W9-BCU-872

MAGILL'S
SCIENCE ANNUAL
2001

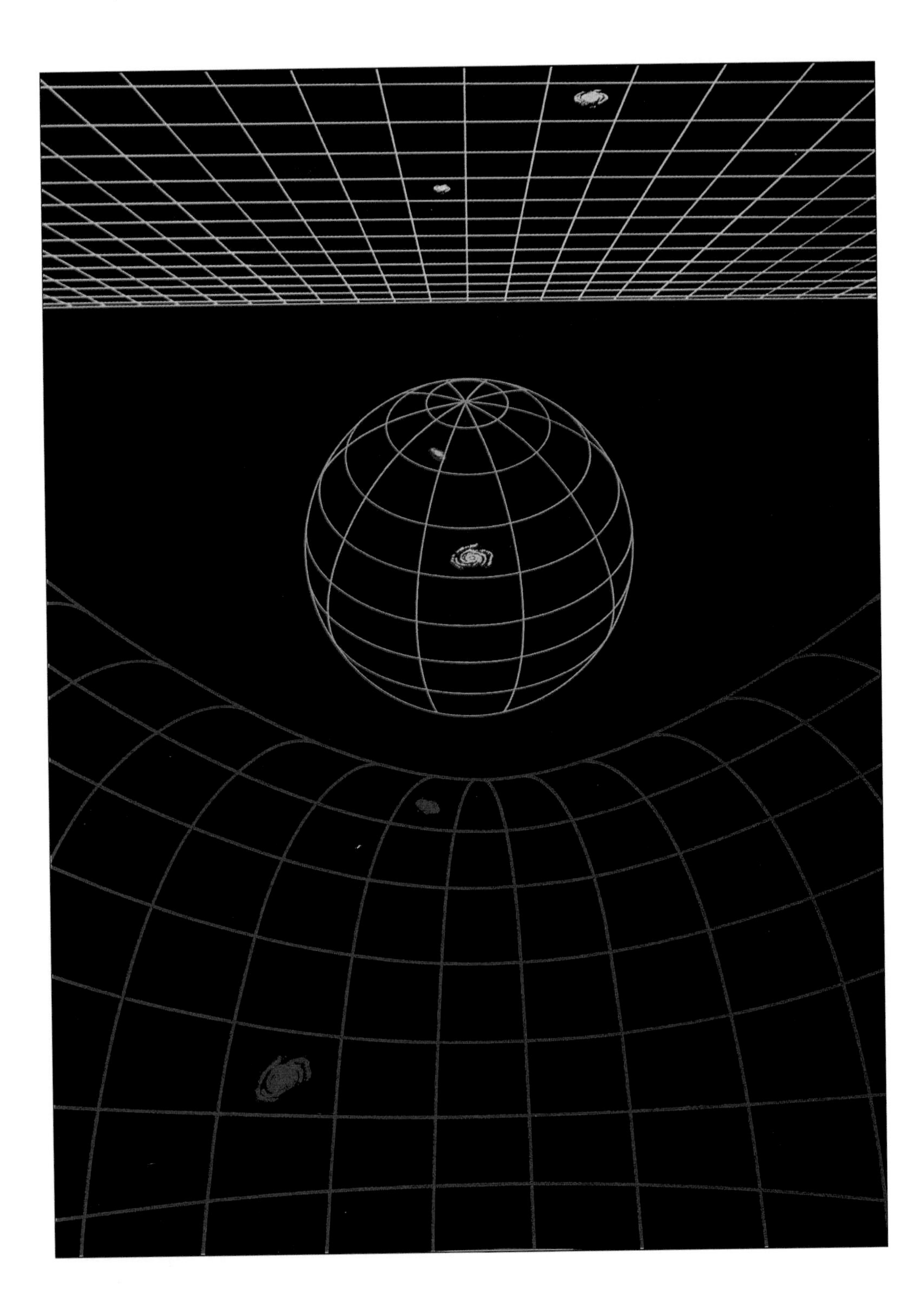

Essays on major scientific breakthroughs and events of 2000

Editor
Joseph L. Spradley
Physics Department
Wheaton College

Salem Press, Inc.
Pasadena, California Hackensack, New Jersey

Managing editor: Christina J. Moose
Project editor: Rowena Wilden
Research supervisor: Jeffry Jensen
Research assistant: Jeff Stephens
Acquisitions editor: Mark Rehn
Photograph editor: Philip Bader
Page layout: James Hutson
Cover and page design: Moritz Design

∞ The paper used in these volumes conforms to the American National Standard for Permanence of Paper for Printed Library Materials, Z39.48-1992 (R1997).

ISBN 0-89356-975-5

FIRST PRINTING

PRINTED IN THE UNITED STATES OF AMERICA

Publisher's Note

Salem Press's *Magill's Science Annual, 2001,* is the first in an annual series of books showcasing the year's most important developments in major scientific fields. This year's annual presents the most prominent scientific events and discoveries in the year 2000 for the fields of Astronomy and Space, Computer Science and Information Technology, Earth Sciences and the Environment, Life Sciences and Genetics, Mathematics and Economics, Medicine and Health, Physics, and Applied Technology.

Each of the eight scientific fields forms a chapter that begins with "The Year in Review," a summation of the notable breakthroughs, discoveries, and events in the field during the year, followed by five to sixteen essays presenting in-depth coverage of selected topics among the year's events. The authors' names appear at the start of each chapter, and if more than one person contributed to a chapter, each essay in the chapter is followed by the author's initials. The essays were selected by experts in each discipline based on their timeliness, their lasting worth in the history of science, and their significance for the general public as well as for researchers. The reader will find in these pages some of the year's familiar stories, from the completion of the sequencing of the human genome, to the discovery of evidence of water on Mars, to the ruling that Microsoft was a monopoly in the antitrust suit brought by the Department of Justice, and to commerce and crime on the Internet. Each chapter ends with a listing of resources for students and teachers that contains books, CD-ROMs, and videos released during the year as well as useful Web sites.

The annual contains more than one hundred photographs, including images of distant galaxies, the International Space Station, black holes, a mockup of a "space hotel," an MP3 files player, Jeff Bezos of Amazon.com, Intel's new computer processing chip, cloned cows and pigs, the growing ozone hole, and a wildfire near Los Alamos National Laboratory. About twenty-four illustrations, sidebars, graphs, charts, and tables illustrate processes, clarify points, and provide related information.

Among the features of this annual are a retrospective on 2000 by Editor Joseph L. Spradley of Wheaton College, "Looking Back"—a summation of the year's greatest events and achievements in science, and his projections for the year 2001, "Looking Forward"—in which he notes the areas in which new scientific breakthroughs are mostly likely to occur. A chronological list of the year's scientific achievements appears in the form of a time line. Those scientists who were honored in 2000 are listed in the section on science awards, with a brief description of their achievements. Some forty-seven scientists who died during the year are memorialized in brief obituaries. Supplementary sources of information on science can be found in "Books of the Year," an appendix featuring the year's best new books. "Journals and Magazines" lists the major science periodicals. Another appendix, "Audiovisual and CD-ROM Resources" surveys the year's best videos, CD-ROMs, and DVDs. Science-related organizations, agencies, and archives are addressed in "Organizations, Agencies, and Archives." Finally, "Web Sites," a list of science Web sites, all in existence as of December, 2000, describes the best sources of science information on the Internet. All weights and measures appear in both their metric and U.S. conventional system forms in the text; however, the annual also contains a metric conversion table. An index, which will be cumulative starting with *Magill's Science Annual, 2002*, completes the annual.

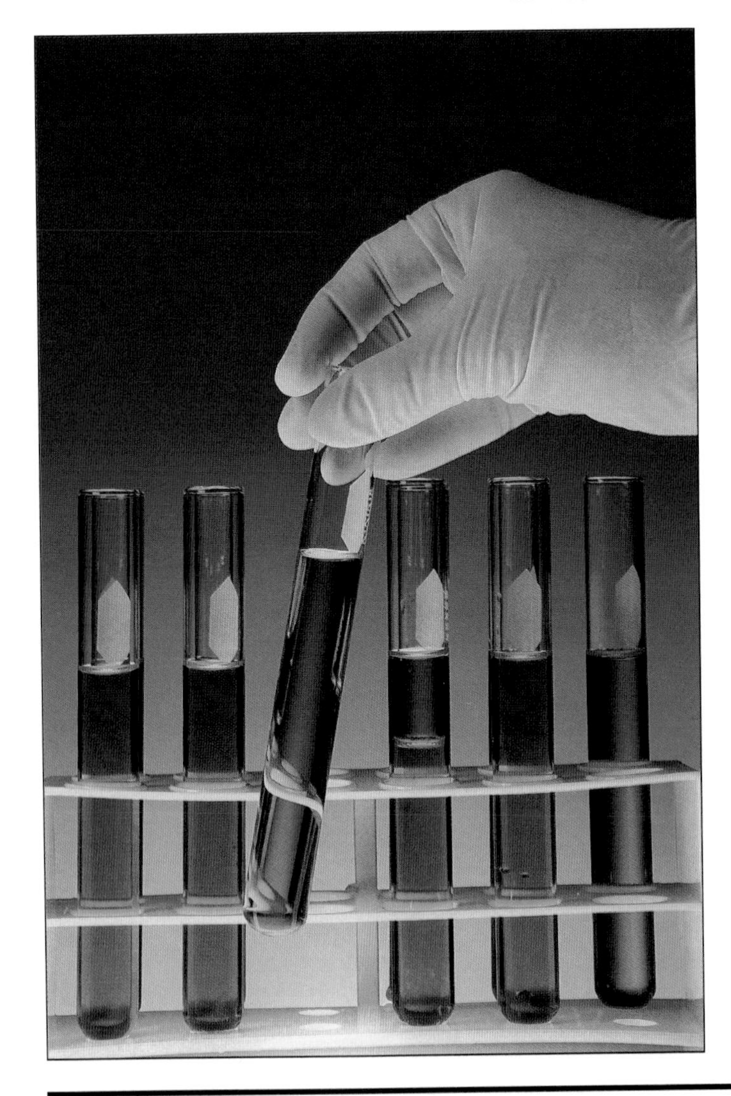

Salem Press would like to thank all those who contributed to *Magill's Science Annual, 2001*, particularly Joseph L. Spradley, Wheaton College, for his counsel as editor. Other authors' names, affiliations, and contributions are listed on the following page.

Contributors

Bezaleel S. Benjamin
University of Kansas

Alvin K. Benson
Brigham Young University

Margaret F. Boorstein
C. W. Post College of Long Island University

John T. Burns
Bethany College

Zachary Franco
Community School of Naples, Florida

Hans G. Graetzer
South Dakota State University

Thomas C. Jefferson, M.D.
Independent Scholar

Michael J. Rafa
Brooke High School

Charles W. Rogers
Southwestern Oklahoma State University

Elizabeth Schafer
Independent Scholar

Jane Marie Smith
Slippery Rock University

Roger Smith
Independent Scholar

Cheryl M. Speer
Bethany College

Joseph L. Spradley
Wheaton College

William R. Wharton
Wheaton College

George M. Whitson III
University of Texas at Tyler

Rowena Wildin
Independent Scholar

Contents

MAGILL'S
SCIENCE ANNUAL
2001

Looking Back
Science and Technology in 2000

➤*Joseph L. Spradley*

At the start of 2000, discoveries and innovations in science and technology were dominated by large-scale collaborations and highly complex equipment. This marked a continuation of a trend toward big science begun in the second half of the twentieth century. This trend included such huge enterprises as the development of nuclear energy and giant accelerators by the Atomic Energy Commission, the moon landings of the Apollo program, and unmanned space explorations by the National Aeronautics and Space Administration (NASA). It led to the electronics and computer revolutions by giant companies such as Intel and Microsoft and the triumphs of molecular biology by the National Institutes of Health (NIH) and international drug companies.

The progress and momentum of big science over the last fifty years was clearly evident in the major advances and uses of science and technology in 2000. Multinational cooperation led to the start of continual occupation of the International Space Station and the prospect of new scientific and medical studies made possible by it. Giant telescopes and complex spacecraft probed the universe, exploring from the nearest asteroids to the most distant galaxies. Satellites and supercomputers monitored natural disasters and environmental problems more closely than ever before, with warnings and potential solutions to match. International cooperation and new information technologies led to major progress in several gene-mapping programs, including the historic announcement of the first draft of the human genome. Advances in genetics, along with new cell research and cloning techniques, opened up new challenges in medicine.

These developments were all greatly aided by the rapid development of computer and communication technologies. Faster and smaller computers with smarter software, increased memory, and better networking contributed to improved scientific equipment and medical techniques. From the largest particle accelerators to the smallest nanotechnologies, computers were indispensable in finding new forms of matter and new kinds of forces. These in turn offered new prospects for continuing the computer revolution. Computer advances and rapid improvements in the Internet also combined to close the gap between big science and the average person interested in its results.

Astronomy and Space: Progress and Discoveries

The International Space Station (ISS), dubbed Alpha by its first crew, became the most visible symbol of space research when continual occupation began on October 31, 2000, even though it had little current scientific significance compared with other achievements in space during the year. It is a good example of the problems and possibilities of expensive multinational science programs. Built by a consortium of sixteen nations under the direction of NASA, the $60 billion project started eight years ago. However, it ran into large cost overruns and years of delays caused by financial problems experienced by the Russian space agency, which will bring the final price tag to about $90 billion.

The crew of the Alpha space station, consisting of American mission commander Captain William M. Shepherd and Russian astronauts Yuri Gidzenko and Sergei Krikalev, finally began their four-month mission some three years later than originally planned after five years of train-

Occupation of the International Space Station began in October, 2000. (NASA CORE/ Lorain County JVS)

ing together. After setting up basic systems for water processing, oxygen generation, carbon-dioxide removal, food preparation, suction toilet operation, and communication links, they installed the indispensable computers and networks to control the various functions of the station.

At present, the station consists of three pressurized modules and a variety of appendages. The three main units include the Russian-made Zvezda control and living module docked to the Soyuz craft that serves as its escape pod, the Zarya cargo module launched in November, 1998, and the American-made Unity module, which serves as the central docking port for future shuttle missions. The station will grow over the next five years from an 80-ton structure stretching over 143 feet into a 480-ton labyrinth of laboratories and other equipment spanning more than a city block. In November, 2000, an unmanned Progress cargo craft delivered 1,300 pounds of supplies and returned with waste material. In December, the shuttle *Endeavour* delivered a huge solar-panel array, giving the ISS new visibility as the third-brightest object in the night sky after the Moon and Venus.

Although the ISS opens up interesting possibilities for scientific and commercial research in the future, it does not begin to match the dazzling achievements in 2000 of a host of unmanned spacecraft. Of the many space probes launched by NASA, the European Space Agency (ESA), and Japan's Institute for Space and Astronautical Sciences over the last decade, several began to make impressive discoveries during the last year at a fraction of the cost of the ISS program. The Near Earth Asteroid Rendezvous (NEAR) mission in its orbital embrace of the asteroid Eros determined its elemental and physical structure and found the elusive connection between S-type asteroids and chondrite meteorites. The Galileo mission to Jupiter focused on its larger moons during the year, revealing the most extensive volcanic activity in the solar system on Io and providing persuasive evidence for an ocean under the icy surface of Europa. After several disappointing failures of spacecraft sent to Mars, the Mars Global Surveyor surpassed all expectations with surprising evidence for ancient underground rivers, recent surface water seepage, and possibly even lakes.

The station will grow from an 80-ton structure stretching over 143 feet into a 480-ton labyrinth of laboratories and other equipment spanning more than a city block.

A variety of sophisticated new telescopes developed over the last decade, both ground-based and in orbit, probed the universe with unprecedented results during the last year. A new generation of giant telescopes in Hawaii and Chile discovered a dozen more extrasolar planets from the tiny wobbling of a few nearby stars, using new spectroscopic techniques and massive computer analyses. An international fleet of space observatories launched over the last decade was ready for the year 2000 sunspot maximum, giving new insight into the nature of solar storms and early warnings of their results on earth. These included the U.S.-European Solar Heliospheric Observatory (SOHO) and Ulysses space ships, Japan's Yokoh satellite, and NASA's Advanced Composition Explorer (ACE), WIND spacecraft, and Transitional Region and Coronal Explorer (TRACE) spacecraft.

Space observatories designed to detect electromagnetic waves over the entire spectrum, from gamma rays to radio waves, made several breakthroughs during the year. The orbiting Compton Gamma Ray Observatory and other space and ground-based instruments began to unravel the mysteries of powerful gamma-ray bursters and show their relation to distant supernova explosions. The Chandra X-Ray Observatory along with other X-ray and infrared detectors have revealed the birthing of new galaxies and the existence of supermassive black holes at the centers of many galaxies. Radio telescopes around the world provided evidence that the expansion of the universe is accelerating, giving rise to the possibility of a new cosmological force acting at galactic distances.

Earth and Environment: Developments and Disasters

Closer at hand than the exploration of space, a plethora of computers, earth satellites, and growing communication networks also monitored global events on earth. Computers again played the central role as they continued to proliferate and accelerate. Improvements in chip design introduced in 2000 by companies such as Intel and AMD led to relatively inexpensive microcomputers with processor speeds of one gigahertz (a billion simple operations per second), approaching the speeds of supercomputers. NASA's Jet Propulsion Laboratory discovered new possibilities for using "pairs of entangled photons" in laser light to produce even faster and smaller computer chips.

Computer-related applications advanced on other fronts as well. Improvements in blue laser beam technology opened the way for further advances in high-density optical storage techniques that can focus light to read mul-

Fire destroys a tree and home in Los Alamos, New Mexico. Numerous wildfires consumed thousands of acres and taxed firefighting resources in the southern and western United States during the summer of 2000. (AP/Wide World Photos)

tiple rings of a single spiral track on CDs and DVDs. Rapid increases during 2000 in the availability of high-speed Internet access by satellite, cable, optical, and wireless systems has improved computer networking for research purposes and public access to information. Activities such as e-commerce and e-trading expanded with the aid of new mathematical techniques and encryption standards.

Several government agencies were active in large-scale programs to document, analyze, and predict global disasters and their future effects. The National Climatic Data Center (NCDC) and the United States Geological Survey closely monitored the record-breaking drought of 2000, including the devastating wildfires during the summer months. The National Oceanic and Atmospheric Administration (NOAA) was able to partially correlate the drought with predictions based on widespread atmospheric-oceanic linkages associated with the La Niña phenomenon of colder-than-average temperatures in the southeastern Pacific Ocean. The National Drought and Mitigation Center (NDMC) tracked the impacts of the drought across the United States and made recommendations for federal aid.

Manmade disasters, both actual and potential, were closely studied during the year, especially the growing ozone hole over Antarctica and the possibility of global warming. Again, satellites and computers played an essential role in surveying and analyzing these threats to human welfare. In September, 2000, satellite imagery associated with NASA's Total Ozone Mapping Spectrometer (TOMS) revealed that the area of ozone depletion in the stratosphere had grown to its largest size ever, increasing the dangers of ultraviolet radiation from the sun. NOAA reported that the depleted area over Antarctica had grown to about 20 times its value in 1981. However, measurements by NOAA since 1994 show that the chlorofluorocarbons (CFCs), believed to be mainly responsible for ozone depletion, declined in the troposphere (lower layers of the atmosphere). This is an encouraging result of a 1987 international agreement to phase out CFCs, but computer models suggest that twenty to forty years might be needed before concentrations reach pre-1980 levels, and even more time for ozone levels in the stratosphere to increase.

Global warming research led to somewhat ambiguous results in 2000. Most atmospheric scientists agree that surface temperatures of the earth increased about 1 degree Fahrenheit over the last one hundred years, and some data suggests that the rate of increase might now be as high as 3.5 degrees per century. The National Research Council of the National Academy of Sciences announced in January, 2000, that an analysis of satellite data for the past twenty years showed surface warming. However, it also revealed less warming and even some cooling in the middle to upper levels of the troposphere where most weather is produced, suggesting to some scientists that global warming computer models are flawed because they did not predict this result.

Computer models suggest that twenty to forty years might be needed before concentrations of chlorofluorocarbons reach pre-1980 levels

On the other hand, studies of satellite imagery during the summer of 2000 showed that melting of the North Polar ice cap over the past twenty years had decreased its area by about 6 percent and its thickness by 40 percent, posing a long-term threat to many polar species. In July, 2000, a Texas A&M University computer analysis suggested that human factors are more important than natural factors in causing global warming over the last century. Several solutions to this problem were discussed during the year, such as planting more trees to remove excess carbon dioxide from the atmosphere, and reducing other greenhouse gases such as methane by changing livestock feed mixtures and draining rice fields more often. Sadly, an international conference in November was unable to reach agreement on targets for reducing greenhouse gases.

Advances in the Study of Life

Several dramatic advances were announced during 2000 in the study of the smallest units of life, mostly by international collaborations of scientists. Biological studies included gene-mapping projects, cloning programs, and cell research. Paradoxically, in most cases, the smaller the object of study, the larger the number of researchers involved, with a corresponding increase in the complexity of computer support needed and the size of the equipment required. In many cases, medical and technological benefits associated with these advances were clearly evident.

The map of the entire genetic code forms an instruction manual of human life longer than two hundred telephone books.

Perhaps the most dramatic disclosure of 2000 was the June 26 joint announcement of the completion of the first draft of the human genome several years ahead of schedule by Francis Collins, director of the NIH's Human Genome Project, and J. Craig Venter, founder of Celera Genomics. This map of the entire genetic code is based on the four letters representing the DNA chemicals (nucleotide bases) that make up the more than 3 billion paired bases in some 100,000 genes, forming an instruction manual of human life longer than two hundred telephone books. It was originally scheduled for completion in 2005, but the finished version is now expected by 2003 at a total cost of about $3 billion. Paving the way for the human genome and augmenting its interpretation was the March announcement of the completion of the fruit fly genome by Celera Genomics, together with the Berkeley and European *Drosophila* genome projects.

The Human Genome Project is a publicly funded consortium made up of four large gene-sequencing centers in the United States, as well as the Sanger Center in England and labs in France, Germany, China, and Japan. It has worked for a decade with more than 1,100 scientists at a cost so far of $250 million, using a combination of robotics, computers, and automated DNA-sequencing machines. Starting with blood and sperm cells from a half dozen anonymous individuals, it separated out the twenty-three pairs of chromosomes and clipped bits of DNA from each. It then identified the sequence of DNA units in each bit and matched them up to form the sequences of all the genes in each chromosome. Actually, the first draft includes about 85 percent of the genome, sequenced four times to give an error rate of less than one in one hundred bases. The finished version will

On June 26, 2000, the Human Genome Project announced that it had sequenced about 90 percent of the human genome. Andre Rosenthal, a member of the German Human Genome Project, explains human genome sequencing at a conference in Berlin. (AP/Wide World Photos)

be sequenced about ten times for an error rate of one in ten thousand bases.

Celera Genomics entered the race just two years ago with strong private financing and a shortcut approach called the "whole-genome shotgun" strategy. Instead of working piece by piece, Celera shredded the entire genome of one man and sequenced the fragments three times with three hundred automated DNA sequencers. To fill in gaps, it also sequenced parts of the genomes of three women and a second man of a different ethnicity. Accuracy was improved beyond that of the Human Genome Project by adding the consortium's data to Celera's as the information was made available to the public on the Internet. Celera then fed the data into the company's supercomputers to connect the fragments in their proper order, forming a "consensus" genome. Its computer algorithms ensured that the order of Celera's genome was better than the supposedly preestablished order of the Human Genome Project's genome, thus giving it commercial value for drug companies looking for better data to design drugs and diagnostic tests.

As the Human Genome Project nears conclusion, the hard work of "annotation" begins, which attempts to correlate specific genes with their functions. Here again, computers and special software are essential. By scanning billions of bases, computer programs can compare sequences to known genes and translate these genes to related proteins with known functions. Other programs compare genome data with that from "model" organisms such as the fruit fly, in which 60 percent of the 289 known human disease genes have equivalents. That is why Celera's March, 2000, completion of the fruit fly genome with its 120-million paired bases is so important, as well as its work on the mouse genome with about 90 percent of its proteins similar to human proteins. Because 99.9 percent of the human genome is believed to be common to all humans, drug companies are especially interested in the other 0.1 percent of genes, often related to diseases, that can be targeted by gene-specific drugs.

The first complete plant genome was completed in December, 2000, in a joint U.S.-European effort. The five chromosomes containing 25,500 genes of the *Arabidopsis thaliana* (thale cress) were mapped in a five-year effort. The plant's biological simplicity and rapid growth make it easy to manipulate in the laboratory, making it a model genome for studying ways to improve staple crops such as rice, corn, and wheat.

Additional progress was reported during 2000 in cellular and cloning research with both medical and ethical implications. Several groups obtained atomic-level views of the action of ribonucleic acid (RNA) within the protein factory of the cell, called the ribosome, showing how proteins are assembled from amino acids within the cell. Human stem cells derived from embryos or fetal tissue are progenitors of many kinds of cells. They offer great hope for improved treatment of diabetes, Parkinson's, and other diseases, but antiabortion groups have lobbied hard to block federal funding for such research. In August, 2000, the NIH announced new guidelines allowing NIH-funded researchers to work with embryonic stem cells, but only if privately funded researchers have established the cell lines from surplus frozen embryos created for fertility treatment. Among other conditions, the decision to donate embryos must be separated from fertility treatment or any financial or other compensation. Preliminary success was also reported in September in transplanting stem cells from the bone marrow of seven lupus patients and reinserting them later to stimulate the immune system. Six of the seven severe cases appear to be cured.

Pigs have the highest priority for cloning to supply organs for transplantation to humans as they are the closest physiological matches to humans.

Cloning research also continued to advance in 2000, with the successful cloning of pigs in both Scotland and Japan by the use of new techniques. Pigs have the highest priority for cloning to supply organs for transplantation to humans as they are the closest physiological matches to humans. In March, PPL Therapeutics, which first cloned sheep in 1997, announced success with a double nuclear transfer method that produced five cloned piglets. Japan's National Institute of Animal Industry published its results first in September, which were based on a microinjection technique. However, these results were tempered by news that pig retroviruses can infect human cells, which will require further research to avoid the danger of another retrovirus pandemic.

Probing Deeper into Matter

The study of matter at ever-smaller dimensions continues to lead to giant budgets, large-scale programs, and huge equipment. In 2000, these studies included molecular nanotechnologies, new forms of subatomic matter, and evidence for new elementary particles. Several advances in the growing field of nanotechnology illustrate this trend, including a $500 million presidential research initiative for 2001. Materials with dimensions of one to one hundred nanometers (billionths of a meter), ten to a thousand times larger than an atom, have a high ratio of surface atoms that make them highly reactive. Recent Japanese studies show that nanoscale gold particles are excellent catalysts for partial oxidation reactions that could contribute to the $210 billion catalyst market. The use of alternating nanolayers of magnetic and nonmagnetic metal films that produce a giant magnetoresistance (GMR) effect are currently revolutionizing the $140 billion computer memory business. Consolidated nanostructures made from pure metals with nanosize grains are several times stronger than large-grained metals, with great potential for a variety of useful structures and devices, including molecular computing prospects.

CERN's large hadron collider (LHC) will be about seven times more powerful than Fermilab's Tevatron by 2005.

At the subatomic level, an international team at the Fermi National Accelerator Laboratory in Illinois, home to the world's largest proton accelerator with its 6.3-kilometer Tevatron ring, announced in July, 2000, the first direct evidence for the tau neutrino. The tau neutrino is the final particle in the standard model of elementary particles and completes the lepton family of electron, muon and tau particle, and their associated neutrinos. The standard model has been highly successful in accounting for the particles and forces of nature, except for gravity. One of its predictions appeared to be confirmed earlier in the year when physicists at the European Organization for Nuclear Research (CERN) laboratory announced that their six-kilometer super proton synchrotron (SPS) ring had produced a new state of matter called a quark-gluon plasma. They smashed lead ions into gold atoms, producing free quarks and gluons in a state that existed near the beginning of the universe. In September, tantalizing but incomplete evidence that threatens the standard model appeared at CERN in experiments with its twenty-seven-kilometer large electron-positron collider (LEP) ring.

At CERN, electron-positron collisions at the high-energy limit of the LEP collider produced 228 pairs of tau particles rather than some 170 pairs predicted by the standard model. This is consistent with a rival supersymmetry theory that doubles the number of elementary particles and links them with particles that carry forces, including gravity and the long-predicted Higgs particle that explains how particles attain their observed masses. In October, before enough evidence could be accumulated to verify the Higgs particle, the LEP collider had to be closed down for a long-scheduled reconstruction that will use the twenty-seven-kilometer ring to house the large hadron collider (LHC). The LHC will be about seven times more powerful than Fermilab's Tevatron by 2005. In the meantime, the best hope for discovering the Higgs particle will be the newly upgraded Tevatron, which would open up a new era in the understanding of matter and its origins.

The year 2000 has demonstrated the continuing growth of scientific knowledge, especially the accelerating trend toward big science with its large collaborations, complex equipment, and heavy dependence on computers. From the largest structures of space to the planets, the earth, and its environment, and down to the smallest units of life and matter, major progress and important discoveries filled the year. These advances form the foundation for many applications in technology and medicine and point the way to new understandings of the universe and its many mysteries.

Looking Forward
What to Watch for in 2001

➤ *Joseph L. Spradley*

Many opinions have been voiced about the actual beginning of the twenty-first century. Mathematical purists have insisted that 2000 is the last year of the twentieth century and that the year 2001 marks the beginning of the twenty-first century. This is an appropriate view for looking ahead to possible scientific and technological advances in 2001 as these will mostly build on the theoretical framework of the twentieth century as exemplified in the achievements of the year 2000. This framework includes quantum and atomic theories at the deepest levels of matter, molecular genetics in biological and medical studies, the global perspectives of atmospheric and oceanic sciences, and the use of space technologies and theories of cosmology to study the solar system and the universe beyond.

Extrapolations of current scientific work suggest a number of innovations and applications to watch for during 2001. Souped-up particle accelerators could find more evidence for the Higgs particle at energies that would support a supersymmetry theory of elementary particles. Advances in atomic and molecular studies suggest the possibility of new types of lasers with many applications. Refinement of the human genome and that of other model organisms will lead to further correlations of specific genes with their biological functions and the design of new drugs to treat genetic diseases. A better understanding of the relations between the oceans and the atmosphere will lead to continued improvements in weather prediction and new solutions to the dangers of global warming. A new generation of space probes and telescopes will add to human knowledge of the possibilities for life in the solar system and on extrasolar planets and provide a better understanding of the birth and developing structure of the universe.

Elementary Particles and Laser Physics

The transition from the twentieth century to the twenty-first is especially meaningful in the search for new elementary particles at the high energies associated with the birth of the universe. The standard model of elementary particles came close to completion in 2000, but evidence also emerged that might replace the standard model with a new supersymmetry theory and a host of new particles. The standard model provides a theoretical framework for all the known fundamental particles, which include six leptons (such as electrons and neutrinos), six quarks (such as components of protons and neutrons), and the particles that transmit forces between them (such as photons and gluons). The one missing link in the standard model is its prediction of a force particle at energies just beyond the current stage of particle accelerators, called the Higgs boson, which should interact with matter particles to produce their observed masses.

The first organic lasers sandwich high-purity tetracene crystals between field-effect transistors, exciting them to amplify a yellowish green beam of light.

Evidence for Higgs particles began to emerge in several events at the high-energy limit of the large electron-positron collider (LEP) at the European Organization for Nuclear Research (CERN) laboratory in Geneva, Switzerland, before it had to be shut down in October, 2000. The evidence was not sufficient to lay claim to the Higgs. However, it strongly suggests the existence of the Higgs at lower energies than those predicted by the standard model but consistent with supersymmetry theory. A major upgrade in 2000 of the Tevatron accelerator at the Fermi National Accelerator Laboratory in Batavia, Illinois, provides it with high enough energy to reveal the Higgs. Guided by the LEP data, it could confirm the LEP evidence for the Higgs particle and open the door to the search for supersymmetry particles.

Two new laser techniques developed in 2000 offer many possible applications in the years ahead. In July, Lucent Technologies announced the first organic lasers, which sandwich high-purity tetracene crystals between field-effect transistors, exciting them to amplify a yellowish green beam of light. Other organic crystals offer the possibility of cheaper lasers and a new palette of colors. Physicists at the Massachusetts Institute of Technology (MIT) and in Japan and Germany demonstrated the first lasers that amplify a beam of atoms using a quantum-coupled system of atoms near absolute zero temperature known as a Bose-Einstein condensate. These matter-wave lasers suggest thousandfold improvements in resolution over optical lasers, comparable with

that of electron microscopes over optical ones. Because the atomic laser beams focus to nanometer spot sizes, they offer new possibilities for accurate sensors and high-resolution deposition of atoms on surfaces to fabricate nanostructure devices and computer circuits.

Gene Mapping and Drug Designing

With the Human Genome Project nearing an end, genetic studies are beginning to shift toward the mammoth task of interpreting how gene sequences are expressed in protein synthesis, bodily functions, and diseases. This is expected to provide a more rational approach to diagnosing diseases and designing drugs. One of the first tasks will be to compare the human genome with model genomes of a few animals, whose functions can be studied in the laboratory and related to their gene sequences. The mouse and rat genomes, with some 90 percent of proteins similar to human proteins, will be completed within a year and comparisons with the human genome have already begun. Lobbying has also begun for a chimpanzee genome project as studies have already shown that the genes of chimpanzees have about a 99 percent similarity with the human genome. This is attractive to drug companies, which would like to find genetic differences that might account for why chimps seem to be free of certain human ailments.

Rational drug designing will depend heavily on how well gene sequences can be related to the proteins they synthesize and their various functions in the body. The key to understanding the differences that produce diseases are the one-in-a-thousand variations in the human genome, called SNPs (single nucleotide polymorphisms). Researchers at the SNP Consortium, a nonprofit partnership of top universities and thirteen of the largest drug companies in the world, have found a million of the estimated 10 million SNPs so far. Identification of several thousand SNP patterns in genes with known connections to diseases have already led to preliminary but limited successes in developing drugs for asthma, hemophilia, and leukemia. However, most diseases involve interactions between many genes and the many more possible SNP patterns they contain. As more connections between SNPs and diseases become clear, more drugs will be developed to correct the genetic spelling errors underlying diseases such as Alzheimer's and cancer.

The mouse and rat genomes, with some 90 percent of proteins similar to human proteins, will be completed within a year.

The year 2001 should begin to see the benefits of two other genome programs. The *Plasmodium* genome of the mosquito parasite that causes malaria is near completion, and malaria researchers are beginning to identify targets for new drugs from these data. Sequencers are also gearing up to attack the genome of the *Anopheles* mosquito that transmits malaria, opening up new avenues for antimalarial drugs and even the possibility of designing a malaria-resistant mosquito to replace natural populations. Another program to watch is the International Rice Genome Sequencing Project, which, along with Monsanto researchers, could begin to greatly improve rice production.

Weather Prediction and Global Warming

Accumulating data on oscillating ocean temperatures and atmospheric pressures have led to some successes in long-term weather predictions, such as those associated with the El Niño warming and La Niña cooling in the Pacific Ocean from 1997 to 2000. An increased rate of global warming of 3 to 4 degrees Fahrenheit per century is now evident, but complex computer models associated with its

This insignia depicts the U.S. laboratory, Destiny, just before its attachment to the International Space Station. (NASA)

causes need to be improved. There is some evidence of warming in the South Atlantic that may be altering ocean currents. Paradoxically, one concern about global warming is the possibility of colder weather in Europe because of a shift in the Atlantic Ocean currents that warm Northern Europe. If that happens, temperatures could fall as much as 20 degrees Fahrenheit over the next decade, leading to conditions similar to the "little ice age" in the seventeenth century.

Unfortunately, new evidence of human contribution to global warming was not sufficient to produce agreement at a November, 2000, international meeting in The Hague to finalize rules of the 1997 Kyoto global-climate treaty for reducing greenhouse gases. One problem involved how well industrialized countries could meet their carbon dioxide emission targets by relying on carbon "sinks" such as forests and crops to absorb carbon dioxide. Although failure of the Kyoto protocol is a disappointment, it might spur efforts to find new ways to increase carbon dioxide absorption on land and in the oceans. It might also lead to a new focus on ways to reduce other greenhouse gases such as methane from farm animals and rice paddies and soot-laden aerosols from diesel engines and agricultural burning. A good start has already been made in the United States under the Clean Air Act; and the Montreal protocol to reduce chlorofluorocarbons has resulted in a measurable reduction in this pollutant and indications that the hole in the ozone layer will begin to diminish.

One concern about global warming is the possibility of colder weather in Europe because of a shift in the Atlantic Ocean currents that warm Northern Europe.

Initiatives in Space and Advances in Astronomy

Although it will be difficult to match the many space-probe successes during 2000, several new and continuing efforts should yield interesting results. The International Space Station begins its scientific programs early in 2001 after delivery and docking of the U.S. Laboratory Module to the orbiting Alpha station on February 10. This 4.3-meter (14.1-foot) diameter by 8.5-meter (27.8-foot) long module, with rack locations for thirteen experiments, will provide a permanent science institute for long-duration microgravity research. Some of the first research will involve the effects of long-term exposure on humans in microgravity and earth observations to study environmental changes and ways to improve weather forecasting systems. Life science studies could lead to a better under-

The Mars Odyssey orbiter is scheduled for launch in April, 2001, and is expected to reach Mars in October. (NASA)

standing of organic processes and the development of new drugs. Industrial experiments may lead to lighter, stronger metals, purer crystal growth, and more powerful computer chips.

The search for possibilities of life beyond the earth will focus on Mars and the continuing study of extrasolar planets during 2001. The Mars Global Surveyor mission will continue to send high-resolution photos of the surface of Mars, and further analysis should add to the existing evidence of ancient rivers, oceans, lakes, and even signs of recent groundwater seepage. The 2001 Mars Odyssey orbiter was scheduled for launch in April and is expected to arrive at Mars in October, 2001. Its instruments will study the kinds of minerals on the surface and measure subsurface hydrogen as evidence for water and ice buried below the surface. This mission will also prepare for the launch of twin rovers to Mars in 2003. Meanwhile, there is every reason to expect more discoveries of extrasolar planets, which should give a better idea of whether all have rapid or eccentric orbits or if any have planetary arrangements that are similar to the solar system and can foster life.

The mysteries of gamma-ray bursts and the structure of the universe should become clearer during 2001. Gamma-ray bursts have been identified as hypernova explosions in the early universe, releasing more energy in one second than the sun will emit in its 10-billion-year lifetime. Recent observations by the Beppo-SAX X-ray satellite and the Chandra X-Ray Observatory satellite of iron emission in the spectrum of two gamma-ray bursts support theories that they result from the collapse of supermassive stars into black holes. The October, 2000, launching of MIT's High Energy Transient Explorer 2 (HETE-2) will help to refine these theories by providing accurate positional data for about one burst per week, quickly alerting large telescopes on earth to observe their afterglow. Meanwhile, the National Aeronautics and Space Administration (NASA) plans a spring 2001 launch of the Microwave Anisotropy Probe (MAP) satellite into a halo orbit 1.5 million miles further out from the Sun than Earth. It will provide a much higher resolution mapping of the sky than the 1992 Cosmic Background Explorer (COBE), shedding light on the shape of the universe, the formation of the first galaxies, and the existence of exotic dark matter.

The year 2000 has indeed provided a good foundation for predicting discoveries and perhaps even breakthroughs in 2001. Confirmation of the Higgs particle at Fermilab would be such a breakthrough, opening the door to many new supersymmetry particles. New laser materials should lead to many new kinds of lasers and new possibilities in nanotechnology. Completion of several genome projects marks the beginning of a new understanding of biological functions and diseases and prepares the way for exciting new drugs. Improved Earth monitoring systems will reveal further implications of global warming and possible solutions. Continuing studies of Mars, nearby extrasolar planetary systems, and the universe at large will reveal new possibilities for life and the unusual, if not unique, nature of planet Earth.

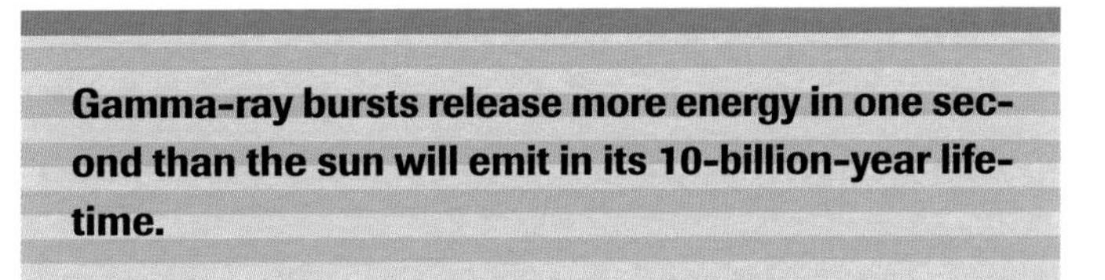

1 · Astronomy and Space

The Year in Review

➤*Joseph L. Spradley*
Physics Department
Wheaton College

➤*William R. Wharton*
Wheaton College

Astronomical discoveries in 2000 range from evidence for recent groundwater seepage on Mars to an apparent acceleration in the expansion of the universe. They include the discoveries of about ten new extrasolar planets and hundreds of new X-ray sources, many of which may be supermassive black holes at the centers of galaxies. Research leading to many of these discoveries was greatly aided by a flotilla of satellites and space probes launched during the 1990's. Working in tandem with ground-based telescopes, a great variety of specialized space observatories have begun to probe every corner of the solar system and the universe beyond with instruments that detect the entire spectrum of radiation and particle energies.

In addition to the well-publicized Hubble Space Telescope, these observatories include space probes orbiting the Sun, Earth, Mars, Jupiter, and for the first time ever, an asteroid. They carry a bewildering array of instruments, each specially designed to detect some region of the electromagnetic spectrum, including radio waves, visible light, infrared radiation, ultraviolet radiation, X rays, and gamma rays as well as charged particles, magnetic fields, and gravitational fields. They are providing a vast quantity of high-resolution photographs and extraterrestrial data for astronomers to analyze.

Planetary and Solar Discoveries

Early in the year, a milestone was reached when for the first time, an artificial satellite was placed into orbit around an asteroid. The Near Earth Asteroid Rendezvous (NEAR) Shoemaker satellite was eased into an orbit around the asteroid Eros, fittingly on Valentine's Day. Dramatic photographs began to reveal characteristics of the asteroid's surface with a resolution that showed boulder trails down the sides of impact craters. Analysis of X rays from the surface of Eros, caused by several fortuitous solar-flare eruptions during the year, solved an old mystery by revealing that S-type asteroids and chondrite meteorites have the same elemental composition despite differences in color apparently caused by space weathering.

The year 2000 coincided with the sunspot maximum in the eleven-year solar cycle, not only aiding the NEAR mission but also shedding new light on solar and geomagnetic storms. With a fleet of solar observatories in various orbits, scientists were prepared as never before to observe and even predict solar storms and thus issue warnings and make preparations for their disrupting effects on Earth. The Solar Heliospheric Observatory (SOHO) satellite obtained data for the important discovery that the high-speed solar wind comes from the cracks between convection cells in the surface of the Sun.

During 2000, the Galileo spacecraft made its final flybys of three of Jupiter's larger moons: Europa on January 31, Io on February 22, and Ganymede on May 20 and December 28. Magnetic measurements near Europa on successive flybys revealed a changing magnetic field that adds further confirmation to the presence of an ocean under its icy surface. Jupiter's magnetic field, rotating with the giant planet, apparently induces electrical currents in a salty ocean beneath Europa's icy surface, producing its observed magnetism with a synchronous fluctuation. The closest flyby ever of Io reinforced its reputation as the most active moon in the solar system with an estimated three hundred periodically erupting volcanoes.

The most dramatic discoveries of the year came unexpectedly from the Mars Global Surveyor. Planet-wide altitude and gravitational data revealed southern highlands with a thick crust like that of Earth's continents, and northern lowlands with a thin crust like that under Earth's oceans. The data also suggested underground water channels that might once have carried huge flows of water into an ancient northern ocean. High-resolution images provided even more surprising evidence for recent groundwater seepage and runoff, defying all earlier ideas about Mars's frozen surface and providing new impetus for the search for primitive life-forms on Mars.

The continuing search for extrasolar planets around Sun-like stars led to the discovery of the first six Saturn-size planets, giving hope that eventually Earth-like planets may be discovered. These new discoveries include two planets with minimum masses even less than that of Saturn, discovered with the Keck telescopes in Hawaii, and three near the size of Saturn discovered at the La Silla Observatory in Chile. One of the Saturn-size planets appears to have a smaller companion in a multiplanet system. However, all the newly discovered planets continue to exhibit either very elongated orbits or very rapid periods of a few days, neither of which offers a good environment for life.

Stellar and Galactic Discoveries

During 2000, discoveries beyond the solar system were also greatly aided by space probes, unhindered by Earth's atmosphere. The mystery of gamma-ray bursts, first observed in the late 1960's, began yielding to the cooperative efforts of a variety of observational instruments. Over the last decade, some 2,500 of these highly energetic bursts of high-frequency electromagnetic radiation have been recorded by the Compton Gamma Ray Observatory, which was finally taken out of orbit and destroyed on reentry into Earth's atmosphere on June 4, 2000. More recently, several new instruments, both in space and on Earth, have combined to study these events over the entire spectrum. These bursts are among the most powerful and distant in the universe and now appear to be associated with distant supernova explosions.

In 2000, two new orbiting X-ray telescopes began operation, identifying many types of objects for the first time as X-ray sources. In a single image of a star-forming region of the Orion nebula, the Chandra X-Ray Observatory has resolved a thousand young stars emitting X rays. Other X-ray sources being studied by Chandra and the X-Ray Multi-Mirror space telescope include the remnants of supernova explosions, huge and distant gas clouds giving birth to new galaxies, and millions of very young and distant galaxies with X-ray cores that are apparently massive black holes.

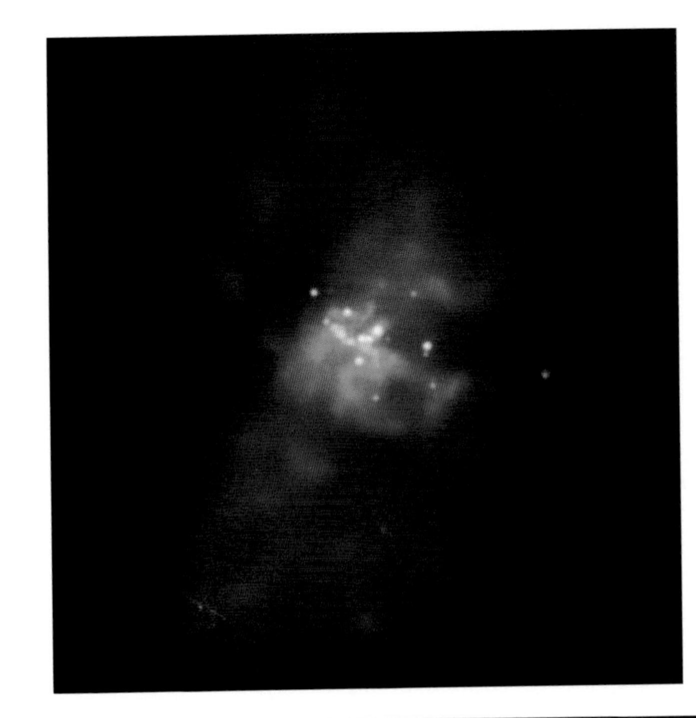

The arrow points to a middle mass black hole, an object whose gravitational field holds more than 500 times the mass of the Sun within a volume the size of the Moon. Unlike supermassive black holes, which are at the centers of galaxies, this black hole is about 600 light-years from the center of galaxy M82. (NASA/CXC/SAO)

The young galactic X-ray sources, many of which qualify as quasars (quasi-stellar radio sources believed to be infant galaxies), are not the only new candidates for black holes. Recently a few black hole remnants of supernova explosions have been identified by their gravitational effect on light passing by them. Combining X-ray and other data has made it possible to identify dozens of supermassive black holes at the centers of nearby galaxies ranging in size from just over a million to a few billion stellar masses. These and other galactic studies are beginning to provide an understanding of the birth and growth of galaxies, showing how their structures evolve with time from quasars to elliptical galaxies and to giant spirals like our and nearby galaxies.

Over the last decade, some 2,500 highly energetic bursts of high-frequency electromagnetic radiation were recorded by the Compton Gamma Ray Observatory.

Recent studies of distant supernovas by U.S. and Australian collaborators provide new evidence for a runaway universe that is expanding at an accelerating rate. This evidence implies that the universe was expanding more slowly in the past, leading to a 10 percent to 15 percent increase in the estimated age of the universe. This result is more consistent with the estimated ages of the Galaxy and the oldest stars and suggests that the universe will not eventually collapse. An accelerating universe also indicates that some kind of unknown repulsive cosmological force or vacuum energy is acting to overcome the gravitational attraction between galaxies.

Although the discoveries in astronomy and space science during 2000 were remarkable, they only begin to suggest the possibilities that are emerging for a new understanding of the intriguing universe in which humans live. New studies of asteroids and solar storms are important not only for the insights they provide about our solar system but also for the practical capabilities that are being developed for warning and protection against space threats to Earth. New evidence of extrasolar planets, water on Mars, and geological activity on the moons of Jupiter is beginning to shed light on the possibilities of life elsewhere in the universe. As the new fleet of space probes and the new generation of ground-based telescopes explore the outer reaches of the universe, the very structure and destiny of the universe are taking new shapes and revealing new patterns of birth and development.

J. L. S.

The Accelerating Universe

➤ Data from supernovas and background radiation show that the rate of expansion is increasing and the universe is flat.

During the last two years, an assortment of data consistently suggests that the universe has a flat geometry and is expanding at an increasing rate. Data were released in spring, 2000, from two balloon experiments, Boomerang and Maxima, carrying microwave detectors examining the cosmic background radiation. This radiation decoupled from the matter in the universe in the distant past, thereby revealing the early structure of the universe. This radiation shows regions that are slightly warmer or compressed subtending an angular range of about 0.9 degree, the size predicted in a flat universe that was 300,000 years old. A universe curved like a sphere or a saddle would have larger or smaller hot spots, respectively. A dynamically flat universe requires a specific total amount of mass plus energy density, each of which will influence the expansion of the universe.

Expanding Universe

Within a distance of about one billion light-years of Earth, the universe is observed to be uniformly expanding. The recessional velocity, v, of each object is proportional to its distance, d, from Earth, $d/v = 1/H = 12\pm2$ billion years (12±2 Gyr). This is simply another version of the standard equation, distance divided by speed equals the time to travel that distance. Assuming unchanging velocities into the past, the expression for time in the equation, the inverse of the Hubble constant, provides the date when all these objects were together, marking the big bang origin of the universe. Of course, the expansion rate may not have been constant, thereby providing an erroneous date for the big bang. Mass has a gravitational attraction, which slows the expansion. Recent data indicate the expansion is speeding up, thereby requiring a repulsive force. In the 1920's, when scientist Albert Einstein believed in a static (unchanging) universe, he postulated such a force to counterbalance the gravitational attraction. This force, called the cosmological constant, Λ, is usually interpreted as a vacuum energy density with a

negative pressure pushing the universe outward.

Knowledge of an acceleration of the expansion of the universe is important for several reasons. It provides a more accurate date for the big bang and a clearer idea of whether the universe will expand forever, ending in a cold heat-death, or collapse into a hot crunch. Combined with other data, it gives a better understanding of the composition and energy/mass density of the universe. It will also provide data to test various cosmological models of the universe.

The mass density, estimated from measurements of the total mass in large clusters of galaxies, is only 30 percent of what is needed to make a flat universe. This mass is estimated from the motions of these galaxies, and most of it is dark matter, invisible to telescopes. For a flat universe, the insufficient mass density must be complemented with

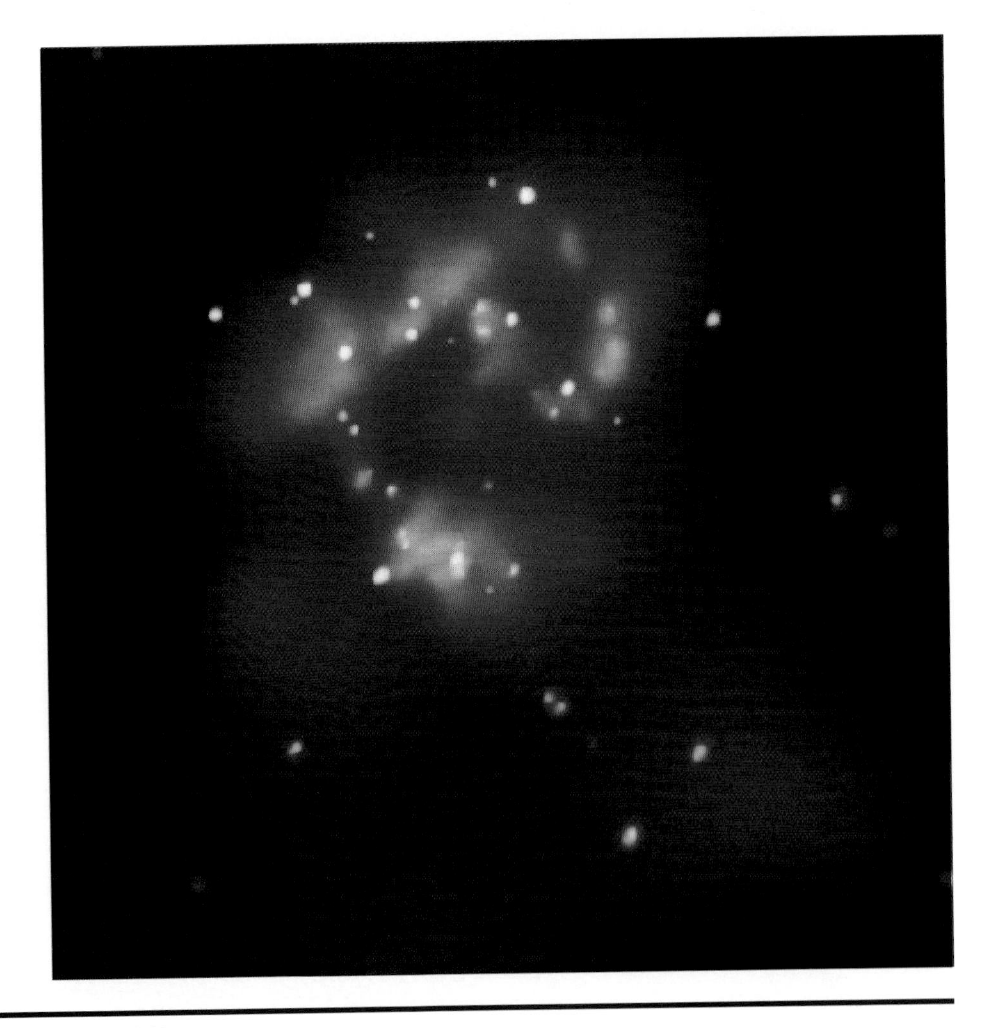

Superbubbles created by thousands of supernovas and pointlike sources created by neutron stars and black holes can be seen in this image of colliding galaxies from Chandra X-Ray Observatory. Images of supernovas helped scientists show that the universe was expanding at an accelerated rate. (NASA)

a vacuum energy density, which also causes an accelerating expansion of the universe.

The recessional velocity is caused by the stretching of space between objects. This stretching is referred to as an increase of the distance scale of the universe. The wavelengths, λ, of electromagnetic spectral lines from a distant object are compared with the same spectra on Earth. Their shift, $\Delta\lambda$, to larger values (redshift) is a measure of the increased distance scale of the universe. For example, if redshift $z = \Delta\lambda/\lambda = 0.83$ from an object 6 billion light-years away, the fundamental distance scale of the universe has increased 83 percent during the last 6 billion years that the radiation was traveling to Earth. By comparing the expansion rate over the last 6 billion years with the expansion rate over the last 1 billion years, it is possible to deduce whether it is accelerating or staying constant. To know the expansion rate, first measure the distance, which for the distant object is 6 billion light-years.

Evidence for Acceleration

The most common method for measuring astronomical distances is the use of so-called standard candles, or objects whose luminosity (emitted light intensity) is uniformly known. Because the brightness of a standard candle decreases as the inverse of the distance squared from the observer, its brightness indicates its distance from the observer. Cepheid variable stars are reliable standard candles for measuring distances to nearby galaxies but are not luminous enough to see in more distant galaxies. Supernova explosions are 100,000 times more luminous for a short period of time. Type Ia supernovas have a fairly regular luminosity that can be known within 25 percent uncertainty, thereby giving their distance to about 12.5 percent uncertainty. There is also a systematic uncertainty for all types of standard candles, which will not be important in this discussion. Type Ia supernovas are carbon-oxygen white dwarfs, with little hydrogen, that gravitationally implode after sufficient mass is accreted onto their surface. They then burn into ^{56}Ni, which decays, producing much of their luminosity.

Two international collaborators—the Supernova Cosmology Project, led by Saul Perlmutter of the United States, and the High-*z* Supernova Search Team, led by Brian Schmidt of Australia—have combined to analyze about one hundred type Ia supernovas with *z* ranging from 0.2 up to 1. The supernova at $z = 1$ occurred when the universe was 35 percent of its present age and its light traveled about 8 billion light-years to reach Earth. The search for high-z supernovas involves a telescope with a charge coupled device (CCD) imager taking ten-minute exposures of patches of the sky, each capturing about five thousand galaxies. A few weeks later, another image is taken of the same patch in the sky. A subtraction of the first image from the second can reveal the appearance of a new light source in the sky.

Once the new light source is identified as a supernova, the powerful Keck telescope takes a spectrum of it. Models in which the universe is expanding at a constant rate predict that almost all of the high-z supernovas should be brighter than observed. Specifically, they are 25 percent dimmer than expected at $z = 0.5$. This suggests that they are 12.5 percent farther away than a $z = 0.5$ supernova should be. With the electromagnetic radiation traveling a longer distance to Earth, the expansion of the universe has more time to redshift the supernova spectra. A longer time for the shift, z, means the universe was expanding more slowly in the distant past than in the recent past; that is, the expansion rate is accelerating.

Implications of Acceleration

Other explanations for dimming the high-z supernovas seem unlikely. Cosmic dust could possibly absorb some of the radiation, but this would normally absorb more strongly in the blue, thereby reddening the spectra. No such effect is seen. An accelerating expansion of the universe means that there must be a repulsive force pushing it apart. The fit to the high-z supernova data is adjusted to give the amounts of mass and vacuum energy density in the universe. The fit gives densities that are consistent with a dynamically flat universe. The fit attributes 30 percent of this density to mass and 70 percent to the vacuum energy. This agrees remarkably well with the other cosmological data.

Another welcomed result of these data is that they solve the age crisis. The age of the oldest stars, found in globular clusters in the Galaxy, are 11 to 14 Gyr old, close to the age of 12±2 Gyr calculated for the universe without any vacuum energy. Also the age of the Galaxy is determined to be 10 to 20 Gyr from the radiometric dating of isotopes produced in its stars. An accelerating universe with vacuum energy, which was expanding more slowly in the past, adds 1.5 Gyr to the age required to reach its present distance scale. With these new results, the best estimate for the age of the universe is 13.5±2 Gyr. This should give enough time for the first stars to form, presumably within the first billion years of the universe. Finally, if the vacuum energy density remains constant, the acceleration of the expansion will increase in the future, resulting in a cold heat-death for our universe sooner than would have been predicted earlier. Nevertheless, this cold heat-death would occur far into the future.

W. R. W.

Evolution of Distant Galaxies

➤*Analyses of data from the Hubble and other space telescopes reveal the demographics of galaxies in the distant past and the rapid rate of star formation within them.*

During 2000, several thousand galaxies at distances that date them more than halfway back to the big bang were systematically analyzed. Galaxies play a central part in most aspects of the universe and are home for the birth, life, and death of most of the stars. The formation and evolution of galaxies are not as well understood as that of the stars. The farther a galaxy is from Earth, the older it is. An examination of the very early galaxies reveals both their history and the history of the universe. Whereas the behavior of stars depends primarily on their mass, the behavior of galaxies is much more complicated.

Because the universe is expanding, light from very distant galaxies is stretched to much larger wavelengths. This so-called redshift, z, is defined as the fractional change in wavelength, $z = \Delta\lambda/\lambda$. Objects with $z = 1$ are halfway back to the big bang origin. For example, $z = 3$ and $z = 5$

A massive cluster of galaxies called Abell 2218, some 2 billion light-years from Earth. In this Hubble Space Telescope image, the arc-shaped patterns are distorted images of very distant galaxies, five to ten times more distant than Abell, which acts as a magnifying lens. (NASA, A. Fruchter and the ERO Team)

are 84 percent and 91 percent, respectively, back to the big bang. The approximate equation for the time of an object with redshift z is $(1 + z)^{-1.5}$ * (age of universe). The redshift z of very distant galaxies can be measured by observing the wavelength shift of the so-called Lyman break. This wavelength marks the boundary at which light starts to be absorbed by cool hydrogen gas. As of May, 2000, more than one thousand galaxies with $z > 2$ have been measured.

In December, 1995, and September, 1998, the Hubble Space Telescope (HST) took ten-day deep-field images of small regions in the sky near the north and south celestial poles respectively. Each Hubble deep field (HDF) recorded more than 3,000 distant galaxies, some of which had their redshifts measured using spectrometers on the 10-meter (33-foot) Keck telescope in Hawaii. The distant galaxies have their ultraviolet and visible light shifted into the infrared. These galaxies were examined again in 1999 using a new infrared camera on the Hubble Space Telescope, called the Near Infrared Camera and Multi-Object Spectrometer (NICMOS). A comparison of the morphology (structure and size) and common rest frame spectral type (color and luminosity) was made of the distant galaxies of the young universe with the nearby galaxies of the present universe. A common rest frame spectrum was corrected for its redshift so that it represented the actual spectrum emitted by the galaxy.

Some spiral galaxies have loose curving arms of gas, dust, and stars, and others have tightly wound lanes that spiral down to the galaxy's hub in the center.

Ancient Galaxy Morphology

Analyses revealed that large ordinary spiral galaxies similar to the Milky Way are found out to $z = 1.25$. Some spirals have loose curving arms of gas, dust, and stars, and others have tightly wound lanes that spiral down to the galaxy's hub in the center. When viewed edge-on, spirals appear flat like a pancake. Spirals cannot easily survive in the dynamic environment of a dense galaxy cluster. Not only was the universe smaller and denser in the past, but also there were more galaxies. Analysis of the Hubble Space Telescope data found that collisions and mergers of galaxies are more prevalent for $z > 1$. In one instance, as many as five galaxies were observed merging into one. It has been suggested that spirals are more mature galaxies that have evolved after mergers, but this is not yet understood.

Red giant elliptical galaxies are found out to $z = 1.8$. Galaxies become redder as their stars get older. This means that the most distant ellipticals had the bulk of their stars form during the first two billion years after the big bang. Most galaxies beyond $z = 2$ appear to be irregular, structurally disturbed systems. Elliptical galaxies are round or egg-shaped depending on how quickly they rotate. The cause of either spirals or ellipticals rotating is not understood but could result from the merging of smaller irregular galaxies.

Even more puzzling is why and how galaxies form in the first place and what determines their size. The oldest, most energetic objects observed in the universe are quasars, which could be the seeds around which galaxies form. Quasars have been seen as far out as $z = 5.5$ and are known to be small. They are believed to be highly energetic material swirling around and into a black hole. Similarly, many galaxies close enough to be examined appear to have a massive black hole at the center. The Hubble Space Telescope's images of many quasars show a galaxy surrounding each one. However, recent pictures suggest that some quasars may form before galaxies, thereby being a seed for an enveloping galaxy.

Star Production Rates

It is important to examine distant galaxies in the Hubble deep fields at longer wavelength to look through any dust, which probably exists. Both visible and ultraviolet light is absorbed by dust and reemitted as infrared. Infrared from such distances would be Doppler shifted to wavelengths between 0.1 and 1.0 millimeter (0.004 and 0.04 inch) The best instrument for examining the Hubble deep fields at these wavelengths is the Submillimeter Common User Bolometer Array (SCUBA), used on the James Clerk Maxwell telescope in Hawaii. SCUBA examined an area of the Hubble deep fields containing hundreds of galaxies. Only five spots of radiation were identified in the SCUBA images, each near a specific galaxy in the deep-field optical image. Four of these galaxies lie in the redshift range $2 < z < 4$. Because of SCUBA's inferior resolution, some of the spots may contain more than one galaxy. These spots were extremely bright at submillimeter wavelengths, indicating that these galaxies were producing stars at ten to one hundred times the rate suggested by the visible light, which is absorbed by dust.

Interesting conclusions have been reached by combining the SCUBA data with the survey by the European Infrared Space Observatory (ISO) and the submillimeter radiation mapping of the whole sky using the Far Infrared Absolute Spectrophotometer (FIRAS) on the Cosmic Background Explorer (COBE). It looks as if half of the starlight ever emitted has been absorbed by dust and then radiated again in the infrared. Furthermore, the first half of the universe's history had a much greater conversion of cosmic gas to stars than in recent times. Galaxies in the early universe were producing stars at rates up to several hundred times as fast as in galaxies such as the Milky Way. Furthermore, the Keck telescope has found a large fraction of gas clouds with $z > 3$ having already been contaminated by carbon that could only have been made in stars. The stellar birthrate may have peaked less than one billion years after the big bang.

This stellar birthrate makes some sense because the early universe had a high concentration of gas and dust, the materials from which stars form. The mystery is the dynamics of this process and why many galaxies formed so early and quickly. The maps of the early universe are too limited to know whether galaxies formed in clusters. Such clustering should show up in future measurements of the cosmic background radiation, which must pass through these clusters. Hubble's medium deep survey shows significant galaxy clustering as far back as seven billion years. Another puzzle is how galaxies such as the Milky Way can still be creating, on average, a new star each year without running out of their gas supply. Recent studies of nearby massive gas clouds indicate that they are providing a continual inflow of material into the Galaxy.

W. R. W.

Demographics of Black Holes

➤ The discovery of eight new black holes suggests that they exist in abundance.

Astronomers at the June, 2000, meeting of the American Astronomical Association reported the discovery of eight new supermassive black holes, bringing the total to more than thirty. A black hole by definition is invisible because no matter or radiation can escape from it, so all evidence for the existence of black holes is indirect. During the last thirty years, considerable evidence has accumulated that black holes exist in abundance. They exist as either stellar black holes, forming as one member of a binary star system, or as very massive cores near the centers of many, if not most, galaxies. With the best array ever of optical, X-ray, infrared, and radio telescopes, detailed new data are being accumulated on demographics of black holes and the processes occurring around them.

A black hole by definition is invisible because no matter or radiation can escape from it, so all evidence for the existence of black holes is indirect.

The most energetic objects in the universe are quasars, which are thought to be massive black holes gravitationally pulling in huge quantities of material. Quasars are abundant at large distances, which, because of the great travel time of the radiation from them to Earth, places them in the distant past. Recently a quasar was observed with a redshift of 580 percent, which according to the big bang model places it at a time when the universe was only 5.6 percent of its current age. In early times, the universe was much smaller and denser, so much more material was available to power a black hole. As material is sucked in toward the black hole, it spirals into a rapidly rotating disk. The friction, or viscosity, of this so-called accretion disk heats the gas to millions of degrees centigrade and ionizes it. The rapidly rotating ions (charged particles) produce enormous magnetic fields. Once the gas is within the so-called Schwarzschild radius, it is lost forever, thereby creating a doughnut hole in the accretion disk.

The radiation pressure from the luminous accretion disk can become so large that the supply of gas into the black hole is cut off. This luminosity cutoff is called the Eddington limit. The huge currents in the accretion disk can cause magnetic field lines to be twisted along the axis of rotation, which is perpendicular to the disk. Excess matter, which because of the radiation pressure is prevented from entering the black hole, is funneled along these magnetic field lines. The strong magnetic forces eject this material as two high-powered back-to-back jets. Velocities in these jets have been measured up to 99 percent of the speed of

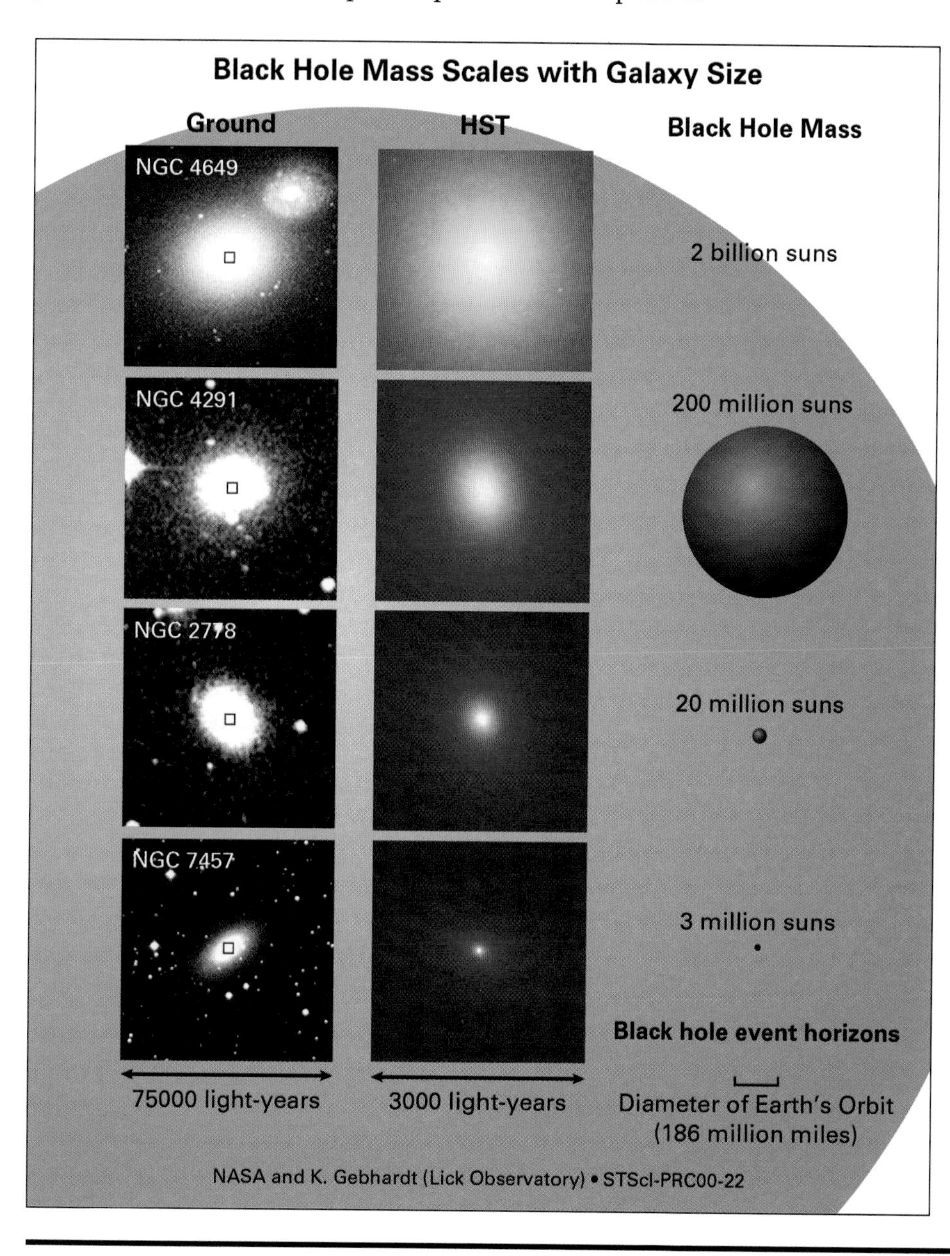

A comparison of black-and-white images of the hearts of four elliptical galaxies taken by ground-based telescopes (left column) and close-up images taken by the Hubble Space Telescope (right column) enabled astronomers to determine that the more massive a galaxy's central bulge of stars, the larger its black hole. (NASA and Karl Gebhardt, Lick Observatory)

light. Unlike the small accretion disks, these jets can stretch to great distances. Both the jets and the accretion disk are strong emitters of radiation, especially X rays, at most wavelengths.

Demographics of Massive Black Holes

The Hubble Space Telescope imaging spectrograph (STIS) can precisely measure the speed of gas and stars around a black hole, thereby determining its mass. Masses of thirty-three black holes at the center of galaxies are found to be from 1 million to 2.4 billion solar masses. Surprisingly, these masses are linearly proportional to the mass of a galaxy's central bulge but unrelated to the size of any disks in these galaxies. For example, the Milky Way with a small bulge has a central black hole of a few million solar masses, whereas giant elliptical galaxies harbor black holes up to a thousand times more massive. This suggests that black holes in galaxies with small bulges are undernourished. All these black holes probably developed along with their host galaxies from infancy. The disks around spiral galaxies, such as the Milky Way, appear to have formed later.

The Chandra X-Ray Observatory has identified the sources of most of the X rays in the universe. About 80 percent of the X rays hitting the Earth come from about 70 million sources, most of which are so dim in visible light that their existence was previously unknown. About one-third of the sources are cores of faint distant galaxies. These cores do not shine in visible light. The probable explanation is that these are quasars, that is, massive black holes, whose visible light from the accretion disk is blocked by a cloud of gas and dust. Another third of the sources do not have a visible galaxy associated with them. These are thought to be cores of even more distant galaxies, from which the visible light of the whole galaxy is blocked by gas and dust. This is a likely scenario for the early universe, in which gas was dense and abundant. Chandra's observations suggest that massive black holes were abundant and formed early in the history of the universe. Along with the STIS data, the Chandra data suggest that they might have been seeds around which new galaxies formed.

Stellar Black Holes

When massive stars die, they undergo a supernova explosion, leaving behind either a neutron star or a black hole. Neutron stars also have strong gravitational fields,

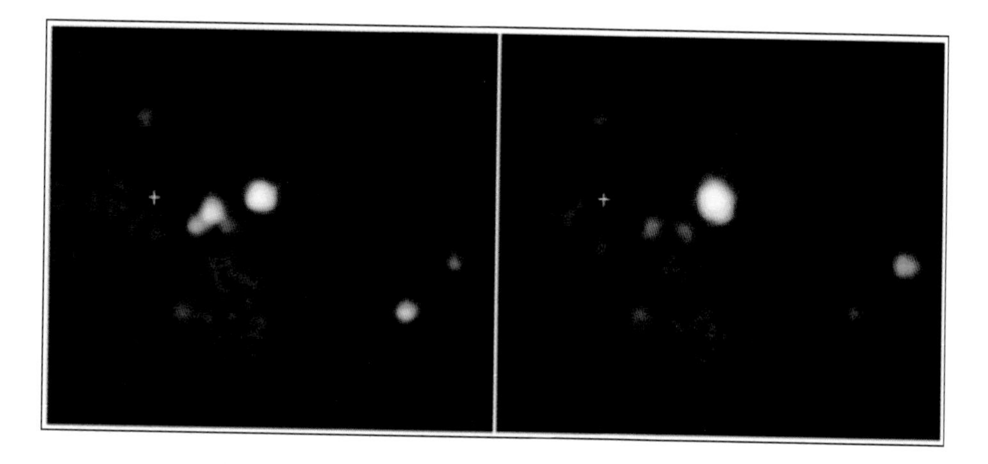

Chandra X-Ray observations of galaxy M82 revealed a bright source (near the center of the left image) that increased in intensity over three months (right image). This increase, along with short-term flickering, are evidence of a black hole—this one outside the nucleus of a galaxy. (NASA/CXC/SAO)

which may result in accretion disks and bipolar jets similar to black holes. Reliable theoretical calculations determine that neutron stars must have a mass smaller than three solar masses. The most reliable way to distinguish a black hole from a neutron star is to measure its mass. Neither of these objects can be easily seen unless a source of material is flowing to them. Until recently, this restricted all stellar black hole candidates to binary stars in which the black hole is collecting material from its companion star. About a dozen such stellar black holes have been tentatively identified in the Galaxy, with masses ranging from four to fifteen solar masses.

Early in 2000, David Bennett of the University of Notre Dame reported the discovery of two dark, isolated black holes, with no noticeable accretion disk or bipolar jets. These black holes were discovered as microlenses when passing directly in front of a background star. The gravitational field of the black hole acts as a lens, bending and magnifying the light from the distant star behind it. The two lensing events lasted five hundred and eight hundred days respectively, which is longer than Earth's orbital period around the Sun. Earth's motion changes the angle of sight from one side of the Sun to the other, thereby affecting the microlensing. Using the Earth-Sun distance, the distance to the lensing objects was calculated to be several thousand light-years. In addition, each of their masses was found to be six times the mass of the Sun, too great to be a neutron star. Only black holes could be so close and massive without emitting enough light to be seen. This discovery suggests that many more black holes are likely to exist.

W. R. W.

The Supernova Connection to Gamma-Ray Bursts

➤ *The study of gamma-ray bursts from brief, powerful, distant events suggests that some are very strong supernovas, called hypernovas.*

Earth is showered more than once each day with brief, strong bursts of gamma rays that last from a fraction of a second to a few minutes and rarely more than an hour. Current research is attempting to understand the sources of these gamma-ray bursts (GRBs), which are some of the most powerful and distant events in the universe.

Beginning with the gamma-ray burst GRB970228, named according to the date it hit Earth, February 28, 1997, astronomers have located afterglows emanating from some of these events. Gamma-ray telescopes can locate the source of each burst to a circular region of the sky with radius greater than two moon diameters, containing many thousands of galaxies. X-ray detectors, which detect the afterglow, can pinpoint sources in the sky up to one hundred times more accurately, thereby allowing optical telescopes to tentatively identify the host star and/or galaxy of the gamma-ray burst.

The most distant gamma-ray burst observed so far occurred on January 31, 2000, about 12 billion light-years away. Gamma rays are the most energetic photons, with X rays the next most energetic category. Much of the intensity of a gamma-ray burst is concentrated between 100,000 electronvolts (100 kiloelectronvolts, or KeV) and 1 million electronvolts (or 1 megaelectronvolts, or MeV), with some gamma rays as high as 50 billion electronvolts (50 gigaelectronvolts, or GeV). One MeV is the energy of an electron accelerated through one million volts.

Gamma-ray bursts were first seen in the late 1960's by the U.S. Department of Defense while it was attempting to monitor suspected nuclear detonations by the Soviet Union in outer space. In April, 1991, the space shuttle orbiter *Atlantis* launched the Compton Gamma Ray Observatory, which included the Burst and Transient Source Experiment (BATSE). Until the controlled destruction of the Compton Gamma Ray Observatory on June 3, 2000, BATSE had identified about 2,500 gamma-ray bursts distributed uniformly across the sky. This isotropic distribution ruled out the Milky Way or other nearby galaxies as being their primary source.

X-Ray and Visible Light Studies of GRB

In April, 1996, the Italian-Dutch satellite Beppo-SAX was launched from Cape Canaveral. It is named after the Italian physicist Giuseppe (Beppo) Occhialini, and SAX is an abbreviation for an Italian name meaning X-ray astronomy satellite. This satellite has a gamma-ray burst monitor (GRBM) that is sensitive to photons between 60 and 600 KeV with a 1-millisecond temporal resolution and wide field camera (WFC) X-ray detectors with a sensitivity range of 2 to 30 KeV. Together, these two instruments identify the gamma-ray burst and locate its position in the sky to an accuracy of about one-fifteenth of a degree or 4 arc minutes (4′). The scheduling of observations using the many instruments aboard Beppo-SAX allows for target of opportunity (TOO) alerts when a gamma-ray burst is detected. Observations are suspended while lower-energy X-ray telescopes are pointed at the gamma-ray burst six to sixteen hours after the burst. More than 80 percent of the gamma-ray bursts show a fading X-ray afterglow, which these lower-energy (0.1 to 10 KeV) telescopes can pinpoint to about 1 arc minute (1′). Within weeks or months, the Hubble Space Telescope and powerful land-based telescopes can also look at visible light at this location.

Twenty-one hours after Beppo-SAX located GRB970228 in the sky, the 4.2-meter William Herschel Telescope in the Canary Islands saw its pointlike visible counterpart. It dimmed so much that a week later, this telescope could no longer see it. However, the Hubble Space Telescope was able to see it on both March 26 and September 5, as well as its fuzzy, very distant host galaxy. That it took so long to fade is evidence that this event is some type of relativistic fireball in which large amounts of material are ejected at very high speeds.

Afterglows of other gamma-ray bursts have been studied. Their electromagnetic spectra indicate large redshifts (stretching of the wavelengths), which is attributed to their great distance and travel time in a continually expanding universe. GRB000131 has the largest observed redshift of 4.5, indicating it is about 12 billion light-years away, when the universe was only 8 percent of its current age. To see such distant events means they are extremely powerful, one hundred times the energy emitted in a supernova explosion. A vivid example of this is GRB990123, observed on January 23, 1999. It is the only GRB whose visible light was seen simultaneously with the gamma-ray burst. Its redshift of 2.6 indicates it was about 9 billion light-years away, yet its visible light was bright enough to see in a pair of handheld binoculars. Its intrinsic power was 10^{16} (or ten million billion) times that of the Sun. Only skill

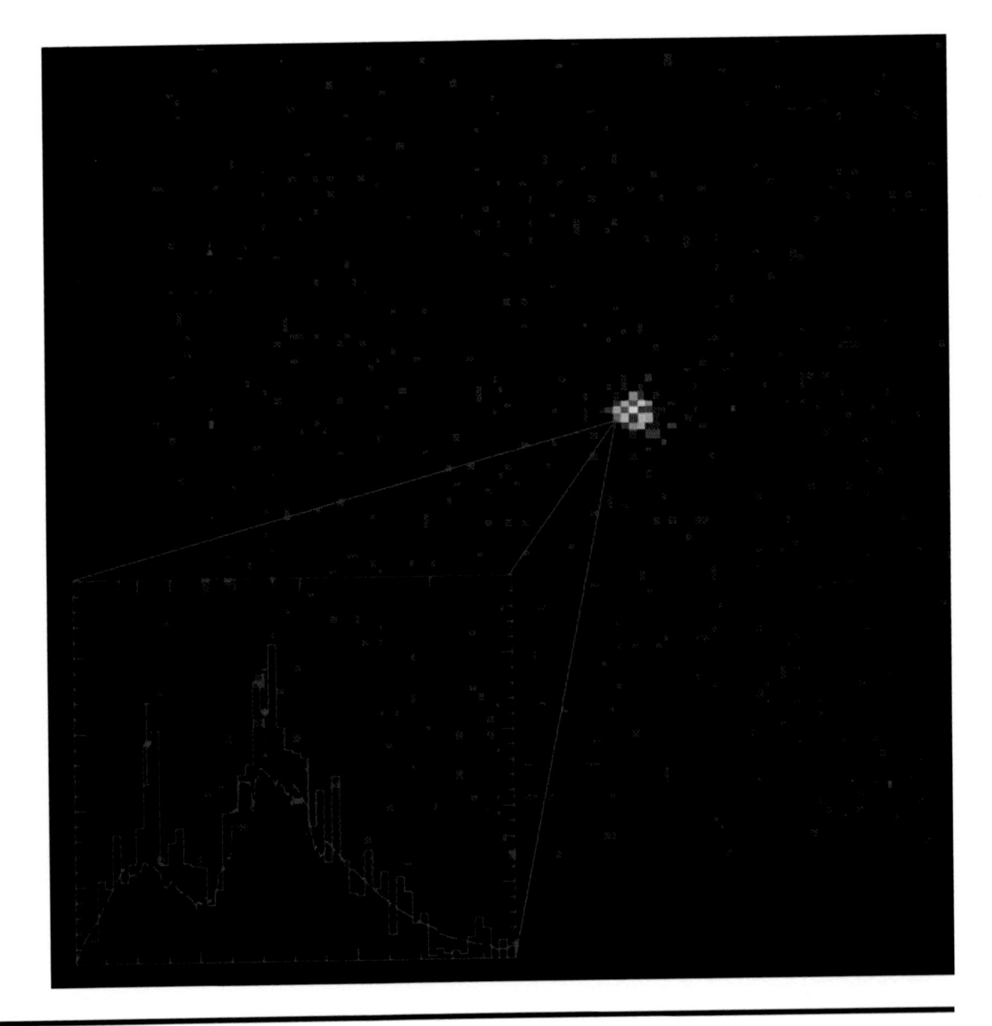

This composite image from the Chandra X-ray Observatory, which detects the afterglow of gamma-ray bursts, demonstrated that Zeta Orionis, one of three belt stars in the constellation of Orion, is a binary star system. (NASA/CSC/Waldron & Cassinelli)

and new technology allowed astronomers to capture this brief flash of visible light.

Normally data collected by orbiting telescopes are stored onboard and downloaded every hour or two. When the tape recorders associated with BATSE failed, the National Aeronautics and Space Administration's Goddard Space Flight Center implemented a real-time data retrieval system called the Gamma-Ray Burst Coordinate Network (GCN). The BATSE data were automatically intercepted at Goddard, a rough gamma-ray burst position calculated, and within seconds, the burst's location was distributed over the Internet to eager observers around the world. The Robotic Optical Transient Source Experiment (ROTSE) is designed to point quickly to a gamma-ray burst location, which it does not need to know very accurately. The response was so quick that a mere 22 seconds after the burst GRB990123 had begun, ROTSE had received the GCN signal and took its first optical image of the event. Because the burst lasted 110 seconds, it was able to take three optical images.

Earth is showered more than once each day with brief, strong bursts of gamma rays that last from a fraction of a second to a few minutes.

Another example of cooperation is the detection of GRB00301C on March 1, 2000. The Earth-orbiting RXTE, Sun-orbiting Ulysses, and asteroid-orbiting Near Earth Asteroid Rendezvous satellite saw this 10-second burst. Within days, six major telescopes were looking at its visible, infrared, and radio waves. The Hubble Space Telescope measured its redshift near 2, placing it two-thirds of the distance across the observable universe. Sudden brightening of the afterglow occurred evenly at all frequencies. This suggests that fiery bubbles, blown off from the gamma-ray burst, passed directly behind a massive object, such as a star. The star would gravitationally focus the light, thereby brightening it uniformly at all frequencies.

Supernovas and Gamma-Ray Bursts

The most similar events to gamma-ray bursts are supernova explosions, which occur after a massive star has converted the light-element nuclear fuel in its core into heavier elements. With nuclear reactions burned out, a sufficient mass will trigger a sudden gravitational collapse followed by recoil, which blows away the outer part of the star. A small neutron star or black hole is left at the center. The total number of supernova explosions in the universe is estimated to be ten thousand times the number of gamma-ray bursts per unit of time, but these explosions are believed to be only one-hundredth as energetic as the strongest gamma-ray burst. Supernova explosions vary depending on detailed properties of the exploding star, and gamma-ray bursts experience even more variation, which suggests a rich morphology.

Some evidence indicates that some of the gamma-ray bursts are also supernova explosions. The best association is GRB980425 with supernova SN1998bw. These events occurred so close together in both time and location in the sky that the odds against accidental alignment are estimated to be one in ten thousand. SN1998bw was very bright, showed very large velocities, and had a very bright radio source. This suggests that this supernova is really more energetic than commonly known supernovas and could be the hypothetical "hypernova," something much stronger than a supernova. Images taken in June, 2000, by the Hubble Space Telescope show that the location of this burst is a great star-forming region, which is a likely place for supernovas and hypernovas.

After detecting a gamma-ray burst, the satellite Swift should be able to direct high-quality X-ray telescopes toward the event in a matter of seconds.

One major difficulty with associating SN1998bw with GRB980425 is that the X-ray afterglow of the gamma-ray burst observed with Beppo-SAX's lower-energy X-ray telescopes near the location of SN1998bw was 100,000 times weaker than the initial X rays and was varying very little over a ten-day period. In another example, the exceptionally luminous SN1997cy is aligned with GRB970514 with a 1 percent chance of association. These are the two best associations.

One of the best observational evidences for the existence of hypernovas is the suggestion of two hypernova remnants in the nearby galaxy M101. One of these so-called nebulas, MF83, has radius greater than 430 light-years, making it one of the largest supernova remnants known. The other, NGC5471B, is rapidly expanding. Both have X-ray luminosity ten times brighter than the brightest supernova remnants. It is suggested that hypernovas result in the formation of a black hole from an extremely massive star or involve the merger of two very dense objects, such as neutron stars. M101 has more vigorous ongoing star formation than other nearby galaxies, which may explain why it has two relatively rare hypernova remnants with ages less than a million years.

Estimates of the energy content of a gamma-ray burst are made assuming isotropy, which means the energy is being released uniformly in all directions. If Earth is seeing a collimation or jet stream, the total energy of a typical gamma-ray burst should be a factor ten smaller than the original estimate. In this case, the most energetic gamma-ray burst would be only a factor of ten greater than the supernova. The best circumstantial evidence that gamma-ray bursts are some type of collimated supernova is that the afterglow of a gamma-ray burst usually dims faster than a supernova. One expects this dimming to be faster as a collimated flow slows and spreads laterally. For collimation to exist, there cannot be a thick hydrogen cloud around the event. SN1998bw, which is the best candidate for a gamma-ray burst, is a type 1c supernova, meaning that it has a negligible hydrogen and helium envelope.

Beppo-SAX was scheduled to be shut down in April, 2001. The year 2003 should see the launching of satellite Swift, so-named because of its fast-moving detection system. After detecting a gamma-ray burst, it should be able to direct high-quality X-ray telescopes toward the event in a matter of seconds. Further in the future will be the launching of the Gamma-Ray Large Area Space Telescope (GLAST), which will be able to measure photons between 10 and 300,000 MeV.

W. R W.

Solar Storms at the Sunspot Maximum

➤ The eleven-year cycle of solar activity peaks, influencing electrical systems and communications on Earth.

The year 2000 marked the peak of the Sun's eleven-year cycle of sunspots and associated solar storms. More than ever before, this increased solar activity was both a challenge and an opportunity for science and its high-technology products. Sun storms can produce surges of charged particles that are dangerous to astronauts and can disrupt satellites, power transmission grids, radio communications, and other sensitive electrical systems. The National Aeronautics and Space Administration (NASA) and other international space agencies prepared well for this millennial solar maximum with a fleet of space solar observatories and by placing solar-wind detectors on several other space probes. These preparations have paid off in a wealth of new information about the Sun and a growing ability to make invaluable predictions of solar storms.

The first large solar magnetic storm of the year occurred on April 6, 2000, producing a colorful auroral display that lit up the night sky at many northern latitudes. This storm and its geomagnetic counterpart provided a showcase for the new fleet of space weather stations located at various points in the inner solar system. The first detection of the storm was by the Advanced Composition Explorer (ACE) spacecraft positioned between the Sun and Earth. ACE detected a shock wave in the solar wind accelerated by a magnetic storm on the Sun to between 375 and 600 kilometers (233 and 372 miles) per second. This blast of protons and other atomic particles hit Earth's magnetic field less than an hour later, producing a category 4 geomagnetic storm on NASA's new space weather scale (category 5 is the highest). Images from the U.S.-European Solar Heliospheric Observatory (SOHO) revealed the origins of this outburst to be a solar flare that produced a large ejection from the corona of hot gases surrounding the Sun two days before ACE detected the resulting shock wave.

SOHO detected an even larger solar storm on July 14, 2000, the biggest solar-radiation storm since October, 1994. The next day, the solar wind velocity soared to nearly

1,000 kilometers (621 miles) per second, producing bright northern lights as far south as 40 degrees latitude and disabling Japan's Advanced Satellite for Cosmology and Astrophysics (ASCA) satellite. Two days later, a powerful shock wave struck Earth's magnetic field and triggered a category 5 geomagnetic storm, disrupting some satellites and short-wave radio communications and prompting a call to delay a Russian space launch. On August 9, a coronal mass ejection on the Sun produced a shock wave that hit Earth's magnetosphere on August 11, producing a category 3 geomagnetic storm and northern lights that could be seen as far south as Los Angeles. On September 22, the largest sunspot in nine years came into view, covering an area a dozen times the surface of the entire Earth.

Solar Storms and Their Effects

Solar storms are believed to be driven by energy release from the Sun's magnetic field and produce a variety of effects. Solar flares are rapid increases in brightness in active regions near sunspots and produce intense radiation, including X rays that often exceed those of the entire Sun. Sunspots have a dark central region with lower temperatures than the surrounding solar surface, probably caused by the action of their strong magnetic fields inhibiting the convection of hot gases to the surface. They last from a few days to a few months, rotating with the Sun's twenty-seven-day period, and varying in number over an eleven-year period that is directly related to solar activity. Solar prominences are loops of magnetic fields with hot gases trapped inside, which sometimes erupt and rise off the Sun's surface. Coronal mass ejections (CMEs), discovered in the 1970's, are explosive events on the surface of the Sun. They can propel as much as ten trillion kilograms of material up through the super-hot corona and into the solar wind, which can reach Earth a few hours later and influence geomagnetic activity.

In August, a coronal mass ejection on the Sun produced a shock wave that hit Earth's magnetosphere, producing a category 3 geomagnetic storm and northern lights that could be seen as far south as Los Angeles.

Little was known about solar storms and their effects on Earth until after the invention of telegraphy in 1841 and the telephone in the 1870's. During solar storms, charged particles from the Sun produce changes in Earth's magnetic field that are strong enough to induce powerful electrical currents in wires. Long telegraph and telephone lines began to experience surges, some large enough to nearly electrocute telegraph operators. At the Royal Greenwich Observatory in 1879, William Ellis made the first solar storm forecast when he announced a correlation between sunspots and strong auroral activity (northern lights) produced by charged solar particles deflected along Earth's magnetic field into northern latitudes. He warned about possible magnetic storm damage during the next maximum in the Sun's eleven-year sunspot cycle. Confirmation followed in 1881 when reports of electrical surges began to accumulate again in Boston and London.

During the March, 1940, magnetic storm, voltages up to 2,600 volts were recorded in the Atlantic cable, and electrical services were disrupted from New England to the upper Midwest. A February, 1958, storm caused severe interruptions of telephone service on Western Union's North Atlantic telegraph cable and blew a power transforimer in British Columbia. In August of 1972, another storm produced 25,000-volt surges in South Dakota and Wisconsin and another power transformer exploded in British Columbia. The most dramatic storm occurred on March 13, 1989, during the peak of the last sunspot cycle, when power was shut down for nine hours in much of Quebec.

A satellite outage occurred on May 17, 1998, shutting down service for 80 percent of North American pagers and credit card transactions.

Satellites are especially vulnerable to solar storms. The Geostationary Operational Environmental Satellite-7 (GOES-7) weather satellite lost half of its solar cells when high-energy protons were released from the Sun during the March 13, 1989, storm, reducing its operating life span by half. A malfunction in the GOES-8 weather satellite followed a solar storm on January 11, 1997, and shortly thereafter AT&T's Telstar 401 satellite experienced a massive power failure. Satellites appear to be most vulnerable to high-energy electrons produced by solar storm activity, especially when they are injected into the geosynchronous orbits in which satellites match Earth's rotation. A dramatic satellite outage on PanAmSat's Galaxy IV satellite occurred on May 17, 1998, during the peak of such electron activity, shutting down service for 80 percent of North American pagers and credit card transactions.

Detection and Discoveries

The new solar space observatories are well positioned for early detection and warning of solar storms and pro-

vide a variety of instruments for studying their causes and effects. The Ulysses spacecraft is a joint effort of the European Space Agency (ESA) and NASA. It was launched on October 6, 1990, toward Jupiter and used Jupiter's large gravitational field to accelerate it into a nearly polar six-year orbit around the Sun, reaching high northern latitudes above the Sun at the maximum of solar activity. The Yohkoh satellite was launched from Kagoshima, Japan, on August 31, 1991, as a project of Japan's Institute for Space and Astronautical Sciences, and includes instruments from the United States and Great Britain. It is designed to study X rays and gamma rays from the Sun produced by solar flares.

NASA's WIND spacecraft, launched November 1, 1994, operates in a small circular "halo" orbit 1.5 million kilometers out in space between Earth and the Sun where their gravitational forces are balanced. It is designed to measure the solar wind and Earth's magnetosphere. The joint NASA and ESA Solar Heliospheric Observatory (SOHO) was launched December 2, 1995, into a similar halo orbit and was designed to study the internal structure of the Sun, the heating of the solar corona, and the origin of the solar wind. NASA's Advanced Composition Explorer (ACE) mission was launched August 25, 1997, from the Kennedy Space Center in Florida into a large elliptical orbit at the balance point between Earth and Sun. It provides near real-time solar wind data and can give an advance warning of geomagnetic storms about an hour before they reach Earth.

The Transitional Region and Coronal Explorer (TRACE) spacecraft was launched on March 30, 1998, from Vandenberg Air Force Base in California. It has four telescopes to study ultraviolet radiation from the Sun in conjunction with SOHO, especially in the transition region between the 6,000-degree surface of the Sun and the corona with temperatures up to 16 million degrees. An announcement in September, 2000, indicated that TRACE had obtained ultraviolet images of the large arcs of energetic gas ejections called coronal loops. These images of the loops that make up the solar corona revealed that they are heated to nearly 2 million degrees Celsius at the base of the loop, contrary to a previous theory that thought they are heated uniformly.

In one of the more important discoveries resulting from this new focus on the Sun, the SOHO spacecraft has been credited with finding the source of the high-speed solar wind, which travels up to 3 million kilometers (1.8 million miles) per hour, or about twice the speed of its low-speed counterpart. Large convection cells below the surface of the Sun are associated with strong magnetic fields, which can accelerate hot charged particles. Hot gases in the solar wind source regions emit ultraviolet light, which was detected by SOHO's instruments, revealing that the high-speed solar wind came from cracks between the convection cells.

Solar physicists analyzing data from SOHO have found that waves moving through the Sun from explosive events on its surface make it possible to detect disturbances that start on the far side of the Sun and later rotate into view. This gives up to two weeks of warning about potential solar storm activity that could affect Earth. Such advance warnings would be especially valuable to protect astronauts in a space shuttle or on the International Space Station.

J. L. S.

Orbiting X-Ray Telescopes Map the Nonvisible Universe

➤ *Chandra and XMM-Newton reveal unexpected properties of both small and large objects.*

X rays are thousands of times more energetic than photons of visible light, which allows them to penetrate opaque clouds of interstellar gas and dust. The most energetic and hottest objects in the universe produce large quantities of X rays. These objects range in size from newly formed stars to larger supernova remnants and black holes gulping up huge quantities of material and to gigantic clouds of hot tenuous gas with sizes stretching millions of light-years. Many features of these objects are visible to humans only with X rays. However X rays do not easily penetrate Earth's atmosphere. Recently, two new orbiting X-ray telescopes have advanced X-ray astronomy by largely the same degree as the Hubble Space Telescope advanced visible astronomy. The Chandra X-Ray Observatory started regular science observation late in 1999, and the X-Ray Multi-Mirror space telescope (XMM-Newton) began regular observing in June, 2000.

The space shuttle *Columbia* launched Chandra, named after Nobel Laureate Subrahmanyan Chandrasekhar, on July 23, 1999. It is the largest and heaviest satellite ever launched by the space shuttle. After five firings of its own integral propulsion system, Chandra was placed in a highly elliptical orbit stretching more than one-third of the distance to the Moon. The fraction of the sky covered by Earth is small over most of Chandra's 63.5-hour orbit, allowing for long uninterrupted observations lasting more than two days. The XMM-Newton, named in honor of Sir Isaac Newton, was launched December 10, 1999, by the European Space Agency on an Ariane 5 rocket. Its 48-hour elliptical orbit is somewhat smaller than Chandra's. X rays are much more difficult to reflect by mirrors or refract by lenses than is visible light. X rays will reflect only at steep grazing angles. Therefore, the heart of Chandra's telescope resembles a huge set of polished, curved, open-ended barrels, nested inside one another. XMM has three separate barrels, each holding fifty-eight concentric cylindrical mirrors, together totaling the surface area of a tennis court.

The Chandra X-Ray Observatory's high-resolution images allow scientists to gain a better understanding of space. This image captures Sirius A, the brightest star in the northern sky, and its near neighbor, Sirius B, a much dimmer star that is harder to see with an optical telescope. (NASA/SAO/CXC)

Because of the steep angles, the effective cross-sectional area of XMM varies between a mere 0.43 and 0.18 square meter (0.5 and 0.22 square yard), depending on the X-ray energy. However, this effective collection capacity is six to twenty times that of Chandra.

Chandra has spatial resolution of 0.5 arc second, a factor of ten better than XMM and capable of reading 1-centimeter (0.4-inch) newspaper print at a distance of half a mile. Both telescopes have an assortment of excellent spectrometers providing detailed energy spectra of X-ray sources. These allow studies of structure, not only in X-ray intensity but also in temperature, velocity, and chemical composition. XMM has a better spectrometer, but it must be cooled, and the coolant will run out in two years. XMM can detect the highest-energy X rays, up to 700 kiloelectronvolts, which is a factor of seventy larger than the highest-energy X rays detected by Chandra. An electron accelerated through 1,000 volts acquires 1 kiloelectronvolt of energy. Both telescopes may operate up to the year 2010. Another first-class X-ray telescope, ASTRO-E, launched on February 10, 2000, by the Japanese, was not put into a sustainable orbit because of a rocket failure.

Brown Dwarfs and Young Stars

In December, 1999, Chandra studied continuously for twelve hours a brown dwarf 16.2 light-years away. This brown dwarf is the size of Jupiter but much more massive. The mass of every brown dwarf, however, is insufficient to ignite into a star. For more than nine hours, Chandra saw nothing, but then suddenly it observed a visible flare lasting two hours. Comparable in strength to low-level solar flares, it surprised the astronomers. The brown dwarf lacks a hot ionized atmosphere in which solar flares occur, but it most likely has a strong magnetic field that generated this flare. The flare probably originated in a hot ionized gas below its surface.

The closest massive star-forming region to Earth is the Orion nebula, 1,500 light-years away. In a single image of the nebula, Chandra resolved a thousand faint X-ray-emitting stars. With sensitivity twenty times better than previous X-ray telescopes, Chandra was able to resolve individual stars in their natal cloud of gas and dust. Young stars are known to be much brighter in X rays than older stars. The source of the X rays is thought to be violent flares in strong magnetic fields near the surfaces of young stars. The physical causes and evolution of this activity are not understood. It is hoped that this new data will shed light on this phenomenon.

A massive star sheds a shell of material thousands of years before undergoing a supernova explosion. The supernova explosion creates a shock wave that, after ten or more years, collides with this material, heating it to millions of degrees, thereby creating a supernova remnant. Chandra in conjunction with the Hubble Space Telescope (HST) is for the first time witnessing the birth of a supernova remnant in the study of supernova SN1987A. The Hubble Space Telescope is observing gradually brightening hot spots as the shock wave hits the shell of material. Chandra is observing the much hotter gas behind the shock wave, which can be seen only with X rays.

Both Chandra and XMM-Newton are mapping nearby supernova remnants. XMM-Newton has provided a detailed study of the Tycho supernova, named for Danish astronomer Tycho Brahe, who saw it explode in 1572. The distributions and expansion velocities of chemical elements such as sulfur, calcium, and iron are measured. There is clear evidence that some elements formed in different parts of the exploding star and subsequently mixed.

X-Ray Hot Clouds, Distant Galaxies

About 5 billion light-years away is a cluster of one hundred galaxies with the giant elliptical galaxy 3C295 at its center. Chandra reveals that these galaxies are embedded in a cloud of 50-million-degree-Celsius gas with a diameter of about 2 million light-years. This X-ray cloud contains enough material to create another thousand galaxies. A much more complicated cluster is Abell 2142, with a diameter of 6 million light-years in which hundreds of galaxies are embedded. Chandra is able to measure variations of temperature, density, and pressure of this cluster's atmosphere with unprecedented resolution. Chandra finds a relatively cool 50-million-degree-Celsius central region embedded in a large elongated cloud of 70-million-degree-Celsius gas, all of which is surrounded by a faint (tenuous) 100-million-degree-Celsius gas. Apparently this system has resulted from the collision and merger of two smaller clusters.

Chandra has also studied the Antennae galaxies, a recent merger of two galaxies 50 million light-years away. With unprecedented detail, the images show superbubbles 5,000 light-years in diameter, consisting of extremely hot gas enriched with oxygen, iron, and other heavy elements. Also seen are ninety-six pointlike sources, thought to be neutron stars and black holes. Presumably the collision of gas and dust as the galaxies merged created shock wave compressions leading to the rapid birth of millions of massive stars. Massive stars have relatively short life spans, ending in supernova explosions. These supernovas provided the heat and heavy elements for the bubbles, and the supernova remnants are the neutron stars and black holes, each with hot material swirling around it.

During the early days of space exploration, a pervasive glow of X rays from all directions was observed. With Chandra and XMM, astronomers can resolve about 80 percent of this glow into specific sources. About two-thirds of these are attributed to tens of millions of distant galaxies that cannot be seen with visible light but whose cores shine bright in X rays. These galactic nuclei are thought to be so-called type 2 quasars with an excess of high-energy X rays. A cloud around the quasar filters out the visible light and low-energy X rays. Quasars are thought to be black holes, and this new discovery greatly increases the number of black holes in the universe.

W. R. W.

More Extrasolar Planets

➤ *Ten planets are discovered in surprising orbits outside the solar system.*

Discovery of the first two extrasolar planets smaller than Saturn was announced on March 29, 2000. This announcement was followed in May and August by reports of some eight new extrasolar planets discovered in the Southern Hemisphere, including four near the mass of Saturn. Some thirty planets discovered in the last five years in orbits around nearby sunlike stars have exceeded the size of Jupiter, which is larger than the combined mass of all the other planets in its solar system at 318 times the mass of Earth. More than a dozen are smaller than Jupiter, and three of the new planets are smaller than Saturn, the second largest planet in the solar system at 90 Earth masses. The smallest of the new planets may be as small as only 40 Earth masses, showing that astronomical techniques have become sensitive enough to detect a planetary system as small as the one containing Earth.

The discoveries announced March 29 were made by astronomers Geoffrey Marcy of the University of California, Berkeley, and Paul Butler of the Carnegie Institution, Washington, D.C. The two new planets smaller than Saturn were found in a database of more than a thousand sunlike stars by using a revised computer program that corrects for irregularities in their equipment at the 10-meter (33-foot) Keck telescope on Mauna Kea, Hawaii. They are roughly one-fifth and one-third the size of Jupiter. The first is 109 light-years from Earth in the constellation Monoceros, orbiting the star HD 43675. Like most of the other extrasolar planets, it is very close to its host star with a period of only 3 days. The second and smaller planet is 117 light-years away in the constellation Cetus, orbiting the star 79 Ceti with a period of 74 days. It has a mass of only about two-thirds that of Saturn and an orbit about two-thirds that of Mercury.

Some thirty planets discovered in the last five years in orbits around nearby sunlike stars have exceeded the size of Jupiter.

The discoveries announced in May were made by a

Swiss team headed by Michael Mayor at the La Silla Observatory in Chile. The smallest of their six new planets (around the star HD 168746) was about one-fourth the size of Jupiter (about 80 percent the mass of Saturn) with an orbit of 6.4 days. Two of the others were slightly more massive than Saturn. One of these two (around HD 83443) had the shortest orbital period to date, slightly less than 3 days. The other three had minimum masses of about one, two, and three times that of Jupiter and were in eccentric orbits of 119, 443, and 230 days.

In August, a second planet around HD 83443 was announced with a minimum mass as small as one-half that of Saturn and a period of 30 days. Another announcement in August by William Cochran at the McDonald Observatory of the University of Texas was a planet around the star Epsilon Eridani, the nearest yet at only 10.5 light-years from Earth, with period and mass similar to those of Jupiter but a highly eccentric orbit that appears to preclude any Earth-like planets in the system.

Search Methods

Marcy and Butler have discovered about two-thirds of the extrasolar planets detected since the first planet orbiting a sunlike star was found by Mayor and Didier Queloz in 1995 around the star 51 Pegasi about 40 light-years away. These discoveries were made possible by a new generation

of computers and optical instruments to analyze the light from nearby stars. Because planets are about a billion times fainter than the host star, they are virtually impossible to see directly. An indirect method involves searching for a tiny wobble in the motion of a star as it and any companions it may have orbit about their common center of mass. Although the gravitational interaction between a star and planet is too small to observe directly, the back-and-forth velocity of the star increases and decreases the wavelength of its light, causing an alternating Doppler shift toward the red and blue ends of its spectrum.

The amount of a star's Doppler shift determines its velocity as it wobbles. The shift in the wavelength caused by a Jupiter-size planet is only about one part in ten million. The periodic variation in wavelength reveals the period of a planet's orbital motion. The velocity of the star and the period of its motion can be analyzed to determine the radius of the orbit and the minimum mass of the planet. However, the unknown inclination of its orbit allows for a larger wobble than its apparent radial motion and, therefore, a larger mass up to a factor of about two. The shape of the periodic variation curve reveals the shape of the orbit. A circular orbit produces a perfect sine wave, and an eccentric orbit produces an irregular variation that can be analyzed by computer to determine the orbital shape.

Because planets are about a billion times fainter than the host star, they are virtually impossible to see directly. An indirect method involves searching for a tiny wobble in the motion of a star.

Detecting these changes requires the use of a sensitive spectrometer to measure wavelengths. The light first passes through an iodine-vapor cell, which absorbs some of it, producing a dark line at a well-known wavelength superimposed on the spectrum. Shifts in the wavelength of the light can be measured relative to the standard iodine line to obtain a precise measurement of the motion of the star. The score of Jupiter-size planets discovered by Marcy and Butler involved velocities as small as 8 meters (26 feet) per second, less than the 12-meter-per-second (39-foot-per-second) motions of the Sun caused by Jupiter. Scientists can detect star wobbles of only 3 meters (10 feet) per second, making it nearly possible to detect Neptune-size planets. Although this is still not precise enough to detect Earth-size planets, it could reveal planetary systems with outer gas planets having longer orbits of more than about ten years, similar to our solar system.

Planets condense from materials in the disk produced by the collapse and rotation of a newly forming star and are believed to have an upper limit in size of about ten Jupiter masses. In standard theories of planetary formation, matter in the protoplanetary disk collides and coagulates into planetesimals ranging up to 10 kilometers (6 miles) in size. These planetesimals attract each other by gravity to trigger a sequence of mergers that produces the inner rocky terrestrial planets and the outer rock-and-ice cores that seed the giant gas planets in the cooler regions farther from the central star.

Scientists can detect star wobbles of only 3 meters (10 feet) per second, making it nearly possible to detect Neptune-size planets.

Because giant planets require such a large amount of material, they should form only in the cooler regions several astronomical units from their host star (1 AU = Earth-Sun distance of 150 million kilometers, or 93 million miles). Only in these outer expanses of the disk (greater than about 5 AU) is it cool enough for ice to form out of water molecules, roughly tripling the amount of solid material available for planet making. When the ice-and-rock core reaches about 10 Earth masses, it begins to attract huge amounts of hydrogen and helium gases and expands to about 1 Jupiter mass (318 Earth masses) until its gravity can begin to tear a gap in the disk that feeds it, thus stopping its growth. With this model, theorists were successful in accounting for the solar-system sequence of inner rocky planets (Mercury to Mars) and outer gas planets (Jupiter to Neptune) beyond 5 AU.

Results and Prospects

Extrasolar planet discoveries around sunlike stars have revealed two new and unexpected types of planetary objects: hot-Jupiter planets with small circular orbits and eccentric-Jupiter planets with elongated orbits. The masses of these thirty-odd planets vary almost continuously from roughly two-tenths of Jupiter's mass to eleven times the mass of Jupiter. Hot-Jupiter planets have periods from 3 days up to about 40 days with surface temperatures in excess of 1,000 degrees Celsius (1,832 degrees Fahrenheit). Eccentric-Jupiter planets have periods from about 60 days to 4.5 years, with orbits between 0.2 and 2.5 AU. Only two multiplanet systems have been detected. Three planets have been found around the star Upsilon Andromedae with periods of 4.6 days, 241 days, and 3.5 years, and two planets orbit HD 83443 with periods of 3 days and 30 days.

The discovery of extrasolar planets around sunlike stars may at first seem to offer new hope for the existence of planetary systems that would support extraterrestrial life. However, the unexpected nature of these planets has raised new doubts about the possibility that any of them might harbor life. The bizarre nature of the newly discovered planets has initiated a new generation of theories about planet formation and the uniqueness of our solar system. Evidence so far seems to indicate that our solar system is highly unusual in its life-supporting planetary arrangement.

Revised theories suggest that giant gas planets might have formed beyond 5 AU (with periods of about 12 years) in a dense protoplanetary disk, which then slowed them down and caused them to spiral inward. Such a process would obliterate any small, inner terrestrial planets congenial to life as it is known on Earth. New theories suggest that two or more super-Jupiters forming from a dense protoplanetary disk might then interact with each other gravitationally, causing some to be thrown into eccentric orbits or even tossed free of the star. Such eccentric giants would gravitationally disturb and eventually collide with smaller inner planets, again precluding life-supporting planets such as Earth.

Although the new Saturn-size planets are like their larger cousins in having very rapid orbits closer to their host stars than the nearest planet to the Sun, they do offer hope that even smaller planets can be detected. The National Aeronautics and Space Administration's Space Interferometry Mission in 2006 will make it possible to detect Earth-size planets. However, they may not be able to support life without giant gas planets in larger orbits such as those of our solar system. Recent studies have shown that Jupiter in its present stable orbit beyond Earth sweeps up most killer asteroids and comets, reducing the Earth-collision rate about a thousand times. None of the extrasolar planets found so far has the kind of orbit that would protect inner planets enough to permit the development of life, let alone its higher forms.

J. L. S.

The Global Surveyor Finds Hidden Mars

➤ *Data provide evidence of ancient underground rivers and more recent surface seepage on the third planet from the Sun.*

After the double disappointment of two failed missions to Mars, the National Aeronautics and Space Administration (NASA) finally began to reap a waterfall of new information in 2000 from its orbiting Mars Global Surveyor (MGS). The 1999 losses of the $125 million Mars Climate Orbiter and the $165 million Mars Polar Lander were eclipsed by the stunning results from the Global Surveyor, designed to map the surface of the Red Planet. The Global Surveyor has provided evidence of vast underground rivers flowing from the southern highlands toward the northern plains a few billion years ago and even more surprising indications of groundwater seepage within the last few million years. Photographs released as part of a report in December revealed layered outcrops similar to those formed on Earth within bodies of water.

The Mars Global Surveyor arrived at Mars in 1997, shortly after the dramatic Mars Pathfinder landing with its Sojourner rover on July 4, 1997. However, the Global Surveyor received little attention, especially because its mapping mission was delayed by a year because of damage to some of its hardware during its 370-million-kilometer (229-million-mile) flight from Earth. After an initial insertion into a highly elliptical 11.6-hour orbit, it was eventually brought down to a circular 2-hour orbit in which its camera, operated by San Diego-based Malin Space Science Systems, began to reveal unprecedented details with a resolution ranging down to about 1 meter (slightly more than 3 feet). The evidence for water on Mars in both the distant and near past increases the possibility that life also might have existed at one time on Earth's neighbor.

The first flyby of Mars, by Mariner 4 in 1965, revealed a dead, moonlike planet that appeared to be disturbed only by impact craters. Mariner 9 and two Viking spacecraft in the 1970's revealed dry riverbeds and volcanic mountains that suggested ancient geologic activity some three or four billion years ago. Further studies of Martian craters and meteorites in the

1980's showed that many lava plains were only a little older than one billion years, with volcanic cones, such as Olympus Mons, as young as a few hundred million years. The Mars Global Surveyor mission, supported by new dating of Martian meteorites, is revealing an active Mars with recent volcanic activity and flowing water over much of its geologic history and perhaps continuing into the present time.

Underground Rivers to an Ancient Ocean

Results from the Global Surveyor's laser altimeter, which has mapped precise altitudes across the entire planet, were published in the March 10, 2000, issue of the journal *Science*. The research team, led by Maria Zuber of the Massachusetts Institute of Technology, used this altitude data along with an analysis of gravitational measurements showing variations in the thickness of the Martian crust to reveal a new picture of the planet's history. These global readings indicate higher altitudes in the north, where the crust ranges up to 70 kilometers (43 miles) deep, and in northern lowlands with a crust of a fairly uniform thickness of some 35 kilometers (22 miles). Mars appears to have two distinct crustal zones, which resemble the thicker crust of Earth's continents and the thinner oceanic crust. These differences show that billions of years ago, when Mars was warmer and more Earth-like, water could have flowed downhill from south to north, forming a northern ocean.

Zuber and her colleagues have found a new clue in their gravity data showing how water might have been transported from south to north in spite of the lack of visual evidence for extensive water channels. Gravity-field readings indicate broad strips of lower density beneath the surface of the southern highlands some 200 kilometers (124 miles) wide and 1 to 3 kilometers (0.6 to 1.9 miles) deep, stretching more than 1,500 kilometers (932 miles) in length. These features suggest underground flood channels that branch out into the northern lowland plains. Researchers believe that these purported channels were cut into the surface by ancient floodwaters flowing from south to north and then covered over by relatively light sediment. The scale of these buried channels indicates that they could have transported sufficient quantities of water to form an ancient ocean in the northern lowlands.

This mosaic of two images taken by the Mars Global Surveyor reveals deep channels running down a crater wall in the planet's southern hemisphere. Scientists believe these channels are evidence of water on Mars. (AP/Wide World Photos/ NASA)

One confirmation of the idea that water flowed from southern highlands and collected in a northern ocean comes from additional Global Surveyor altimeter measurements. These data show a large, smooth-floored basin surrounding the north polar region with a shelf around it, similar to coastal shelves that form from delta deposits along the shores of seas on Earth. The altimeter measurements showed that this shelf was at a uniform level around the basin, suggesting a "sea level" of an ancient shoreline.

One model suggests that water flowing into the northern lowland region might have been under high pressure from the weight of overlying sediment. It could then have burst from the ground in catastrophic floods, cutting the larger outflow channels such as the Ares Vallis north-draining riverbed, where the Pathfinder probe landed in 1997. Some of these channels begin abruptly, without smaller runoff channels to feed them, indicating that the water feeding these channels came from underground rather than surface runoff. Global Surveyor photographs reveal rows of collapsed pits before the start of some channels, suggesting that they extend underground in the upstream direction, and the ground collapsed here and there into the subsurface flood channel.

Billions of years ago, when Mars was warmer and more Earth-like, water could have flowed downhill from south to north, forming a northern ocean.

Evidence of relatively recent geologic activity on Mars also comes from about a dozen Martian meteorites, identified by their unique elemental composition. Radioactive dating and mineralogical studies indicate that some of these meteorites solidified from molten lava only about 100 to 700 million years ago, within the last 10 percent of the planet's lifetime. Several of these younger meteorites have rustlike minerals or carbonate deposits inside them that most likely formed by the action of liquid water. Global Surveyor photos have also shown young lava flows in the Elysium Planitia region of Mars, where analysis of impact craters reveals less than 1 percent as many craters as found in older regions.

Groundwater Seepage Possibilities

One of the most dramatic and surprising results from the Mars Global Surveyor mission was published in the June 30, 2000, issue of *Science*. Prepublication announcement of the article entitled "Evidence for Recent Groundwater Seepage and Surface Runoff on Mars" by Michael Malin and Kenneth Edgett of Malin Space Science Systems in San Diego made headlines around the world. The possibility of liquid water flowing recently on the surface of Mars was both surprising and confusing because the extremely low pressures and temperatures (averaging –50 degrees Celsius, or –58 degrees Fahrenheit) should result in rapid vaporization or freezing of any water reaching the surface.

The evidence for water flowing on the surface of Mars came from about 150 high-resolution images of 120 locales out of more than 65,000 pictures that were reviewed. These photographs show gullies on the sides of crater walls and valleys, which look as though emerging water cut into these steep slopes and flowed down to form a channel-ridden apron of debris. These "aproned alcoves" are so free of impact craters that they must be very young, forming any time between one million years ago and the present. Compounding the mystery of these images that apparently resulted from flowing water is that they appear above 30-degree latitudes, mostly in the southern highlands, and usually on the colder pole-facing slopes (about 50 percent face south and 20 percent north).

There is general agreement that a fluid, most likely water, produced these Earth-like features as it emerged from an aquifer cut by a crater or valley. How such water was kept warm enough to flow on or near the surface is a mystery, because there is no evidence of volcanic heat sources and current theories of the interior of Mars lack an adequate geothermal source. Although they cannot explain how water reaches a crater wall, Malin and Edgett suggest that equator-facing slopes are usually warmed enough by the sun so that any surface runoff is evaporated fast enough to avoid most erosion. On the colder, pole-facing slopes, water freezes and forms an ice barrier. If aquifer pressure builds enough, it would eventually break the barrier and release a burst of water to form the gullies and aprons before it evaporates.

The presence of ground ice shows that Mars is more geologically active than previously thought and increases the possibility that some primitive life-forms may exist beneath the surface of Mars.

Several scientists have pointed out that the cold pole-facing slopes were among the warmest places on Mars four or five million years ago because of the chaotic wobbling of the poles of Mars as calculated from planetary dynamics. When its axis had a large tilt, Mars might have warmed enough to evaporate ice from its southern polar cap, pro-

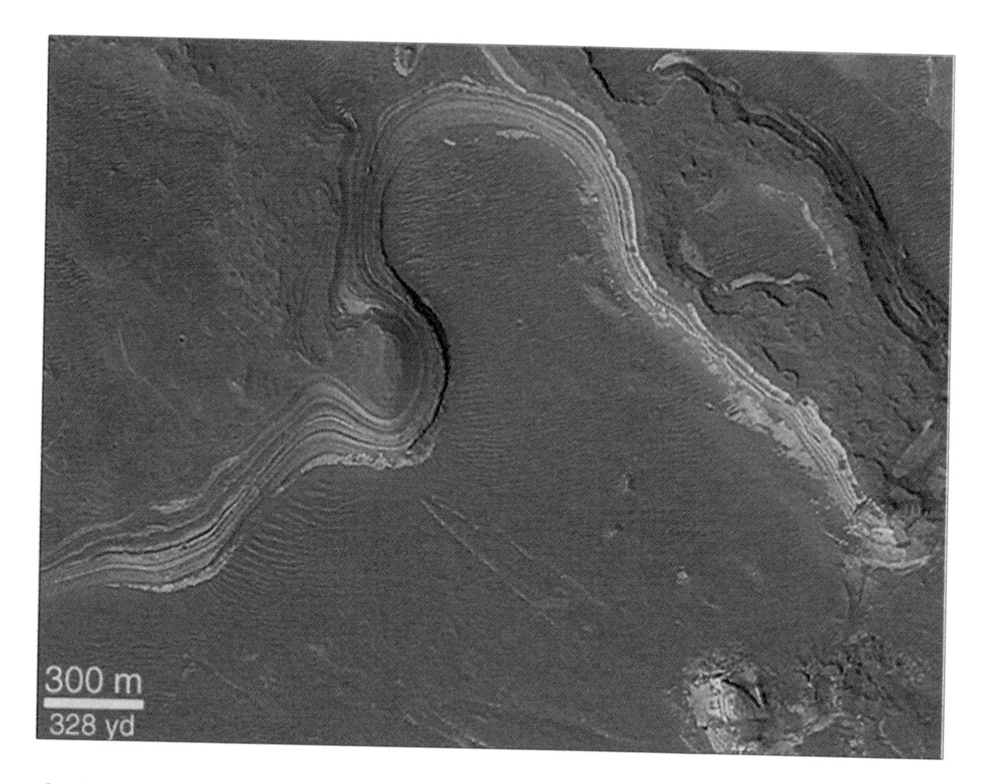

This Global Surveyor image of Holden Crater shows layered sedimentary rock, thought to have been produced by water. (NASA/JPL/ Malin Space Science Systems)

ducing a greenhouse effect that trapped the Sun's heat. A large tilt would especially warm pole-facing slopes, which would be in full sunlight through long summers, melting any water ice frozen into rock layers and producing surface runoff.

In the December 8, 2000, issue of *Science*, Malin and Edgett published a study analyzing photographs from the Global Surveyor that showed layered geographical outcrops across the bottom of the Martian canyon Valles Marineris. On Earth, these layers are typically formed by sediments deposited by water. Malin and Edgett believe the layering indicates that the surface of Mars may have been covered with lakes.

These layered outcrops and the gullies found earlier point to the presence of water on Mars, possibly billions of years ago. The presence of ground ice and the fact that it got loose in recent times shows that Mars is much more geologically active than previously thought. It also increases the possibility that some primitive life-forms may exist beneath the surface of Mars.

J. L. S.

The Galileo Mission

➤ *Galileo successfully ends four years of work investigating volcanoes and oceans on Jupiter's moons and begins its Millennium Mission.*

After more than a decade in space, the Galileo spacecraft made its final flyby of Jupiter's icy moon Europa on January 3, 2000, and passed closer to its fiery moon Io than ever before on February 22, 2000. The National Aeronautics and Space Administration's (NASA) Galileo mission began when astronaut Shannon Lucid launched the Galileo spacecraft from the space shuttle *Atlantis* on October 18, 1989, after a three-year delay caused by the 1986 *Challenger* shuttle accident. Six years and several equipment failures later, Jupiter and its moons began to reveal their diverse structures and some of the most geologically interesting features in the solar system, including the best possibility yet for the existence of life outside Earth.

On January 31, 2000, the Galileo successfully completed a two-year extended mission that followed its two-year primary mission. NASA decided to extend the Galileo mission a second time when the spacecraft continued to function well even after being exposed to twice the level of radiation it was designed to withstand. The radiation did cause some problems with spacecraft instruments but not enough to diminish the value of operating the spacecraft for another year. The new extended mission, called the Galileo Millennium Mission, included flybys on May 20 and December 28, 2000, of Ganymede, the largest moon in the solar system. This allowed for joint observations of Jupiter with the Cassini spacecraft when it passed Jupiter on its way to Saturn in December, 2000.

Ice and Water on Europa

When the Galileo spacecraft made its final Europa flyby on January 3, 2000, it provided strong new evidence that a liquid ocean lies beneath Europa's icy crust, confirming data from several earlier close encounters. Although it did not pass as close as the 200-kilometer (124-mile) flyby on December 16, 1997, it was able to make crucial magnetic measurements from an altitude of 351 kilometers (218 miles). It was programmed to see if the moon's magnetic north pole had changed from previous encounters. Galileo's magnetometer instrument did observe that the magnetic field had changed and indeed appears to reverse about every 5.5 hours.

The change in Europa's magnetic field matches the 5.5-hour change in the direction of Jupiter's magnetic field at Europa's position, which can produce electrical currents in a conductor such as an ocean. Those currents generate a magnetic field similar to Earth's magnetic field but with its magnetic poles near Europa's equator and shifting in conjunction with Jupiter's field to cause a complete reversal every 5.5 hours. These changes in the direction of Europa's magnetic field are consistent with a shell of electrically conducting material, such as a salty, liquid ocean. It is unlikely that the electric currents on Europa flow through solid surface ice, because ice is not a good electrical conductor. However, melted ice containing salts is a good conductor.

Although currents could flow in partially melted ice beneath Europa's icy crust, that is unlikely because Europa is hotter toward its interior, where the ice is probably completely melted. These findings are consistent with previous Europa images showing tortured surface features apparently formed when surface ice broke and rearranged itself while floating on a sea below. The changing magnetic field appears to confirm the existence of an ocean beneath the surface somewhere in the outer 100-kilometer (62-mile) layer of ice and water.

The evidence from the Galileo orbiter appears to suggest that Europa may have three of the basic ingredients necessary for carbon-based life: organic molecules, liquid water, and an energy source. Earlier flybys of Jupiter's large moons, Callisto and Ganymede, revealed the presence of complex organic molecules on their surfaces, indicating that such molecules are likely to be present on Europa. Although Europa's surface temperature is a frigid –130 degrees Celsius (–202 degrees Fahrenheit), it experiences enough tidal flexing from gravitational interaction with Jupiter and with its neighboring moons to generate significant internal heating. Europa experiences gravitational interactions, called resonances, in which every two orbits of Io around Jupiter are matched by one orbit of Europa, and every two orbits of Europa are matched by one orbit of Ganymede. This phenomenon stretches and squeezes Europa as it passes closest to Io at one end of its elliptical orbit, where it is also closest to Ganymede on every other orbit.

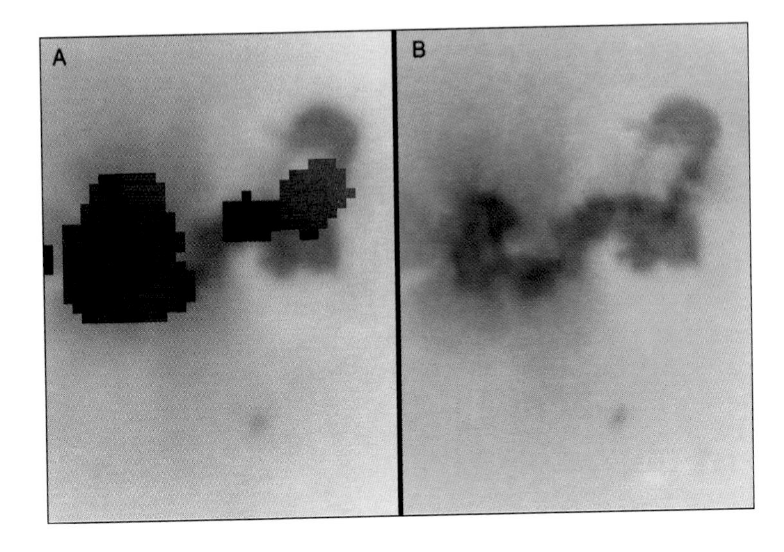

Eastern and western hot spots appear on the Prometheus volcano on Jupiter's moon Io in this color temperature map (left) generated with data obtained by a spectrometer on Galileo. An earlier Galileo image (right) shows an 80-kilometer (50-mile) lava flow. (JPL)

Volcanoes and Lava on Io

The Galileo spacecraft completed its third and closest-ever flyby of Jupiter's fiery moon Io on February 22, 2000. Io experiences even greater tidal flexing than Europa from its gravitational interaction with Jupiter, coming closest to the giant planet at one end of its elliptical orbit every forty-two hours and closest to Europa at the other end of its orbit every eighty-four hours. Volcanic action caused by this stretching and squeezing produces gases that interact with Jupiter's magnetism and radiation, leaving a stream of charged particles in Io's wake. As Galileo was approaching within 200 kilometers (124 miles) of Io, intense radiation triggered two computer resets about four hours apart; however, onboard software registered these as false alarms and the flyby continued unaffected. Mission planners scheduled this closest encounter near the end of Galileo's extended missions so that any damage to the spacecraft would come after most data had been collected.

Despite these radiation problems near Io, the spacecraft completed all its planned activities and successfully transmitted its data to Earth. These data included observations designed to study changes in Io's volcanoes since previous flybys in 1999. Loki, the most powerful volcano in our solar system, was observed near the beginning and end of one of its major periodic eruptions. Dramatic photographs from the three Io flybys have revealed oceans of multicolored lava, superheated geysers spewing sulfur gases and molten rocks, and vast plains of yellow sulfur with streaks of green and spots of bright red. More than one hundred volcanoes have been counted out of an estimated three hundred, including mile-high lava fountains, making Io the most volcanically active body in the solar system. Its surface is unusually young because of continual resurfacing by more than one hundred times the lava produced on Earth, making it the only body in our solar system with virtually no impact craters.

Io is about the size of Earth's moon, but it is a far stranger world. Away from its volcanic vents, temperatures drop below 150 kelvins; however, its lava beds reach 2,000 kelvins, more than 500 degrees hotter than terrestrial lava. Tidal flexing lifts its surface up and down by 30 to 100 meters (98 to 328 feet) every forty-two hours. Its tenuous atmosphere is mostly sulfur dioxide, a poisonous gas. It is losing about one ton of volcanic vapors every second into space, which has caused it to shrink by about 2 miles (3 kilometers) over the 4.5 billion years since it formed. During this time, its surface has completely melted and remelted at least 400 times. It has mountains up to 16 kilometers (9.9 miles) high, requiring a solid crust above its molten core of at least 30 kilometers (18.6 miles) for their support.

Photographs of Io have revealed oceans of multicolored lava, superheated geysers spewing sulfur gases and molten rocks, and vast plains of yellow sulfur with streaks of green and spots of bright red.

The close encounter of Galileo with Io gave it an energy boost that redirected the spacecraft into a more elongated orbit stretching some 11 million kilometers (6.8 million miles) from Jupiter and increasing its orbital period to three months. This brought it close to Ganymede on orbit number twenty-eight in May and back near Ganymede on orbit number thirty in December. A stream of dust stretches from the vicinity of Jupiter out some 300 million kilometers (186 million miles) into the solar system, more than twice the distance between Earth and the Sun. Galileo's dust detector has shown that the dust impacts peak about every forty-two hours, indicating that Io is its main source. The December, 2000, joint observation of Jupiter by the Galileo and Cassini spacecraft provided a unique opportunity to study the Jovian dust streams with dust detectors on both spacecraft. The interaction of this dust with Jupiter's magnetic field can then be traced from events beginning at Galileo to their subsequent results at Cassini.

J. L. S.

The Near Earth Asteroid Rendezvous Mission

➤A satellite is eased into orbit around the asteroid Eros, marking the first successful placement of a spacecraft in orbit around an asteroid.

The Near Earth Asteroid Rendezvous (NEAR) spacecraft was eased into orbit around the asteroid 433 Eros on February 14, 2000, after a four-year journey from Earth. This first successful orbiting of an asteroid by a spacecraft required a 57-second engine burn to slow it to about 5 kilometers (3 miles) per hour relative to Eros. It was a fitting embrace of the 13-by-13-by-33-kilometer (8-by-8-by-20-mile) potato-shaped object named after the Greek god of love Eros, son of Aphrodite, on Valentine's Day. Beginning with a large 22-day orbit 321 by 366 kilometers (199 by 227 miles) wide, NEAR received a 22-second engine burn on March 3 that nudged it into a 204-kilometer (126-mile) 10-day orbit around Eros. On April 1, it began a 6-week descent toward a 50-kilometer (31-mile) 1.2-day orbit and its 10-month mission to image, measure, and analyze the entire surface of the rotating space rock. On October 26, it passed within about 6 kilometers (3.7 miles) of the surface on its way to a 200-kilometer (124-mile) 9.4-day orbit.

The NEAR mission was the first launch in the Discovery Program of the National Aeronautics and Space Administration, a NASA initiative for small planetary missions. The $224 million project is managed for NASA by the Johns Hopkins University's Applied Physics Laboratory with the help of the Jet Propulsion Laboratory's Deep Space Network. The program scientist is Andrew Cheng and the mission manager is Robert Farquhar. At the end of its mission, the NEAR spacecraft was crash-landed on the asteroid on February 12, 2001. Scientists obtained some close-range images of the asteroid that reveal a smooth surface with small rocks, rivulets, a large rock, and few craters.

The 5.6-meter-wide (18.4-feet-wide) spacecraft flew past the asteroid Mathilde in June of 1997 and was originally scheduled to start orbiting Eros in December of 1998, but engine problems required a course change and another orbit of the Sun before the spacecraft's successful rendezvous with Eros. Eros is about 265 million kilometers

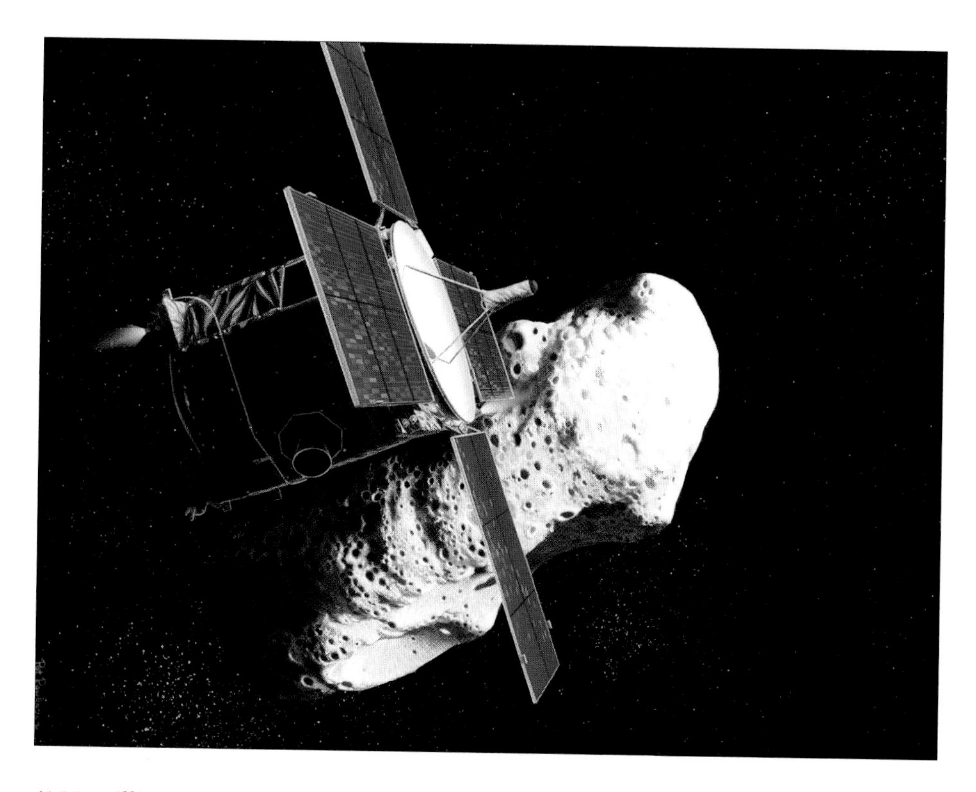

The Near Earth Asteroid Rendezvous (NEAR) spacecraft orbits the asteroid Eros in this artist's rendering. (NASA)

(165 million miles) from Earth in a changing orbit that is expected to cross Earth's orbit in the next few million years. The chance of its hitting Earth within the next 100 million years is approximately 5 percent.

Near-Earth Asteroids

Asteroids are the numerous small bodies in orbits around the Sun, ranging in distance from inside Earth's orbit to beyond Saturn but primarily between Mars and Jupiter in the main asteroid belt from 2.1 to 3.4 astronomical units (1 AU is the Earth-Sun distance of 150 million kilometers, or 93 million miles). The first two asteroids were discovered in 1801 and 1803 and are also the largest, 914-kilometer (568-mile) Ceres and 522-kilometer (324-mile) Pallas, each with a 4.6-year orbit at an average distance of 2.77 AU from the Sun. About 230 asteroids, named after Greek and Roman gods, have diameters greater than 100 kilometers (62 miles), and thousands of others are smaller. The largest are spherical in shape, and the smaller ones usually have irregular shapes. The combined matter of all the asteroids is considerably smaller than the smallest planet.

Near-Earth asteroids probably originated in the main belt, where a combination of collisions and gravitational interactions has pushed them into the inner solar system. Eros was the first one to be discovered when its elliptical orbit was found to stretch from inside the orbit of Mars at 1.13 AU (about 20 million kilometers, or 12 million miles, from Earth's orbit) to beyond Mars at 1.78 AU. It was discovered in 1898 and was designated 433 Eros (the number indicating the order in which it was cataloged). In 1932, the asteroid 1862 Apollo was found passing inside Earth's orbit. Near-Earth asteroids are grouped into three types named after prototype asteroids. Amors pass inside the orbit of Mars but do not cross Earth's orbit; Apollos cross Earth's orbit; and the rare Atens have orbits that are smaller than Earth's orbit.

The largest near-Earth asteroid is the Amor-type asteroid 1036 Ganymede, with a diameter of about 30 kilometers (19 miles), which comes within about 0.2 AU of Earth's orbit, followed by Eros, which reaches 0.13 AU of Earth's orbit. The largest Apollo asteroid is 1627 Ivar, with a diameter of about 8 kilometers (5 miles), similar in size to the asteroid that hit Earth 65 million years ago and believed by many scientists to have caused the extinction of

the dinosaurs. About two hundred near-Earth asteroids have been cataloged, but estimates indicate the existence of some seven hundred Apollo asteroids between 1 and 10 kilometers (0.6 to 6 miles) in size with the probability of colliding with Earth about once every 200,000 years. Many smaller asteroids have been observed, with sizes down to about 10 meters (33 feet), including one in 1991 at half the distance to the Moon. These asteroids, which hit Earth about every thousand years, include the one that caused Meteor Crater in Arizona about 50,000 years ago.

Results of the NEAR Mission

The NEAR spacecraft is now designated as NEAR Shoemaker in honor of the late Eugene Shoemaker, who pioneered the study of asteroid impacts. Even before reaching Eros, it had gained important information when it flew within about 1,200 kilometers (746 miles) of the asteroid 253 Mathilde in 1997. This 59-by-47-kilometer (37-by-29-mile) object was found to have a low density, about half that of Eros, and is believed to be similar to a compacted rubble pile. It has two large craters, with one covering about 62 percent of the asteroid's mean diameter. On its way to Jupiter, the Galileo spacecraft flew by the 12-by-16 kilometer (7-by-10-mile) asteroid 951 Gaspra in 1991 and the 55-kilometer (34-mile) asteroid 243 Ida in 1993. The latter was found to have a 1 kilometer (0.6-mile) moon circling it in a 100-kilometer (62-mile) orbit.

As NEAR Shoemaker began circling Eros, high-resolution images began to surprise scientists with the abundance of ridges, craters, boulders, and even chains of craters. The images revealed numerous boulders as small as 50 meters (164 feet) across and long ridges extending several kilometers along the surface, which is nearly covered with craters smaller than 2 kilometers (1.2 miles) in diameter. Many of the crater walls have bright markings from loose material, called regolith, probably the result of impact cratering. One larger crater some 5.5 kilometers (3.4 miles) wide has bright streaks on its walls and a boulder that appears to have rolled into the crater in line with the asteroid's gravitational force. Because gravity on Eros is only about one-thousandth that on Earth, an astronaut could easily jump out of the crater, but gravity is still strong enough to make boulders roll downhill.

NEAR Shoemaker carries a laser rangefinder to measure the asteroid's surface profile. This data will help determine if the surface features are from erosion, fault lines, or other events. The spacecraft's orbit made it possible to calculate the mass of Eros and thus its average density of 2.4 grams per cubic centimeter, similar to Earth's crust, suggesting that it is more compacted than Mathilde and nearly solid except for its surface, which has experienced fragmentation from collisions. Its rotation rate of once every five and one-half hours is not fast enough to pull it apart even if it were a loosely aggregated rubble pile. In fact, only asteroids less than 200 meters (660 feet) in size are found rotating faster than once every two and one-half hours, suggesting that they are solid and that larger ones are held together by their weak gravity.

The NEAR spacecraft's instruments also include a magnetometer and three spectrometers to analyze the composition of elements that make up the surface of Eros. The spacecraft was able to use its X-ray spectrometer to begin the first X-ray detection from an asteroid while it was still 212 kilometers (132 miles) above the surface. This early detection of X rays was the result of a brilliant solar flare on March 2 that provided a 600-second window of opportunity for the X-ray spectrometer to detect X rays from four times farther away than expected. The solar X-ray burst caused elements on the asteroid to emit fluorescent X rays that were measured by the spectrometer, revealing the presence of magnesium, iron, and silicon.

A brilliant solar flare on March 2 provided a 600-second window of opportunity for the X-ray spectrometer to detect X rays from four times farther away than expected.

The NEAR mission seems to be close to solving one of the oldest mysteries in planetary science, the relation between asteroids and meteorites. The most common meteorites, known as ordinary chondrites, do not appear to come from the most common asteroids, S-types such as Eros, because their colors do not match. However, a major solar flare on May 4 stimulated X-ray fluorescence received by the NEAR spectrometer from the surface of Eros, revealing an elemental composition in the same range as that of chondrites. It now appears that micrometeorite impacts and the solar wind cause a reddening of S-type asteroids that masks their similarity to chondrite meteorites.

The pitted surface of Eros was in contrast to what most planetary scientists expected. Before the NEAR mission, it was thought that a collision with another asteroid would have catapulted Eros somewhat rapidly into its near-Earth orbit, leaving it less blemished by impacts than asteroids remaining in the more populated main belt. The battered surface that has now been revealed suggests a slower passage out of the main belt following the impact that tore it from its parent body, requiring hundreds of millions of years to approach Earth and probably even longer before it reaches Earth.

Several missions have been planned to follow up the success of the Near Earth Asteroid Rendezvous, including two that will collect samples and return them to Earth. NASA's Stardust spacecraft, launched in February, 1999, toward Comet Wild 2, is expected to return in 2006 with dust from the tail of the comet. The Japanese space agency's MUSES-C space probe will be sent in 2002 to the asteroid Nereus to collect material and to release a NASA-built nanorover to hop across the surface. Over the next decade, several other comets will be visited, and asteroid flybys are also planned. These missions should help clarify the relationship between meteorites and their celestial sources and provide a better understanding of the origins of the solar system.

J. L. S

Resources for Students and Teachers

Books

Comins, Neil, and William Kaufmann III. *Discovering the Universe.* 5th ed. New York: W. H. Freeman and Company, 2000. This college textbook comes with a CD-ROM featuring the *Starry Night* planetarium software and a special issue of *Scientific American,* "Magnificent Cosmos," describing the latest discoveries in astronomy. Print supplements include an instructor's manual and resource guide, overhead transparencies and slides, a media activities book, and a computerized test bank. A Web site to accompany the text is continually updated with the latest astronomical findings: www.whfreeman.com/ astronomy.

Fix, John D. *Astronomy, Journey to the Cosmic Frontier.* 2d ed. New York: McGraw-Hill, 2001. This college textbook includes a free interactive CD-ROM with chapter summaries, quizzes, animations, simulations, and planetarium software. Also available from the publisher are an instructor's manual and test item file, a microtest computerized testing system, a transparency set with two hundred overhead transparencies, a visual resource library with two hundred images to produce PowerPoint presentations, and a Web site for both instructors and students (www.mhhe.com/fix).

Fraknoi, Andrew, David Morrison, and Sidney Wolff. *Voyages Through the Universe.* 2d ed. 2 vols. Orlando, Fla.: Saunders College Publishing, 2000. This college textbook comes with a CD-ROM that includes the *Redshift* planetarium software, seven hundred full-screen photographs, and access to the *Penquin Dictionary of Astronomy* with more than two thousand entries. Ancillaries include *Saunders Internet Guide for Astronomy* by David Bruning, Randy Reddick, and Elliot King; *The Voyages Instructor's Manual/Test Bank,* a computerized test bank; and an astronomy transparency collection. The Web site is voyages@saunderscollege.com.

Wheeler, J. Craig. *Cosmic Catastrophes: Supernovae, Gamma-Ray Bursts, and Adventures in Hyperspace.* Cambridge, England: Cambridge University Press, 2000. Examines topics such as black holes and gamma-ray sources and the most recent theories such as string theory, quantum gravity, and wormholes.

CD-ROMs and Videos

Astronomy: The Earth and Beyond. CD-ROM. Films for the Humanities and Sciences, 2000. This CD presents core elements of astronomy and earth science curriculums, including information on the origin of the universe and the life and death of a star and a guided tour of the solar system. An eclipse simulator can be downloaded from an affiliated Web site.

The Beginning. Video. Lights in the Sky series. Phoenix Multimedia, 2000. The beginning of the universe is the subject of this video, the first in a series on the history of astronomy. Other videos in this series include *Meteors, Asteroids, and Comets; The Sun and Stars; Planets and Moons: Our Solar System; and Space Travel: The History.*

Stargaze: Hubble's View of the Universe. DVD. Alpha, 2000. This DVD presents views of the universe from the Hubble Space Telescope along with narration in a widescreen format.

The Universe: A Guided Tour. DVD. Films for the Humanities and Sciences, 2000. *The Complete Cosmos* video series is presented in digital video disc. The solar system, astronomy, and space explorations are covered in twenty-five ten-minute programs. Subjects include the Hubble Space Telescope, black holes, and the big bang theory.

Web Sites

Amazing Space
http://amazing-space.stsci.edu/
Space Telescope Science Institute
This site provides educational activities for both teachers and children, answers commonly asked questions, and has a wide selection of topics including black holes.

Astronomy News and Links
http://home.t-online.de/home/Sweimer/astro.html
Provides news on the latest developments in the field of astronomy.

Astronomy Picture of the Day Archive
http://antwrp.gsfc.nasa.gov/apod/archivepix.html
National Aeronautics and Space Administration (NASA)
Features the astronomy picture of the day. Each picture has a caption with links to more information.

Black Holes FAQ
http://cfpa.berkeley.edu/BHfaq.html
Center for Particle Astrophysics, NSF Science and Technology
Provides answers to commonly asked questions about black holes.

Chandra X-ray Observatory
http://chandra.harvard.edu/pub.html
http://chandra.nasa.gov/
Harvard University and National Aeronautics and Space Administration (NASA), respectively
These sites provide public information and news on the Chandra X-ray telescope. Allows users to e-mail astronomy postcards.

European Southern Observatory in Chile
http://www.eso.org/
Provides information on the intergovernmental European astronomy organization and its projects and research. Includes news and visuals such as videos and posters.

The Extrasolar Planets Encyclopedia
http://www.obspm.fr/encycl/encycl.html
This French extrasolar planet home page gives the latest news on discoveries of other planetary systems. There are links to a full catalog of all discoveries as well as publications and tutorials about detection methods.

Galileo Mission
http://nssdc.gsfc.nasa.gov/planetary/galileo.html
http://www.jpl.nasa.gov/galileo/index.html
National Aeronautics and Space Administration (NASA)
The sites for the Galileo Mission offer links to recent photos of Jupiter and its moons from the Galileo Jupiter orbiter. They also include movies, news, and educational activities related to the Galileo Mission.

Gamma-Ray Bursts
http://pao.gsfc.nasa.gov/gsfc/SpaceSci/gamma/gamma.htm
National Aeronautics and Space Administration (NASA) Goddard Space Flight Center
The home page for gamma-ray burst detection and discussion provides much information and many resources.

Hubble Space Telescope
http://hubble.esa.int/
http://www.stsci.edu/
European Space Agency (ESO), Space Telescope Science Institute (STSI), respectively
These sites provide information about the Hubble Space Telescope, including free multimedia playback software to view the images it captures. Movies are also available. The STSI site offers Hubble deep-field south images, showing thousands of very distant galaxies.

Jet Propulsion Laboratory (JPL)
http://www.jpl.nasa.gov
National Aeronautics and Space Administration (NASA), California Institute of Technology
Contains information on JPL and its activities as well as the latest news from NASA on the Galileo, Cassini, and Mars Global Surveyor spacecraft and the Hubble Space Telescope.

Mars Exploration Program
http://mars.jpl.nasa.gov/index.html
National Aeronautics and Space Administration (NASA), Jet Propulsion Laboratory (JPL)
Features the latest photographs of Mars and information on all Mars exploration, including future missions such as the Twin Rover project.

Mars Global Surveyor
http://mars.jpl.nasa.gov/mgs/index.html
National Aeronautics and Space Administration (NASA), Jet Propulsion Laboratory (JPL)
Offers many images from the Mars orbiter and recent news, including the photos indicating the presence of recent water seepage from canyon and crater walls on Mars.

Microwave Anistropy Probe (MAP)
http://map.gsfc.nasa.gov/html/web_site.html
National Aeronautics and Space Administration (NASA)
Provides basic concepts in cosmology and information about the MAP mission.

Near Earth Asteroid Rendezvous (NEAR) Mission
http://near.jhuapl.edu/NEAR/
The Johns Hopkins University, Applied Physics Laboratories
This site has links to a wide variety of photos and movies of Eros taken from the NEAR spacecraft at distances ranging from two hundred kilometers to as close as six kilometers.

Planetary Science at the National Space Science Data Center
http://nssdc.gsfc.nasa.gov/planetary/
National Aeronautics and Space Administration (NASA)
The planetary science home page highlights current activities and observations in the solar system.

Science/Astronomy
http://space.com/scienceastronomy
The Space.com site devoted to providing news and information on astronomy and space.

The Search for Extrasolar Planets
http://exoplanets.org/index.html
Department of Astronomy, University of California, Berkeley
The home page for the primary American extrasolar planet search team details the hunt for extrasolar planets.

Solar Heliospheric Observatory (SOHO)
http://sohowww.nascom.nasa.gov
National Aeronautics and Space Administration (NASA), European Space Agency
The home page for the Solar Heliospheric Observatory (SOHO) spacecraft features the most recent photos of solar activity.

SpaceWeather
http://www.spaceweather.com
Features solar wind and sunspot data and solar storm predictions and summaries based on National Aeronautics and Space Administration (NASA) and National Oceanic and Atmospheric Administration (NOAA) information.

Supernova Cosmology Project
http://panisse.lbl.gov/public/
Lawrence Berkeley National Laboratory
The home page for high redshift supernova data, with good background information and links to other sources.

XMM-Newton, XMM-Newton Science Information Center
http://xmm.esa.int/
http://xmm.vilspa.esa.es/
European Space Agency (ESA), Villafranca Satellite Tracking Station
These sites contain information on the mission of the XMM-Newton X-ray telescope space observatory and the science behind it.

J. L. S. and W. R. W.

2 · Computer Science and Information Technology

The Year in Review

➤ *George M. Whitson III*
Computer Science Department
University of Texas at Tyler

Every year, a great deal of progress is made in computer science and information technology, and the year 2000 was no exception. Today's computers are faster and less expensive than ever. Computer software meets more user needs, and the Internet continues to change how people access information. However, the big story for the year was that social, legal, and security issues have become more important than technical innovations in computer science and information technology.

Hardware Continues to Improve

A few years ago, advances in computer technology were synonymous with improvements in computer hardware. However, in 2000, hardware innovations were eclipsed in importance by legal issues. Still, an enormous number of important advances in computer hardware took place within the year. Certainly, the delivery of 1-gigahertz processors was one of the major hardware achievements. Both Intel and AMD now offer 1-gigahertz processors, and advertisements from retail computer companies such as Dell and Gateway feature 1-gigahertz computers at reasonable prices. Although a careful study of the technology used in the 1-gigahertz processors shows no great technological breakthroughs, getting all that power on a single chip is still quite an achievement. Faster system buses and memory further enhance the overall speed of today's motherboards. The new Pentium IV and Athlon processors, with an announced system bus speed of up to 400 megahertz and high-speed memory, make today's computing systems more than twice as fast as those of last year.

One of the most important events in 2000 was the use of Extensible Markup Language (XML) data description language in Microsoft's Office 2000.

Hard disk density technologies give today's storage devices capacities that were unheard of until recently. Disks with more than 50-gigabyte capacity have already been announced, and the improved intelligent drive electronics (IDE) interfaces have increased the disk to central processing unit (CPU) transfer speed. Optical storage devices also improved in 2000. The new 8X CD-RW (compact rewritable disc) drives are fast enough to allow them to be legitimate backup devices. Although standardization issues continue to slow the development of digital video disc (DVD) drives, progress was also made in this area. The price of moderate - resolution digital cameras dropped to as little as one hundred dollars in 2000, and more and more high-quality digital cameras appeared on the market. Even a device as simple as the mouse saw many improvements, with optical mice, remote mice, and even finger mice appearing.

Cable modems are now available in most major cities. Digital subscriber line (DSL) modems in 2000 were deployed much more rapidly and had fewer problems than expected. The appearance of these two high-speed Internet access methods resulted in considerable competition that benefited those desiring high-speed access. Cable modem costs dropped in anticipation of the competition from DSL, then DSL's initial prices dropped to better compete with cable modems, and satellite access prices also dropped because satellite was no longer the primary choice in high-speed access. Most people can now have fairly reliable high-speed Internet access using one of these three technologies. As this becomes better known by customers, improved high-speed Internet access will be seen as one of the big hardware events of year.

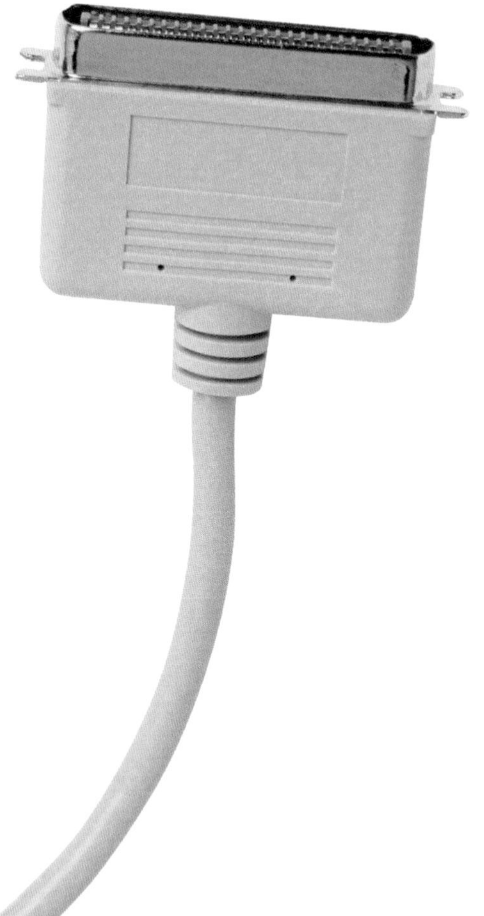

Smarter, Internet-Enabled Software

New and improved software products appeared throughout 2000. Most of these were designed to be used in user-friendly Windows environments. Major improvements occurred in word-processing, Internet-access, and database software. A number of specialized graphics packages also had major upgrades. Macromedia's Fireworks was especially notable in its support for developing graphics for the Internet, and Visio added a lot of support for the Unified Modeling Language (UML). One of the most important events in 2000 was the use of Extensible Markup Language (XML) data description language in Microsoft's Office 2000. Making XML and Hypertext Markup Language (HTML) the base format for all the Office 2000 documents gave a major boost to XML as the standard data description language for the Internet. In 2000, object-oriented programming continued to improve its position as the dominant form of programming. UML continued to demonstrate its importance as a systems analysis and design tool for all forms of object-oriented programming. Whether users are writing a C++ program with lots of classes or a Visual Basic program with lots of ActiveX controls, they can use UML as their design tool. Both Visio and Rational Software (in its Rational Rose) made substantial improvements to support UML this year.

Artificial intelligence provided many improvements to computing software this year. Intelligent agents were increasingly used in software on the Internet to answer questions, to look up information in databases, and to develop a profile of the Web site users. Neural networks were often used as part of intelligent agent programs because of their ability to learn how to respond to users on the fly. Many people now own scanners and use them on a regular basis. All the optical character recognition (OCR) software programs were improved this year. It is now possible to scan a newspaper article into a word processor reasonably well, retaining the columns, formatting, and graphics. Neural networks software is also a key component of most OCR software because it is used to identify the letters of the scanned text.

Almost no one pays for an Internet browser anymore, and it has never been easier to navigate the Internet.

Upgrades to the three major voice recognition software products appeared this year, and the new versions work considerably better than the earlier versions. Users can accurately dictate not only into their favorite word processing program but also directly into the most popular e-mail clients and Internet browsers. The purchase of Dragon by Lernout & Hauspie this year means that the research capabilities of the two largest voice recognition software companies will be combined, and this undoubtedly will result in major improvements in voice recognition software in the near future.

Although almost no one pays for an Internet browser anymore, the improvements in Internet Explorer 5 and its competitors in 2000 benefited all users. Almost all Internet documents can now be printed without problems, and it has never been easier to navigate the Internet. Standard e-mail and Internet e-mail programs also improved, along with Internet relay chat (IRC) chat room software, so that everyone can use the Internet as easily and reliably as a word processor.

The main events in the computer industry in 2000 involved legal issues, not hardware developments. A bicyclist rides by Redmond, Washington, offices of Microsoft, which was ruled a monopoly by a U.S. district judge in an antitrust case brought by the Department of Justice. (AP/Wide World Photos)

Legal Issues Arise

The computer industry encountered major legal issues for the first time in 2000. Almost no week went by without some major legal case being front-page news. The federal government now has a cyber crime group, as do many states. Items in the news included several mass thefts of credit card numbers and major frauds committed at various popular Internet auctions. Congress has held hearings about regulating the Internet. One question of interest to both Congress and state governments is how to collect taxes on merchandise being sold over the Internet Another question Congress considered this year is whether the Internet's content should be regulated, particularly sites that contain sexually explicit or socially offensive material. Cyber crime touched even software giant Microsoft, which had its source code stolen by some Internet thieves. Internet crime, its prevention and prosecution, were big issues in 2000 and will be even more important in the future.

Of course, the most important legal story in the computer industry in 2000 was the Department of Justice's antitrust lawsuit against Microsoft. In the first round of this lawsuit, the Department of Justice prevailed, finding Microsoft to be a monopoly. If the decision of the District of Columbia District Court stands, Microsoft will be broken up into two companies, one developing operating system software and the other developing all other software, and each company will be severely restricted in what it can do in developing new software. Most observers believe that the District of Columbia Court of Appeals or the U.S. Supreme Court will modify the original decision so that Microsoft will be able to continue to develop personal computer software, with some major restrictions on its business practices, but the fact that Microsoft was successfully sued by the Department of Justice caught the attention everybody in the computer industry.

Personal Computer Security

A majority of personal computers in homes across the nation are also used to access the Internet. This has greatly increased the need for better security for these computers. However, security can be added only by decreasing the actual operating speed of the computer, controlling the peripheral devices that the computer supports, and forcing the user to be more security conscious (for example, regularly typing in passwords). Unix and Windows 2000 can provide a fairly secure environment, but many of the most popular operating systems do not. To combat the security problem for Internet access, a number of computer companies provided personal firewalls in 2000 to control attacks on home computers over the Internet. More and more computer viruses appeared, and they have never been more dangerous. In addition to the traditional viruses that can be introduced from a file on a floppy disk and destroy the user's hard drive, a number of new viruses appeared in 2000 that could attack a user's computer through an e-mail program. Antivirus software companies continue to improve their software to combat virus attacks and, as long as users regularly update their virus definitions and keep their antivirus monitor programs active, they have a good chance of avoiding problems.

Education Goes High Tech

Teachers have been criticized for being slow to adopt new technology to enhance their teaching. Some teachers still spend most of their time presenting verbal lectures with an occasional trip to the blackboard to draw a diagram. This year, almost every university encouraged its teachers to develop a Web site to complement their courses and to consider using some high-tech tools in their teaching. Many instructors now hand out a CD (compact disc) at the beginning of class that they created themselves with a CD-RW (rewritable compact disc). In addition to showing rented videotapes of a topic, some instructors now create their own videos in their schools' multimedia studios. It is not unusual today to find teachers using scanners, voice recognition software, or digital cameras to prepare materials for their courses. More and more teachers today are using high-tech tools.

Although virtual universities are still in their infancy, there is little doubt that this form of education is going to be successful.

New and improved versions of many of the HTML editors were released in 2000. It has never been easier for instructors to develop Web sites to support their courses or to develop complete Web-based distance education courses. In 2000, virtual universities became a marketplace reality. A year earlier, a visit to a virtual university's home page would have turned up empty links and promises of courses and programs being developed. Now, students can choose from many virtual universities where they can take Web-based distance education courses and complete entire majors. Although virtual universities are still in their infancy, there is little doubt that this form of education is going to be successful.

Convergence

Many computer researchers recognized that the basic method of sharing data is similar to having a telephone conversation or watching television. Although they differ in the level of interactivity, these data-sharing activities have a user and a provider connected by some type of medium (usually wiring). In 2000, progress was made in the development of hardware, software, and standards, enabling the appearance of individual devices that can process data, voice, and video. It is quite common today to use a portable computer to work with data, have a telephone conversation, listen to an MP3 (MPEG-1 layer 3) song, or watch a video. Although the days when people can use a wristwatchlike device for all types of communications are still not here, people can listen to MP3 songs on their cell phones if they want.

The Arrival of the 1-Gigahertz Processors

➤ *Relatively inexpensive microprocessors execute operations at a speed of more than one gigahertz per second.*

One of the most important components of a computer is the central processing unit (CPU), which performs the arithmetic and logical operations of a computer. The CPU also exercises overall control of the computer. It has an internal clock that determines the number of discrete operations that the computer can perform per second. Each cycle is called a hertz (hz), which means that a 1-megahertz (Mhz) computer executes a million simple operations per second. More complex computer operations, such as addition, multiplication, and division, are completed in one or more clock cycles.

Early computers were called complex instruction set computers (CISC) because they had arithmetic instructions that included loading the data as a part of the instruction and required many clock cycles to complete. Most computers today are called reduced instruction set computers (RISC) because their arithmetic instructions do not load the data as a part of the instruction and are completed in a small number of clock cycles. In fact, most RISC CPUs complete an integer addition in a single clock cycle and complete more complex floating-point division instructions in a half a dozen clock cycles. A CPU uses other techniques to decrease the number of clock cycles required for arithmetic and logical operations so that modern computers come close to executing one operation for each clock cycle. In 2000, a number of microprocessors functioned at a speed of more than 1 gigahertz (Ghz) per second. In other words, they performed nearly one billion arithmetic and logical operations per second. Although supercomputers achieved this speed long ago, the fact that a typical microcomputer can now achieve a similar speed with a thousand dollar chip is truly amazing.

Some of the most important components of the CPU that determine its speed are the material used as a conductor for the chip's "wires" and the CPU's ability to pipeline arithmetic and logical instructions. An aluminum layer is generally used to create a wire on a chip to transfer data from one part of a chip to another. Other substances could be used to speed up a chip's internal bus. For example, gold and silver have been used for workstations but are a little expensive per chip. Intel, IBM, and Motorola have announced that they will use copper rather than aluminum, which they predict will produce a great increase in chip speed and a decrease to 1.3 microns in the size of a transistor on a chip.

The arithmetic and logic units of a CPU can be made up of subunits. In a pipeline unit, each subunit can be working on different data. For example, if an adder has four subunits, then it can be processing four different data items, one in each subunit, rather than a single item. Except for during the startup and closedown of this adder pipeline, properly organized data would be processed four times as fast as with a non-pipeline adder. Some CPUs also support chaining the various pipeline units, which can further increase the speed of the CPU. Pipelining is one of the major improvements in CPU technology that has led to the 1-gigahertz microprocessor CPUs.

The speed of the CPU is related to a number of general features of a computer's architecture, including the bandwidth and speed of the memory system, the bandwidth and speed of the system for the bus, and the size, speed, and bandwidth of the cache subsystem. The CPU needs to load data from the computer's memory to do its computations and to store the data when it is finished. The memory system can be designed, for an increase in cost, to do these transfers more rapidly. For example, the Rambus technology announced in 2000 for the Pentium III and 820 chip set marks the first time that the advanced supercomputer memory technology of a memory controller was to be used this extensively in microcomputers. The system bus ties together the CPU, the main memory, the graphics subsystem, and sometimes the cache memory. To take advantage of the higher transfer speed of the individual components, the system bus needs to have the maximum bandwidth and speed. Until recently, most system buses were 66 megahertz, had 64 data lines, and operated with a transfer rate of less than 1 megabyte per second. The main bus for the Pentium III is 66, 100, or 133 megahertz with 64 data lines and a data transfer rate of several megabytes per second. The AMD Athlon system bus is 100 or 200 megahertz with a transfer rate of 0.8 to 1.6 mega-

Copper will produce a great increase in chip speed and a decrease to 1.3 microns in the size of a transistor on a chip.

bytes per second. Also, adding the right amount and type of cache is very important to improving the speed of the CPU. In November, Intel introduced the Pentium IV processor, with an announced system bus speed of up to 400 megahertz and high-speed memory.

Improvements in Floating-Point Arithmetic

Early microcomputers did only logical and integer arithmetic instructions in hardware. All floating for instructions was done in software. As the microprocessor manufacturers introduced more sophisticated processors such as the Intel 286, they began to do floating-point arithmetic in hardware with a coprocessor. The early coprocessors were very primitive, using a stack architecture for doing arithmetic. Because floating-point operations can take a lot of clock cycles, improving the floating-point unit of the CPU was the key to increasing the overall performance of the CPU. When Intel introduced its i860 and Pentium microprocessors, its floating-point units started to look just as sophisticated as those of a Cray supercomputer. The current crop of 1-gigahertz microprocessors being produced by Intel and AMD have greatly improved floating-point units and are very fast processors for applications in areas such as image processing and speech rec-

In late November, 2000, Intel added another processor to its lineup: the Pentium IV. (Intel)

ognition. Some high points for Intel processors in the development of floating-point arithmetic units of a CPU are shown in the table.

INTEL PROCESSORS AND FLOATING-POINT SUPPORT	
Processor	**Floating-Point Support**
8088	All floating-point arithmetic done in software.
80286	Floating-point arithmetic done with a numerical coprocessor using a simple stack architecture.
80386	Numerical coprocessor integrated onto the main chip. Floating-point arithmetic still done with a simple stack architecture.
Pentium I	Floating-point arithmetic done with traditional components including a floating-point adder and multiplier.
Pentium II	Support for up to 12-stage pipeline units including-floating point operations is added to the CPU.
Pentium III	Support for up to 20-stage pipeline including floating for operations is added to the CPU. Support for Single Instruction Multiple Data (SIMD) for some floating-point operations is added to the CPU.

In 2000, AMD led the field with its Athlon processor, which has the most developed floating-point capabilities of a microprocessor. It has three pipelined arithmetic units to support floating-point operations, including an adder and a multiplier. It also has some support for single instruction multiple data (SIMD) capability. The company claims that the Athlon 600-megahertz computer can return as many as four 32-bit floating-point operations per clock cycle, giving a peak performance of 2.4 gigaflops, or 2.4 billion floating-point operations per second.

Although the best microprocessors still are not as powerful as the CPU boards of the supercomputers, companies such as Intel and AMD are making great strides in moving the key floating-point operations onto their CPU chips. These chips can be combined in some parallel computer architectures to make truly powerful computers that actually execute in the teraflop range, or 1 trillion floating operations per second.

Dynamic Look-Ahead Architectures

As long as the CPU does not have to branch, it can use pipelining to achieve substantial speed improvements. Short loops whose instructions do not have data dependencies can achieve great increases in speed as a result of pipelining. However, branches tend to reduce the speed advantages of pipelining. To improve the ability of the CPU to pipeline instructions when a branching instruction is encountered, many processors attempt to build machine code on all the branches of a decision so that no matter which branch is taken, the pipeline will not be broken. Moving this type of CPU optimization onto microprocessor chips is one of the big recent technological breakthroughs achieved by chip manufacturers. Both Intel and AMD support multiple-branch code prediction and multiple-execution pipes in their latest microprocessors and claim that the next generation of microprocessor chips will be even better.

The large number of cache architectures is one of the more interesting developments in CPU design in 2000.

Another problem for keeping the instruction pipeline full is instructions whose data depends on other instructions in the pipeline. An instruction with data dependencies cannot be placed into the pipeline because the data values needed are not known when the instruction needs to be loaded. To solve this problem, processors try to reorder instructions by discovering these data dependencies before the instruction is placed into the pipeline. Research on this advanced data-flow analysis is difficult and ongoing. It used to be strictly the province of the large research universities and supercomputer companies, but now some of the microcomputer CPU chip manufacturers are implementing new data-flow analysis into their chips as rapidly as anyone.

Cache Crucial for Performance.

Most computer manufacturers have to balance the cost of a component versus its speed. For example, spending more money on the memory subsystem and the system bus can increase the overall performance of the CPU, but of course this greatly increases the cost of the computer. Cache memory is a very high-performance and expensive kind of memory. Cache memory contains copies of what is in the real memory of the computer. By using the data in the cache memory, the computer can increase its speed of operation, but it needs to periodically update the real memory.

Throughout the history of computers, CPU manufacturers have experimented with various cache architectures. Most modern microcomputer CPU chips have a small cache located on the chip that operates at the chip's speed. The on-chip cache is from 16 to 128 kilobytes and generally is divided into two caches of equal size, one containing instructions and the other containing data. This on-chip cache is called the L1 cache. Most computers also have an L2 cache, and some even have an L3 cache. Each type of cache memory has a different speed and cost. An important element in building the CPU of a computer is to use the various levels of cache memory to build the fastest possible computer at a reasonable cost. Modern computers do an extremely effective job of combining the L1, L2, and L3 caches to increase the overall speed of the CPU.

Comparing the Intel Pentium III and the AMD Athlon gives a fairly good picture of the current options for cache memory. The Pentium III has only 16 kilobytes of L1 instruction cache and 16 kilobytes of L1 data cache, while the Athlon has 64 kilobytes of each type of cache. The Pentium III has 256 kilobytes of L2 cache that operates at 1,000 megahertz, and the Athlon has 512 kilobytes of L2 cache that operates at 333 megahertz. The algorithm used to map the Pentium's L2 cache to real memory is a little better than that used by the Athlon. Experts are still arguing over whether the Pentium III has a better overall cache architecture than the Athlon, and with the differences in their architectures, it is easy to see why disagreement exists. On the other hand, Intel's Celeron has no L1 cache memory at all, and Intel's multiprocessor cache architecture has up to 8 megabytes of L3 cache. The large number of cache architectures is one of the more interesting developments in CPU design in 2000, and even more innovations are likely in the future.

More Transistors on a Chip?

A recent discovery in quantum physics may allow chips to be designed with smaller transistors. Scientists at the National Aeronautics and Space Administration's Jet Propulsion Laboratory think that they can use "pairs of entangled photons" to create streams of laser light particles to manufacture smaller transistors on computer chips. The new technology would allow chips to have transistors approximately one-third the size of existing transistors and thus triple the power of modern chips. The use of copper for chip connections to increase on-chip speed and decrease transistor size is further proof that innovations in chip manufacture are alive and well in 2000.

USB, Firewire, and Other Microcomputer Buses

➤ *Interconnect speed between frontside and backside microcomputer buses doubles from 133 to 266 megabytes per second, and nearly all computer manufacturers adopt USB standard.*

A microcomputer bus is the hardware and software needed to connect a number of devices on the motherboard via a shared media. Only two devices can share the bus at one time. If more than two devices request access to the bus at the same time, the bus needs a method of determining which two devices it will allow to use it. The most important feature of a microcomputer bus is the number of bits per second that can be transferred over that bus. A feature that is a close second is the protocol that the bus uses to start sending data over the bus, determine that all the data have been successfully sent, and support the interaction of the bus with the other buses on the motherboard.

Early minicomputers and microcomputers had a single bus and relatively simple bus access protocols. A typical modern microcomputer has many buses and a very complex set of protocols for keeping all the buses operating. A microcomputer bus is implemented on a motherboard by a chipset and a number of "wires" embedded in the motherboard. The chipset synchronizes the simultaneous requests for access to a common wire.

The most important feature of a microcomputer bus is the number of bits per second that can be transferred over that bus.

The motherboard of a modern microcomputer has two main buses, the frontside and backside bus. The frontside bus supports all the input and output devices, including hard disk drives, a keyboard, the mouse, printers, and network cards. The backside bus connects the CPU (central processing unit), main memory, and graphics adapter and has many other buses, such as a SCSI (small computer system interface) and ISA (industry standard architecture) buses, connected to it. The frontside and backside buses also need to be connected with a bus between buses. Until recently, a 33-megahertz PCI (peripheral component interconnect) bus, which transfers data at a rate of 133 megabytes per second, connected the buses. The latest chipsets that are used to implement the

A TYPICAL MOTHERBOARD

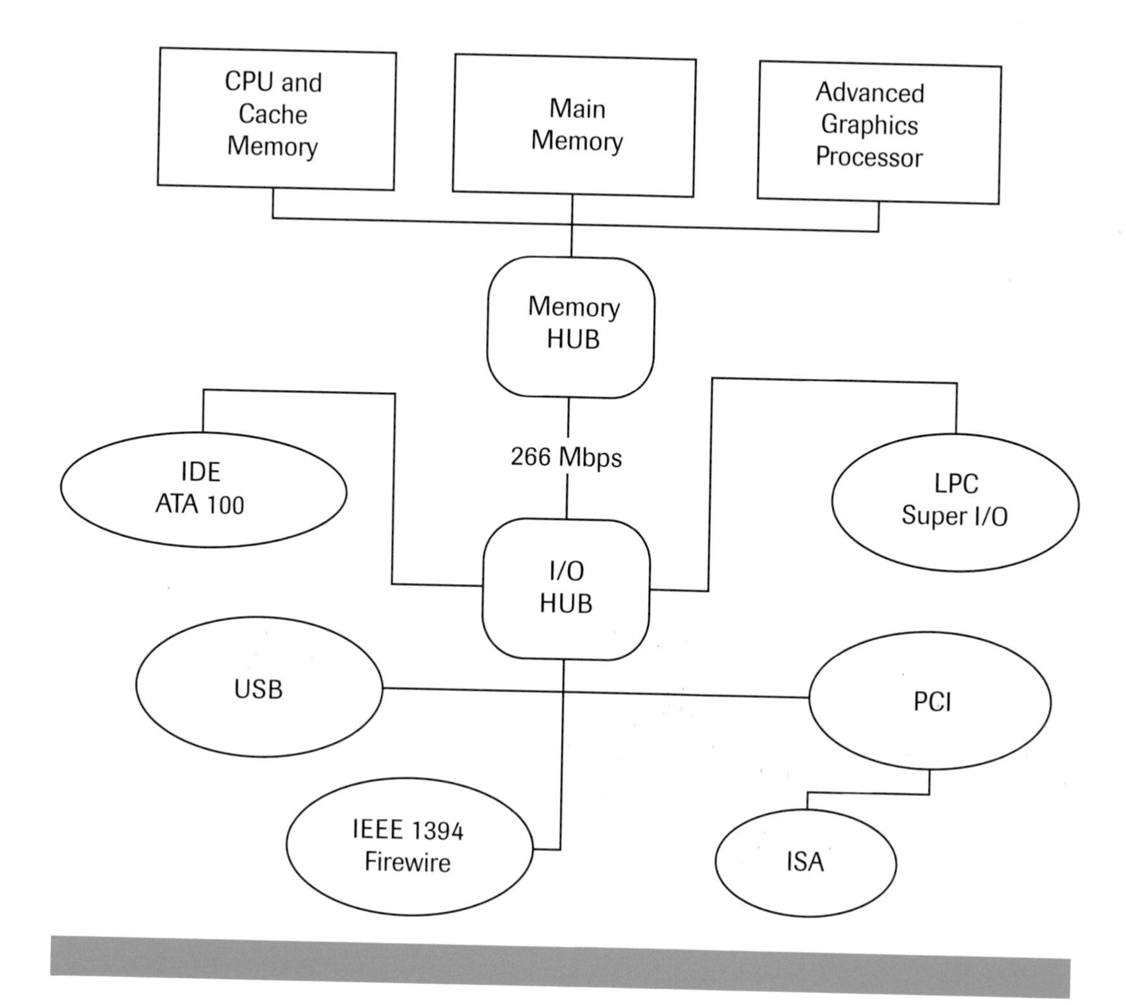

bus structure of the motherboard have increased the transfer rate between the frontside and backside buses to 266 megabytes per second. This doubling of the interconnect speed is one of the major improvements in microcomputer buses in 2000.

The architecture of modern microcomputers is much more complex than that of the first microcomputers. In the early days, buses needed to deal with input and output from only a few devices such as keyboards, printers, floppy disks, and hard disks. A modern-day microcomputer has to deal with input from a wide range of devices, including digital cameras, sound cards, and modems. In addition, plans are underway to interconnect microcomputers and

a wide range of other devices in users' homes. For this to happen, the buses of the microcomputer have to be integrated with the buses used to connect to the other devices in a house. The traditional Ethernet network will not be adequate to support the complete interconnection of all the devices in a home, so buses such as the universal serial bus (USB) and Firewire are being developed not only to improve the bus structure of the microcomputer but also to support connecting the microcomputer and all the devices in a house.

As processors achieve speeds of one gigahertz, the bus structure of the motherboard needs to improve to take full advantage of the speed of the processors. The central idea of the microcomputer architecture being developed is to increase the speed of the buses and to better synchronize how the buses work together. A memory hub implements the frontside bus, and an I/O (input/output) hub implements the backside bus. The hubs synchronize the many other buses of the motherboard.

A Closer Look at the Motherboard Buses

The CPU is connected to the cache memory by a very high speed bus. Typically, the CPU chip contains a small L1 cache, and a larger L2 cache is connected to the CPU chip by a short local bus. The synchronization of these buses is usually handled entirely by the CPU chip and is transparent to the rest of the system. As the advent of the Pentium IV and comparable processors from AMD and Via demonstrates, the speed of the cache buses will continue to increase for the foreseeable future.

The access time and the bandwidth of memory continue to improve. Whether the world settles on SDRAM (synchronous dynamic random access memory) or Rambus memory technology, the speed with which data can be accessed from and stored to memory will continue to increase for the foreseeable future. Computer manufacturers express some concern that the memory bus will not be able to keep up with the other buses in providing the usual doubling of speed each year with only a small increase in cost, but so far the memory bus has kept up.

AGP (advanced graphics processor) 4X will more than double the current AGP 2X speed, providing a transfer rate of more than one gigabyte per second between the AGP card and the main memory. In addition, AGP 4X will have its own high-speed buffer memory to support games and high-quality graphics applications.

The Pentium IV and comparable processors demonstrate that the speed of the cache buses will continue to increase for the foreseeable future.

Several improvements are being made to the intelligent drive electronics (IDE) bus. One of the most advanced is the ATA (AT attachment) 100 interface that will more than double the current transfer rate to the ATA 33 devices. The LPC (low point count) Super I/O bus will provide support for all the legacy devices such as a PS/2 keyboard, a PS/2 mouse, serial ports, parallel ports, game ports, and floppy disks. The PCI bus will be the main bus for add-on boards and will have a bridge to allow the continued use of the ISA bus. USB 2.0, new in 2000, will be the major new bus for video, scanners, cameras, and the like for IBM-type PCs. Firewire will also be available, but it is not clear if it will be a major bus in the near future for IBM-type PCs. Of course, Firewire will be important for those using a Mac.

Improvements to the PCI Bus

The ISA bus, with its sixteen interrupts, was developed for the original IBM PC's peripheral bus. Many have predicted its demise, but it is still here and will probably be around for some time. Other microcomputer peripheral buses have appeared and been used for a while, including the SCSI, EISA (extended industry standard architecture), VESA (Video Electronics Standards Association), and MCA (micro channel architecture) buses. At present, the major bus for connecting peripherals to a microcomputer is the PCI bus. The PCI bus was introduced by Intel and then turned over to a standards organization. Still, most of the current implementations of the PCI bus are modeled after those of the Intel 400 series chipsets. These buses are compatible with the ISA bus, sharing its sixteen interrupts; have five PCI slots with all but a shared ISA slot supporting bus mastering; have a 33-megahertz bus speed with a 133-megabyte-per-second data transfer rate; and support two devices sharing one interrupt for a properly written device traveler. This means that the problems associated with using the sixteen ISA interrupts have been solved for the PCI bus.

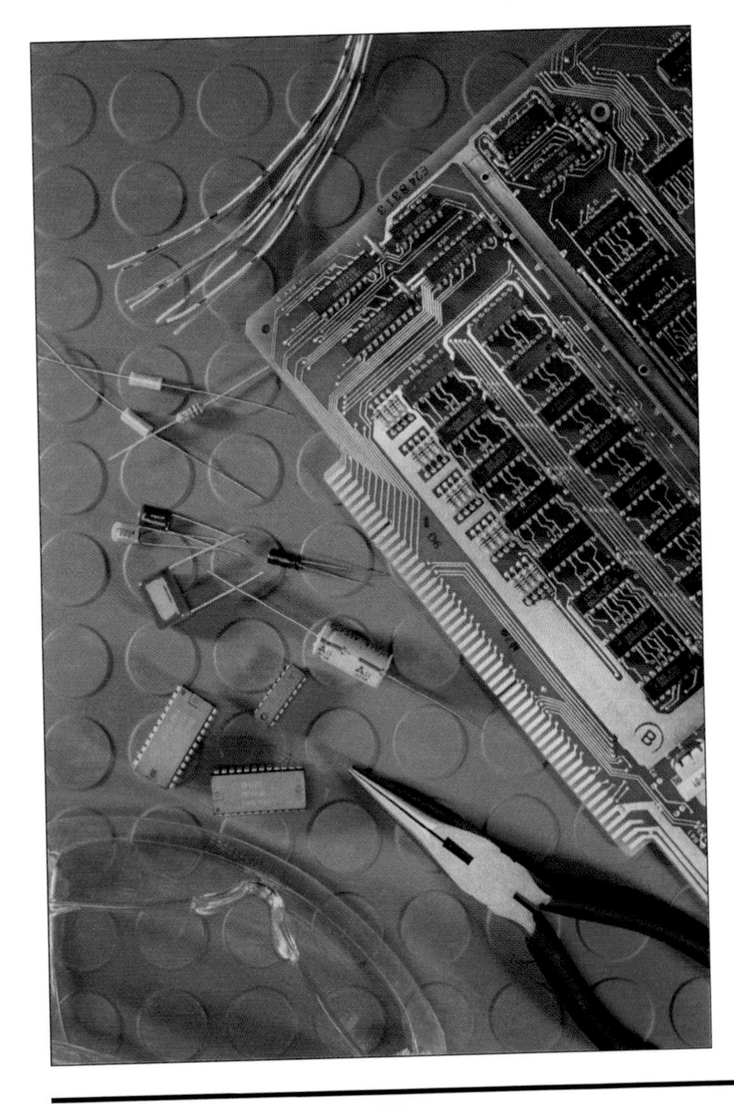

Recent improvements to the PCI bus include increasing the bus speed to 66 and 133 megahertz, supporting more than five PCI slots, supporting the old PCI bus structure that treats the frontside and backside buses as extra peripherals, and supporting the new hub architecture that uses the PCI bus exclusively as a peripheral bus. The major improvements to the PCI bus in 2000 are the increases in speed and the successful implementation of ISA interrupt sharing.

USBs

The USB was developed to provide features to the microcomputer bus architecture that were not supported by other buses. The three main reasons for the creation of USB are to provide a bus that can connect to other devices such as a telephone, to simplify the process of adding the devices to a microcomputer, and to increase the number of devices that can be added to a microcomputer by eliminating the need for ISA interrupts.

The USB is a hierarchical star, or tree. It has a root hub that generally has two connectors. A device or another hub can be attached to the root hub, and more devices can be attached to any added hubs. A USB cable has four wires, two of which are for power and the other two for sending and receiving data. One of the advantages of USB over earlier buses is the ability of a USB cable to transfer sufficient power to operate some devices. In a computer that contains the root hub, a control program manages setting up connections between the host computer and USB devices. The control program constantly polls all the USB devices to see which need to transfer data between the host computer and the device. USB supports asynchronous and isochronous data transfers. An asynchronous data transfer is the usual packet way of transferring data over a bus. An isochronous data transfer is designed to support the transfer of voice or video data by allocating a time slot and then transferring the data within the allotted time.

An isochronous data transfer is designed to support the transfer of voice or video data by allocating a time slot and then transferring the data within the allotted time.

The original USB standard was version 1.1. For this version, the rate of transfer is 12 megabytes per second. The USB 1.1 standard supports multiple pipes. Each pipe is actually a minibus, so USB can support data transfers between the host computer and several devices simultaneously.

In 2000, two big developments occurred in USBs. First, the USB standard has been implemented by almost all computer manufacturers. Almost every computer peripheral manufacturer now makes USB models. Computer owners are using USB scanners, CD-ROMs (compact disc-read-only memory), modems, and many other devices successfully. USB has moved from an interesting standard to a bus that every computer needs.

Second, USB 2.0 was released. USB 1.1 was a reasonably good standard, but as with all new technologies, a lot of small problems popped up. The USB 2.0 standard should fix most of the small problems and produce a peripheral bus that lives up to the early expectations of the USB developers. Of course, the main feature of the USB 2.0 standard that excites everyone in the computer industry is

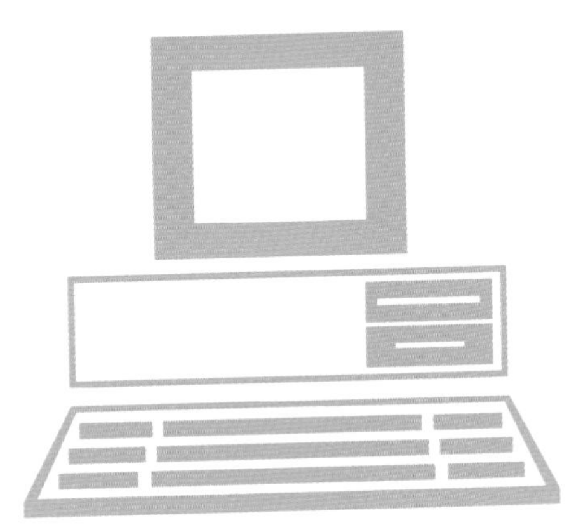

its 640-megabyte-per-second transfer rate. With this rate, many are considering USB as the most likely candidate for a bus to connect a microcomputer to other devices in the home. It is too early to tell if USB 2.0 or Firewire will be the home bus of the future, but the early success of USB 1.1 gives the USB standard a better chance to succeed than had been anticipated by computer experts.

Firewire vs. USB 2.0.

The Institute of Electrical and Electronics Engineers (IEEE) 1394 bus, or Firewire, is very similar to the USB bus; however, some major differences exist. A Firewire cable has six wires, two power lines, and two twisted-pair data lines. Firewire uses a slightly more distributed bus assignment protocol with a cycle master rather than polling. Its overall bandwidth is 100, 200, or 400 megabytes per second. This is much more bandwidth than USB 1.1 had, but not as much as USB 2.0 has. The Firewire cabling system supports daisy-chaining as well as a hierarchical star. Firewire has been used almost exclusively for high-end video applications. Although most chipsets support Firewire at the I/O hub level, they do not support Firewire connectors on the motherboard. Instead, the Firewire bus is generally added through an add-on card. Until recently, this greatly slowed the speed of the Firewire bus because it had to go through the PCI bus, but the Intel 800 series chipsets and their I/O hub eliminate this problem. Systems using the Intel 800 chipset will get the full speed of Firewire with an add-on card.

Firewire is the bus of choice for Mac users. When Firewire was first announced, everyone considered it the microcomputer bus of the future. It did a good job as the peripherals bus and was well designed to connect to a host of other devices, including telephones, air conditioners, and refrigerators. Firewire may well fulfill its promise, but the high cost of its early implementation and its slow acceptance by the IBM-type PC motherboard manufacturers have resulted in a slow beginning for the bus. Firewire is used extensively on Macs, but at the moment, it is simply a SCSI replacement for Macs and PCs. It remains to be seen if Firewire will become the new bus standard or suffer the same fate as the SCSI bus.

Intel and the Competition

Intel processors and chipsets are the current standard for microcomputers. However, AMD has produced the AMD 750 chipset to support motherboards based on its

Athlon processor, and the success of the Athlon has resulted in substantial use of the 750 chipset. The current AMD chipsets are pretty much Intel clones, but AMD has announced that it plans to have a 400-megahertz frontside bus to support the follow-on processors to the Athlon. Considering the current success of the Athlon processor with the AMD 750 chipset, it is not unreasonable to expect a bus structure from AMD that will challenge and perhaps surpass those from Intel. Taiwan has long been a producer of motherboards and associated chipsets. A new company, Via, has produced a chipset that is being used on a number of motherboards, and it cannot be ignored as a real competitor to the chipsets of Intel and AMD. No great architectural differences exist between the Via systems and those of AMD and Intel, but in the past, some Taiwanese systems showed features superior to those of systems produced by Intel and AMD, so it would not be surprising to see a new Via system that establishes the standards for bus architectures.

Optical Storage

➤ *Improvements in laser beam technology show promise of leading to the development of a laser that can record on and read from multiple concentric rings of a single spiral track. In addition, technological improvements allow recordable digital video discs to use almost all their storage space and to fall in price.*

Although optical storage discs appear to be made entirely of plastic, they actually consist of a number of layers. The outer layers are used only to protect the inner layers. One of the inner layers consists of a substance that has the ability to absorb and reflect light differently so that an optical reader can determine whether a particular point on the disc represents a zero or a one. The first optical storage device was the audio CD (compact disc). It was developed by the music industry as a medium for storing digital music.

Shortly after audio CDs became popular, the computer industry recognized that the optical storage technology being used for audio CDs could easily be adapted to store data for a computer. Because optical CDs use one long spiral track to store data, computer manufacturers could apply the packet technology of computer networks and store files on the spiral track as a sequence of packets. The resulting optical storage disc, called a CD-ROM (compact disc-read-only memory), can hold 650 megabytes of data. The data are written to the CD-ROM once and then read over and over. The original CD-ROM technology used a laser to place physical pits representing a zero or a one on the track during the write phase. A CD reader measured the deflection of a laser light beam focused on the pit using a mirror and determined whether the pit was a zero or a one. CD technology is used to store data and digital audio. Almost everyone has come to use CD-ROMs with a computer to install software, retrieve information from an electronic dictionary, or play a game.

Optical discs used to store only data and audio files fall under the CD technology. As digital video increased in importance, both the entertainment and computer industries looked for ways to use optical storage for combinations of data, digital audio, and digital video. All these optical storage technologies are referred to as digital video disc (DVD) technology. All DVDs have a single spiral track that stores data, digital audio, and digital video in packets. The distance between tracks, the distance between pits, and

A COMPARISON OF THE TYPES OF CD TECHNOLOGY	General Description	Approximate Capacity (megabytes)
CD	Stores digital audio files.	650 MB
CD-ROM	Stores data and digital audio files. Laser creates physical pits. Uses write-once technology.	650 MB
CD-R	Stores data and digital audio files. Laser creates pits in film layer. Uses write-once technology.	650 MB
CD-R/W	Stores data and digital audio files. Laser creates pits in a media material by generating a phase change in the material's structure from amorphic to crystalline. Uses multiple-write technology.	650 MB

the minimum length of a pit are all smaller for a DVD than for a CD, so a DVD can store much more information than a CD of the same size. Also, recent advances in DVD technology allow up to four layers, each having its own spiral track, to store data.

Improvements in Laser Beam Technology

Both CDs and DVDs have a single spiral track that stores data as packets. In comparison with CDs, DVDs use much improved techniques to physically store data on an optical disc. These techniques are based on a number of different ideas. The single spiral track requires a minimum distance between two "concentric" rings of the spiral. The minimum distance for a DVD is half of that needed for a CD. To write to and read from a DVD with tracks that are closer together, the computing industry needed to develop better laser beam focusing technology to create pits on the disc and better laser beam focusing technology and mirror technology to recognize the pits on the disc. The pits on a DVD can be much closer together than the pits on a CD, and the minimum length of a pit can be shorter. Again, better laser beam focusing technology needed to be developed to support this improved storage capacity for DVD optical drives.

The length of the spiral track, the distance between concentric rings, the distance between pits, and the minimum length of a pit establish a particular optical disc for-

A COMPARISON OF THE TYPES OF DVD TECHNOLOGY

	General Description	Approximate Capacity (gigabytes)
DVD-video	Stores digital video files as MPEG with an interactive directory capability.	4.7 GB (1 layer) to 17 GB (4 layers)
DVD-ROM	Stores data, digital audio, and digital video files. Uses write-once technology. May provide better protection for data than DVD-Rs.	4.7 GB (1 layer) to 17 GB (4 layers)
DVD-R	Stores data, digital audio, and digital video files. Laser creates pits in film layer. Uses write-once technology.	4.7GB (1 layer) to 17 GB (4 layers)
DVD-RAM	Stores data, digital audio, and digital video files. Laser creates pits in media layer using a wide variety of technologies. Uses multiple-write technology.	4.7 GB (1 layer) to 17 GB (4 layers)

mat, whether CD or DVD. One of the major improvements in optical disc technology that has been discussed in 2000 has been to consider developing new disc formats that can place the concentric rings of the spiral track closer together, place the pits closer together, and allow the minimum length of a pit to be shorter. Although no great breakthroughs have occurred to compel a new optical disc format, the improvements in laser beam focusing technology that have occurred in 2000 are promising and indicate that new formats may be forthcoming in the near future.

Improvements in laser beam technology in 2000, including the blue laser technology, indicate that in the near future the computer industry may develop a laser that can record on and read from multiple concentric rings of the single spiral track. The technology needed for this is very similar to the technology needed to create a higher density optical disc format, but the blue laser technology could work with current CD and DVD formats. This is primarily a DVD development, so in addition to working with several concentric rings on a single track, the blue laser technology might work with spiral tracks on several layers.

More Data, More Accurately

Recent developments in DVD technology have resulted in the ability of some DVD readers to read information on two different layers. The media used on the layers have different reflective properties so that a single laser can recognize the pits on each of the layers. The standard for a DVD defines the minimum thickness of the disc to be half that of a traditional CD. DVDs have two identical sides so that they physically resemble CDs, but most DVD-videos and DVD-ROMs do not use the second side. However, some products containing DVDs with data on both sides of the disc have begun to appear on the market. Of course, existing DVD readers have a laser that works on one side only, so in order to use the data, digital audio, or digital video on the other side of the disc, the user must open the DVD drive and turn the disc over. It is important to note that each layer of a DVD has a format that is similar to the standard DVD format. Many computer manufacturers are experimenting with using more than two layers per side on a DVD, but no one has announced a reader for more than two layers.

All the optical disc formats decompose data, digital audio, and digital video files into packets to be stored on a portion of the spiral track of the disc. Because the operation of composing a file into packets is subject to error, an encoding scheme is used to detect and correct these errors. This is, of course, the same type of technology used in sending files over a network. Research to improve both the error-detecting and error-correcting capabilities of the encoding scheme is ongoing. Improvements in either area increase both the amount of data that can be stored on an optical disc and the speed at which the data can be stored and read. The DVD encoding scheme is much better than the original CD encoding scheme. In 2000, a number of companies and research institutions have discussed creating a new optical disc format with an even better encoding scheme.

The universal disc format (UDF) is a DVD standard that has been proposed by the Optical Storage Technology Association (OSTA) as a standard for all DVD-type discs so that any UDF-formatted DVD can be read in any similarly formatted drive. The UDF standard also defines how a UDF-formatted DVD can be read by different operating systems; thus a single format would work for Windows, Unix, and Mac operating systems. This standard appears to be headed for adoption by all the optical disc manufacturers. Although standards may not seem as exciting as having multiple tracks, they are just as important, and UDF may well be the most important improvement in optical discs in 2000.

DVD-Rs and DVD-RAMs

Much has been written about the work being done on write-once DVD-R (digital video disc-recordable) discs and multiple-write DVD-RAM discs in 2000. The DVD-R discs being developed use a technology similar to that of the CD-R (compact disc recordable) drives. At the start of the year, DVD-R drives were expensive and wasted about two gigabytes of storage (only using 2.6 gigabytes of the 4.7 gigabytes available). As the end of the year, DVD-R drives were using almost all the 4.7 gigabytes available and had become much more affordable.

Many in the computer industry are eagerly awaiting the arrival of a DVD-RAM drive standard. At the moment, a number of companies have developed DVD-RAM drives that support multiple writes to a DVD. The technology used for DVD-RAMs is still developing, although it clearly is similar to the technology being used to create CD-RWs (compact disc-rewritable). Standards are so lacking in the DVD-RAM arena that two different physical optical discs have been developed. The year 2000 saw much development in DVD-RAM drives, but it would be premature to say that anything really works.

CD-RW Drives Get Really Fast

Late in 2000, the tried and tested CD-RW drives got a lot faster. Typical of the new faster writers is the HP 9500. This drive, a 12X CD writer, can create a standard CD, copying a full 650 megabytes of data to the new CD, in six minutes. The most significant improvement for this drive is the write speed for CD-RW discs. The 9500 is an 8X CD-RW writer that copies data at twice the rate of the Hewlett Packard (HP) 9100. The slowest operation of a CD-RW drive is the initialization of a new CD-RW disc. HP claims that CD-RW discs can now be formatted in minutes. Other drives are claiming even faster CD-RW write speeds, so it appears that CD-RW drives will soon be as easy to use as floppies. The only downside is the initial cost of the discs. A CD-RW disc is about $4, which is more than twice as expensive as earlier 4X discs, but as these drives become more popular, the price of the medium should drop.

Here Comes the High-Tech Mouse

➤ *Mice evolve, becoming optical and cordless; handheld, ring, and tactile versions are also released.*

When a graphics user interface such as Microsoft Windows is employed, the computer screen is divided into a set of rows and columns. The intersection of a row and a column is called a pixel. A pointing device is a physical device that identifies (points to) a pixel on the computer screen. The pointing device is represented on the computer screen by a cursor and internally in the computer by its row number and column number. A pointing device is used in computer programs to select items on the computer screen. In addition to selecting items on the screen, a pointing device can also position and rotate items and draw paths.

A large number of pointing devices are currently in use. These include graphic tablets, light pens, touch pads or trackballs, and mice. Users can press a pen on a rectangular graphic tablet to point to a pixel. Alternatively, they can shine a light pen on a light-sensitive computer screen to point to a pixel, and if the computer has a touch-sensitive screen, they can dispense with the pen and press on the screen with the tip of their finger. To point to a pixel using a touch pad or track ball, users move a finger across the touch pad or roll the trackball. To point to a pixel using a mouse, users move the mouse on a mouse pad.

Although most people think of Microsoft as a software company, it has also been a leader in developing some graphics user interface hardware.

By far the most popular pointing device is the mouse. Although exactly who first introduced the mouse is not clear, everyone agrees that the mouse first became famous in 1981 as a pointing device of the Xerox 8010 Star. The standard two-button mouse used in 2000 looks very much like the original Star mouse. The Xerox Star is a good example of the arguments that can develop over the best pointing device. Some members of the Star team went to work for Apple and helped develop the Macintosh computer. Reportedly, one of the major reasons that some of the Star scientists went to Apple was their conviction that a mouse

should have a single button. As Microsoft began to develop its Windows operating system (Windows 2.0), the company adopted a two-button mouse. When the Unix community added graphics user interfaces, it selected a three-button mouse. Many Windows' mice have four buttons, two of which are programmable. A lot of research has been done to determine the best number of primary mouse buttons. This research looks at questions such as how many errors a novice operator makes, how quickly a person can create documents, and how much physical stress results from the selection of a one-button, two-button, or three-button mouse.

Few improvements were made in mouse technology until the last few years. As the Internet increased in popularity, many mouse manufacturers attempted to enhance a standard mouse to improve its usability with a browser. Of course, improving the mouse for Internet usage also improved mouse usage in regular applications. As 2000 came to a close, many high-tech mouse technologies had been refined and their performance improved. In addition, an unusually large number of novel mice were developed in 2000. A few years ago, a high-tech mouse was one with a cute mouse cover and wiggly cursor. Now, a high-tech mouse probably has an optical sensor, no wires, a wheel, and lots of buttons.

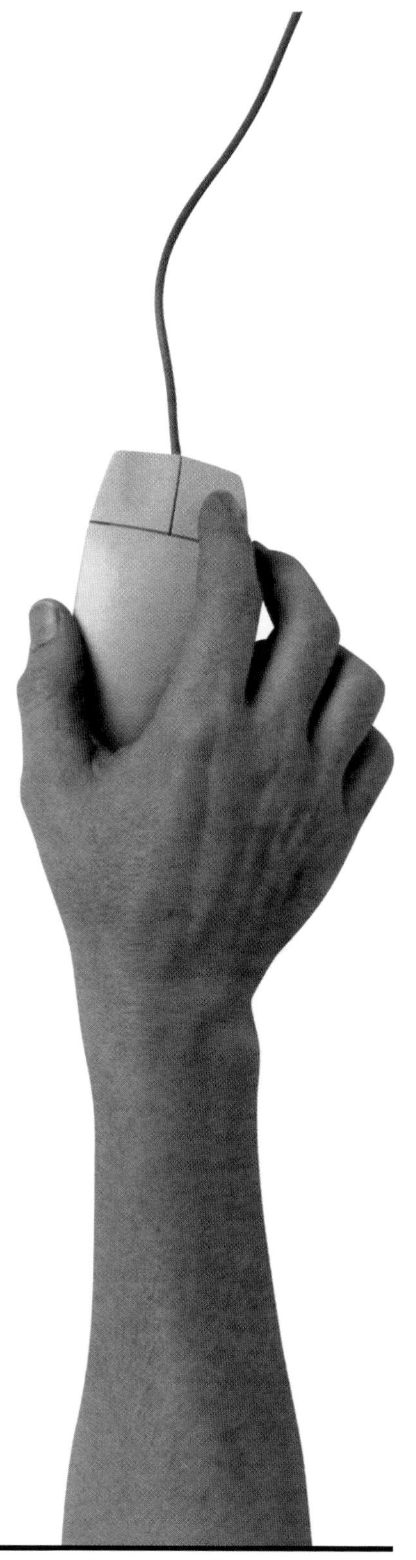

The Optical Mouse

Although most people think of Microsoft as a software company, it has also been a leader in developing some graphics user interface hardware. The Microsoft keyboard and the Microsoft mouse are often seen as leaders in both quality and innovation. Microsoft was among the first to add a wheel for vertical scrolling and has enabled all its applications to take advantage of this scrolling capability.

Everyone who has owned a traditional mouse has suffered as the mouse ball mechanism slowly stiffened up and ultimately became inoperative because of dust and lint on the mouse pad. Even worse, sooner or later the mouse ball itself gets a little warped and just does not work very well. The mouse ball gets dirty, even if a new mouse pad is purchased every month. To solve this problem, a number of companies initiated the development of an optical mouse. Microsoft's version of the optical mouse appeared early and is still one of the best around. Everyone who has visited a computer store recently has seen the Microsoft demonstration of its optical mouse working on a mirror. Unfortunately, it still

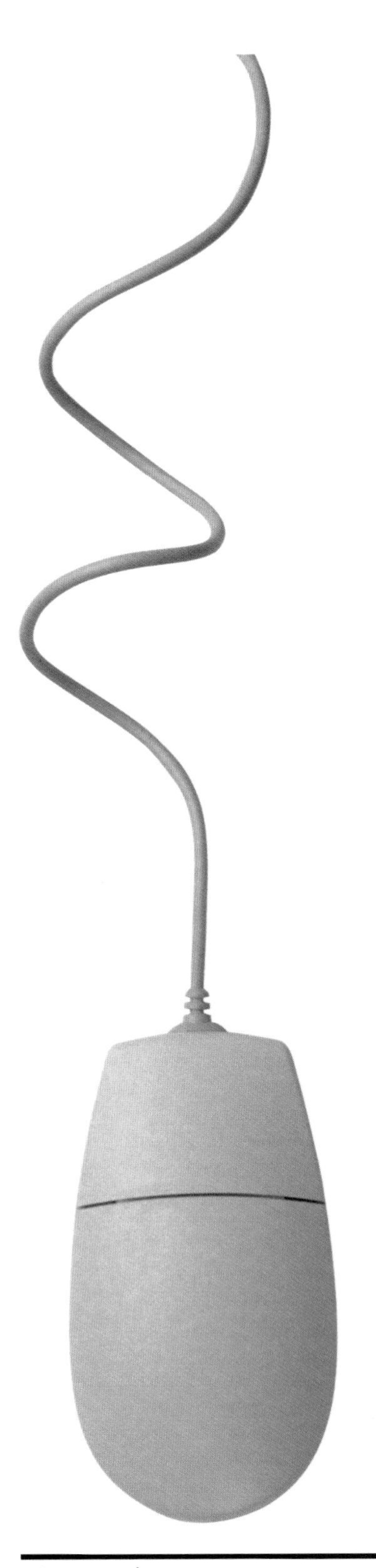

does not work all that well on a pure black mouse pad, but it works well if one with a blue pattern is used.

The Microsoft optical mouse uses a tiny complementary metal oxide semiconductor (CMOS) digital camera to take 1,500 pictures per second of the surface beneath the mouse. A digital signal processor analyzes these pictures and determines the mouse movement on the screen by picking up picture differences. The Microsoft optical mouse has a small computer, running at 18 million instructions per second (MIPS), to keep everything running smoothly, which would have been a good central processing unit (CPU) ten years ago.

Although Microsoft has been the leader in optical mouse development, many other companies, notably Logitech, also have a number of optical mice. The optical mice save users the bother of cleaning their mouse balls every few months, which is a good reason for upgrading, but other than that, they are much like regular mice.

Total Mouse Freedom

Logitech has been in the mouse-making business for a number of years. In fact, most high-quality computers use either a Logitech or a Microsoft mouse. Logitech has been a leader in developing novel mice. It was among the first companies to feature the trackball mouse and has improved on the Microsoft vertical scrolling wheel by adding a horizontal scrolling wheel on many of its mice. It provides lots of mice for Macintosh computers, including a number of transparent and tiny-mouse models.

Logitech was one of the first companies to use high-frequency radio technology to create a cordless mouse. Because the Logitech mouse uses high-frequency radio waves to transfer mouse actions to the computer, it does not have to be pointed directly at the computer. In fact, the mouse can be up to six feet from the receiver. The radio waves are multidirectional and can be picked up by the receiver just like any radio waves. The mouse comes with a receiver that plugs into a USB port or PS/2 mouse connector. The receiver is a little large but does a good job at picking up the radio signals. The receiver also has some shielding capability so that users' favorite FM radio stations do not move the mouse cursor. Because the mouse is attached to the computer, it needs batteries, but as with many such devices, the batteries are inexpensive and last a long time. One of the more interesting Logitech cordless mice is the model that has a trackball on the left side, a wheel, and an optical mouse sensor.

Microsoft and most other mouse manufacturers have been a little slow to introduce cordless mice, but recently Microsoft introduced its first cordless mouse that looks and acts a lot like those of Logitech.

Many products have been developed in the search for a better mouse, including Comtel Korea's combination mouse and telephone shown at Comdex Spring 2000. (AP/Wide World Photos)

The Ultimate High-Tech Mouse.

An interesting mouse introduced in 2000 is a small handheld mouse called the One Finger Mouse. The user holds the mouse in one hand and positions it by rolling a trackball with a thumb. He or she keeps a trigger finger on a button for quick selections but can easily click either of the other two buttons on top of the mouse with the thumb.

The new two-button ring mouse fits on a finger and frees users from holding the mouse. It has two buttons and a trackball that operate like those on a regular mouse. Although users might try a different approach, the prescribed method for using the ring mouse is to put the mouse on the index finger and manipulate the trackball and buttons with the thumb. The ring mouse is a wireless mouse that uses triangulated transmission with both ultrasonic sound and infrared technology.

In September, 2000, Logitech announced a mouse that provided tactile sensations when positioned at various places on the computer screen. In addition to the normal mouse features, the Logitech tactile mouse contains a small motor that can provide different sensations as the mouse passes over various parts of the screen. This enhances the standard graphic user interface by producing different vibrations for different objects. For example, if the mouse is passed over a rough surface, it produces a strong vibration. On the other hand, if the mouse is passed over a basketball, it would produce a rubbery sensation. If the mouse were passed over a piece of pie, it would yield a mushy sensation. Some people may feel that the tactile mouse is just one more novelty, but it actually adds a great deal to the user's experience. The technology used by the Logitech tactile mouse is licensed from a third party, so many other tactile mice are likely to be produced. The addition of a tactile sense is the first major graphics user interface improvement to the mouse in years.

An Ergonomic High-Tech Mouse?

Although mouse technology has advanced, an old low-tech problem has gotten worse. Many people are reporting physical problems associated with mouse usage. Most of the improvements in mouse technology this year did little to create a more ergonomic mouse. Experts estimate that up to 30 percent of the time a person spends creating a document with a word processor involves the use of a mouse. Moving one's arm from the keyboard to the mouse and back and making sure that one's arm is always in the correct position is difficult. In 2000, researchers investigating where to place a mouse to reduce the tension on the arm muscles found that the best place was as close as possible to the keyboard. Obviously, having a keyboard with an integrated mouse should create the most ergonomic combination of mouse and keyboard. A number of combinations incorporating this idea are available, the most common being the inclusion of a trackball in the keyboard.

A novel way to create a more ergonomic environment is to use a combination of keyboard, mouse, and voice recognition software to input data. Many of the actions usually done with a mouse could then be done with a voice command, such as selecting text and deleting it. One company developed a light pen with an embedded microphone; a combination of traditional pointing techniques and voice commands controlled this pointing device. Another idea for reducing muscle strain that surfaced in 2000 was the finger mouse, but because it does not provide all the capabilities of a regular mouse, its use is limited as an ergonomic solution. Many advancements were made in mouse technology this year, but building a better ergonomic mouse was not one of them.

Voice Recognition Software

➤ Technology leader Dragon is acquired by business-oriented Lernout & Hauspie; Phillips releases a handheld voice recognition device that is a combination of a high-quality noise-canceling microphone, a speaker, and a trackball mouse.

Voice recognition software allows someone to input text into a computer program by speaking into a microphone. Voice recognition software exists for a wide range of natural languages from American English to Mandarin Chinese. The key to voice recognition algorithms is to break up each spoken word into a sequence of sounds called phonemes. Each natural language has its own set of phonemes. American English has about 1,600 phonemes. Each of these phonemes corresponds to a single letter (vowel or consonant) or a series of letters that yield a unique sound. Regardless of the written language being produced, voice recognition software accepts verbal input from a speaker, decomposes it into a sequence of phonemes, and constructs written words by combining the phonemes according to the language's grammar and the software's vocabulary.

To use voice recognition software, users input speech into the system, usually by speaking into a microphone. Alternatively, users can use a special mobile recorder to create a voice file that is processed by the voice recognition software after being loaded in the computer. Voice signals are analog signals and need to be converted into digital data. The special software and hardware that do this are referred to as digital signal processing (DSP) technology. The major component of voice recognition software is that which pulls out the phonemes from the stream of sounds. Much of the real research on voice recognition software deals with this component. Usually a voice recognition software system will have a vocabulary consisting of words and the phonemes needed to pronounce the words. The voice recognition software will also have some grammatical rules about how phonemes can be combined to form words and phrases or short sentences. The artificial intelligence phoneme predictor module of a typical voice recognition software package is usually centered on either neu-

Voice recognition software accepts verbal input, decomposes it into a sequence of phonemes, and constructs written words by combining the phonemes.

Among the devices incorporating voice recognition software is the Xybernaut Mobile Assistant, a mobile computer with a head- or side-mounted display, voice recognition, and a 233-megahertz processor, shown at Comdex 2000. (AP/Wide World Photos)

ral networks or hidden Markov model (HMM) systems. The most popular neural network approach appears to be using a number of back-propagation neural networks to identify words from phonemes and short sentences. The most popular use of HMMs is to apply a number of tests to phonemes to produce a translation to words that has the highest probability of being correct.

The three major types of organizations involved in developing voice recognition software are university research laboratories, industry research laboratories, and voice recognition software companies. A number of universities, including the Massachusetts Institute of Technology, have laboratories devoted to artificial intelligence research. Voice recognition has long been a major problem in artificial intelligence, so these university laboratories are constantly producing new systems. Although a lot of innovation takes place at the university laboratories, most of the voice recognition systems produced there do not work well in the real world. However, the laboratories' innovative solutions, produced with small sets of data, provide hints that industry can use in developing practical full-fledged systems. Several large companies, including Microsoft and IBM, have special laboratories devoted to voice recognition research. In fact, Microsoft claims that voice recognition is its most important current project. These labs produce both practical and theoretical results that complement the work of the university laboratories.

In addition, all major commercial voice recognition software companies (including those with research laboratories devoted to voice recognition) are actively developing new systems. Dragon, Lernout & Hauspie (L&H), IBM, and Phillips are producing new products at a rapid rate. Many of these products contain major enhancements to existing voice recognition technology. All the research on voice recognition software comes together in these companies' products because their systems are being marketed to users.

Developments in Voice Recognition

In 2000, the leading technology company, Dragon, was acquired by L&H, a larger, more versatile business-oriented company. Dragon, with its Speaking Naturally, has been the leader in introducing user-friendly and accurate voice recognition software. About 60 percent of the voice recognition software systems sold last year were from Dragon. L&H produces a wide range of products, including

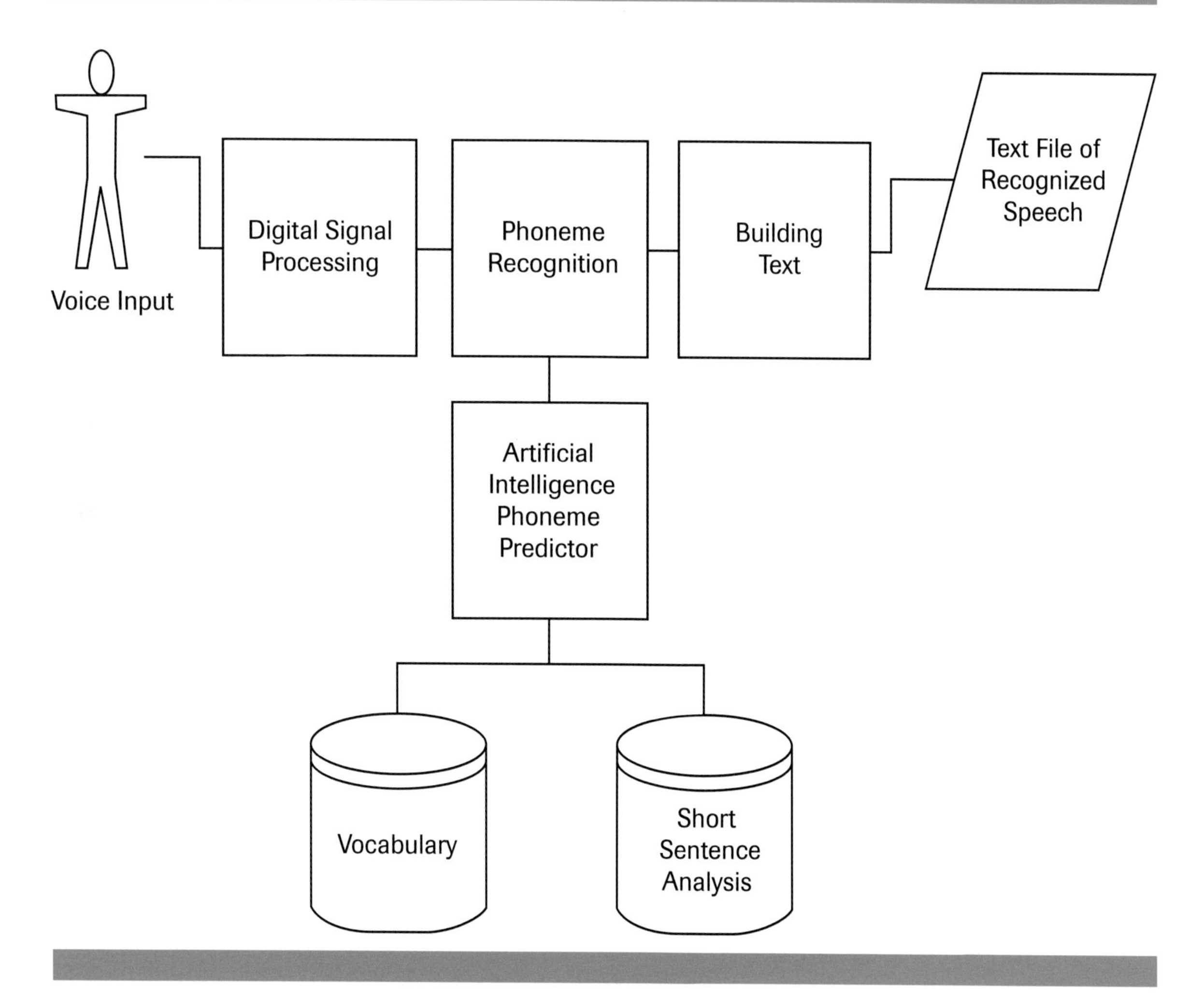

professional Kurzweil voice recognition systems, Voice Xpress personal systems, and many Internet device voice recognition systems. Combining the research capability of the two companies should result in substantial improvements in the current voice recognition software systems. Also, L&H has had a close relationship with Microsoft. As a result of this partnership, Microsoft is ceasing to add voice recognition capability to all personal computer operating systems and is using a standard application program interface (API) to support applications development. Both Dragon and L&H had some very good voice recognition systems for special areas. The combined company now offers some high-end products for fields such as legal voice recognition, medical voice recognition, and handicapped voice recognition.

Voice recognition software does more than simply translate what the user dictates into text. It also can be used to give commands to applications. For example, when using a word processor such as Word or WordPerfect, users can give voice commands to save the file, a boon to those who have difficulty using a mouse or keyboard. The three major providers of this type of voice recognition software are Dragon, L&H, and IBM. Each of these companies released an upgrade of its basic software in 2000, and all three added a large number of applications to those for which they provided command support. Dragon's Speaking Naturally and L&H's Voice Xpress now support Microsoft Office, WordPerfect, and a wide variety of Internet applications. IBM's ViaVoice product supports most of Microsoft Office and some Internet applications. Voice recognition software is among the most complex and troublesome software that can be run on a personal computer. The modules required to support commands in applications often crash computers. The new versions of the software all claim to be more stable.

With the cost of DSP chips dropping every day, many companies are likely to be producing USB microphones with an included sound card in the near future.

The traditional input device for voice recognition software is a high-quality noise-canceling microphone that works with a microcomputer's sound card. Traditional microphones, such as VXI's Parrot series and those of Anderea, continue to improve. Recently, many new input devices for speech recognition software have appeared. One interesting new microphone is based on the latest universal serial bus (USB) standard. Rather than connecting to a sound card and using the DSP processing of the sound card, the USB microphones of VXI and IBM include a small sound card as a part of a microphone. Those using voice recognition software know all too well how tricky it can be to integrate a microphone with a sound card. Getting all the necessary sound card capability included with a microphone should greatly increase the accuracy of the voice to phoneme translation. The cost of the USB microphone with its own sound card is not a great deal more than that of a traditional microphone, and the microphone does not interfere with the use of the traditional sound card for other purposes. With the cost of DSP chips dropping every day, many companies are likely to be producing USB microphones with an included sound card in the near future.

Another important new input device for voice recognition software is the mobile recorder. A mobile recorder is simply a recorder with some type of flash card memory that saves the users' dictation. Most of the mobile recorders have considerable editing capabilities, so users can immediately make some changes to the dictation. When users return to their computers, they simply connect the mobile recorder to their computer with a thin cable and download a file that can then be further edited with the voice recognition software. Although existing mobile recording devices and file editing software need much improvement, they show much promise.

Possibly the most interesting new input device for speech recognition software introduced in 2000 is the SpeechMike Pro from Phillips. This is a handheld device that has a high-quality noise-canceling microphone, a speaker, and a trackball mouse. It looks more like a remote control for a television set than a microphone but actually is an interesting combination of the devices a person uses to dictate to a computer. Although it is possible never to touch a mouse or keyboard while dictating, most users actually make substantial use of the mouse to assist the voice commands of the voice recognition software. Also, most voice recognition software has a playback feature so that users can listen to see if what they dictated was correctly entered into the system. Having a single device that supports the mouse and the playback functions of the voice recognition software as well as a microphone is a useful improvement to the usual method of dictation at a computer. The only thing missing is a keyboard.

Think Small, Talk Tall

The first computers were mainframes that filled a room. Microcomputers covered a desktop, and laptops occupied a lap. Some of the latest computing devices can be worn on the wrist. Although the small size of these computers is appealing, input and output can be a problem. It is rather hard to put a keyboard on one's wrist. Everyone realizes that input and output devices for the super-small computers of tomorrow need to be different from input and output devices for existing small computers. A number of people are experimenting with getting much of the input for small computing devices from voice recognition software. Voice recognition software is not perfect; developers are still in the early stages of replacing keyboard and mouse input with voice input, but a number of the systems produced recently show promise.

An interesting example of one of the new small computers that uses voice recognition software as a major form of input is the Qbe Personal Computing Tablet. The Qbe Tablet is a little larger than a palmtop computer, but smaller than a laptop. Its processor, memory, and storage devices are much like those of the palmtop. It has a 13.3-inch touch-screen display that almost entirely covers the computer. It has an embedded mouse like most palmtop and laptop computers, but its standard keyboard is a display featuring keys that are clicked with a special touch pen. The Qbe Tablet has built-in voice recognition software that is accessed with a standard headset/microphone. The Qbe Tablet works well as a portable computer except for the difficulty of transporting the headset and the touch pen. If an ear-mounted microphone and touch pen could be embedded in the basic Qbe Tablet, it would be portable and might become a significant type of small computer. Many personal digital assistants (PDAs) are also incorporating voice recognition software to either replace or complement keyboard input. As soon as the voice recognition software is perfected and placed on the chip, users will be talking to their computing devices and listening to the answers.

Many personal digital assistants (PDAs) are also incorporating voice recognition software to either replace or complement keyboard input.

Lots of companies are adding voice recognition capability to their software. With the dominance of Windows for microcomputer operating systems, much of this software is being aimed at the Windows operating system. In the traditional approach to software development, each company develops a little library of functions to implement its voice recognition software. To keep everything running smoothly, the operating system should include a library of speech recognition software routines that can be used by most vendors. The vendors may add some library functions of their own, but they will use the operating system's voice recognition functions whenever possible. To

encourage developers to use its library, a smart operating systems company provides a free, or low-cost, software development kit that makes it easy to use the speech recognition libraries of the operating system. As one might expect, Microsoft has expended considerable effort developing the Microsoft Speech Software Development Kit (MSSDK) so that the applications using speech recognition software on Windows systems will work harmoniously together.

Software Design with the Unified Modeling Language

➤ *UML becomes the universal object-oriented graphical notation, and software engineers and scientists develop partial adaptations of UML for Web applications.*

Unified Modeling Language (UML) is a graphical notation that can be used in the analysis and design of object-oriented systems. As object-oriented programming has developed over the years, a number of methods to analyze and design object-oriented systems have also developed, each with its own graphical notation. Three of the most popular early object-oriented analysis and design methods were the Booch method, the object modeling technique, and object-oriented software engineering. The Booch method was very good at helping a programmer discover and describe the classes of a problem and how they were related by inheritance, dependence, and composition. The object modeling technique spent more time than the Booch method in considering the objects instantiated from classes and how data moved from object to object. Object-oriented software engineering decomposed the solution of the entire object-oriented problem into a number of simpler use-cases. Each use-case involved having services provided to an actor, the client, by one or more other actors, the service providers.

Three of the most popular early object-oriented analysis and design methods were the Booch method, the object modeling technique, and object-oriented software engineering.

About 1995, the developers of these three methods, Grady Booch, James Rumbaugh, and Ivar Jacobson respectively, of Rational Software Corporation, began to develop a single graphical notation that could be used for all three of their methods. They selected the Unified Modeling Language as the name of this graphical notation. The developers envisioned UML as being not only a graphical notation for their methods but also sufficiently general to serve as the primary graphical notation for all object-oriented analysis and design methods. Although UML is much too complex to be summarized in a few short paragraphs, some of its most important features are that it captures all the class modeling capability of the Booch method and does not attempt to integrate database modeling into the graphical notation. It also has very good use-case modeling and state diagrams to show object relationships. In addition, UML uses sequence diagrams to illustrate the steps of a use-case, as well as deployment and component diagrams to model distributed objects. The current version does not support Internet applications.

It is clear that object-oriented programming has arrived. Small modules can still be developed with procedural programming techniques, but large systems all need to be based on object-oriented analysis and design techniques. When developing individual programs to run on a single computer, object-oriented programs usually concentrate on developing well-thought-out classes. In this situation, UML is excellent as a design notation because of its strong class modeling capability. When developing distributed programs to run on several computers, UML is also an excellent design notation because of its ability to represent run-time objects on these computers. Many large applications are now being developed with UML notation, and many extensions of UML are being developed.

Given a well-designed object-oriented program, all the user has to do to create a corresponding set of class diagrams is click a button.

UML 1.3 Becomes De Facto Standard

UML 0.8 was the first version of the Unified Modeling Language, and UML 1.3 is the latest. Each of the versions of UML provided incremental changes and bug fixes. With version 1.3, UML is now a stable product that can be used successfully to develop most object-oriented systems. Microsoft was an early adopter of the UML notation and has its own tool set for implementing UML called Visual Modeler. Visio 2000 Professional also has an outstanding implementation of the UML graphical notation. Of course, Rational Software has a full set of tools to aid Java, C++, and Visual Basic programmers, and because this is where UML started, the company has excellent tools such as Rational Rose. All these tools generate nice diagrams, often directly from the code. Given a well-designed object-oriented program, all the user has to do to create a corresponding set of class diagrams is click a button.

When UML was first proposed as a universal object-oriented graphical notation, it was only one of many such projects. Many people were trying to introduce a graphical notation that would be adopted by all object-oriented programmers. All of these, except UML, failed. As 2000 comes to a close, almost everyone in the computer industry recognizes that the Unified Modeling Language is the only graphical notation for object-oriented programming.

Extending UML to Real-Time Applications

Some of the most complex computing systems are the real-time embedded systems such as the Airborne Warning and Control System (AWACS) used to detect missile and airplane attacks against the North Atlantic Treaty Organization (NATO). These systems are expensive, so a computer model of any proposed AWACS system is generated before it is manufactured. Such a computer model involves both real-time computing and distributed computing. One of the best ways to describe a real-time distributed computing system is with a series of finite-state diagrams. The entire state of the system can be understood at any time by looking at the state of all the diagrams. The state diagrams of UML are actually finite-state diagrams with a few extra features. NATO software engineers are using UML to design the next AWACS system. Because the current version of UML was not designed to model real-time distributed systems, the Object Management Group (OMG) has set up a group to study how best to integrate modeling real-time distributed systems into UML.

As 2000 comes to a close, almost everyone in the computer industry recognizes that the Unified Modeling Language is the only graphical notation for object-oriented programming.

UML allows extensions to the original purpose of the graphical notation in three ways: stereotypes, constraints, and required tags. A stereotype is one of the standard graphical elements of UML, like a class symbol, that are given special meanings when used with adornments (icons). If the conditions that are usually required to create a standard graphical element of UML are limited (constraints)—for example, never allowing a class to have a method—then a new element is added to the system. Also, UML allows the addition of an attribute name and value pair (required tags) to any graphical element.

By redefining some of the graphical elements used in the UML state diagrams, the OMG has created a subset of UML that can be used to create finite-state diagrams to adequately represent real-time distributed systems. More than fifty software engineers are now using UML on the NATO Midterm Modernization Program to create the next version of AWACS. The OMG predicts that this is just the start of using UML on real-time distributed systems in both the military and the computer industries.

Other UML Uses

In 2000, Rational Software announced a product that can be used to create Extensible Markup Language (XML) documents using extensions to UML. UML classes represent the XML tags. The structured hierarchy of the XML tags is captured by association and uses relationships between classes. This modeling technique is useful in understanding portions of the XML hierarchy because it replaces the indentation used to indicate the XML hierarchy with a two-dimensional directed graph in UML. Because this is a relatively new development, it remains to be seen just how much help the conversion from linear code to a directed graph will be, but for the moment, a number people seem to find it useful.

UML provides a graphical notation that is useful in discovering a solution to an object-oriented problem and documenting that solution. A number of people believe that UML will also improve the ability of software engineers to discover abstract solutions to a class of object-oriented problems that will make it easier to develop solutions for specific problems. The classic example is the development of Windows programs using an application framework such as Visual C++ with Microsoft foundations classes (MFC).

A pattern in object-oriented programming is an abstract solution to a class of problems. For example, one pattern is a general set of UML diagrams to describe a customer's relationship with a bank. An application framework starts with a pattern, then adds tools to assist in developing specific solutions based on the pattern. For example, in developing a set of programs for a new bank to use on its computer systems, a company might integrate the UML banking pattern with a graphical user interface system and a complete database model. The entire package could be customized to produce the computer programs for the new bank in much less time than would be required if the entire code had to be developed from scratch.

When UML first started in 1995, most programmers were happy to have it to assist in the development of their object-oriented programs and to provide graphical documentation to help in later modifications. In 2000, as programmers become more comfortable with UML, many patterns and applications frameworks have appeared on the market.

A Web Modeling Tool

When the Internet was first developed, users simply downloaded Hypertext Markup Language (HTML) pages and viewed them in their browsers. Later, forms allowed users to provide information to a Web server so that it could be saved on the server. In both cases, the Web was simply a mechanism for transferring information from one computer to another. Web developers realized that the Internet could be used to run complete distributed applications just as mainframe computers had in an earlier era. A Web application is a set of Web pages and programs distributed on both the client computer and the server that work together to complete a business task. Most users are familiar with Web applications such as online stores. The technology needed to create Web applications has been developing at a rapid pace. Most programmers are racing deadlines, so they have devoted little time to thinking about documenting a Web application with a graphical notation. As might be expected, the lack of documentation becomes a major problem when developers need to modify existing Web applications. Although UML was not specifically developed to model Web applications, UML extensions have been used by some to provide quite satisfactory graphical models.

A Web application is usually centered on Web pages that either contain code or invoke code. In creating the UML model for a Web application, the first step is to identify those pages that execute code on the server and those that execute code on the client machine under control of a browser. Complicating matters is the fact that some pages contain code, or references to code, that execute on both. In this case, pages are used twice in creating graphical models of the Web application. When using UML to model a page containing code that will execute on the Web server, the user selects a class with an adornment (icon) to represent the page and other classes that have a used or associated relationship with the Web page to represent Java code, ActiveX controls, Java Beans, VBScript code, and redirected pages. To model a page containing code that will execute on the client machine, the user also uses a class with an adornment (a different one, of course). The Java code, Java Script, ActiveX controls, Java Beans, Flash components, and other client side-code are

A Web application is a set of Web pages and programs distributed on both the client computer and the server that work together to complete a business task.

represented as classes that have uses and associated relationships with the Web page. Sequence diagrams and use-cases are also useful in describing interactive sessions and business logic. The first major steps in adapting UML to model Web applications started in 1999, and in 2000, more and more software engineers and scientists published partial solutions to the problem of representing Web applications with UML. With the increased importance of Web applications illustrated by the proliferation of e-commerce sites, a complete adaptation of UML for Web applications is likely to be developed soon.

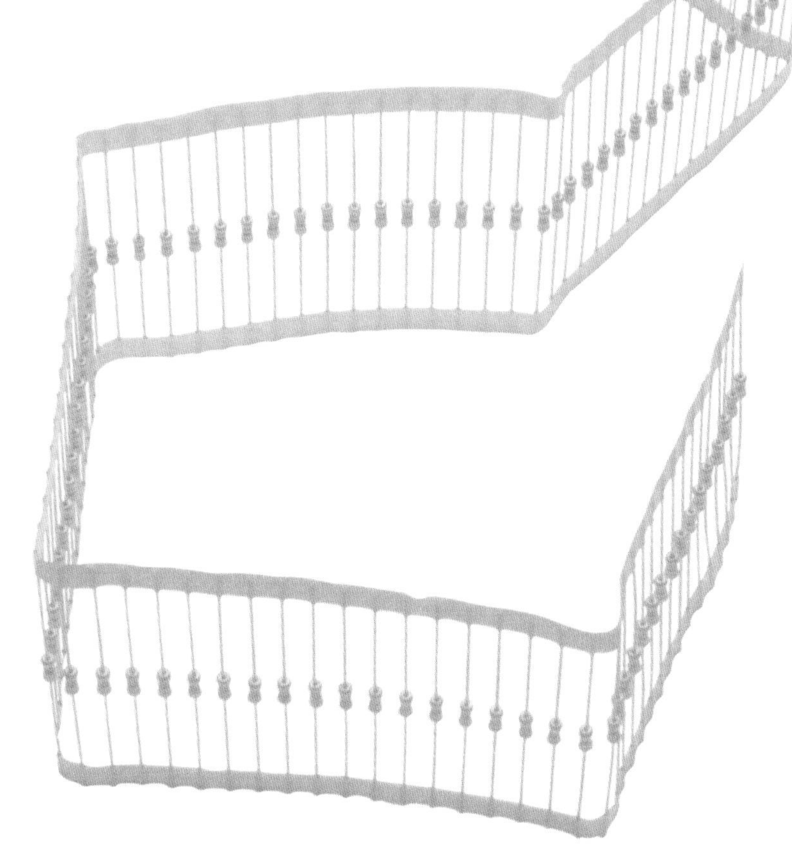

Intelligent Agents

➤ *New techniques allow Web search engines to do better searches, and Microsoft adds voice-activation capacity to one of its intelligent agents.*

Intelligent agents, sometimes called software agents, are computer programs that interact with humans much as another human would. The creation of intelligent agents is as old as the field of artificial intelligence. Many of the early artificial intelligence programs were really nothing more than intelligent agents. In recent years, the development of intelligent agents has become more specialized. Today, an intelligent agent has three distinct characteristics: autonomy, the capacity for cooperation, and the ability to learn. First, an intelligent agent must be autonomous. This means that the computer program that represents the intelligent agent must be able to run on its own regardless of what happens in its environment.

Second, an intelligent agent must be able to cooperate with other intelligent agents. Because an intelligent agent is nothing more than a specialized computer program, this means that these programs must be able to operate as distributed programs. As with all distributed programs, the intelligent agent program must be able to synchronize with other intelligent agent programs. In addition, it must be able to share information with other intelligent agent programs.

Third, an intelligent agent must have the ability to learn. Programs that have the ability to learn are almost the definition of artificial intelligence programs. Intelligent agent programs can use any of the many learning techniques of artificial intelligence, including rule-based expert systems and artificial neural networks as well as more traditional means.

There are many examples of intelligent agents. Progress in their development is most easily understood by looking at some of the best-known general categories of examples, which include information, collaborative, interface, and mobile agents. Information agents are those that collect information locally and forward it to other agents that organize all the information into a meaningful database. Among the most important of the information agents are the Web search agents. Nearly every Internet user knows how to do a general search of the Internet for information. To create a database of information about a search subject such as automobiles, a Web search engine asks many agents to search the Internet for information about automobiles and to report this information back to the general search engine. Search engines such as MSN Search, Yahoo, and Lycos all develop their databases of information about a search subject by delegating parts of the search to intelligent searching agents.

Collaborative agents are a set of computer programs that work together to accomplish a single goal. Collaborative agents often do not work very well independently but do a great job of working together as individual pieces of the program. A typical example of collaborative agents would be a database query that invokes a number of separate queries of different da-

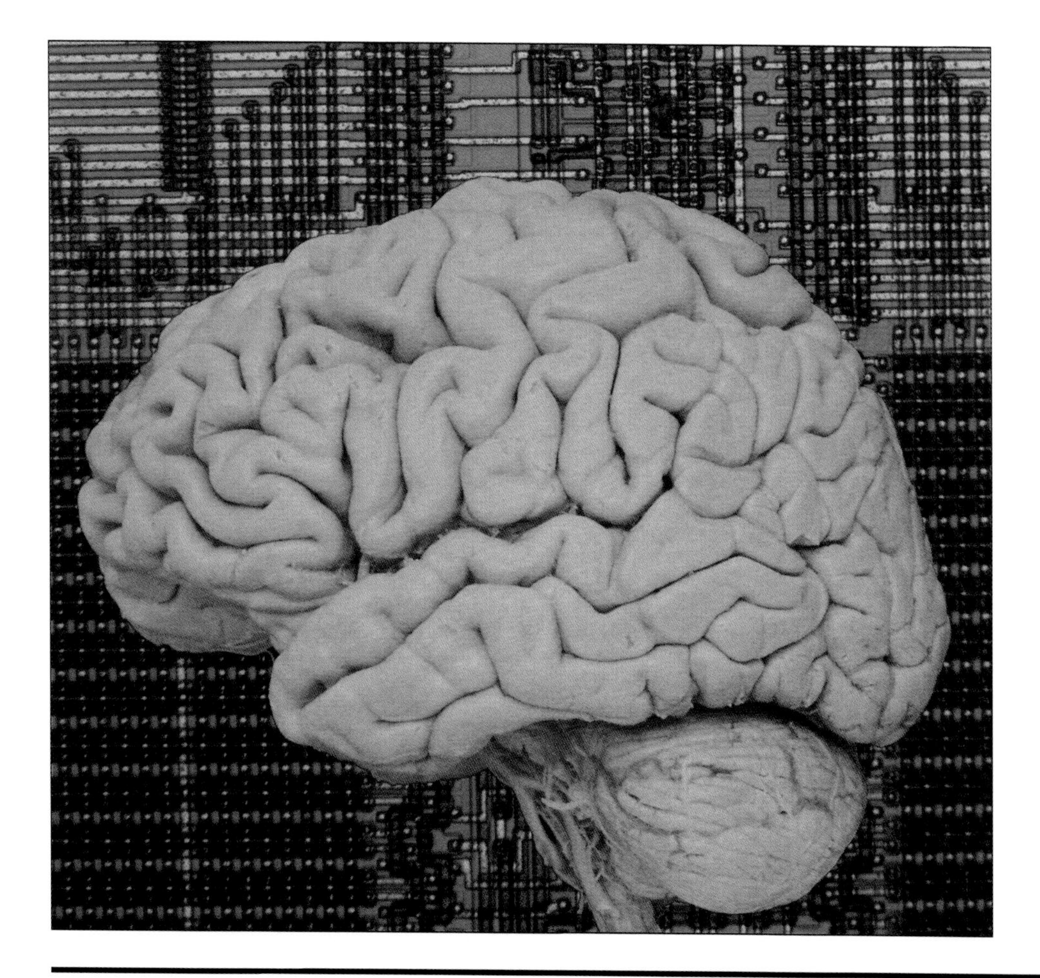

tabases and then combines the information to return to the user. Interface agents work in the background of a graphical user interface to assist someone in using the interface. A classic example of an interface agent is the famous paper clip help in Microsoft Word. Microsoft Word watches what users are doing as they type in their documents, and if it spots an error, it suggests that the users do something else. Mobile agents are those that can move around. For computer software, this means that the program will run in different places. A good example of a mobile agent is a Java script program that is downloaded by the user and then returns information to a Web site.

Intelligent agents are usually part of a complex system. They interact with humans and other computer programs much as a human with access to a high-powered computer would. Many claims are made every year about advances in intelligent agent software, but real progress in this area tends to be a little slower.

Web Search Engines

In 2000, many improvements were made in Web search engines. Until recently, most search engines used a simple approach to creating databases of URLs (uniform resource locators) containing valuable information. A typical algorithm used by a search engine to create a database might be as follows: First, delegate the search to a number of specialized agents. For example, the general search engine might delegate medical searches to a medical agent. Next, the specialized agent sends a mobile agent to a Web site. Once the mobile agent arrives at the Web site, it checks the metatags at the site for index information. Then, the mobile agent performs some type of intelligent analysis on the site. For example, current search engines usually do a simple word count on the keywords of a document and rate the document's importance based on the number of occurrences of the keywords in the document. Next, the mobile agent checks all the URLs to other pages at the site to see if some of their pages contain useful information. Finally, the mobile agent moves to a new site selected from one of the external URLs of the current site.

Determining the value of a page is important for a good search engine. A new idea that some search engines are reported to be implementing is to collect information about the number of links to a page from a defined class of pages. Then the document's value is based on both the number of occurrences of the keywords and the number of links to the page. Assuming that the search engine does a good job determining how many links a page has, this clearly would be an improvement over current techniques of rating pages because it would give a higher rating to pages that are more popular and therefore should contain more important information.

Another technique that has recently been added to improve the methods used to ascertain the value of a document containing keywords is to determine if the document uses the keywords in important phrases or other syntactic elements. For example, a general search engine might delegate building a search database for a medical term such as "cancer" to a specialized intelligent agent that oversees searching for URLs at sites containing information about cancer. This search engine would obtain a URL of a site that is a good candidate to have pages with useful information about cancer. A mobile agent would be sent to the site and would check the metatags to see if the keywords for the home page included pertinent cancer terms. It would also check the page to see the number of times these terms appeared on the page. The specialized intelligent agent would then look to see if the keywords appeared in important phrases. For example, it might determine the number of times "lung cancer" appears as a phrase on a site so that the site could respond to more specialized searches for lung cancer as well as general searches about cancer.

An intelligent agent has three distinct characteristics: autonomy, the capacity for cooperation, and the ability to learn.

Many search engines are improving their databases to contain related sets of keywords. A mobile agent that visits a Web page looking for a primary keyword can then create a set of related keywords on the page. This makes the search engine more helpful to users who cannot initially think of the right keyword but type in something related. When this primary keyword is entered, the search engine returns the associated set of keywords, along with the links of the primary keyword, and the user can select a new and hopefully better primary keyword. Continuing with the cancer example, if the user types in the word "cancer," then the search engine might return a set of links to general cancer sites and a list of other keywords or combinations of keywords so that the user can refine the search. Typically, this is done by having the user select an advanced search option and then viewing a number of suggested search phrases such as "lung cancer."

Microsoft Interface Agents

An agent that many computer users have grown to know is the help agent of Microsoft Office 97 and 2000. If users continue to click the wrong button while creating a Word document, they often see a paper clip jump and offer to help. Although users disagree about how intelligent the paper clip really is, it is certainly a well-known intelligent agent. The program providing the paper clip and its help was written in Visual Basic. As with many useful Visual Basic additions to Microsoft Office, someone figured out a way to enter a computer via the Internet and use the intelligent paper clip to copy files to the user's hard drive. The techniques used were similar to other macro viruses. In this case, the virus was called Dildog, and its inventor simply told Microsoft about the hole in the intelligent paper clip program rather than actually launching the virus. Later Microsoft added a patch that could be downloaded to fix the problem, and the paper clip agent continues to help Word users.

Making it easy to use an intelligent agent as part of the graphical user interface is a major improvement to intelligent agent technology.

Another Microsoft intelligent agent is Microsoft Agent, which allows a program developer to easily add an animated character like the help paper clip of Microsoft Office to any application. Just as with the paper clip, users must perform certain acts to start the animated character such as selecting a menu item after marking something in the text that does not work with the menu item. This intelligent agent can be activated by input from the keyboard, the mouse, or a microphone. Adding a voice-activation capability to Microsoft Agent was the major addition to version 2.0, which Microsoft released in 2000. The heart of the intelligent agent software is an ActiveX control. With the agent control, a program developer need only provide a description of the behavior of the intelligent agent in order to actually implement it. Making it easy to use an intelligent agent as part of the graphical user interface is a major improvement to intelligent agent technology. Of course, those not fond of Microsoft Office's intelligent paper clip may not be any happier with an intelligent help agent added to other applications, and the intelligent agents created using Microsoft Agent may be as subject to virus attacks as the paper clip was. Nevertheless, making it easy to add interface agents to a graphical user interface is a major step forward.

An Intelligent Virus Agent

Like many institutions, Sandia National Laboratories was looking for a good way to detect viruses that arrived at Sandia before its antivirus software could be updated to catch them. If the new virus was of a known type, then Sandia wanted to detect and isolate the virus. The approach used by Sandia was to place intelligent agents on its computers to watch for viruses that were similar to viruses that have been seen recently.

The classifications of viruses that Sandia developed agents for are I/O port checking viruses, programs that probe the system information, programs that activate on system events, denial of service attacks that begin to overuse system resources, and programs that attempt to start from e-mail. If Sandia makes this program work, it would be a major addition to most antivirus programs. Reports are promising, but it is likely to be a while before a complete antivirus agent is available.

Fighting Robots

The software that drives a robot is by definition an intelligent agent. The intelligent agent is the brain of the robot. One of the more interesting applications of robot technology that occurred during 2000 was the development of the television show on the Comedy Channel called *BattleBots*. The show features some house robots that take on participants' robots in a fight to the death on a 48-foot square. Although probably no great new technology was developed for the show, it certainly has demonstrated the current advances in robot technology.

Using XML to Organize Data

➤ *The computer industry recognizes the need to combine HTML and XML, and major companies involved in producing handheld computers and Internet devices agree to use XML files to transfer data.*

An Internet browser downloads and displays a Hypertext Markup Language (HTML) page at the request of a user. The HTML page contains tags that tell the browser how to display the content contained in the page. In the current implementation of HTML, the content is contained all over the page as unstructured text, quite differently from how it is contained in word processing and database files. In these files, the content is usually structured in a hierarchical fashion. However, while the content is nicely structured, it is usually in a proprietary format that is difficult to use in any application other than the one used to create the file. Extensible Markup Language (XML) was developed so that Web

content could be placed in files that supported structured data and were also portable.

A typical example of some structured Web content is contained in the description of a hamburger. A hamburger has a number of attributes that are best described with structured data as opposed to simply writing a linear description. To describe a typical hamburger from the universe of all hamburgers one could use the XML that follows.

```
<?xml version="1.0"?>
<hamburgers>
 <hamburger Cheese ="yes" Onions="yes">
  <name>CheeseBurger</name>
  <options>Super Pickle</options>
  </fresh>
  <description>Quarter Pounder.</description>
  <price>1.99</price>
 </hamburger>
</hamburgers>
```

As is evident from this example, an XML description of structured data is similar to an HTML file that is downloaded to your browser. The XML description starts with a tag that indicates what version of XML is being used. This tag does not require a terminator such as </?xml>. The rest of the description of the hamburger consists of a set of tags that are either empty, including </fresh>, or always terminated, including <options>Super Pickle </options>. As with HTML, any XML tag can have parameters, such as Cheese="yes." However, there are some important differences between HTML and XML. XML is case sensitive, and in some situations, white space is important. However, the most important difference between HTML and XML is that all the XML tags are user-defined. Because all the tags are user-defined, XML is a markup language that is easily extended, thus the name extensible.

During 2000, explosive growth took place in development of applications of XML and in XML technology. A visit to the Web site of W3C, the major organization that approves standards for markup languages, reveals literally dozens of new markup languages related to XML. All major vendors are now embracing XML as the markup language or file format for creating content in portable Web documents. Although the use of XML as the native format for the Microsoft Office 2000 application suite is clearly the best-known use of XML, other companies such as Sun and Oracle have announced their support for XML. The difficulty is not in cataloging many promising developments in XML but in selecting those that will last. Some of the most useful and promising technologies are described in this essay, but in this rapidly changing field, predicting the dominant future technology is tricky.

Combining HTML and XML

XML is a markup language used to describe how the data, or content, of a Web page is structured, whereas HTML is used almost exclusively to describe how to display a Web page in a browser. HTML does almost nothing to describe how the data is structured. Although the display and structure of the data could be described in separate files, it is better to have a single file format describing both the structure of the data and how to display it. Despite the clear benefits of such a file format, the speed with which the Internet has been developing prohibited its early adoption. The most important advancement in XML technology in 2000 is the recognition that it is time to combine HTML and XML.

Although some people had heard of XML before the 1999 release of Microsoft Office 2000, few realized the importance of this language until they experienced it in Word 2000. Most users of Word 2000 were shocked when they saved a Word document to HTML and found the file was two to three times as large as the same file saved to HTML in Word 97. The difference in the two HTML files was a lot of XML. After viewing XML for a year, most users can look at a Word 2000 file saved as HTML and think of it as "simple." The Word 2000 HTML file format places almost all the XML description of the data in HTML comments. Although this works for Microsoft, it does not necessarily lay out a blueprint for a standard format to be approved by W3C. Worse, it actually lays out a blueprint for every company to develop a proprietary way of combining XML and HTML in a single document. Not surprisingly, lots of companies have followed this blueprint.

The most important advancement in XML technology in 2000 is the recognition that it is time to combine HTML and XML.

The W3C has proposed a new markup language called Extensible Hypertext Markup Language (XHTML) as the markup language that will officially combine HTML and XML. The development of the language is still in its early stages and will probably be subject to much change, but the key ideas of the language are fairly simple. The language will have tags that describe both how to display and how to structure data. XHTML will also contain a metalanguage that describes tags. Some tags, such as the data-formatting tags of HTML, will probably have a fixed meaning. Other tags, described in a metalanguage, will describe how the data are structured. The final version of XHTML should also support extensible display tags. The following is a simple example of an XHTML document currently on the W3C site.

```
<html xmlns="http://www.w3.org/1999/xhtml" xml:lang="en" lang="en">
 <head>
  <title>A Math Example</title>
 </head>
 <body>
  <p>The following is MathML markup:</p>
  <math xmlns="http://www.w3.org/1998/Math/MathML">
   <apply> <log/>
    <logbase>
     <cn> 3 </cn>
    </logbase>
    <ci> x </ci>
   </apply>
  </math>
 </body>
</html>
```

An XHTML-compliant browser will recognize that this is an XHTML document by reading the first line above. It will display the HTML just as an ordinary browser does. When it encounters the XML name space tag, it will recognize that it needs to display the text surrounded by this tag using the math vocabulary, ending up with something that looks like $\log_3(x)$.

Although an XHTML file is large, it is also portable and, when fully implemented, will eliminate the need for proprietary file formats. Never again will a Word file not open in WordPerfect.

Communicating Made Easier

In the early days of the Internet, a browser simply received an HTML file and displayed it. This was great for the scientists who were the early users of the Internet and simply wanted to exchange information quickly. As more people started using the Internet, and graphical user interfaces became popular, everyone raced to add active content such as animation and multimedia to an HTML page. One of the earliest ways of adding this active content was using a Java applet. Later developers used either an ActiveX control or a Java Bean to add the active content. Unfortunately, hidden behind the simple, attractive names were some complex technologies that were highly incompatible. A number of computer companies have submitted a plan to the W3C to allow an HTML file to call an object in a platform in an architecturally independent way. The name of the technology being proposed to access objects from an HTML page is the Simple Object Access Protocol (SOAP). When finally approved, SOAP should make accessing Java applets, ActiveX controls, and Java Beans from an HTML file simple and reliable. SOAP might also cover many other active content techniques being used such as Macromedia's Flash.

In 2000, almost all companies planning to participate in the creation of handheld computers and Internet devices agreed that XML files should be used to transfer data between different stations on the Internet.

An XML document can be created in any text editor. In addition, a number of specialized text editors make it easy to create an XML document by using automatic indentation and color coding for the tags of the document. A special viewer is used to view an XML document. In addition to the many shareware XHTML viewers available, Microsoft's Internet Explorer 5 supports viewing XHTML files. Because it is easy to create an XML document and to view it, many businesses are now using XML documents to exchange corporate information. To better support business-to-business communications, a number of companies have created special tags that they and their customers agree to use in XML documents. These special sets of tags are called XML vocabularies, while the hierarchy of the tags is called a schema. By standardizing the tags in the document, companies create a standard to be used in transferring their corporate information. Groups of individuals are also creating specialized vocabularies. An example of a specialized vocabulary that is currently enjoying considerable popularity can be seen in the XHTML sample presented earlier. Here mathematicians and scientists have agreed on a common set of tags to be used in describing mathematical content in an XML page. Another example of a successful vocabulary is Microsoft's BizTalk vocabulary. Microsoft has defined a set of tags that it wants its customers to use in business-to-business transfers of information.

Connecting with XML

Until recently, most people accessed an HTML page in a browser on a microcomputer. However, the Internet is increasingly becoming a means of connecting all sorts of devices. In addition to microcomputers, Web servers, and other computers, the Internet will connect handheld devices and Internet appliances. The handheld devices will include Palm computers, Windows CE computers, advanced cell phones, and other computing hardware. The Internet devices will include everything from refrigerators to air conditioners. It is easy to design a wiring system and networking software to connect this wide range of devices over the Internet. Because a large number of companies will be making these devices, some common file format needs to be developed so that they can all share information. In 2000, almost all companies planning to participate in the creation of handheld computers and Internet devices agreed that XML files should be used to transfer data between different stations on the Internet. No doubt some vendors will still argue about the desirability of using XML files to transfer data between different stations on the Internet. Nevertheless, getting a number of the large companies to agree that XML should be the data description language for the files used by Internet applications is a good first step.

Revving Up the Internet
The New High-Speed Access

➤ *High-speed Internet access becomes readily available to users.*

The original network that became the Internet consisted of a large number of mainframe computers connected to share files and computer programs. A workstation accessed the network directly using an operating-system-supported terminal. Some of the researchers who used the Internet in their offices also wanted to be able to use it at home. To support this, researchers developed the modem, which could convert the digital signals used by computers into analog signals that would traverse standard telephone lines.

In the mainframe era of computing, the standard speed for a modem was about 300 baud, or about 300 bits per second, which by modern standards is very slow. As personal computers became popular, modem technology improved, and modems with speeds of 56,000 (56K) bits per second were developed. Certainly, 56,000 bits per second is a lot better than 300, but with graphically oriented browsers, it is not nearly fast enough.

Some standard techniques can be used to improve the speed at which data move over the Internet. Some of these standard techniques are digital compression, encoding, and use of different media. In digital compression, the idea is to run the bits of a file that is to be sent over the Internet through an algorithm that produces a smaller number of bits with no loss of information. When the compressed file is received, the reverse algorithm is run to reconstruct the original file. In encoding, the bits of a file are encoded onto the analog signal to maximize the bits that are encoded in a fixed length of transport media. Encoding can also utilize previously unused bandwidths of the transport media. When the file is received, the analog signal is decoded and the bits are recovered. Some media, such as fiber-optic cable, can move larger amounts of data than others, such as copper cable. Of course, changing the medium may also greatly increase the cost of sending these data.

Four technologies—satellite, mobile, cable modem, and asymmetric digital subscriber line (ADSL)—have been developed recently to provide high-speed Internet access. A user can access the Internet by using a satellite dish as the first link from the computer to the Internet. Standard satellite radio signals are used to receive data. In mobile technology, a user accesses the Internet by using a modem attached to either a cell phone or a radio transmitter. Data are transmitted using frequencies outside the traditional voice frequencies. With cable modems, a user accesses the Internet by using the coaxial and fiber-optic cable of a cable television company as the first link from the computer to the Internet. Data are sent using frequencies outside the traditional video frequencies and received over an unused video channel. With ADSL, a user accesses the Internet over a standard copper telephone wire as the first link from the computer to the Internet. Data are transmitted using frequencies outside the traditional voice frequencies.

FROM THE SERVER TO THE PC

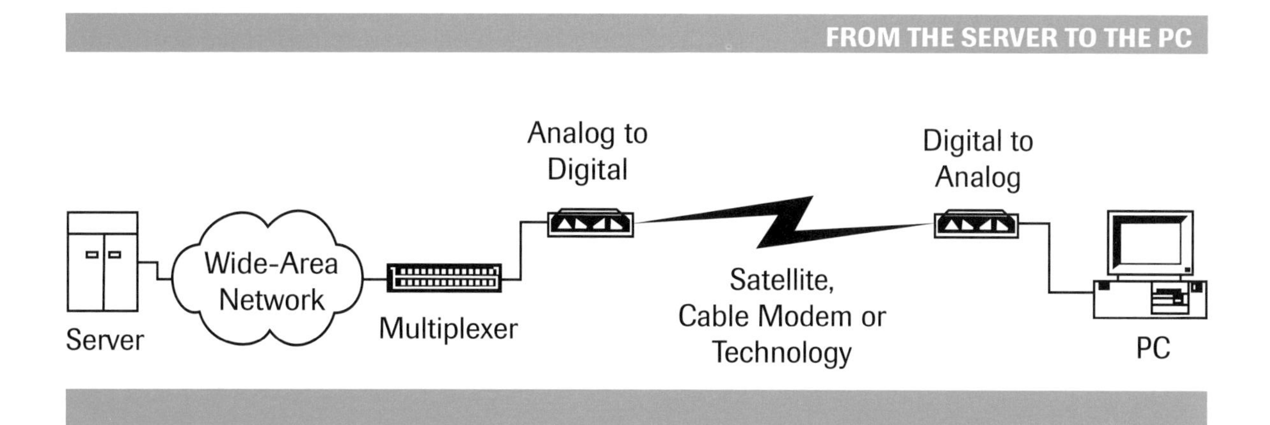

A single diagram can be used to describe all these technologies. All four technologies access a Web server over a wide-area network (WAN). In each case, some type of device is needed by the computer to convert the digital data of the computer into analog signals that travel over copper wire, coaxial cable, fiber-optic cable, or air. The signaling used by the three technologies is different for each technology, as is the method of combining the data sent by multiple home users to a central office. What is most interesting about the new high-speed Internet access technologies is that they all use high-frequency radio waves for their signaling.

The questions of how fast the various methods of accessing the Internet are and what it costs to connect are nearly impossible to answer because of the variety of options available. Also, improvements in the transport speed of the various media take place every day.

SPEED AND COST OF INTERNET ACCESS TECHNOLOGY

	Download Speed	Upload Speed	Setup Cost	Monthly Cost
56K Modem	56Kbps	33.6Kbps	$100	$20
Cable Modem	64Kbps to 14Mbps	64Kbps to 14Mbps	$0 to $500	$25 to $60
ADSL Modem	1.5Mbps to 8Mbps	512Kbps to 1Mbps	$0 to $500	$50 to $70
Satellite	64Kbps to 400Kbps	33.6Kbps to 400Kbps	$200 to $1,500	$40
Cell Phone	64Kbps to 11Mbps	64Kbps to 11Mbps	$300 to $500	$20 to $50

In 2000, the big story in high-speed Internet access is its availability, which has increased dramatically. Users who live in medium-size cities may have four choices for high-speed access, whereas a year ago they probably had only satellite access. As the number of choices for high-speed access increased, the cost went down. Most people can get high-speed Internet access for less than $40 a month and an installation fee under $200. The theoretical increases in speed for satellite and cable modems are not as important this year as availability, but the higher data transfer rates announced this year will be important in the future as more products come to market using the higher speeds.

Internet Access via Satellite

Satellite access for television has been around for a long time. Rather than purchasing a satellite dish and the associated cabling system to watch television shows, users can upgrade their existing satellite systems to support Internet access. A typical example of satellite access is DirecPC, which has a download speed of 400,000 bits per second and an upload speed of 33,600 bits per second (using a 56K modem). DirecPC has a complex payment plan, but users get more or less complete Internet service for $40 a month plus installation charges.

The changes in Internet access via satellite in 2000 have not been so much in technology as in price. The cost of Internet access via satellite has dropped because of competition from other methods of high-speed access.

Mobile Internet Access

Handheld devices such as a personal digital assistant (PDA) are now very popular. As these devices have grown in popularity, manufacturers have designed a wide variety of ways to access the Internet through a cell phone system. One approach is to add a modem to the handheld device, attach it to a cell phone, and then access an Internet service provider through a cell phone system. Sometimes a radio transmitter is attached to a handheld device and accesses an Internet service provider directly over the cell phone system. By using high-frequency radio waves, some of these devices can achieve high data transfer rates. The technology used by the handheld devices for increasing the Internet transport speed is similar to the ADSL technology. In both cases, the key component is the encoding method used to pack the digital data onto the radio signals. This technology is well developed and has an Institute of Electrical and Electronics Engineers 802 LAN (local-area network) standard (IEEE 802.11). Recently, the IEEE modified the 802.11 standard to include complementary-code-key encoding, which increased the transport speeds to 5.5 million and 11 million bits per second rather than the earlier 1 million and 2 million bits per second. Of course, many providers do not come close to these new speeds, but the introduction of the higher speeds means that these speeds will soon be the standard speeds.

PDAs have grown in popularity, and manufacturers have designed a wide variety of ways to access the Internet through a cell phone system.

A wide variety of small devices access the Internet through cell phone systems, and the availability of inexpensive high-speed connection hardware for these devices greatly enhances them. A PDA that can be consulted for a daily schedule is a novelty, but a PDA that can also work with e-mail and the Web is a real working tool.

Cable Modems

A modern cable television system is constructed from a combination of fiber-optic and coaxial cables. Most systems support about 110 television channels and some upload bandwidth for interactive television. Some of the television channels can be used for transferring data rather than television content. Most modern cable modem Internet access systems use a television channel for downloading data and some of the interactive television bandwidth for uploading data. With the technology developed over the past five years, the typical upload speed is 128,000 bits per second and the typical download speed is from 64,000 bits per second to 1,024,000 bits per second. The keys to increasing the speed for cable modem Internet access are upgrading the fiber-optic cable system so that the basic transmission rates are increased and enhancing the encoding scheme for the cable modem so that more data can be packed in the carrier radio waves.

Many cable television systems are improving their cable plants so that they can support high-speed cable modem access. Improvements in the encoding technology are also taking place at a rapid rate. Terayon, a leading manufacturer of cable modems, has announced that its new cable modem is capable of downloading data at a rate of 14 million bits per second. Terayon achieves this high download rate by using a new proprietary encoding technology called S-CDMA. According to Terayon, its S-CDMA encoding scheme supports data transfer rates of 14 million bits per second for both uploading and downloading using standard 5-megahertz television channels. Terayon also claims that the S-CDMA encoding works with regular coaxial cable so that less rewiring of the cable plant is required. Of course, using proprietary technology always has problems because it is difficult to integrate proprietary cable modem systems into larger networking environments. Nevertheless, the speed of the Terayon cable modem is impressive and should improve the competitiveness of the cable modem against the ADSL modems.

The keys to increasing the speed for cable modem Internet access are upgrading the fiber-optic cable system and enhancing the encoding scheme.

Multiple cable modems are attached to the same cable, so they need a protocol to allow them to share the cable. Only one packet at a time can pass through a given cable modem. If two packets arrive at the cable modem at the same time, a collision results, and all the data in both packets is lost. Cable modems have a sophisticated access protocol that not only allows multiple users on a line but also allows users to be allocated different transport speeds. Some users get only 64 thousand bits per second, while others get 14 million bits per second. Companies such as Terayon continue to improve the access protocols of cable modems to support flexible, secure, and fast data transmissions.

ADSL and ADSL Lite

Asymmetric digital subscriber line (ADSL) is a high-speed Internet access technology that can send compressed data over traditional telephone lines. ADSL needs to be fairly close to the central office of the service provider and is sensitive to the quality of the telephone lines, but many people are using ADSL for their high-speed Internet access. Because ADSL uses radio frequencies that are outside the normal voice range, it is possible to use the same telephone line for voice conversations and Internet access.

ADSL goes directly from the user's computer to a service provider's central office. No one else shares the telephone line that is being used for the data transfer. This means that ADSL does not need an access protocol and that the service will not be degraded by multiple users being on a single line at the same time, as is the case with cable modems, satellites, and cell phones. At the central office, a multiplexer/router called a digital subscriber line access multiplexer (DSLAM) connects the user's computer to the Internet. For an ADSL system to support both voice and data communications, the home user and central office need a special device to split the voice and data frequencies. Pure ADSL can achieve download speeds of 8 million bits per second and upload speeds of 1 million bits per second.

Because ADSL uses radio frequencies that are outside the normal voice range, it is possible to use the same telephone line for voice conversations and Internet access.

The cost of the telephone splitters needed for full ADSL is substantial, and the equipment usually requires professional installation. The ADSL community recognized the need for a less expensive version. They also wanted a version of ADSL that conformed to an international standard so customers could pick up an ADSL modem at a local computer store, pop it into the computer, and be operating quickly. Therefore, they developed ADSL Lite, which does not need a telephone splitter. In addition, from a user's point of view, an ADSL Lite modem was just like an ordinary 56K modem. ADSL Lite has a download speed of 1.5 million bits per second and an upload speed of 512 thousand bits per second.

At the beginning of 2000, no one expected either ADSL or ADSL Lite to work. Everyone was predicting that installation and line-quality problems would keep this technology from being available for several years. To everyone's surprise, both ADSL and ADSL Lite systems have exploded onto the high-speed Internet access market. Almost every

Internet service provider is now advertising ADSL services. As ADSL services proliferated, the cost of cable modem, satellite, and cell phone access dropped quickly. As this happened, the initial cost of the ADSL service also dropped, but it remains a little more expensive than some of the other services. The early arrival of reliable ADSL Internet access in 2000 helped make all forms of high-speed Internet access more affordable to customers.

What High-Speed Access Means for the User

Speed really does make a difference for a home user of the Internet. As browser speed goes up, people tend to use the Internet as a tool in their daily work. They start checking the Internet first for the weather, sports, and late breaking news. Sending and receiving large attachments in e-mail is done with little or no trouble. The Internet becomes just one more tool for people to use. In addition to making the use of e-mail and a browser second nature, high-speed Internet access also enhances a number of applications including e-commerce, distance learning, telecommuting, image transfer, video conferencing, and multimedia.

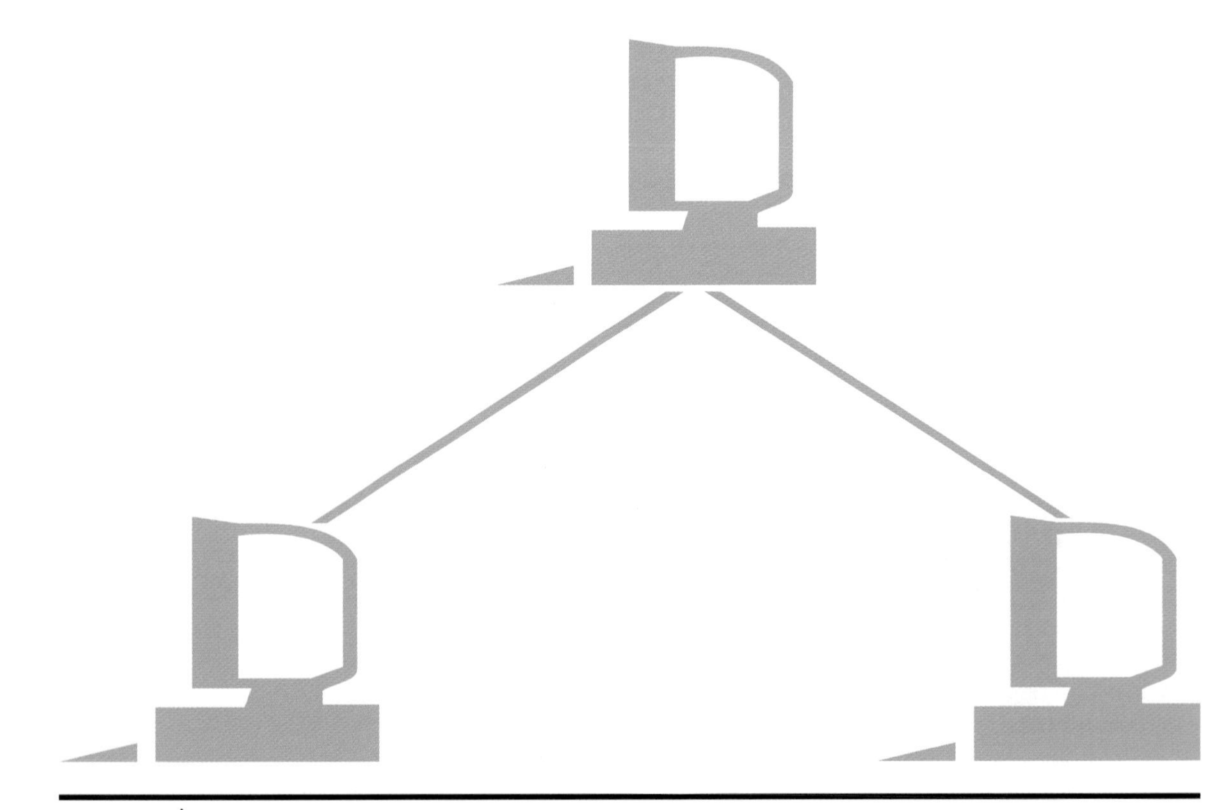

Digital Music from the Internet

➤ *Secure Digital Music Initiative concentrates on standardization and protecting artists' rights.*

Microcomputers have been used to play music since they were first introduced. Until recently, the music played consisted of three types: music CD-ROMs (compact disc-read-only memory), MIDI (musical instrument digital interface) compositions using either FM (frequency modulation) synthesis or wave tables, and digital WAV (waveform audio format) files.

To create a digital WAV file, the analog musical source has to be converted into a digital music file conforming to the WAV file format. This is generally done by using a microphone attached to a sound card of the computer. As the sinelike analog sound waves pass through the sound card, their amplitudes are sampled and recorded as digital data in a WAV file. When an audio player plays the file using the sound card, a replica of the original music is played. A standard WAV file uses a sampling rate of 44.1 kilohertz, or 44,100 digital samples per second, and stores the amplitudes as 16-bit numbers. Although a WAV file gives a faithful recording of the original music, it is very large and, even with compression, too large to be downloaded over the Internet.

The first MPEG (moving pictures expert group) standard developed for digital video also created a digital audio standard. This new audio file format included substantial compression so that the overall file size was smaller than a WAV file. Many sound cards were developed that play WAV files, MIDI files, and the new MPEG audio files. This opened the door for Web developers to think about new file formats for audio files downloaded over the Internet. The MP3 (MPEG-1 layer 3) digital audio file format was developed from the MPEG 3 digital video format by dropping off the video.

The early MP3 developers wanted to be able to play music contained on audio CD-ROMs over the Internet. To do this, they needed to take a song on an audio CD, convert it to a WAV file, then convert the WAV file to an MP3 file. Many tools were developed to create MP3 files from audio

CD-ROMs, and many MP3 players were developed to play this music. The Internet community immediately accepted MP3 music, and users created a vast number of MP3 sites from which songs can be downloaded. With the original MP3 format, it was impossible to control access to and copying of the file. This meant that most MP3 files that were created from CDs were illegal and generally not acceptable to the music industry as an Internet file format. The latest version of MP3 includes watermarks, a technique used to limit access to and copying of a file, but recent court cases indicate the music industry is still somewhat skeptical of the MP3 format. Regardless of the outcome of the court battles, the MP3 format is significant because it started the current Internet digital music explosion.

MP3 Files Still the Best?

Digital music files have a number of formats, each of which has the following components: a file descriptor describing the file type and its security control information; public information describing the music in the file, the file format, and other useful data; the music files as sampled digital files created from the original audio music; and watermarks and other embedded security information to control who can open the file and whether the file can be copied.

An MP3 file currently has no security information. It can encode a stream of audio into a file to be downloaded at 128 kilobytes with no substantial loss in quality. Because the standard Internet download speed is 56 kilobytes, it is also possible to convert an audio music file into a 64-kilobyte form with very little loss in quality. Those who download and play MP3 files regularly and compare the results to a standard WAV file usually find that the MP3 file sounds better. When one says that a particular file format "sounds better," this means that a group of expert listeners classifies the results as better.

When MP3 files first began to appear on the Internet, the large music companies tried to ignore their existence. As MP3 became "the" file format for digital music on the Internet, a number of competing file formats developed.

The big feature of most of the new file formats was a method of controlling who could open the file and whether the file could be copied. Six major competing file formats have been developed. All these formats provide comparable quality, but some sound better with male voices, some sound better with female voices, and some sound better with a particular type of music. The five new file formats have better compression than the original MP3, and all of them are being used.

The six major file formats are MP3, a2b, Liquid Audio, RealAudio, QuickTime, and Windows Media. MP3 is the original digital music file format for the Internet and still the most popular. The new proprietary format a2b provides good security and an improved compression algorithm. Therefore, a2b files are smaller than MP3 files. Liquid Audio is another new digital music file format with good security and an improved compression algorithm. RealAudio is the digital music file format for RealNetworks. It also provides adequate security and compression. In tests, it appears to give the "best" quality of any of the formats. QuickTime is Apple's digital file format. Windows Media is Microsoft's current name for its digital music file format included as part of its NetShow streaming media server. It has an ambitious and developing security model and excellent compression. Its quality appears to be just a little below RealAudio. A different player is required for most of the file formats other than MP3. Microsoft's Media Player is the most versatile, and its latest version plays only MP3, Windows Media, and QuickTime files.

Musical Standards for the Internet

With the number of file formats likely to increase in the next few years, Internet users might despair over the dozens of players they will need if they want to be able to listen to all the digital music on the Internet. The Secure Digital Music Initiative (SDMI), the standards organization for music over the Internet, is trying to create a standard that would allow any player to play any file format in a few years and to ensure that all file formats can provide adequate security in the near future. However, the number of file formats probably will continue to grow because the varying file formats create files of different sizes and quality. Two competing factors control the file size and its quality. First, better quality can be obtained by having higher digital sampling rates, but this results in more amplitudes being placed in a file. Second, the human ear cannot detect

many sounds, so many compression algorithms simply omit sounds that are not seen as important. Of course, although the sounds may not be important to many, they just might be what adds that something extra to provide a high-quality music file.

Downloading and playing digital music has clearly caught on with Internet users, but two major problems must be solved in order for digital music to be truly successful. The first problem is to find some method of protecting the intellectual property rights of the musicians and music companies. This means that the creators of the digital music downloaded from the Internet need to be able to get a reasonable fee for their music. To do this, a digital music file needs a file header that contains some information about the security attributes of the digital music file and some embedded watermarks that control who can play and copy the file. The second problem that needs to be solved is to make sure that every digital music player can play every digital music file format.

Several companies have created proprietary digital file formats and players that fully protect the intellectual property rights of digital music. Some music providers have made arrangements with these companies to encode their digital music in the proprietary format. However, none of these companies has been very successful up to now. Microsoft has proposed a general file format that would support any of the proprietary file formats, but it is too early to know if this will catch on. None of the efforts by individual companies has really gained wide support in the Internet community.

To develop a complete standard for digital music delivered over the Internet, SDMI proposes that its file standard must be open, interoperable, financially practical, upgradable, and testable. It must also have the capacity for control of who plays and copies any file. Although an open standard will not eliminate proprietary digital music file formats, it will require each file format to conform to a standard. All files will contain a content description, the musical content, and control information. The standard must be interoperable so that any approved player can play all file formats. All approved digital file formats must support the ability to control who can play the file and who can copy it. Special provisions will be made for the current MP3 files so that all approved players will play standard MP3 files. The standard will be practical in keeping the costs associated with conforming to the standard reasonable. The standard will be ungradable so that new versions will always be backward compatible. The standard will be testable so that a list of conforming products can be developed.

The SMDI standard will apply to both microcomputers and the mass of portable devices that can also play digital music. Efforts to get the music production companies, the music device developers, and the Internet music download sites to agree on developing a common standard is the major event in digital music for the Internet in 2000.

When developing a player for digital music downloaded over the Internet, a software developer needs to write a program that accepts a music file and then plays the file on the computer's sound system. Better and lower priced software can be developed if an operating system company provides a programming interface for the developer. Microsoft has announced that it will support the SMDI standard for its Windows operating systems. Developers who want to develop a new player for Microsoft Windows that conforms to the SDMI standard will be able to use some Microsoft libraries in their work. Future versions of Microsoft's Media Player will conform to the SMDI standard. Microsoft's support of SDMI should help in improving the quality and reliability of Windows digital music players for Internet digital music files.

Some people think that what really got MP3 going was the development of the Diamond Rio MP3 player. This was a digital music player, not a complete computer, designed to play MP3 files downloaded from the Internet. In addition to the original Diamond Rio player, many other handheld players have been developed for all the new digital music file formats. Recently, a company announced that its cell phone would not only have e-mail capability but would also have the ability to play MP3 digital music files. So far, no one has announced a wristwatch MP3 player, but it may not be far off. The new SDMI standard will greatly assist the development of handheld digital music players because the standardized file formats will allow the development of chips to support digital music.

Music for Fees?

In September, MP3.com lost a lawsuit that could result in its paying $250 million dollars to the Universal Music Group (UMG), the world's largest recording company, for copyright violations for downloading UMG's music. At

The Diamond Rio MP3 player, which weighs less than three ounces, is designed to play CD-quality digital music files downloaded from the Internet. (AP/ Wide World Photos)

about the same time, Vitaminic, a European MP3 download site, announced that it was going to sign agreements with as many music companies as possible to avoid lawsuits over copyrights. Shortly after this, MP3 was reported to be negotiating with a number of companies, including UMG, for agreements to allow a small fee to be charged for downloading music from the MP3 site. At the end of 2000, the days of free Internet music appeared to be limited.

Microsoft Versus the Department of Justice

➤ *The Department of Justice wins the first round of an antitrust lawsuit against software giant Microsoft.*

All citizens of the United States experience government regulation of various aspects of their lives. They stop at stop signs and use their Social Security numbers to make purchases and to pay their taxes. However, some aspects of life, including religion and speech, are relatively free from government regulation.

Until recently, the high-tech computer industry was largely unregulated. Some chip factories had their waste disposal methods scrutinized, and supercomputers could not be sold abroad easily, but by and large, the computer industry was unregulated. In 2000, things changed. The government is no longer reticent about regulating the computer industry. Some of the factors behind the new move toward more government regulation of the computer industry are the popularity of computers, the growing importance of Internet access, and the increasing number of people who make their living in the computer industry.

In 2000, using a computer is no longer something that only scientists do; many people have come to feel that they have a pretty good understanding of what a computer is and how to use it. Also, the Internet has become more than just a new technology for accessing information. It is changing the way people live and do business. Access to the Internet is as important as access to a car and a telephone. Most access to the Internet is via a personal computer, and this being the case, it is important to ensure that there are lots of choices for Internet access from a personal computer. In addition, many people now make their living in the computer industry. Keeping the computer industry viable in all fifty states is important. The computer industry is important not only as an industry but also as an essential resource for all industries in the information age.

In 2000, the government made numerous attempts to regulate the computer industry, but none got more attention or is more important than the antitrust case against Microsoft.

Some regulation of the computer industry is inevitable. Cyber crime is just getting started, and laws are needed to assist in controlling this new type of crime. In many parts of the country, the computer industry is the most significant industry in the region. Elected officials will be just as protective of their local computer industry as they are of older industries such as steel manufacturers. The exact amount and type of regulation are just beginning to be determined and will clearly change a lot over the next few years, but it is also clear that there will be more regulation of the computer industry in the future than there has been in the past.

In 2000, the government made numerous attempts to regulate the computer industry, but none got more attention or is more important than the antitrust case brought by the Department of Justice against Microsoft. This case is important for several reasons. First, Microsoft's development of a working distributed operating system based on its proprietary run-time object architecture is one of the most significant achievements in the computer field. Since the industry's early days, the debate has raged over whether an open architecture for distrib-

uted computing was better than a proprietary architecture with an open interface. In either case, it is essential that the system really work. Microsoft has demonstrated that its distributed architecture works. It remains to be seen if an open architecture will really work. It is unlikely that Microsoft will continue to enhance its distributed operating system if the Department of Justice lawsuit prevails. The question is whether the rest of the computer industry would continue the work that Microsoft has begun and move to build a distributed operating system that really works and or if it will let this important research area falter.

Second, the computer industry has always had giants and dwarfs. At one time, the IBM mainframe was the only computer used by large businesses. At its zenith, Cray Research produced almost all commercial supercomputers. Changes in the computer industry have always corrected

inequities between companies. IBM is now just one of the many companies providing business computing, and Cray is owned by Silicon Graphics. Last year's giants often are this year's Chapter 11's. The question is whether the Department of Justice's assistance is really necessary to increase competition for Intel-based personal computer operating systems. If the Department of Justice's lawsuit succeeds, other lawsuits are likely to be filed, and this will result in more government regulation of the entire computer industry.

Third, the computer industry, like all businesses, has some very aggressive business practices. Giving discounts on volume purchases, creating special partnerships, and selectively sharing trade secrets is currently common practice in the computer industry. The Microsoft case raises questions of whether all companies need to use the same business practices and whether a company should be required to change its business practices when it reaches a certain size or controls a certain market share. Assuming a company should be required to change its business practices, the question is whether this change should be brought about by legislation or by judges' decisions.

If the Department of Justice's lawsuit succeeds, other lawsuits are likely to be filed, and this will result in more government regulation of the entire computer industry.

As of November, 2000, U.S. District Judge Thomas Penfield Jackson had issued a decision based on a finding of facts that agreed with the Department of Justice's position that Microsoft had a monopoly on Intel-based personal computer operating systems, had used unfair business practices as a result of its monopoly power, and should be broken up into two companies.

The Supreme Court of the United States declined to hear Microsoft's appeal of Judge Jackson's decision. Microsoft plans to file its appeal with the U.S. Court of Appeals for the District of Columbia. Although no one knows how this court will decide the appeal, most think that it will alter, in Microsoft's favor, Judge Jackson's original decision. Regardless of the decision of the appeals court, it is almost certain that either Microsoft or the Department of Justice will appeal to the U.S. Supreme Court. Thus at this point, the Department of Justice has won the first round of what will undoubtedly be a long legal process.

The Government's Position

A number of cases have been filed against Microsoft in the past few years. These include a bulk licensing case, a contempt case, a Quicken case, and the current antitrust case. Microsoft survived the first three cases. It lost the first round of the antitrust case, but won the second round when the U.S. Supreme Court declined to hear the case immediately.

The Department of Justice's position against Microsoft, while very complex, is based on three key points. First, computer professionals disagree as to the definition of an operating system. Some believe that the Unix operating system provides the working model for the definition of an operating system. With Unix, the operating system provides a small set of services for process, memory, and file management. Others believe that the Windows NT operating system provides the working model for the definition of an operating system. With Windows NT, the operating system can contain many services in addition to the basic operating system services provided by Unix. The Department of Justice favors a narrower, Unix-type definition of an operating system.

The Department of Justice believes that Microsoft's dominance of the Intel-based personal computer operating systems is a monopoly and that this has hurt consumers.

Second, a Windows operating system on an Intel-based personal computer is important enough to be subject to antitrust considerations. The Department of Justice believes that Microsoft's dominance of the Intel-based personal computer operating systems is a monopoly and that this has hurt consumers. The Justice Department further believes that the computing industry cannot correct this problem without government intervention and that a legal remedy is required. One solution would be to create at least two viable operating system companies, each producing a Windows operating system. A second solution would be to break up Microsoft into two companies, one concentrating on operating systems (as defined by the Department of Justice) and one concentrating on software. Using the Department of Justice's definition of software, all server-side software would move to the new software company. The Justice Department favors splitting Microsoft into two separate companies.

Third, for ten years, or until real competition exists in the Intel-based personal computer operating systems market, Microsoft cannot use ordinary business practices.

Microsoft Chairman Bill Gates, shown addressing a group of business and government leaders, disagreed with the U.S. district court, which found his company in violation of antitrust laws. (AP/Wide World Photos)

The courts would control all of the actions of the two new companies, created by breaking up Microsoft, during this period. The Department of Justice argues that the antitrust laws were developed precisely for situations such as its case against Microsoft. From the department's point of view, Microsoft has been successful in the development of its operating system not only because of improvements in the quality of the operating system but also because Microsoft has used unfair and predatory business practices.

Each time Microsoft adds more capability to the operating system, some companies go out of business and other companies are created to provide add-on capabilities for the new operating system. Microsoft controls the interface to its operating system and therefore has some control over which companies can successfully develop software for a Microsoft Windows computer.

In the process of developing a final proposal for breaking up Microsoft, a number of options were considered. One reported to be favored by Judge Jackson was to create two competing Windows operating system companies for Intel-based personal computers and a separate software company. After the breakup into three companies, the government would remove itself from controlling the development of the operating system software for Intel-based personal computers. Although this does not eliminate questions such as where the server-side software would be placed and how convergence of data, audio, and video for Windows CE would be handled, many industry experts believe it would be a better solution than that proposed by the Department of Justice.

Microsoft's Position

Microsoft views an operating system as a large and expanding collection of components that provides services to a personal computer user. As technology changes and current versions of a Windows operating system become inefficient and unstable because of add-ons such as compact discs, universal serial buses (USBs) and programs directly accessing the Internet, Microsoft feels that it has a right to modify its operating system to incorporate support for the new technologies.

Microsoft also believes that being a major developer of user software, such as Microsoft Word, allows it to develop better operating systems. In other words, Microsoft believes that vertical integration of the operating system with some of its key products produces better software for the user community. Microsoft also believes that products such as Microsoft Office benefit from their association with an operating system company because they can do a better job of utilizing standard features of the operating system. Microsoft says that every company should have the freedom to innovate.

Microsoft believes that vertical integration of the operating system with some of its key products produces better software for the user community.

Microsoft argues that its business practices are no different from those of its competitors and that if they were less aggressive in the marketplace, they would soon find themselves out of business. They point out that the focus of the lawsuit, Microsoft having a monopoly in Internet access, is clearly not the case. With the acquisition by America Online (AOL) of Netscape and the emergence of many new ways to access the Internet, Microsoft does not have any hope of becoming a monopoly in the Internet access field. In fact, as AOL continues to gain share in the Internet access market, many think that Microsoft might be the only company strong enough to keep AOL from becoming a monopoly in Internet access.

Who Is Right?

Those who believe that the Microsoft Windows operating system provides good value for its price find it hard to see why they should support the government's antitrust lawsuit against Microsoft. The best the government appears to offer is a choice of companies to provide a Windows operating system on Intel-based personal computers, and the worst the government is offering is a return to the incompatibility and low quality of the Windows-based systems before Windows 3.1. However, those who do not like Microsoft Windows and also do not want to use a Mac, a Linux-based PC, a UNIX-based PC, or OS/2 can look forward to the day when there are two competing companies, both providing comparable Microsoft Windows operating systems, assuming the government prevails.

Most analysts agree that because Microsoft is producing high-quality software at a low price, it has gained market dominance in a number of areas. So far, Microsoft's business practices appear to have benefited consumers although they have alienated others in the industry. Any company with Microsoft's share of the market could, at some time in the future, use this market share to charge monopolistic prices. Most believe that this will not happen in the software industry because customers would shift to other products, but it is a possibility that cannot be overlooked. The Department of Justice has now placed itself in the position of regulating one part of the computer software industry. This may make some of Microsoft's competitors happy at the moment, but many analysts think this is not good for either the software industry or consumers in the long run.

Combating the "Love Bug" and Other Viruses

➤ *More than fifty variants of the Love Bug virus attack computers worldwide.*

A computer virus is a program that attacks the computer in which it resides. The attacks can take many forms, but almost all are designed to damage or completely disable the host computer. The key feature of a true computer virus is its ability to replicate itself, just as a biological virus does. Computer users detect some viruses fairly quickly because they overload the computer's memory and disk space with copies of themselves. Other viruses are subtler and simply do a certain amount of damage to the host computer while creating copies of themselves.

The damage done by a computer virus ranges from a mild irritation to the total destruction of the information contained on the computer. Past computer viruses have displayed a message or bitmap on the screen at random times, erased some specific files (such as all the MP3 music files), and modified the partition tables or boot sector information of a hard drive so that the hard drive becomes completely unusable. They have also caused computers to freeze up and stop working. This differs from a badly behaved application that causes the computer to lock up in that the action will be replicated many times. Additionally, computer viruses have filled up the hard drives of computers with copies of themselves or of some useless file.

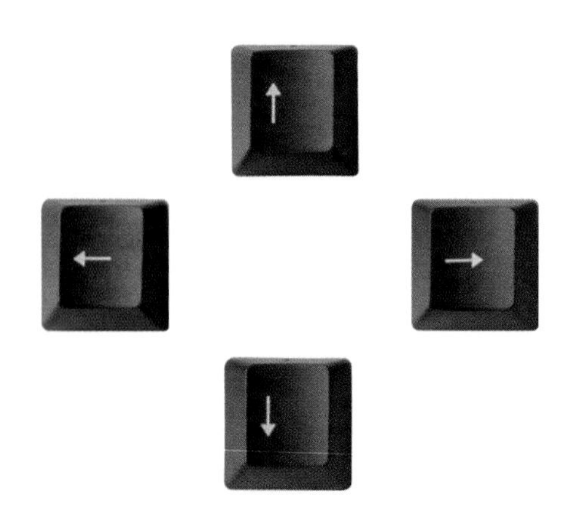

Early viruses generally infected a new computer when someone inserted a floppy disk containing the virus. If users never exchanged data with floppy disks, they could not infect their computers. Later, viruses were caught by downloading an infected file over a network with a file transfer protocol (FTP) program. Because most personal computers were not attached to networks, most computers were still safe from viruses. Today, with so many computers connected by the Internet, no computer is really safe. Computers can get viruses from a large number of servers attached to the Internet or even from e-mail.

In addition to programs that are classified as true computer viruses, many other programs act like computer viruses and therefore are included with them. Three of the most important of these are stealth programs, Trojan horses, and worms. A stealth program waits somewhere in the computer for a period of time and then loads itself into the computer's memory. The actual program that does the damage is then started by the stealth program, and this child program acts just like a virus. This means that the child program generally does considerable damage to the host computer and infects other files or entire drives on the computer. The famous Chernobyl virus is basically a stealth virus. It lurks in the computer until the date of the Chernobyl nuclear accident and then destroys all the data in the computer. Even though the Chernobyl virus has been around for a while, several of the antivirus companies report that it remains quite active. Of course, those antivirus companies have scan and clean software that is good at catching Chernobyl.

The damage done by a computer virus ranges from a mild irritation to the total destruction of the information contained on the computer.

Trojan horses are programs that masquerade as legitimate executable programs but attack their host computers when used. The attack can be viruslike, but in many cases, the Trojan horse simply damages the computer without replicating itself. A Trojan horse program is often spread by a dropper program that has the replication characteristics of a virus. One of the latest Trojan horses is called XalNaga Trojan. When the program is running, it disables a number of the features of the Windows 95/98 user interface, including several of the options on the Start menu.

Worms not only attack an individual computer in a viruslike fashion but also make copies of themselves on many other computers networked to the host computer. A worm is one of the most dangerous viruslike programs around because it can quickly infect all the computers of a company. When a worm arrives at a networked computer, it can start destroying files with no action on the computer's part. Two of the most famous worms are the Melissa virus and the Love Bug virus. Variants of the Love Bug virus were among the most active viruses in 2000.

Attack and Counterattack

Fortunately, computer viruses and their cousins attack computers in a limited number of ways. Most frequently, the virus attaches itself to an executable file, to the boot record of a floppy disk or hard drive, or to a complex file that contains executable code. The most famous exam-

ple of these types of viruses are the macro viruses that attach themselves to Microsoft Office documents. Macro viruses were very popular in 2000 with literally hundreds of new ones appearing. The problem was so bad that Microsoft made slowing down these viruses one of the main Office 97 to Office 2000 enhancements. A virus also can attach itself to a special kind of file such as a ZIP file and infect a computer when the ZIP file is unzipped. In 2000, hackers continued to think of new ways to infect ZIP files, especially those containing Office documents.

Shortly after the first viruses appeared, computer companies began to develop antivirus programs to inspect the crucial files on a computer to see if they were infected. An important part of the strategy of these early antivirus programs was to identify the files that are likely to be infected by their extension or type. This greatly reduced the number of files that the antivirus program had to scan and clean. In fact, the early antivirus programs for PCs using DOS looked only at EXE, BAT, and COM files. The technique used by the early antivirus programs to detect an infected file remains in use. The principal concept is that each infected file has a set of key characteristics that can be converted into a small file that gives the signature of the virus. Once the signature of the virus has been discovered, an antivirus program can calculate the approximate signature of any file and compare it to the signature of the virus to determine if the file is infected. When users run an antivirus program and tell it to scan and clean, it tests the files on the computer and deletes those that appear to be infected. In 2000, companies continued to improve in developing signatures for viruses, which is good because the number of viruses is always increasing.

In addition to checking for files that are infected, an antivirus program checks the partition table and boot sector of any drive on the computer with a set of known signatures to be sure that these areas are also virusfree. Programs loaded in memory may actually be Trojan horses, so they also need a signature check during a scan and clean.

Scan and clean techniques are essential for antivirus programs. However, a great deal of research has also been done by virus detection companies to develop monitoring programs to stop a virus from entering a computer. Many of the recent advances in antivirus software are in the improvement of the monitoring programs.

The McAfee VirusScan 2000 antivirus program, typical of modern antivirus programs, has three types of monitoring programs. They are system scan, e-mail and download scan, and an Internet filter. A system scan watches memory and disk actions while the computer operates and stops computer viruses when they attempt to attack. Some specific activities a monitoring programs watches for are programs attempting to modify the boot sector of a hard drive, staying in memory for a long time without doing anything, or accessing every name in an e-mail address book. E-mail and download scan programs check e-mail as it is opened to see if either the e-mail message or one of the e-mail's attachments is infected. For example, the e-mail monitors take an attachment, place it in a temporary directory, and then run a scan to see if it is infected. Internet filters check for bad Java applets or ActiveX controls. Although these are supposed to be safe from viruses, many successful attacks were launched from Java applets and ActiveX controls in 2000. The progress made in Internet monitoring programs over

The progress made in Internet monitoring programs over the last year has been significant and has made all computers more secure.

the last year has been significant and has made all computers more secure.

Another area of significant improvement in antivirus software in 2000 was in identifying the so-called polymorphic viruses. These viruses modify themselves so that their signatures change constantly, and the antivirus software thus fails to detect them. A lot of the polymorphic viruses are also macro viruses, so all computer users welcome the improvements in polymorphic virus detection. The antivirus companies seem to believe that the creators of polymorphic macro viruses are this year's most prolific virus authors.

The Love Bug and Love Letter Virus

One of the best-known virus programs that appeared in 2000 was the Love Bug virus. The original Love Bug virus was actually a worm attached to an Outlook Express e-mail message as a VBS file. VBS is the extension of an executable VBScript file. When someone opened the attached VBS file, a VBScript program destroyed many of the multimedia files on the host computer and sent an infected e-mail to everyone in the Outlook Express address book. In 2000, more than fifty distinct variants of the Love Bug appeared. Many of the later variants were called Love Letter viruses

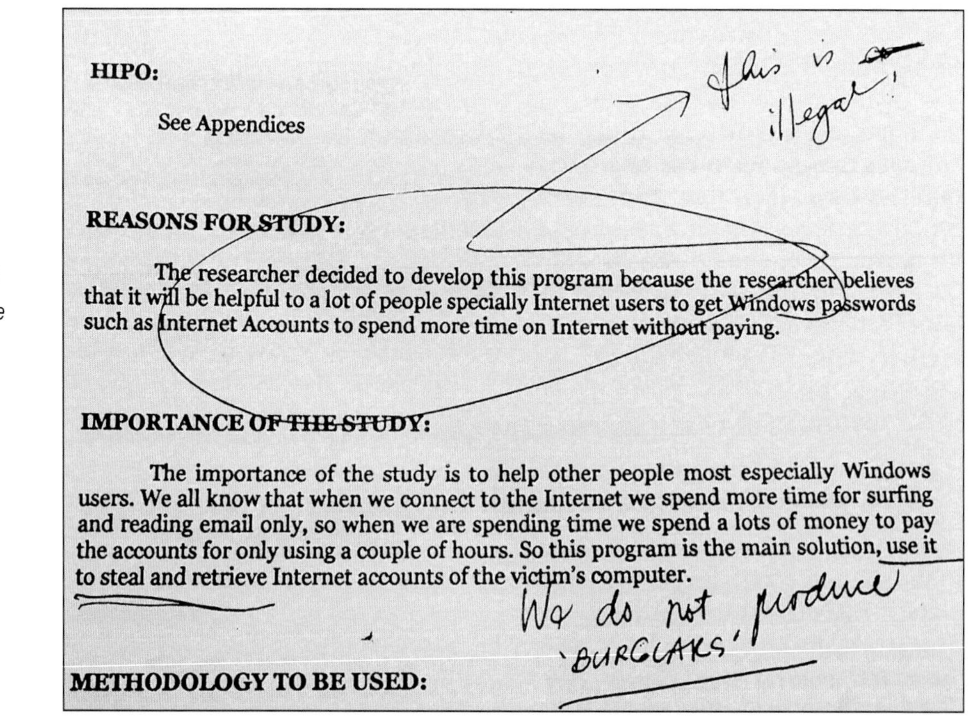

HIPO:

See Appendices

REASONS FOR STUDY:

The researcher decided to develop this program because the researcher believes that it will be helpful to a lot of people specially Internet users to get Windows passwords such as Internet Accounts to spend more time on Internet without paying.

IMPORTANCE OF THE STUDY:

The importance of the study is to help other people most especially Windows users. We all know that when we connect to the Internet we spend more time for surfing and reading email only, so when we are spending time we spend a lots of money to pay the accounts for only using a couple of hours. So this program is the main solution, use it to steal and retrieve Internet accounts of the victim's computer.

METHODOLOGY TO BE USED:

For his thesis, Filipino computer student Onal A. de Guzman proposed to create software very similar to the Love Bug virus. De Guzman's proposal was rejected, but he said that he might have accidentally released the computer virus. (AP/Wide World Photos)

because the subject of the e-mail implied that opening the infected file would display a love letter. One of the versions of the Love Letter virus contained a Trojan horse as well as the standard Love Bug worm. Another recent variant of the Love Letter virus effectively attacked Windows NT's "secure" file system.

The Love Bug virus was reported to have cost computer companies billions of dollars. In addition, many unsuspecting home users had all their multimedia files wiped out by the Love Bug. Most virus investigators suspect that a student in the Philippines developed the Love Bug virus. In fact, the suspected student proposed the development of this virus as his thesis for an advanced degree in computer science (the faculty rejected his proposal). The student was arrested and later released without ever really admitting that he was the author of the original Love Bug. Reportedly, the student has had several job offers because a number companies are now interested in his proposed thesis topic.

Palm Viruses

One of the big events in 2000 was the rapid acceptance of handheld computers such as the Palm and the Windows CE computers. These diminutive devices have a small amount of memory and disk space and are great for checking e-mail. The developers of the Palm computers attempted to sell their computers as super desktop organizers or calculators. As such, they did not have the virus problems associated with laptop computers, which have full operating systems. Unfortunately, in 2000, hackers created a number of viruses that wiped out Palm computers and opened up a whole new market of online virus checks for the antivirus software companies. To scan and clean a Palm, the user can simply log in to an antivirus site and purchase a checkup.

Cyber Crime and Internet Security

➤ *A judge orders digital music exchange site Napster to close, and members of a group including Microsoft and Intel employees are indicted for pirating software. Security and privacy issues come to the forefront when Yahoo warns its free e-mail users that data collected about them will become its to use or sell, and the FBI admits use of an e-mail monitoring system.*

The Internet is a collection of connected computers that support the transfer of information. Internet security is a combination of physical security, hardware, and specialized software that protects the integrity, confidentiality, and availability of the information on the Internet. Cyber crime refers to any criminal activity associated with the Internet. When the Internet was first developed, only the military and some research institutions could gain physical access to it. This physical security of the Internet was satisfactory during its early development. Except for a few spies, almost no one committed cyber crimes. The Internet has expanded and can be used by anyone worldwide, with the proper equipment and connections. Web browsers make the Internet easy to use, and information is being exchanged over the Internet at exponential rates. Although Internet security has also improved, it has not kept pace with the overall expansion of the Internet. Cyber crime is doing well, and Internet security needs to catch up.

Rights Questions Close Napster

It is easy to download files from a Web site, and it is not much harder to place files on the Web site. Thus, one of the real strengths of the Internet is its ability to allow users to exchange files. However, computer files often contain content that is private property and not part of the public domain. If a computer file contains a program (or part of a program) such as Microsoft Word and a person places it on the Internet for someone else to download, then that individual has broken the law because Microsoft owns Word; that is, Word is the intellectual property of Microsoft. If a person places other types of files on a Web site and allows someone else to download one of the files, that individual may or may not be breaking the law, depending on the type of file. For example, most people believe that if they own a copy of a recording (compact disc, vinyl album, tape), it is legal to place a digital music file on an FTP (file transfer protocol) site and allow their friends to download a single copy, especially if they are "exchanging" digital music files. If it is okay for individuals to exchange digital music files that they have purchased, then it must surely be okay for a Web site to facilitate this process.

One of the most successful digital music file exchange sites is called Napster. Napster makes it very easy to exchange MP3 digital music files, so easy that many people created CDs (compact discs) from MP3 (MPEG-1 layer 3) files at Napster. In July, the Recording Industry Association of America filed a lawsuit against Napster, and a judge ordered Napster to close. In August, an appeals court granted a stay that allows Napster to continue to operate. On February 12, 2001, a federal appeals court ruled that Napster must prevent its users from sharing copyrighted material. However, Napster would not be shut down until a lower court reissued its injunction. Napster planned to renew its legal battle. When the final decision is rendered in this case, a major step will have been taken in defining intellectual property rights to content on the Web. Most computer professionals anticipate a final decision that continues to protect the intellectual property rights of the developers of content such as music while still allowing downloads of files from Web sites that are not anyone's intellectual property.

In front of headquarters in Redwood City, California, Napster Chief Executive Officer Hank Barry (right) and cofounder Shawn Fanning address the media after a judge issued a stay allowing the service to stay online, at least temporarily. (AP/Wide World Photos)

Piracy Aboard the Internet

Since the early days of computers, some people have used software without purchasing it. In fact, this is so common that many people do not even think of it as cyber crime. The Internet provides a perfect mechanism for distributing stolen software. As with many forms of cyber crime, the theft of software crosses international borders because it is easy to download the pirated software anywhere in the world. Unfortunately for the cyber criminals, the Internet also aids law enforcement agencies in catching the pirates.

In February, a federal grand jury indicted seventeen defendants from across the United States and Europe for pirating software. These software pirates actually had a company named Pirates with an Attitude. Some of the cyber criminals were employees at Microsoft who helped steal Microsoft software, and some were employees of Intel who helped set up a Web site in Canada to distribute the software. It is estimated that these cyber criminals stole five thousand programs and distributed them throughout the world from their server in Canada. These modern thieves communicated using a chat room and e-mail. Although the Pirates with an Attitude were very clever, they did not count on the Justice Department and other law enforcement agencies using the trails they left on the Internet to aid in their capture. The pirated software was stored on an FTP server, and in addition to finding the server on the Internet, the Justice Department gained access to this server to prove that it contained pirated software. Because cyber criminals who use the Internet to either steal computer software or distribute stolen computer software often do not get caught, it is impossible to tell how many cyber criminals of this type exist. However, 2000 provided many examples of law enforcement agencies capturing these criminals using Internet technology.

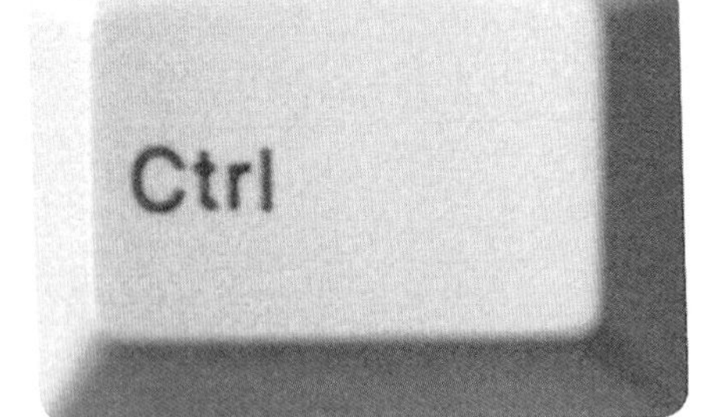

Congress Tackles Privacy and Encryption

Encryption technology has been around a long time. The basic idea of encryption is to take a message and convert it to an encrypted message that cannot be understood by someone who intercepts it. The favorite encryption methods for the Internet involve selecting one or more keys to both encrypt and decrypt the message. In general, the length of the encryption keys determines how hard it is to break the encrypted message, with longer keys giving greater protection. Encryption is essential for some uses of the Internet. For example, whenever a credit card number is passed over the Internet, it needs to be safely encrypted. This is also true for crucial corporate and military data. The technology for using encrypted messages over the Internet exists; however, this technology is not always used for the general good. Many cyber criminals use the Internet to communicate with each other. Obviously they do not want anyone, in particular law enforcement officials, reading their messages. For example, the Federal Bureau of Investigation (FBI) has reported that several terrorist organizations have already used encrypted Internet messages to plan their activities. Law enforcement agencies do not support encryption technologies that do not include a backdoor that allows them to read Internet messages sent by criminals.

In general, the length of the encryption keys determines how hard it is to break the encrypted message, with longer keys giving greater protection.

In 2000, several bills in Congress related to encryption technology. The current standard restricts encryption keys to 56 bits. Both the House and the Senate agree that this should be raised to a 64-bit standard. The industry believes that it needs a 128-bit standard, but a 128-bit standard would produce messages that law enforcement agencies cannot decrypt. Although in 2000, much discussion took place regarding whether privacy of messages is more important than controlling criminal use of the Internet to send messages, no resolution appears likely soon.

Not-So-Private Information

In 2000, in a number of well-publicized cases, the government and industry were found to be collecting personal information about users of the Internet. Each of these cases resulted in information being collected about a user without his or her knowledge or permission.

One of these cases was the discovery that many Internet sites that required users to log in in order to use a service (such as free e-mail) were collecting information about the users and selling this information without the users' knowledge or permission. In August, the well-known Yahoo portal decided to modify its warning to users signing up for free e-mail to make it clear that all information entered by a user became the property of Yahoo and might be shared with Yahoo partners or simply sold to others. In fact, Yahoo indicated that it had already shared information about users of its site with some of its partners.

In another case, several companies that provide a free

downloading service were secretly collecting, categorizing, and using information about users who downloaded files. As users did a "really fast" download, the service collected their URL (uniform resource locators), Internet service provider's IP (Internet protocol) address, and their own computer's Internet identifier. At present, the users of the services receive no warning that this information is being collected.

One of the biggest stories about cyber crime in 2000 was the admission by the FBI that it is using an e-mail monitoring program called Carnivore. Carnivore has a tap that is placed in an Internet service provider's mail server and filters all the e-mail of a suspect from the entire e-mail stream of the Internet service provider. The filtered e-mail is then directed to the FBI's computers. To use Carnivore, the FBI must get a court order. In July of 2000, the press discovered that the FBI was using Carnivore to collect information from the e-mail of suspects. Until this time, no one outside the FBI and the legal system was aware of Carnivore.

Although Carnivore is still being used, the appropriateness of law enforcement agencies monitoring e-mail has become a topic of discussion. Many commercial Internet service providers also use monitoring software on their e-mail servers, as the FBI points out on its Web site for Carnivore. For example, without monitoring software, an Internet service provider has no way of checking to see if its customers are using their e-mail account appropriately. As the use of e-mail monitoring software became public knowledge, many people began to consider the questions it raises regarding the privacy rights of those using e-mail. Although this issue received a lot of attention, few answers have been provided. At present, the only way to be sure that one's e-mail cannot be read by anyone is to encrypt it with a very large key and an encryption algorithm that law enforcement agencies (and others) cannot break.

Progress on International Standards

To prosecute cyber criminals, law enforcement agencies often need to operate in a number of countries; however, the necessary cooperative agreements between countries to allow law enforcement agencies to pursue cyber criminals across national boundaries are largely nonexistent. The Justice Department has been very active in developing a cyber crime division and has developed a number of good general ideas about cyber crime and how to catch cyber criminals. An inspection of its Web site shows that the Department of Justice's approach to stopping cyber crime has three major aspects. First, a law enforcement agency needs to develop numerous technical support tools to monitor criminal activity on the Internet. Second, all law enforcement agencies need to work with various levels of government to improve the laws that are applied to cyber criminals. The existing laws are far behind the technology that the criminals are using. Educating the legislative branch of government as to the need for better laws has become a major part of law enforcement agencies' cyber crime activities. Third, law enforcement agencies need better operational procedures to catch cyber criminals. Therefore, better equipment, training, and coordination are needed by law enforcement agencies to combat cyber criminals.

In March, Attorney General Janet Reno prepared a report "The Electronic Frontier: The Challenge of Unlawful Conduct Involving the Use of the Internet," outlining the United States' approach to catching cyber criminals. A major part of this report was a series of suggestions about how to improve the federal and international laws so that the federal government can catch cyber criminals. Many international conferences have also considered how the countries of the world can modify their laws to make it easier to prosecute cyber crimes that involve multiple nations. Although much remains to be done, 2000 marked the beginning of a serious effort on the part of many countries to work together to capture cyber criminals.

Communications, Computing, and Entertainment

Is It All Good?

➤ *Two companies introduce Internet-enabled refrigerators, the start of a planned line of Internet-enabled appliances, and H.323 gains ground as a standard for media convergence.*

The term "media convergence" has several meanings in the computer field. It is most commonly used to refer to audio, video, and data being transmitted over a common network. The key for these media sharing a network is digital data representation for the high-level transfer protocols. For audio, the transfer protocol might be a simple MP3 (MPEG-1 layer 3) file transfer. For video, it could be a streaming media protocol, and for data, it might be the HTTP (hypertext transfer protocol) for transferring Web pages.

Having audio, video, and data share a common network is not a new idea. Large corporations have operated fiber-optic networks that carry all these media for years. What is new is the idea that TCP/IP (transmission control protocol/Internet protocol) networks can be used as the backbone to move audio, video, and data. In this new concept of interconnectedness, IP packets can flow back and forth from a television station to an interactive television set over the network in the same way that Web information moves back and forth from a browser to a server. As is often the case in the computer field, a popular phrase, "voice over IP," has been adopted to describe this transfer of audio, video, and data over the TCP/IP network. Just as information flows across today's wide-area networks, which are implemented using any combination of satellites, fiber-optic cables, and coaxial cables, the new voice over IP network protocols will work over all wide-area networks.

In a typical voice over IP system, a video supplier such as a television station sends its signal across the backbone network to a television in a home. At the same time, a fax might be arriving at a computer in the home. In addition, a data file could be arriving at the computer, and the microwave oven could be receiving a new set of programming instructions for tonight's dinner. From the homeowner's point of view, all these media have converged into a single logical system.

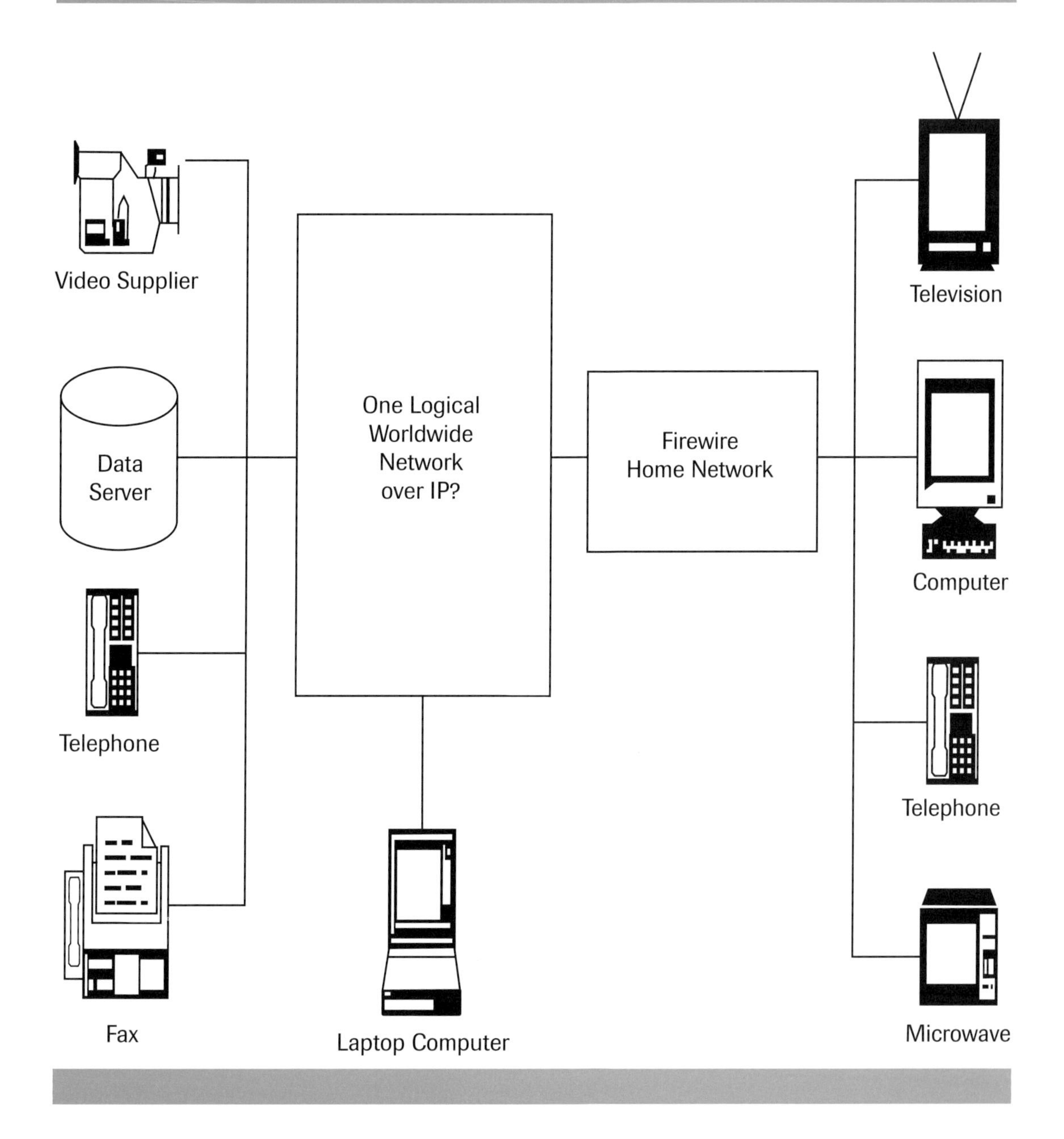

Looking at the totality of products being created for the convergence market, it now appears that voice over IP is the convergence technology that the industry is adopting. Voice over IP assumes that the IP packets are delivered directly to devices on the network. However, the home might well have its own network such as Firewire or uni-

versal serial bus (USB) tying all the devices in the home together. In this case, a small router would convert the IP packets to the home network packets and then forward these to devices such as televisions, phones, and computers, which would then convert the packets back to IP packets. At the close of 2000, the preferred network architecture for convergence seemed likely to be voice over IP going directly from a source to a home device. Nevertheless, the idea of having a home network handle specialized problems such as programming an air conditioner from a computer remained attractive. Although tremendous progress has been made in 2000 in the voice over IP market, less progress has been made in implementing the standardized home networks.

Many people see media convergence as involving the use of a single device such as a laptop computer to process audio, video, and data. A laptop computer can be used to tie into the logical worldwide network and listen to an MP3 song, view a streaming media news story, read e-mail, or upload a Word document describing the day's events. However, the convergence of audio and data in a single device does not require a laptop. Many of the new handheld

Palm computers, intelligent cell phones, and Windows CE computers support audio and data convergence. So far, no one has introduced a handheld computer that supports audio, video, and data convergence, but it probably will not be long before full multimedia handheld devices are available.

The final, and most interesting concept of media convergence is a logical one. Rather than requiring a single device such as a Palm computer to process audio, video, and data, media convergence can be achieved by networking a number of separate devices together in a single room and using them simultaneously. For example, a homeowner might be watching television, talking on the telephone, reading e-mail, and letting the microwave be programmed to cook dinner. The logical convergence of media could be achieved by using a single device or by simply having a number of devices that are networked.

At the close of 2000, the preferred network architecture for convergence seemed likely to be voice over IP going directly from a source to a home device.

In addition to private homes, media convergence based on voice over IP will also have applications in industrial settings. Examples include real-time document collaboration, distance learning, employee training, video conferencing, and video mail. Most of these are not new applications for industry, but the use of voice over IP is new and should lower the cost of supporting media convergence in industry. Voice over IP should also ease the implementation problems often associated with moving to a single network for audio, video, and data for physically distributed organizations. Therefore, voice over IP is increasing the acceptance in industry of media convergence as the standard way of communicating.

Internet Enabled Refrigerators

Two companies, Whirlpool and LG Electronics, have recently announced refrigerators that are Internet enabled. Limited information is available about the network protocols used by the refrigerators, but every indication is that both of the refrigerators are using voice over IP as their basic protocol. Both of the companies also claim that they will soon have a full set of Internet-enabled appliances.

The Whirlpool refrigerator has a mounted Web pad, developed by Qubit Technologies, that can be used as a calendar and to order groceries, check e-mail, and leave notes for family members. A handheld device can be used to control the refrigerator and other Whirlpool Internet appliances remotely. Whirlpool says that it is using Cisco's VoIP technology.

The LG Electronics refrigerator has two fifteen-inch displays, one on each of the two doors of the refrigerator. With these displays, users can check the real-time price of groceries, access the Internet to download a recipe, and obtain information about the contents of the refrigerator. The LG Electronics refrigerator also has a small video camera mounted above one of the displays that allows the homeowner to videoconference as well as to process e-mail. The suggested retail price of the refrigerator is $8,850.

The Internet-enabled refrigerators have an interesting feature for building a grocery list. When an item in the refrigerator is used up, users can scan its bar code and automatically add the item to their grocery list.

The H.323 Multimedia Standard

The computer industry is good at inventing a large number of short cryptic names that describe standards for technologies. Literally hundreds of such standards exist for media convergence. The standards are necessary if a user is to purchase an Internet telephone, hook it into the USB port of a computer, and have it work. One of the most important standards for media convergence is the H.323 standard. It is a comprehensive International Telecommunications Union (ITU) standard for the transfer of audio, video, and data over connectionless networks such as IP-based networks and the Internet. It provides call control, multimedia management, and bandwidth management for point-to-point and multipoint conferences. The H.323 standard defines support for audio and video codes and supports data sharing.

When an item in the Internet-enabled refrigerator is used up, users can scan its bar code and automatically add the item to their grocery list.

Cisco, one of the leaders in router technology, has fully embraced the H.323 standard. Cisco builds routers that have the capability of converting analog signals from voice and video into compressed digital signals in conformance with H.323 that can be sent over the Internet in IP packets. Microsoft also has based its new telephone application program interface, TAPI 3.0, in large part on the H.323 standard. The list of other vendors supporting the standard includes virtually all the major companies in the voice over IP market, so it is a stabilizing force in helping different companies develop compatible Internet devices. For example, Microsoft NetMeeting is a conferencing and collaboration tool designed to be 100 percent compatible with the H.323 standard.

Media Convergence and Automobiles

Modern automobiles have many small computing devices controlling a variety of automotive components. Some of these are even networked. However, recently a flood of media convergence concepts have been developed for automobiles, all of which appear to be based on technologies similar to voice over IP for wide-area networks. For example, Palm and Delphi Automotive Systems have announced plans to support Palm computers in automobiles. Soon drivers and passengers will be able to access maps, weather information, and stock prices through a voice-activated version of Palm computers. General Motors is going slowly on this idea because of safety concerns, but the technology is there. In addition, at the Convergence 2000 automotive electronics conference in Detroit, Michigan, Microsoft introduced its automotive version of Windows CE, version 3. The software will enable automobile manufacturers to converge "communications, information, and entertainment." Both the Palm system and Windows CE 3 are purported to have similar characteristics. Cars will be tied into the worldwide network just as large corporations, small businesses, and homes are.

Distance Education

➤ *Virtual universities become a reality, enabling those far removed from a learning institution to receive an education.*

The first formal education probably took place when a teacher sitting on a rock explained the importance of counting to a student. This type of education is an example of face-to-face education. As time progressed, education became more formal, with many students being taught by a teacher in a classroom. This was still face-to-face education. As transportation and communication systems improved, many realized that learning did not have to take place in a classroom. Students could be removed in space and/or time from a teacher. When distance learning is the primary method of acquiring information, the corresponding educational system is called distance education.

Correspondence courses were the first form of distance education. In correspondence courses, students receive a set of written materials from the instructor to guide them through the course. The written materials usually include a detailed course outline and assignments that are carefully correlated to the textbook. The students interact with the teacher using regular mail. Although many students successfully complete correspondence courses, the written materials are generally not very exciting, and using the regular mail service to correspond is much too slow.

By 2000, most distance education courses had switched to using e-mail and chat rooms for their correspondence. With e-mail, students and teachers can communicate several times a day rather than once every two weeks. As students and teachers gained access to the World Wide Web via the Internet, many distance education courses placed all their written materials on the Internet. This very popular form of distance education is generally referred to as Web-based distance education. Another popular method of distance education is to use two-way television, generally called interactive television, as a means of extending the traditional classroom. Sometimes, distance education courses use a combination of technologies. For example, a Web-based distance education course might schedule weekly

interactive television sessions to allow students to ask questions face to face and also provide written materials on a CD (compact disc) to complement the Web site.

Video Conferencing

Advances in video conferencing technology are improving communication between teachers and students in interactive television classes. When teaching a course using interactive television, the instructor delivers a lecture in a classroom while being filmed with a television camera. In addition to the students in the teacher's classroom, students at remote sites can view the lecture on television monitors. All the students can see the teacher and listen to the lecture. Students at the remote classrooms have access to microphones so that they can ask the instructor questions. The instructor uses a microphone to answer the questions, so that students at the remote classrooms can hear the answer just like the students in the primary classroom. Usually, television cameras are present at the remote classrooms and can focus on the student asking questions. Both the student and the teacher are dis-

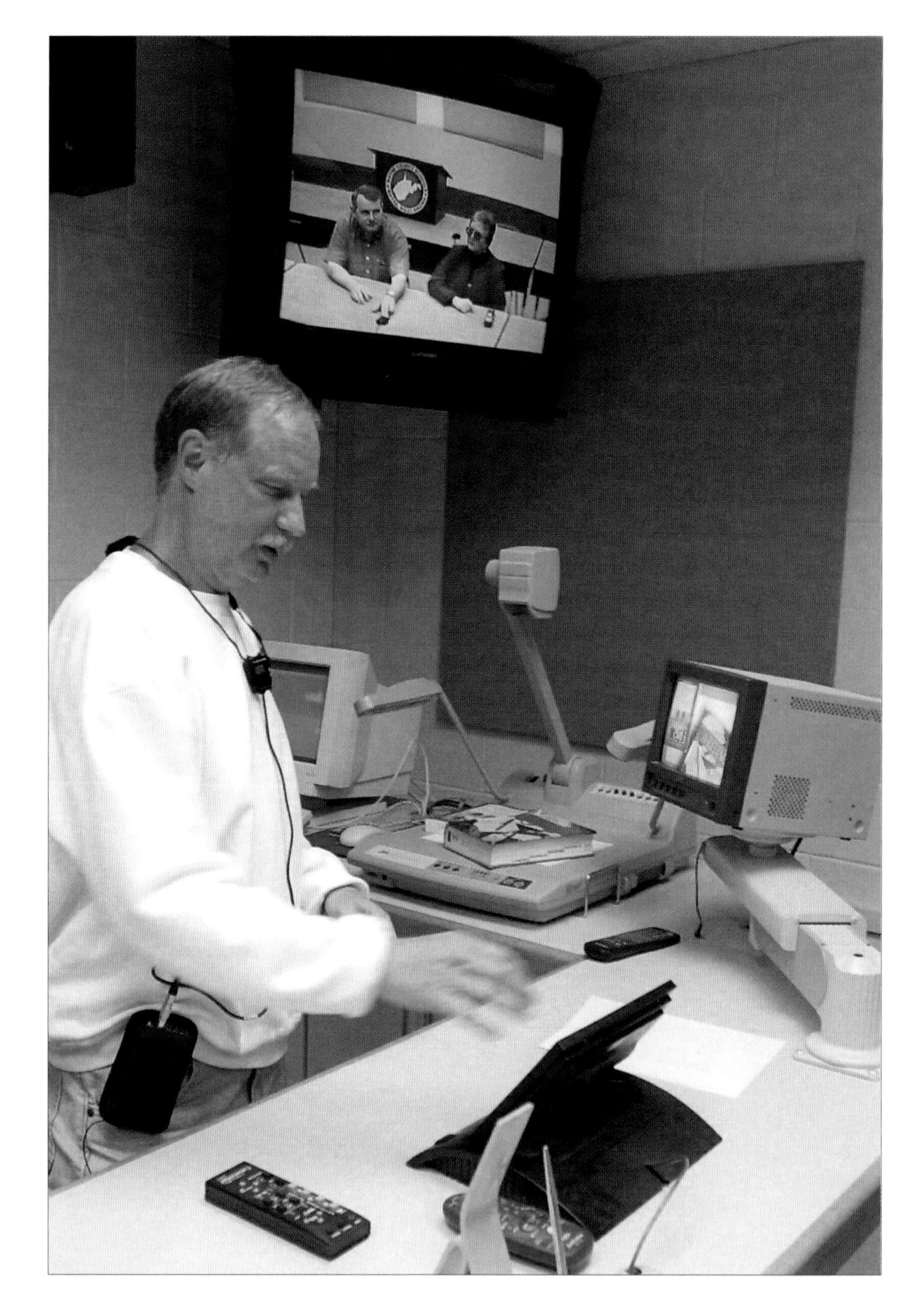

Paul Martin, a high-school teacher in Spring Valley, West Virginia, teaches students at remote sites from the school's distance learning center. (AP/ Wide World Photos)

played on television monitors at all the sites. If everything goes well, it is just like one big classroom.

When interactive television first started, the line speed made it difficult to actually see and hear a conversation between a student at a remote site and the instructor. Sometimes, the lips were moving but there was no sound. Over the last few years, as line speed improved, the concept of having all the remote sites and the primary classroom appear as a single virtual classroom has become nearer to a reality. Also, the cost of microphones, television cameras, and the associated control equipment has dropped dramatically, allowing the use of more microphones and television cameras in both the remote and the primary classrooms. Students often have their own microphones rather than relying on several microphones positioned in a room. Students and teachers can see more than one person at a time in the other classroom through the use of additional television cameras. This allows students and the instructor to view most of the people involved in a discussion at both the remote sites and a primary classroom. This enhanced interactivity has many educators looking at interactive television as a major method of teaching distance education classes. When the technology works well, this form of distance education can put everyone "in the same room."

In spite of many news stories describing virtual universities, students found that most of the Web-based distance education majors were "under construction."

Real Virtual Universities

Over the past few years, virtual universities have been the subject of many discussions. These virtual universities were supposed to have online libraries, numerous majors, and even a student union building. In spite of many news stories describing virtual universities, students found that most of the Web-based distance education majors were "under construction."

In 2000, a number of functioning virtual universities appeared. When students visit a virtual university, they usually are given a tour of the campus. This includes a visit to a library that has many online collections of articles that students can read as background for their course work. They can also stop by the campus bookstore and take a look at the textbooks that are being used for the Web-based distance education courses being taught at the virtual university. If they like what they see, they can request information on how to register as a student at the virtual university. Once registered, students are assigned advisers, who determine their initial schedules of courses. For example, students interested in a master's of business degree can go to dozens of virtual universities.

Streaming Video

Adding multimedia effects to a Web-based distance education course has long been a goal of distance education instructors. However, multimedia files tend to be large, making it difficult to effectively add multimedia to these courses. In particular, it has been virtually impossible to download and play video files in spite of the desirability of adding some of these to a virtual classroom. For example, it is a lot more effective to watch a short video explaining how to save a file rather than read a long and boring HTML (Hypertext Markup Language) page description of how to do this.

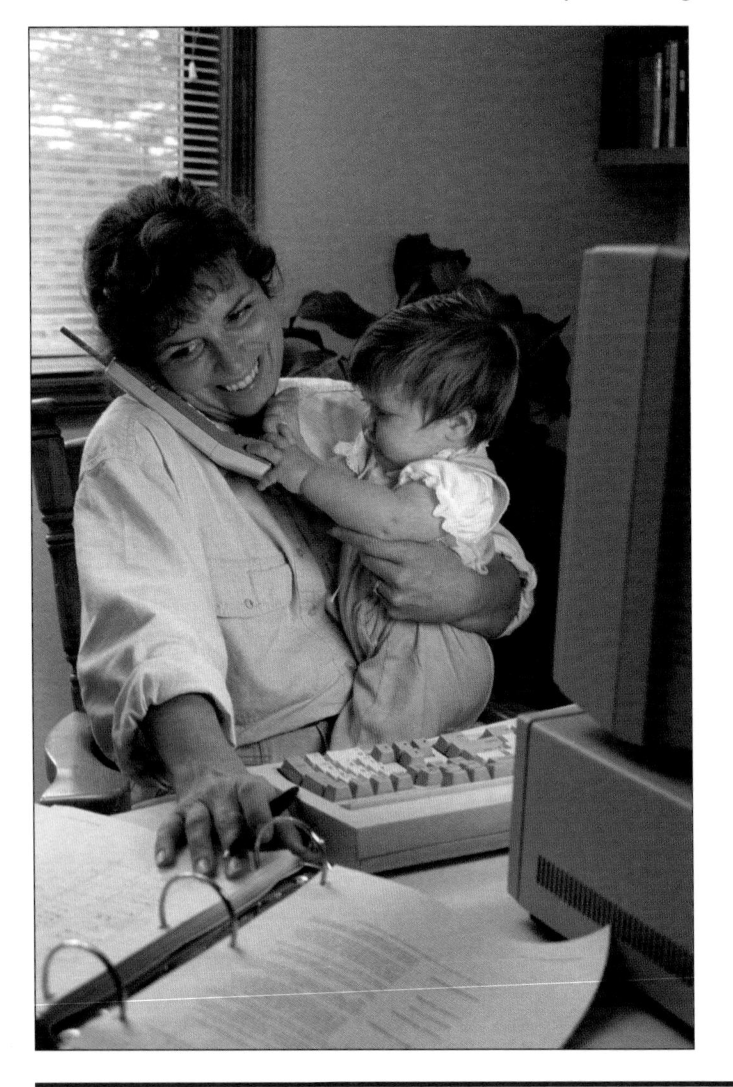

Both Microsoft and the RealPlayer have improved the quality of their streaming media technology. Users with 56-kilobyte modems can watch streaming media videos with the minimal amount of time spent waiting for part of the video to download. Traditional video presentations use video files such as Microsoft's AVI (audio video interleave) files or Apple's old QuickTime files, which need to be completely downloaded before they can play. Many video files are too large to fit into a computer's memory, even if the user is willing to wait for the entire file to download before playing it. With streaming media, the entire video presentation is broken up into a number of small video files. When users play a streaming media video presentation, they actually download a number of small files, one after another, playing them in a seamless fashion. When they play a video, it appears to be one large file. Until recently, the software used to decompose the presentation into small files did not work very well and resulted in many interruptions of the presentation. This resulted in streaming video being mostly a novelty. With improvements in line speed and quality of the software, streaming video is now being added to many Web-based distance education courses. Today, Web-based distance education courses can have many streaming video presentations to add dynamic content.

U33
C66
U34

Resources for Students and Teachers

Books

Downing, Douglas, Michael A. Covington, and Melody Mauldin Covington. *Dictionary of Computer and Internet Terms.* 7th ed. Hauppauge, N.Y.: Barrons Educational Series, 2000. This dictionary contains more than 25,000 items, including a number of historical items. Many of the articles have accompanying diagrams. It also contains a large number of Internet terms.

Feibel, Werner. *The Network Press Encyclopedia of Networking.* 3d ed. San Francisco: Network Press, 2000. The encyclopedia features a comprehensive set of entries on networking hardware, software, operating systems, and associated technologies and contains many tables, charts, pictures, and diagrams. Topics covered include interface cards, cabling, standards, applications, and protocols. The accompanying CD-ROM has a searchable version of the encyclopedia and a copy of the dictionary of networking.

Gates, Bill, with Collins Hemingway. *Business @ the Speed of Thought: Succeeding in the Digital Economy.* New York: Warner Books, 2000. Like him or not, Bill Gates is often ahead of the curve in recognizing where the computer industry is headed, and his latest book is an inexpensive and insightful look at the convergence of computer, communications, and entertainment technologies. The book also presents some of Microsoft's ideas on how business is becoming e-business, as well as an interesting discussion about the relationships between government and the high-tech industry.

Gookin, Dan, and Sandra Hardin Gookin. *Illustrated Computer Dictionary for Dummies.* 4th ed. Foster City, Calif.: IDG Books Worldwide, 2000. Contains humorous definitions and pictures of many computer terms, including "geek."

Kent, Allen, and James G. Williams, eds. *Encyclopedia of Computer Science and Technology.* Vols. 42-43. New York: Marcel Dekker, 2000. This multivolume encyclopedia on a wide variety of computer-related subjects adds new volumes, usually on a single topic, each year. The two volumes published in 2000 are on artificial intelligence and electronic commerce.

Pfaffenberger, Bryan. *Webster's New World Dictionary of Computer Terms*. 8th ed. Foster City, Calif.: IDG Books Worldwide, 2000. This edition has been completely revised, with many new terms specific to the Internet and World Wide Web. It contains more than four thousand entries, including computer terms, acronyms, and jargon.

Ralston, Anthony, Edwin D. Reilly, Jr., and David Hemmendinger, eds. *Encyclopedia of Computer Science*. 4th ed. Condon: Nature Publishing Groups, 2000. Presents fairly detailed articles in the areas of computer hardware, software, and systems; information and data; the mathematics of computing; the theory of computation; and methodologies. About 70 percent of the encyclopedia was rewritten for this new edition. Contains many photographs, diagrams, and tables to complement the articles. A special section covers computer acronyms. Glossary included.

Tulloch, Mitch. *Microsoft Encyclopedia of Networking*. Redmond, Wash.: Microsoft Press, 2000. An exceptionally well-written book describing Microsoft's networking that includes all the standard networking terms needed for Microsoft networking. It contains more than one thousand entries arranged alphabetically with articles that contain accurate definitions and lots of practical examples. Topics covered include protocols, standards, hardware, network operating systems (including Windows 2000), Internet services, distributed applications, and certification programs. An electronic version of the book is contained on the accompanying CD.

CD-ROMs and Videos

Encarta Reference Suite 2000 CD. CD-ROM. Microsoft Corporation, 2000. In addition to entries on general computing terms, the 2000 edition of *Encarta* contains a computer dictionary. Features multimedia presentations, diagrams, and pronunciation aids to make learning about computers enjoyable.

History of Computing: An Encyclopedia of the People and Machines that Made Computer History. CD-ROM. Lexikon Services, 2000. This CD edition of a book by Mark W. Greenia is a history of computing rather than a compendium of the latest advances in the field. It contains many pictures, and diagrams, as well as more than seven thousand terms, chronologies, and bibliographies.

McGraw-Hill Encyclopedia of Networking Electronic Edition.

CD-ROM. Osborne McGraw-Hill, 2000. This is the CD version of a well-known encyclopedia of networking by Thomas Sheldon. It has excellent cross-linking capabilities and many diagrams and pictures. A special feature of this encyclopedia is a list of companies and their products with links to the companies' Web sites.

Technoculture: Finding Our Way in the Terra Incognita. Video. Living in the Brave New World series. Films for the Humanities and Sciences, 2000. The interactions of technology and society are the focus of this video produced by Galafilm in association with Electronic Post Office and CBC Newsworld.

Technoscience: Blurring the Line Between Man and Machine. Video. Living in the Brave New World series. Films for the Humanities and Sciences, 2000. The scientific and social effects produced by interactions between people and computers are the subject of this video produced by Galafilm in association with Electronic Post Office and CBC Newsworld.

Web Sites

The PC Reference
http://www.pctechguide.com/sitemap/sitemap.htm
One of the best Internet sites for information about computer hardware. The discussions and associated diagrams are outstanding.

Webopedia
http://www.webopedia.com/
A very complete Web encyclopedia and dictionary. Using the search engine, it is easy to look up a short definition of almost any computer term. In addition, the site always has some interesting articles about new developments in computer technology.

Whatis?com
http://www.whatis.com/
This computer dictionary has an easy-to-navigate alphabetical menu that quickly leads to a short definition of almost any computer term.

3 · Earth Sciences and the Environment

The Year in Review

➤ *Margaret F. Boorstein*
Department of Earth and Environmental Science
C. W. Post College of Long Island University

The year 2000 was marked by violent and unusual weather, discoveries and new scientific interpretations in earth sciences, and growing discussions about the role of human beings in changing the environment. Although thousands of lives were lost worldwide, the number of deaths and damage from certain natural disasters, such as earthquakes, volcanic eruptions, and hurricanes, was small compared with other years.

Ongoing efforts to understand Earth led to the collection and analysis of much new data. Although some new measurements supported already established theories, others spawned the development of new theories. For example, exploration of the oceans revealed heretofore unknown phenomena as well as remarkable scenes.

The importance of human intervention was a major theme of environmental news. By the end of the year, the U.S. Congress had passed and the president had signed legislation to save the Everglades, correcting damage that began in 1948 with water diversion. Although 2000 was a presidential election year, remarkably little of the election rhetoric concerned the environment.

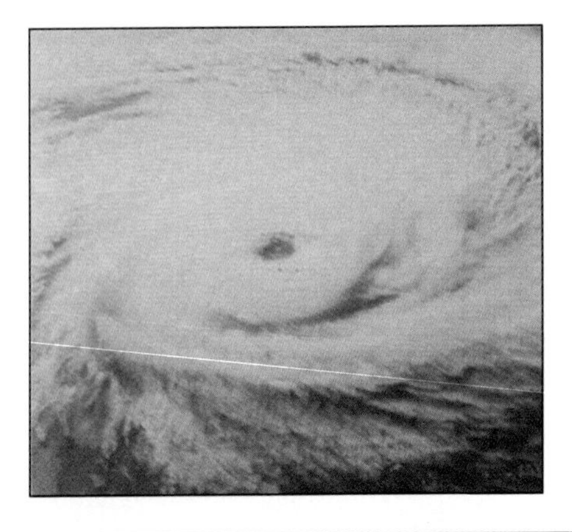

Weather Extremes

In the year 2000, more than one continent suffered from both torrential rain and drought. Typhoons, tropical cyclones, monsoons, and rainstorms caused flooding and mudslides, killing hundreds of people and damaging homes and crops in numerous areas of the world. Rains caused flooding in southeast Africa in February, March, and April. At the same time, Iran, the interior of the Indian subcontinent, Pakistan, and the horn of Africa suffered agricultural drought; millions of people were affected. The outback of Australia, normally dry, was flooded with heavy rains in February and March, including the highly unusual inundation of Uluru (Ayers Rock). A dry inland area was transformed into a sea, flooding towns and isolating visiting tourists. The rains led to a locust plague in April. Flooding and mudslides occurred in Mexico and Guate-

mala in April, while parts of western India, Pakistan, and Afghanistan were very warm and dry. Afghanistan experienced its worst drought in thirty years.

In August, flooding killed more than one hundred people and displaced millions in parts of India, including along the Brahmaputra River and its tributaries as well as Bhutan, Nepal, and Bangladesh. Parts of China, Japan, Brazil, and Cameroon also suffered floods. In August, flooding across India and Bangladesh killed at least seven hundred people, and monsoon rains affected more than four million people in Vietnam, Cambodia, Thailand, and Laos. In contrast, that same month, drought and heat contributed to fires in Bulgaria, Croatia, Albania, and Greece. In October and November, severe rainstorms disrupted traffic in London and forced evacuations in York, England. The storms produced the worst flooding in fifty years in some parts of Western Europe before moving on to Eastern Europe, where they caused more damage.

Weather extremes also affected the United States, where drought scorched large parts of the Southeast and the West, contributing to disastrous wildfires. Hundreds of thousands of acres of forests were destroyed during the

spring and summer. In the fall, humans played a critical role in the blazes in the Appalachians, where suspected arson and careless campers were blamed for fires that spread through dry areas.

The weather seemed to be playing a little joke in August in England, where a shower of dead sprats (a type of fish) fell on Great Yarmouth after a thunderstorm. The meteorological office thought that the fish shower was caused by a waterspout or tornado at sea.

In 2000, a tornado struck Pine Lake, Alberta, Canada, in July, killing seven people. Previously, the last deadly tornado on the Canadian prairies had occurred in 1987. In the United States, the tornado season, with 741 reported tornadoes, was unusually mild. This was the lowest total since 1989 and about two-thirds the total for each of the past two years.

The hurricane season in the Atlantic was relatively mild, with only one making landfall on the eastern seaboard of the United States. In contrast, typhoons caused much damage. For example, in November, Typhoon Xangsane in Taiwan led to the worst flooding in thirty years. The Philippines was hit by Typhoon Bebinca, which killed dozens of people, forced evacuations, and caused landslides. The Philippines had already been hit by damaging rains in February.

Are Humans Intensifying Global Warming?

Atmospheric and oceanographic scientists generally agree that the atmospheric level of carbon dioxide has increased in the twentieth century because of the burning of fossil fuels. In addition, they generally agree that the increased level was at least correlated with and perhaps the cause of the increase of 1 degree Fahrenheit in atmospheric temperature over the same period. Although most of these scientists say that humans were a significant cause of the increased carbon dioxide and temperature levels, some respected scientists disagree.

At the end of 2000, the pendulum of explanation swung to the side of human responsibility when the United Nations Intergovernmental Panel on Climate Change stated that the burning of fossil fuels by humans substantially contributed to the rise in atmospheric temperatures. The report also predicted that the temperature change would be much greater in the coming century. The panel distributed a draft of the report in October so that governments would be prepared for discussions occurring in The Hague in November, 2000, to decide on implementation of the Kyoto protocol. The Kyoto protocol, signed in 1997 by representatives in 150 countries, has not been ratified by any industrial country. The final report by the panel on climate change was to be issued in January, 2001.

Politics and Environmental Issues

The environment was not a major topic of the U.S. presidential campaign of 2000. The main environmental issue was whether it was desirable to develop petroleum resources in the Alaskan wilderness. Important and timely issues, such as potential intensification of the greenhouse effect, the apparent thinning of the stratospheric ozone layer, endangered species legislation, and air and water pollution controls, were barely discussed.

The Kyoto protocol, signed in 1997 by representatives in 150 countries, has not been ratified by any industrial country.

Although no environmental issue dominated the presidential campaign, outgoing Democratic president Bill Clinton made environmental law part of his legacy. He acted to protect hundreds of thousands of acres of federal lands by declaring them to be national monuments. At the end of the year, Clinton signed a law designed to save the Everglades, a 12-million acre "river of grass," consisting of sawgrass and swamp, in southern Florida. The law was passed to undo damage started in 1948 when the Army Corps of Engineers diverted the natural flow of water away from the Everglades to facilitate urban growth in southern Florida. Over the years, animal and plant life in the Everglades was disrupted, with about sixty-eight species in danger of extinction by 2000.

Although no environmental issue dominated the 2000 presidential campaign, outgoing Democratic president Bill Clinton made environmental law part of his legacy.

The new plan, also to be implemented by the Army Corps of Engineers, will catch and store rainwater to facilitate its flow into the Everglades, thus restoring the Everglades and its natural ebb and flow of freshwater. The federal government will spend $7.8 billion (half the cost) over thirty years; Florida will pay the other half. By some means, the waters must still be made available for use by sugarcane growers, cities, and farms of southern Florida as well as utility companies. This is not an easy task. The attachments to this bill authorized $7 billion for improvements on navigable waterways, flood control projects, and environmental projects in a variety of locations around the country.

The Everglades law in many ways exemplified the major environmental perspectives of 2000. It was passed, incorporating much compromise, to undo decades of damage that resulted from a desire to use the natural environment for economic gain combined with a misunderstanding of natural ecosystems. The techniques to be employed in implementing the law were not yet proven and the science behind them could be viewed as uncertain, which meant that the eventual success of an environmentally and financially expensive project was not ensured.

Earthquakes and Volcanoes

The year 2000 was unusual in that although significant earthquakes and volcanic eruptions occurred, none was devastating. Relatively little loss of human life ensued as the volcanic eruptions and earthquakes were either relatively minor or did not occur in densely populated areas. Other significant reasons include greater awareness of the dangers and the willingness of people to evacuate when advised by responsible government officials.

The year marked the twentieth anniversary of the

eruption of Mount St. Helens in Washington state. That eruption and an even more devastating eruption of a volcano, Nevado del Ruiz in Colombia, which resulted in 23,000 deaths in 1985, prompted more government and scientific cooperation in studying about volcanoes and providing warnings for imminent eruptions. The new knowledge appeared to pay off in 2000, when few people died from volcanic eruptions.

Nearly twenty years after Mount St. Helens erupted, on Hummocks Trail, a few miles northwest of the mountain, evergreens rise above the trees felled by the blast. (AP/Wide World Photos)

During the year, earthquake experts developed new theories that became the focus of debate. A great deal of seismic activity occurred in Japan, where the potential for much volcanic and earthquake damage exists. However, scientists were able to predict the volcanic activity sufficiently to evacuate people before the destructive eruptions occurred, and no earthquake caused much significant damage. Japan's largest city, Tokyo, has taken precautions in preparation for a powerful earthquake similar in magnitude to the Kobe earthquake that killed 6,000 people in 1995. Its newer buildings are designed to be earthquake resistant, and its civil defense organization has set up appropriate potential responses. In 2000, however, earthquake experts expressed concern that many older buildings were not earthquake resistant and that the public was not aware of evacuation plans. Experts also predict that an extremely powerful earthquake is likely to occur in the future, perhaps in a location that might cause a tsunami, or tidal wave, and Tokyo and the surrounding region could suffer tremendous and unforeseen damage.

New Explorations

In the latter part of 2000, the National Oceanic and Atmospheric Administration funded an expedition in the Gulf of Mexico by a deep-diving, human-occupied submersible called Alvin. Alvin was originally launched in 1964, but this was its first journey in eight years, and the discoveries were enlightening. Exotic life-forms seen by the scientists included 200-year-old giant tube worms, a blue octopus, and a squid with tentacles that were 10 feet (3 meters) long. Some phenomena were not identifiable, including gray puff balls, which might be colonies of bacteria on the gulf floor. The scientists also observed abyssal storms, or strong currents that suddenly occur and then move. The two-week mission led to the desire to learn even more through future expeditions. One of the areas meriting further study was the ecosystem near bubbling gas and oil seeps because scientists believe that deposits of frozen methane gas hydrates might prove to be a new energy source. Throughout 2000, expeditions and research such as this provided new data in support of already existing facts and theories or for the development of new ideas.

The wide variety of weather-related disasters, new and intensifying debates on global temperature change and its causes, continual examinations of known but not totally understood natural phenomena as well as new discoveries of comparatively unexplored parts of the earth made 2000 a year in which scientists achieved a new understanding of the natural environment. Earth sciences and the environment remained provocative and inviting areas for more study and discovery.

Global Warming
The Debate Heats Up

➤ *New analyses of satellite data fuel discussions in support of and against global warming and suggest ways to mediate its effects. The World Wildlife Fund predicts that 20 percent of species in cold areas could be extinct by 2200.*

The debates about global warming continued on several fronts in 2000. The validity and value of new and old data were discussed; new approaches for mitigation were proffered; possible unforeseen consequences were revealed, but no definitive large-scale actions were taken.

The greenhouse effect is the process through which the sun's energy, largely short-wave radiation, travels through the atmosphere and is absorbed by the earth's surface, which then radiates long-wave or infrared radiation. The atmosphere cannot absorb most of the sun's short-wave energy, but some gases, such as carbon dioxide and water vapor, can absorb long-wave radiation. In this way, the earth's surface directly heats the atmosphere. Without the greenhouse effect, the atmosphere would be too cold to support life.

However, over the last 100 to 125 years, the amount of carbon dioxide has increased from around 290 parts per million to around 370 parts per million. That increase appears to be correlated with, and some feel the cause of, an increase of around 1 degree Fahrenheit over the same period. This apparent increase in atmospheric temperature is wrongly called the greenhouse effect. Instead, it could be called the apparent intensification of the greenhouse effect or global warming.

Old and New Temperature Data

By the end of 2000, most atmospheric scientists agreed that the surface temperatures of the earth increased by around 1 degree Fahrenheit during the twentieth century. Many scientists also agreed that the rate of warming accelerated over the last part of the century to a rate of 3.5 degrees per century. The United Nations Intergovernmental Panel on Climate Change has predicted that over the twenty-first century, the rate will accelerate to 2 to 6 degrees per century. Considering that the temperatures dominating the earth during the last glaciation around 18,000 to 20,000 years ago were only 5 to 9 degrees cooler than present temperatures, the significance of these numbers becomes apparent.

The reliability of these numbers may be questioned because they are based on an assumption that temperature record keeping has been precise and consistent all over the world for a century. Therefore, other sources of data would be useful. In January, 2000, analysis of satellite data from 1979 through 1998 by the National Research Council of the National Academy of Sciences added more fuel to the controversy. The data showed surface warming, but less warming and even some cooling in the middle to upper levels of the troposphere (the lower part of the atmosphere where most of weather occurs). Some respected scientists who think that global warming is not a significant problem used this middle to upper troposphere pattern as support for their position. They also pointed out that the computer models are flawed because they did not predict this inconsistency. Other scientists stated that the surface warming is real and that the difference between surface and middle to upper tropospheric temperatures will disappear over time. Scientists note that despite satellite records, temperature records from upper-air weather balloons have been consistent with surface temperatures for more than four decades.

The greenhouse effect is the process through which the sun's energy, largely short-wave radiation, travels through the atmosphere and is absorbed by the earth's surface, which then radiates long-wave or infrared radiation.

Studies of satellite imagery during the summer of 2000 showed that over the previous twenty years, the North Polar ice cap decreased in area by approximately 6 percent. Also, more open sea existed during the summer of 2000 than previously. The thickness of the ice, as evidenced by submarine sonar data, also appears to have decreased by around 42 percent in the last half of the twentieth century.

Suggestions for Mitigation

Besides increased burning of fossil fuels, large-scale deforestation is another possible cause for the increase in carbon dioxide. The decline in the number of trees has meant less carbon dioxide removed from the air through the process of photosynthesis and less carbon stored in the vegetation. Therefore, one suggestion to lessen global warming has been to plant more trees. However, a study led by Ernst-Detlef Schulze, director of the Max Planck Institute for Biogeochemistry in Germany, reported that young trees do not store as much carbon dioxide as do old-growth forests. The carbon from the air is found not only in the trees themselves but also in the soil and dead organic matter. Removing the old forests will result in the carbon moving back to the atmosphere as carbon dioxide, according to Kevin Gurney of Colorado State University. The German study supports the argument for protecting old-growth forests and also for lessening carbon dioxide emissions from fossil fuels.

Carbon dioxide is not the only greenhouse gas produced by modern industry. Methane is released as a result of coal mining, natural gas and petroleum production, the cultivation of rice, and the digestive processes of livestock. Chlorofluorocarbons (CFCs) and hydrofluorocarbons (HFCs) escape into the air from aerosol cans, coolants, and fire extinguishers. Nitrous oxide is produced by nitrogen fertilizers, manure, waste disposal, and nitrogen-fixing bacteria. Sulfur hexafluoride comes from electrical insulation and asthma treatments. Ozone is created by smog. (Ozone also naturally occurs in the stratosphere around thirty miles above the surface of the earth, where it absorbs ultraviolet radiation from the sun, thus heating the stratosphere and protecting the earth from ultraviolet rays.) Besides these gases, modern industry produces black carbon soot, a solid particulate produced by incomplete combustion.

The World Wildlife Fund has predicted that by the year 2200, perhaps 20 percent of all species living in the cold areas of the earth could be extinct.

Some scientists, including James Hansen of the National Aeronautics and Space Administration (NASA), have advocated working to reduce the emissions of these chemicals rather than carbon dioxide. Reducing carbon dioxide production would be extremely difficult because the primary approach would be burning less fuel, causing inconvenience and hardship to many people and negatively affecting economic activities worldwide. The levels of the other possible greenhouse effect enhancers could in some cases be more easily controlled by new technologies, including improved combustion so that less black carbon soot is produced. Methane produced by agriculture could be lessened by changing the feed mixture given to cows and by draining rice fields more frequently.

The other greenhouse gases have greater global warming potential than carbon dioxide. For the same increase in pollutant level, the increase in temperature is much greater for these gases than for carbon dioxide. Therefore, some scientists argue, reducing the amount of other gases will be more efficient than reducing carbon dioxide emissions. However, others argue that the actual amount of carbon dioxide present and being added is so huge that its levels must be reduced. They are also concerned that a decreased emphasis on carbon dioxide will produce complacency. In addition, the interactions of these chemicals with one another and in the atmosphere are not totally understood, making analysis and prediction complex.

In August, 2000, the administration of U.S. president Bill Clinton suggested an alternative to lessening carbon dioxide emissions in conforming to the 1997 Kyoto protocol. Countries should receive credit for planting forests and crops because these plants will absorb carbon dioxide. The European Union does not like the proposal because Europe does not have as much open space as the United States to grow new plants. Other criticisms include lack of knowledge about how much carbon will be absorbed and for how long.

Newly Recognized Potential Consequences

The World Wildlife Fund, a prominent conservation organization, predicted in August, 2000, that by the year 2200, perhaps 20 percent of all species living in the cold areas of the earth could be extinct, with around 70 percent of their habitat lost. Because the Arctic and neighboring northern areas are predicted to warm most rapidly, species native to those areas would not be able to adapt sufficiently quickly to survive. Not only rare species but also species living in isolated or mountain areas would be vulnerable. Although these predictions cover one hundred years and are conditional on factors that are not yet certain, they still raise another question about the impact of potential temperature change.

One major cause of the warming might be diversion of the warm Gulf Stream waters from their current path to Great Britain and instead to the Canadian Arctic and Greenland. Animals and plants in the Arctic might have a very difficult time adjusting; some plants might have to migrate one hundred times faster than plants did with the end of the last glaciation. Great Britain might be cooler, but warming the fragile ecosystems in the Arctic could be disastrous.

One of the many complexities of analyzing possible climate change is that the flows of the greenhouse gases

The aerial view of Martha's Vineyard (upper) contrasts with the computer-enhanced image (lower) of how many of its beaches and seaside villages will disappear in one hundred years as the sea levels rise because of global warming. (AP/Wide World Photos)

are not completely understood. No one knows for sure how much carbon dioxide is being absorbed by the oceans. If the oceans are absorbing carbon dioxide from the atmosphere, then the actual temperature change might be less than expected. A team from the National Oceanographic Data Center of the National Oceanic and Atmospheric Administration (NOAA), led by Sydney Levitus, analyzed 5.1 million measurements taken from the 1950's to the 1990's that showed the temperature of the top 1.9 miles (3.0 kilometers) of the world ocean increased by 0.1 degree Fahrenheit; for the top 1,000 feet (305 meters), the temperature increased by more than 0.5 degree. As with all studies of temperature change, the question is whether the change is caused by natural or human factors or a combination. Prominent ocean and atmospheric scientists have not yet come to any agreement.

Humans, Nature, and Policy

In July, 2000, Thomas Crowley, a geologist at Texas A&M University, released the results of his computer analysis showing that human factors are more important than natural factors in causing global warming over the last century. His analysis showed that variations in atmospheric temperature for the past 1,000 years generally correlated with variations in solar radiation and volcanic eruptions up until the mid-nineteenth century. Since then temperatures seem to have risen with increased levels of greenhouse gases, and the natural factors accounted for only 25 percent of the warming in the twentieth century. Crowley's analysis has been questioned for a variety of reasons, including its failure to consider the role of oceans and what some scientists consider its inconsistent solar energy data. However, his model appears to support other evidence of human impact on global temperature change and is considered to be valuable to policymakers.

If the oceans are absorbing carbon dioxide from the atmosphere, then the actual temperature change might be less than expected.

Government experts and advisers, industry representatives, and scientists need sufficient information to decide what path or paths the public and their own respective constituencies should take. Although no definitive paths were evidenced in the year 2000, the possible results of taking no action should also be considered.

Pollutants Decline but Ozone Thinning Expands

➤ *Satellites capture a picture of the largest ever ozone hole over Antarctica.*

In September, 2000, satellite imagery showed that the ozone hole over Antarctica was the largest it has ever been since its size has been recorded. Although the concentrations of the chemicals considered mainly responsible for ozone depletion declined in the troposphere, the lowest layer of the atmosphere, and leveled off in the stratosphere above, these images show that the dangers have not gone away and may have even increased.

Most oxygen in the atmosphere is molecular or diatomic oxygen. However, ozone, triatomic oxygen, exists as a naturally occurring gas in large concentrations in the stratosphere, around 12 to 18 miles (19 to 28 kilometers) above the surface of the earth. It absorbs solar ultraviolet radiation, thereby raising the temperature in the stratosphere. For millions of years, the natural chemical reactions of converting oxygen to ozone and back again have resulted in a shield of ozone that served to protect lifeforms from excessive ultraviolet radiation. Although ozone exists in all latitudes of the stratosphere, most ozone is created in the equatorial and tropical stratosphere where sunlight is strongest.

Over the past few decades, the concentration of ozone appears to have diminished in parts of the stratosphere away from the equator, apparently because of the presence of chlorofluorocarbons (CFCs). This decrease in ozone concentration, located seasonally over Antarctica and sometimes the Arctic, is called the ozone hole. It is not really a hole as a complete absence of ozone does not occur. Rather, the coloring of the satellite imagery in the area where the concentration of ozone is low makes the region appear to be a hole.

CFCs, composed of chlorine, fluorine, and carbon, are human-made organic chemicals created in the 1920's as refrigerants. Relatively inexpensive, CFCs gained widespread use as spray can propellants, air conditioning coolants, and in polystyrene plastic foam manufacturing. In 1974, two chemists, Mario Molina and Sherwood Roland, theorized that ultraviolet radiation could break down the

CFCs, freeing a chlorine ion that could then react with stratospheric ozone and decompose it into diatomic molecular oxygen and atomic oxygen. They, along with Paul Crutzen, on whose work they built, won the Nobel Prize in Chemistry in 1995 for this analysis.

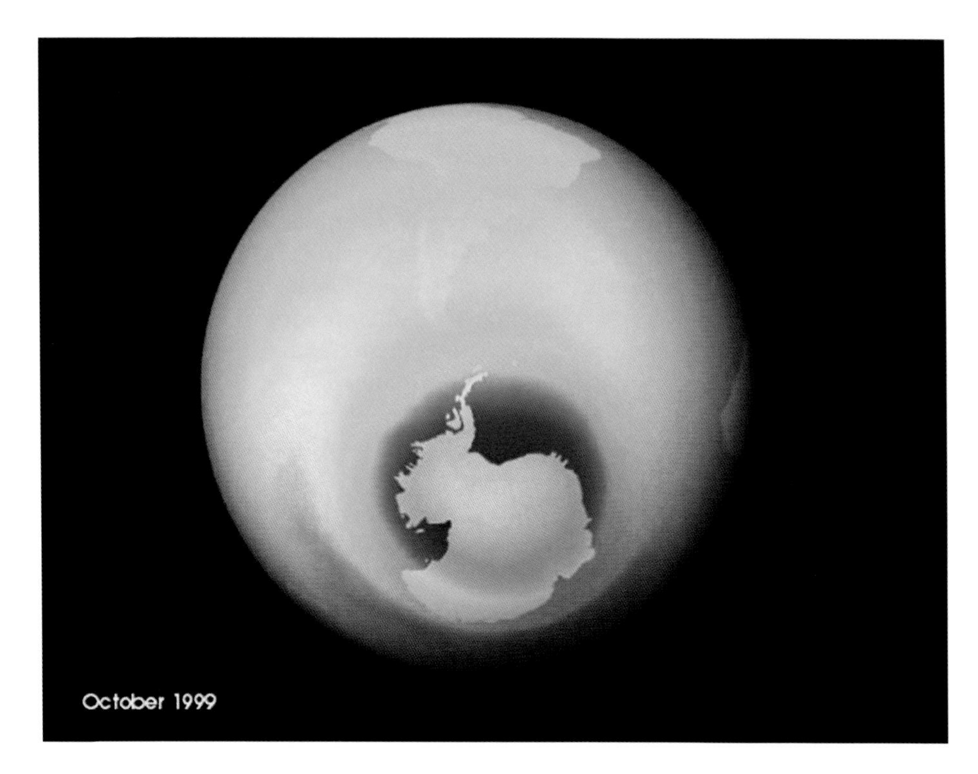

This graphic created using October, 1999, data from the Total Ozone Mapping Spectrometer Earth Probe, clearly shows the growing area of ozone depletion over the Antarctic (shown in blue). (AP/Wide World Photos)

A Dismal Picture

In the 1970's, satellite imagery confirmed surface-based measurements taken since the 1960's by showing a thinning of the concentrations of ozone in the stratosphere, especially in the high northern and southern latitudes. The declines continued into the 1980's and 1990's, although not at a regular rate. Despite the scientific evidence, manufacturers did not want to stop producing CFCs. In 1987 an international agreement, the Montreal protocol, was reached to phase out the chemicals. Perhaps as a result of this protocol, the levels of CFCs in the stratosphere did start to level off around 1990.

In September, 2000, however, NASA's Total Ozone Mapping Spectrometer (TOMS) showed that the size of the ozone-depleted area, or hole, was much larger than it had ever been. NOAA reported that the depleted area over Antarctica was 17.1 million square miles, greater than the area

of North America and much larger than the 900,000 square miles in 1981.

Charles Jackman, an atmospheric modeler at NASA's Goddard Space Flight Center, explained in October, 2000, that CFCs are created on the surface and move very slowly, perhaps taking two years to travel through the troposphere into the stratosphere. Once in the stratosphere, several decades may elapse before the CFC molecules are converted into their destructive form. Then another few years may elapse before the chemical is washed back into the troposphere as hydrogen chloride in rain.

In 1994, scientists at NOAA detected that levels of CFC in the troposphere had declined. Predictions of resulting declines in the stratosphere were made, with Richard McPeters, principal investigator for TOMS, recognizing that computer models indicated that twenty to forty years might pass before CFCs reached their pre-1980 levels. In 2000, even though the concentration of CFCs has leveled off, it is still high, and thus significant ozone depletion should not be surprising.

Temperature, Winds, and Ozone Depletion

The weather in the stratosphere contributes to the level of ozone depletion. The temperatures in the stratosphere become very low during the Antarctic fall and winter as the amount of sunlight is small to nonexistent. Ice crystal clouds, called polar stratospheric clouds, result. The ice crystals accelerate ozone depletion as they offer a surface for chlorine to be changed into catalysts or chemical forms that deplete ozone.

Not only are the temperatures low, but also strong winds move as a whirlpool forming the Antarctic vortex. The vortex is stable and keeps out air that is warmer and contains a higher concentration of ozone. Thus stratospheric air from the equatorial areas and tropics cannot penetrate the vortex. The concentration of ozone within the vortex continues to diminish even into the Antarctic spring. With warmer temperatures in the later spring, the vortex dies through the summer but begins again in the fall.

A vortex may also form over the North Pole, but it does not last as long nor is it as strong or consistent in its location as the South Polar vortex. Atmospheric patterns in the higher latitudes of the northern hemisphere are less stable than over the South Pole. Tall mountains interfere with the upper level winds, causing "atmospheric waves" that often move the vortex to lower latitudes, where temperatures are slightly warmer. An ozone hole is unusual, with the last one occurring in 1997.

Scientists in 2000 considered that a buildup of greenhouse gases might also be contributing to ozone depletion. When located in the troposphere, carbon dioxide and other greenhouse gases appear to have contributed to a rise in temperature, but in the stratosphere, they seem to have a cooling effect. The gases radiate the heat they absorb to space, thereby cooling the surrounding stratosphere. These cooler temperatures may exacerbate the cooling already occurring in the higher latitudes, where the colder air in the stratosphere allows more ice crystals to form, which in turn contributes to more ozone depletion.

The areal extension of ozone depletion and of extremely cold stratospheric air in 2000 spread over the ports of Punta Arenas, Chile, and Ushuaia, Argentina. However, because the sun's rays in September and October, when the hole is largest, are relatively weak, residents do not have to worry about sunburn or cancer dangers, according to Paul Newman, an atmospheric physicist at Goddard Space Flight Center.

The data of 2000 have not led to any conclusive answers about ozone depletion. The leveling off of the concentration of CFCs should, logically, contribute to a diminishment of ozone depletion. However, the diffusion of CFCs is a slow process, and decades may elapse before significant changes occur. As long as CFC levels remain high, a decline in ozone depletion cannot be expected. Complicating matters is the contribution of stratospheric winds and weather in general to temperature levels and to the stability of the Arctic and Antarctic vortices. Regardless of ozone concentrations, temperatures naturally vary significantly from one year to the next, so predictions for the future cannot yet be precise. Predictions of ozone levels are difficult, but the data of 2000 serve to show that amelioration of an environmental problem may be more complicated in actuality than on paper.

When located in the troposphere, carbon dioxide and other greenhouse gases appear to have contributed to a rise in temperature, but in the stratosphere, they seem to have a cooling effect.

Radioactive Waste
Disposing of Geologic and Political Quandaries

➤ *Nuclear waste storage and treatment efforts run into trouble when protests by the Western Shoshone contribute to suspension of a plan to build a national waste storage facility near Yucca Mountain and the Department of Energy turns down a private company's proposal to build a treatment facility at Hanford.*

Two distinct but interrelated problems are involved in the management of nuclear power plant and weapons wastes: leakage from currently stored wastes from nuclear weapons production and the selection of a national site for permanent storage of nonmilitary wastes. Remnants of old power plant components, nuclear weapons, and other radioactive wastes must be disposed of safely to prevent the poisoning of people and animals and the contamination of the natural environment. Proper disposal is extremely difficult because the radioactive substances are difficult to confine, and even in small doses, the wastes are frequently lethal.

In 2000, problems arose in the cleanup of leaking drums containing nuclear wastes at the Hanford site in Washington, which has been storing nuclear wastes from more than fifty years of weapons manufacturing, and contamination was found in the groundwater. Ongoing efforts at cleanup and remediation have made limited progress, and nearby residents have expressed their concern. Also, social concerns and political events caused the government to suspend its plans to construct a permanent storage place for nonmilitary nuclear wastes at Yucca Mountain, about one hundred miles northwest of Las Vegas, Nevada.

Yucca Mountain

Because nuclear wastes are both radioactively and thermally hot, proper storage involves ensuring that the radioactivity and the heat be contained. The site should be geologically stable, with little possibility of earthquakes or volcanoes, and minimal erosion. There should be no occurrences of groundwater, since water is the primary conductor of radioactive wastes. In 1987, the United States Congress, following the advice of the Department of Energy, selected Yucca Mountain as the storage area. The site is unpopulated federal land, and its geological inactivity for the last million years seemed to make it a logical choice. The time line called for a viability assessment and the writ-

ing of a draft Environmental Impact Statement (EIS) in 1998, followed by a final EIS in 2000. If the EIS found the site suitable, licensing would follow in 2002, followed by construction in 2005, and operation in 2010. The government was to build a series of tunnels around 35 miles (56 kilometers) long to store the wastes. As a preliminary step, the government drilled a tunnel about 5 miles (8 kilometers) long for testing purposes. However, in 2000, social and political concerns became so prominent that these proposed actions were suspended.

The Western Shoshone Nation consider themselves caretakers of the land around Yucca Mountain. Although they have hunted animals and gathered plants in this area over the centuries, they do not say they own the land, only that they are caretakers. The Shoshone say that the sole treaty signed by both the United States and the Western Shoshone is the Treaty of Ruby Valley of 1863, which guaranteed safe passage for nontribal peoples through Shoshone lands. The Indian Lands Claims Commission was created in the 1950's to compensate the Indians for land ceded to the United States. The commission decided that 24 million acres (9.7 million hectares) had been trans-

At the Hanford Site, two nuclear chemical operators begin loading drums of radioactive waste into three special stainless steel containers in preparation for shipping some of the waste to the Waste Isolation Pilot Plant in New Mexico. (AP/Wide World Photos)

ferred in 1872 to the United States through gradual encroachment and determined that fifteen cents an acre was sufficient compensation. In 1985, the U.S. Supreme Court upheld the validity of the commission's decision. The Western Shoshone do not recognize the ceding and have been fighting the decisions of the Indian Land Claims Commission and the U.S. Supreme Court. In 2000, they expanded their protests beyond the protection of the Yucca Mountain lands; they are concerned about dangers arising from transporting the wastes across the United States to this site in Nevada. Some Shoshone support making the whole United States nuclearfree.

In April, 2000, U.S. president Bill Clinton vetoed federal legislation that would have required used reactor fuel from electric power plants to be shipped to and stored at Yucca Mountain before the permanent repository was built. The Senate failed to override the veto by a 64-35 vote in May. Whether the wastes are stored locally or moved to the Yucca Mountain site seems to have little impact on the electrical power industry as a whole or on individual power plants, and the location does not pose a safety issue. However, political considerations influenced the actions of the president and Congress as the bill contained two provisions that the Clinton administration viewed as undesirable. The bill would have prevented the

Environmental Protection Agency (EPA) from setting radiation levels until June, 2001, after a new president had taken office, and the secretary of the Department of Energy would no longer have been allowed to increase development fees charged to electrical utilities for the facility.

Recent research indicates that seismic activity is not absent from the Yucca Mountain area. Since 1976, more than 600 seismic events have occurred within a 50-mile (80-kilometer) radius of the site. Most of these events were relatively small in magnitude, but a few were significant. One earthquake in 1992, 5.6 on the Richter scale, damaged the Department of Energy building located around 8 miles (13 kilometers) from the site. The earthquakes were probably caused by faulting that has been occurring for millions of years and that is not likely to stop.

Hanford Site Cleanup

The Hanford site, in southeastern Washington, stores around 60 percent of the nation's radioactive wastes from more than fifty years of nuclear weapons production. The waste includes 190 million curies (54 million gallons, or 204 million liters) of solid and liquid wastes stored in 177 underground tanks as well as 143 million curies in cesium and strontium capsules. The steel and concrete drums, some of which are about four stories high, were supposed to last around twenty years. Sixty-seven have developed leaks, and the others no longer shield their environs from radioactive emissions. To solve the problem, the wastes from the sixty-seven single-walled and leaking tanks were being pumped to new double-walled tanks. However, the transfer was stopped in January, 2000, when welds on the double-walled pipes were found to be defective.

Although government officials have expressed their confidence that no immediate danger to the public is likely, the total amount of radioactive wastes stored in temporary facilities is mind-boggling.

Some of the wastes have been spilled. One million gallons are feared to be in the ground and may be headed to the Columbia River. In February, 2000, a groundwater sample at Hanford contained excessive amount of tritium: 8 million picocuries per liter (0.26 gallon) of water compared with the federal drinking water standard of 20,000 picocuries per liter, or 400 times the accepted limit. The tritium could take anywhere from three to thirty years to reach the river, about 3 miles (5 kilometers) away from the Hanford site.

Plans for wastes currently in the tanks include solidifying the most dangerous wastes into sludge. Even if that were to occur safely, around 50,000 gallons (189,250 liters) of liquid wastes would have to be treated. In May, 2000, the U.S. Department of Energy refused to approve a proposal by a private company, British Nuclear Fuels, to build a high-level waste-treatment facility for $15.2 billion. The wastes would have been melted at thousands of degrees to form glass tubes, which would then be placed in steel and buried. The department's concern was not only with the drastic increments in estimated cost from $6.9 million over eighteen months but also with the proposed financing techniques and management of the company.

The Washington State Department of Ecology, the EPA, and the Department of Energy had agreed in 1989 on deadlines for the cleanup process. However, because those deadlines were continually not being met, the Washington State Department of Ecology decided in March, 2000, to require the Department of Energy to meet new deadlines. These included the issuing of a contract for the construction of a glassification plant by August 31, 2000, with construction to start by July 21, 2001. The plant would be operating by December, 2009, and, by December, 2018, at least 10 percent of the wastes were to be in new tanks. The Department of Energy is looking for alternatives to the British Nuclear Fuels' proposed plant. Managing these wastes is thus complicated not only by the difficulties of containing and coping with radioactivity but also by traditional problems of intergovernmental relations.

Although government officials have expressed their confidence that no immediate danger to the public is likely, the total amount of radioactive wastes stored in temporary facilities is mind-boggling. These dangerous wastes, often in inadequate or aging facilities, are potential sources of accidental exposure and environmental contamination. To develop and operate a program that will store the wastes safely now and for thousands of years to come requires the cooperation of scientists, private corporations, community groups, and government agencies at the local, state, and federal level.

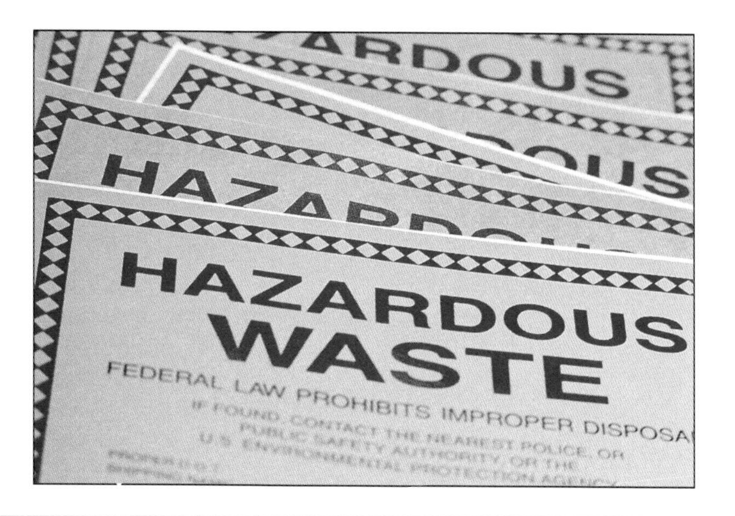

Unplugging the Great Lakes
Declining Water Levels, Exotic Invasions

➤ *Water levels in Lakes Huron, Erie, and Michigan reached their lowest levels since 1964, and the numbers of nonnative fish species in the Great Lakes reached a new high. The International Joint Commission issued a report calling for additional funding and more action to improve the lakes.*

The Great Lakes, located within the heartland of North America and serving as a valuable economic and natural resource, are threatened by declining water levels, invasion by nonnative species, and contamination from atmospheric and sediment sources. Environmental problems are not new in the Great Lakes, a chain of five lakes consisting of Lakes Superior, Michigan, Huron, Erie, and Ontario. Both Canada and the United States, along with provinces and states bordering the lakes, have worked for decades to improve conditions. The efforts continued in 2000 through conferences, legislative funding, and continued governmental monitoring and enforcement of laws and regulations.

Low Water Levels

Normal changes in seasonal water levels of the Great Lakes, the largest body of freshwater in the world, are caused by fluctuations in temperature and precipitation over the year. Lake levels are usually low in the winter and fall when water evaporates from the comparatively warm lakes. In the spring and summer, snowmelt from the surrounding area and precipitation raise water levels. It is normal for precipitation and therefore lake levels to vary over the years. For example, in the early 1970's and in 1986, the levels were very high, and in the 1930's and the mid-1960's, the levels were low. Since 1998, the levels in three of the lakes, Huron, Erie, and Michigan, have been declining, and in 2000, they were at their lowest levels since 1964. This most recent decline is attributed to low precipitation combined with warmer average temperatures during the past few years. Much of the water that would have been in the lake either did not fall as snow or evaporated from the water as a result of the warmer temperatures.

The 1959 completion of the St. Lawrence Seaway, which allowed oceangoing vessels access to the Great Lakes, was a great economic boom but also contributed to invasion of the Great Lakes by nonnative species.

The low water level restricts freight shipment because

Environmental officials are attempting to stop the round goby, a Russian import that has been competing with the native fish in the Great Lakes since its arrival in 1991, from invading the Mississippi River. (AP/Wide World Photos)

ships can run aground in the shallower lakes. Some harbors have been dredged, an expensive process. However, some benefits are possible. Beaches are wider, enabling swimmers to better enjoy the lakeside. The shallower water may facilitate the growth of plants close to shore, providing more food for juvenile sport fish, thus enhancing the fishing industry.

Invasion of the Lakes

Nonnative species have been invading the Great Lakes probably for the last two hundred years but definitely after the 1920's when the Welland Canal was enlarged to permit ships to avoid Niagara Falls and move between Lake Erie and Lake Ontario. The canal allowed fish, including the sea lamprey, and ships, which previously had been unable to climb the falls, to move into Lake Erie from Lake Ontario and bodies of water including the Hudson River and the Erie Canal. Once in Lake Erie, the sea lamprey were able to swim to the other three lakes and contributed to the destruction of the commercial fishing industry by preying on native fish. The 1959 completion of the St. Lawrence Seaway, which allowed oceangoing vessels access to the Great Lakes, was a great economic boom but also contributed to invasion of the Great Lakes by even more species. In the 1960's, after the sea lamprey decimated the trout, the main predator of the lake, a herring-like fish called the alewife, began to proliferate. Millions of alewives began to die each winter from the cold temperatures and food shortages, covering the shores of the lakes.

The rate of invasion appeared to be increasing rapidly in the late 1990's. By 2000, nonnative fish species numbered around 145, up from 139 in 1997, and continued to wreak havoc on the native species. Many, including such fish as the round Goby (fish) and Eurasian ruffe from the Black, Caspian, and Azov Seas, arrived in the 1990's in ballast water from oceangoing vessels. Both the goby and the ruffe are bottom feeders; the ruffe is a competitor of yellow perch and walleyed pike and eats whitefish eggs. The goby eats small fish and eggs as well as clams and mussels, including the dreaded zebra mussel. The zebra mussel entered the lakes in the 1980's, also in ballast water, and has been clogging pipes, thereby seriously inhibiting use of lake waters.

To limit the invasion of species through ballast, the United States, starting in 1993, required and Canada has requested that ships empty their bilge water in the Atlantic Ocean before they enter the Great Lakes. However, around 20 percent to 40 percent of the sludge cannot be pumped out of the ships, resulting in around 100 to 200 tons (91 to 181 metric tons) per ship being transported into the lakes. Efforts to rid the lakes of the fish have been complicated by insufficient knowledge and resistance by a variety of interest groups. Proposals requiring the ultraviolet, biological, or chemical sterilization of the bilge water have been fought by the shipping industry, labor unions, and port interests.

Techniques such as poisoning the larval sites, catching the fish at barriers, and sterilizing and releasing males, all of which have been used against the sea lamprey, so far have not been adopted against these newer invaders. More knowledge needs to be acquired about the life cycles and locations of the fish involved. Environmental concerns have also been cited. New approaches include the development of pheromones, which are odors fish emit as sexual attractants, to provide warnings, or to mark territory. These attempts, as well as restocking of the fish originally inhabiting the lakes, including trout and salmon, are expensive, running into millions of dollars a year.

A History of Contaminants

For decades, tons of sediment contaminated with chemicals such as polychlorinated byphenyls (PCBs), mercury and mercury compounds, lead and lead compounds, and dioxins and nitrogen and phosphorous compounds were carried by runoff or deposited from the atmosphere into the Great Lakes. By 1970, Lake Erie, the shallowest of the lakes, was said to be dead because it was contaminated with pesticides, fertilizers, and other toxic chemicals. The lake's dissolved oxygen level had been diminished by the decomposition of excessive plant life, algae blooms (resulting from the high levels of fertilizer in the lake), and dead fish. In 1972, the United States and Canada embarked on a joint program to clean up the Great Lakes and the surrounding area; in 1978, they agreed to the Great Lakes Water Quality Agreement, which remains in effect.

In the ensuing years, the lakes have improved significantly. Lake Erie, for example, is no longer dead. The levels of contaminants have declined greatly, but the lakes still contain significant, albeit smaller, concentrations of the more dangerous chemicals. These bioaccumulate, or increase in concentration, in the food chain. Thus, a small amount may exist in the lake and in low-level organisms, but as larger and larger animals eat the smaller ones, the concentration of the chemicals in the animals increases. As a result, some fish are considered too dangerous to eat and others can be consumed only in limited quantities. The problem is exacerbated because information about dangerous fish is often disseminated only with fishing licenses; the person who eats the fish is not always the person who has the information. In July, 2000, the International Joint Commission, which oversees the quality of the Great Lakes, in its tenth biennial report, called for additional funding and more action to continue to improve the lakes.

Monumental Events
Clinton Creates Nine National Monuments

➤ *National monuments are created to protect areas with ecological, geological, or historical significance in the United States.*

Acknowledging the fragility of the natural environment, U.S. president Bill Clinton established nine national monuments in 2000 to officially protect hundreds of thousands of acres of land valued for its geological, biological, ecological, and historical features. Unlike other protected areas, such as national parks or national seashores, national monuments do not need Congressional approval for their establishment and can simply be proclaimed by the president. However, not too much can happen without funding, which is the province of Congress. National monuments have a history of controversy. Before President Clinton established the Grande Staircase-Escalante National Monument in 1996, the last president to establish any national monument was Jimmy Carter, who established fifteen in 1978. National monuments cover a wide variety of human and natural resources and as such do not belong to any single federal agency, although most are under the jurisdiction of the National Park Service and the Bureau of Land Management.

Decades of Controversy

The Antiquities Act of 1906 was passed to allow the president of the United States to designate certain antiquities for protection. The original intent of the law was to protect prehistoric cultural features, generally regarded as being small in size. However, as written, the law allowed historic landmarks, prehistoric Indian ruins, and objects of scientific interest to be considered antiquities. Theodore Roosevelt, president when the law was passed, used the scientific interest property of the law to establish national monuments that encompassed geological features and were thousands of acres in size, thus operating contrary to the original spirit of the law.

Other presidents also established large-scale natural monuments, some of which, like the Grand Canyon and Petrified Forest, were subsequently converted by Congress to national parks. Initially, the monuments and their conversion were readily accepted, partially because they were

mostly in Arizona and Alaska, which were territories and therefore did not have representation in Congress. However, beginning in 1943, with the establishment of Jackson Hole National Monument by Franklin Roosevelt, the Antiquities Act came to be viewed as a means to circumvent Congress. Since then, presidents have used the act sparingly, sometimes with congressional cooperation and sometimes without.

In 1978, to preserve lands in Alaska despite strong opposition within that state, President Carter proclaimed fifteen national monuments in Alaska after Congress had adjourned. Two years later, Congress placed most of the Alaskan lands in national parks and preserves, but it also modified the Antiquities Act to curtail "further use of the proclamation authority in Alaska."

The New Monuments

The nine monuments established by President Clinton are meant to protect a variety of important natural or cultural riches, including prehistoric sites, unique geological features, vegetative treasures, and sensitive feeding and nesting habitats. Three were proclaimed on January 11, 2000. The Grand-Canyon-Parashant National Monument, located on the edge of the Grand Canyon, contains geologic features, including Paleozoic and Mesozoic sedimentary rock layers relatively unobscured by vegetation, allowing study and understanding of the geological history of the Colorado plateau. Its biological resources include

The ironwood tree, along with other drought-resistant vegetation of the Sonoran desert, is the heart of the Ironwood Forest National Monument in Arizona. (AP/ Wide World Photos)

endangered, threatened, and sensitive species as well as giant Mojave yucca, mule deer herds, and wild turkey. Archeological remains include prehistoric rock art. In addition, traces of nineteenth century life can be found in the old water tanks and the ruins of ranches, sawmills, and mining operations. Its establishment as a monument will limit damage from new mineral development.

The Agua Fria National Monument contains two mesas as well as archaeological ruins of pueblo communities that existed from 1250 to 1450. Because the monument is within a semidesert grassland with riparian forests running through it, it supports a great diversity of vegetation and wildlife, including the lowland leopard frog and the more widely known mule deer and pronghorn.

National monuments do not need Congressional approval for their establishment and can simply be proclaimed by the president. However, not too much can happen without funding.

The California Coastal National Monument, located off shore and ending where the Continental Shelf meets the Continental Slope, includes crucial parts of the coastal ecosystem. Its geological structures—rocks, islands, exposed reefs, and pinnacles—are home to around 200,000 breeding seabirds and also some mammals. This monument, along with the two previously mentioned, will be managed and administered by the Bureau of Land Management and/or

National Park Service, although California Coastal will be jointly managed with the California State Department of Fish and Game. Most current permits for livestock grazing, hunting, and fishing will be honored, but new mining claims and, in Agua Fria, geothermal leases, will be prohibited.

Other monuments to be managed by the Bureau of Land Management include Canyons of the Ancients in Colorado, Cascade-Siskiyou in Oregon, Ironwood Forest in Arizona, and Hanford Reach in Washington. The Canyons of the Ancients, containing more than twenty thousand archeological sites of the ancestral Northern Pueblo people dating from 450 to 1300, is threatened by tourism and population growth, as well as oil and gas leasing and development. Because the area is a monument, oil and gas leasing and development will continue as long as they do not interfere with the "objects protected by the designation" nor with the rights of the American Indian tribes. New leases will be allowed only if the environment is not to be endangered.

Cascade-Siskiyou National Monument was established to protect an area where several biological, climatological, and geological regions intersect, including the Cascade, Klamath, and Siskiyou ecoregions. In addition to the area's natural features, which include towering fir forests, a great diversity of butterfly species, and manifestations of complex geological history, the monument also contains parts of the Oregon/California Trail.

Ironwood Forest National Monument is biologically and geologically diverse and contains archeological objects dating from 600 to 1450, including rock art. It contains many endangered or threatened species and is home to drought-adapted vegetation of the Sonoran Desert, including ironwood trees, which can live eight hundred years and contribute to the survival of native animals and plants, including desert bighorn sheep, hawks, owls, and even saguaro cactus, by providing frost protection.

The site of Hanford Reach National Monument served as a buffer area in the federal reservation that developed nuclear weapons, limiting development and human use for fifty years. As a result, it contains an a natural and historic legacy preserved by unusual circumstances and is considered irreplaceable. It consists of the last free-flowing, nontidal stretch of the Columbia River with upland shrub-steppe habitats. Its great diversity of flora and fauna includes rare and sensitive plants and even two new recognized plant species, the Umtanum desert buckwheat and the White Bluffs bladderpod, and forty-one newly recognized insect species. The environs contribute to the existence of wintering and migratory birds as well as the spawning grounds of salmon, for which the Columbia River is famous. Like the other monuments, it contains unique archeological and more recent historical remains, as well as significant paleontological and geologic formations, including rhinoceros and camel fossils.

The Grand Sequoia National Monument was established to protect and save the giant sequoia trees and their ecosystem. These trees once grew from California to Colorado to Wyoming; now they live only on the west slope of the Sierra Nevada mountains. Some sequoias are thousands of years old, and they and their ecosystem as a whole need to be protected from too much harvesting within and around the monument. Half of the seventy existing groves were already protected because they are contained in the Sequoia, Kings Canyon, and Yosemite National Parks. The new monument is to protect the remaining groves, which are outside these national parks but within Sequoia National Forest. The monument will be managed by the

Forest Service in the Department of Agriculture with a scientific panel to evaluate the management choices needed to protect the giant sequoias. Existing lumbering permits will be honored for at least two and one-half years, but the eventual goal of stopping the harvest of the sequoias means lumbering activities will have to end.

The Lincoln and Soldiers' Home National Monument in northwest Washington, D.C., includes the U.S. Soldiers' and Airmen's Home and Andersen Cottage, a retreat of President Abraham Lincoln. The Andersen Cottage was a summer retreat for the president and his wife and is where he drafted the Emancipation Proclamation. The Soldiers' Home was established in 1851 for disabled army veterans.

Preservation Versus Traditional Uses

A problem with monument establishment is that while the preservation value may be recognized, the prohibition against normal or traditional use of the lands accompanying monument designation causes hardships for people who were using the lands. To ease the transition, the establishment of these monuments involved public hearings to ensure that local interests were not ignored. Recognizing that the Antiquities Act applies only to federal lands, the proclamations state that private lands within the monuments are not subject to the new restrictions. However, if those private lands are acquired by the federal government, they would become part of the monument.

Similarly, monument establishment has been structured so as to interfere minimally with economic activities. Existing grazing permits will generally continue to be honored unless they are found to be incompatible with preserving the biological resources, and hunting and recreation will be permitted as before. However, off-road travel by mechanized vehicles is prohibited, which in most cases does not change actual practice.

The year 2000 remains remarkable not only because so many monuments were established after so many years but also because the monuments cover such a wide range of biological, geological, archeological, and historical features. The full operation of these monuments will take years and involve much study and compromise to best meet the needs of the human and natural environments.

Cattle Grazing on Federal Lands

➤ *The Supreme Court of the United States ruled that the government may change grazing regulations on federal lands to benefit the public and the environment.*

Cattle grazing on federal lands involves balancing the public good, preservation of the environment, and the interests of private parties. The resulting conflicts have a long history in the United States, involving clashes between early cowboys and Native Americans and between farmers and cattlemen, and are exacerbated by the prominent role that beef plays in the typical American diet. Therefore, grazing and its management, which might appear to be rather mundane activities, involve both high-stakes finances, emotions, and ecological concerns. In May, 2000, the Supreme Court of the United States ruled that the secretary of the interior has the authority to change cattle grazing regulations to benefit the public and restore the health of the range.

The Taylor Act

In an attempt to reverse decades of damage to rangelands caused by cattle and sheep grazing, Congress passed the Taylor Grazing Act of 1934. The task was complicated, and as the Supreme Court noted in its May decision, "the Taylor Act delegated to the Interior Department an enormous administrative task . . . to determine the bounds of the public range, create grazing districts, determine their grazing capacity, and divide that capacity among applicants." By 1937, the Department of the Interior had set the basic rules for allocation of grazing privileges, recognizing that "many ranchers had long maintained herds on their own private lands during part of the year, while allowing their herds to graze farther afield on public lands on other times." The rules gave first preference to stock owners who owned enough private land (or water rights) to support their herds and who had also grazed their stock on public lands during the five years just before the Taylor act became law. Second preference went to stock owners who owned adequate private land for their herds, and third

Cattle grazing and its management involve high-stakes finances, emotions, and ecological concerns.

went to stock owners who did not own enough land but relied on public lands. As the Court noted, this system, in effect, awarded grazing privileges to owners of land and water.

Grazing allocations were determined using animal unit months (AUMs), or the right to obtain the forage needed to sustain one cow (or five sheep) for one month. The Court noted that Congress made the secretary of the interior "the landlord of the public range" with discretion to grant grazing privileges. The department, beginning in 1938, had the power to modify, fail to renew, or cancel a grazing permit or lease. The secretary could cancel permits or leases because of persistent overgrazing of the public lands or failure to comply with the range code, withdraw the land from grazing and devote it to a more valuable or suitable use, and if range depletion occurred, reduce the amount of grazing allowed by partially or wholly suspending AUMS for as long as necessary.

Further Deterioration Leads to Further Action

Despite improved management under the Taylor Act, the rangelands deteriorated. Because grazing capacity was declining, the Department of the Interior Active reduced the number of active AUMs from around 18 million in 1953

A solitary cow stands on federal land north of Reno, Nevada, unaware of the legal turmoil surrounding the government's right to change grazing regulations. (AP/Wide World Photos)

to around 10 million in 1998. Although ranchers did not like to have their allocations decline, the reductions were made proportionately and were restored if and when the grazing capacity increased.

In the 1960's, a Congressional survey of rangelands reported that only 16 percent of the range was in excellent or good condition, 53 percent in fair condition, and 30 percent in poor condition. The U.S. Department of the Interior then decided to raise grazing fees more than 50 percent, from 19 cents to 30 cents per AUM per year. In 1978, Congress passed the Federal Land Policy and Management Act, which called for federal lands to be managed for multiple uses (including fishing, hunting, recreation, watershed management, range, and timber). It also called for sustained yield, defined as management of renewable resources so that their output is steady forever. Congress raised grazing fees for the years 1979 through 1986. The grazing regulations of the Department of the Interior were modified to conform to the new law. Even with the increases, the fees charged by the federal government were lower than those for grazing on privately owned lands.

Department's Actions Lead to Lawsuits

In 1995, the Interior Department established new regulations to speed up restoration of the rangeland, increase the compatibility of the rangeland management program with ecosystem management, streamline some administrative functions, and obtain "fair and reasonable compensation" for the grazing of livestock on public lands. A group of ranching organizations and grazing permit holders, including the Public Lands Council and the National Cattlemen's Beef Association, challenged ten of these new regulations by suing the secretary of the interior.

Challenges to three of the regulations reached the Supreme Court of the United States. The first of these regulations said that the secretary of the interior could change the definition of grazing preference and eliminate reference to AUMs, instead using land-use plans. The secretary also had the power to cancel grazing permits. The second regulation stated that grazing permits could be granted to people other than those active in the livestock business. The third regulation said that improvements made on federal lands would no longer be the property of the improver but of the federal government. The cattle interests were concerned that these new regulations would harm ranchers economically and that the secretary of the interior had violated the Taylor Act. However, the Supreme Court ruled in favor of the secretary in each of the three cases.

The first regulation stipulated that allocating grazing preference would no longer be based on the quantity of forage because AUMs would not be used. Because ranchers had used AUMs as collateral for bank loans, they feared that their economic security would be undermined, especially because the land-use plans that would serve as the basis of allocation were "difficult to predict and easily changed." In its May, 2000, decision, the Court did not agree with the complainants' reasoning and stated that grazing permits were a privilege, not a right. Ranchers therefore were no longer guaranteed grazing privileges for a specific number of livestock.

The ranchers claimed that the second regulation, which granted permits to people not in the livestock business, was "part of a scheme to end livestock grazing on public lands." The ranchers stated that individuals could "acquire a few livestock" and obtain a permit for conservation purposes. However, the Supreme Court ruled that such conservation permits had been ruled illegal by the court of appeals.

The regulation regarding allocation of ownership of range improvements affected ranchers who had built wells, fences, and other structures to benefit their livestock. Although the secretary of the interior had in the past granted ranchers ownership rights to some range improvements, the Court ruled that the secretary of interior could change his or her mind. Any individual who made improvements would have the right to negotiate for compensation.

The purpose of the 1995 amendments was to restore the health of the rangelands in the West. The Supreme Court found that the reform program was within the framework of existing law and reaffirmed the right of the secretary of the interior to use the broad authority granted under the Taylor Act to manage the public's lands. However, individual permit holders retained the right to bring challenges to changes in grazing privileges.

Petroleum Predictions and New Technologies

➤ *United States Geological Survey publishes a report estimating the supply of petroleum, natural gas, and natural gas liquids outside the United States, and petrochemical companies begin a joint project combining enhanced recovery of oil with reduction of greenhouse gases.*

The United States Geological Survey (USGS) released a report in 2000 stating that the amounts of undiscovered oil outside the United States would increase 20 percent over the next twenty-five years but that the amounts of undiscovered natural gas outside the United States would be lower than its previous reports had indicated. The organization's report, *The USGS World Petroleum Assessment 2000*, was based on a study conducted from 1995 to 2000 and predicted change in the thirty-year period from 1995 to 2025 in the amounts of petroleum, natural gas, and natural gas liquids available outside the United States. The USGS study examined information and data exclusive of the United States as the organization had reported in 1995 on a 1994 study of petroleum and gas in the United States. Petroleum, also called crude oil, is a natural liquid composed primarily of hydrocarbon compounds. Natural gas contains mixtures of hydrocarbon gases, primarily methane, but also other, nonhydrocarbon gases. Natural gas liquids include propane, ethane, butane, and other hydrocarbon gases and are liquids under the high pressure and temperature of the rock reservoir in which they exist in the crust of the earth.

Although the USGS assessment found large amounts of energy resources outside the United States, these amounts are finite. The application of environmentally and economically efficient new technologies plays an important role in increasing the availability—and therefore the supply—of the fuels.

Potential Supplies and Past Usage

The world report estimated two measures of these fossil fuels: undiscovered resources and potential reserve growth. Undiscovered resources, or volumes that have not yet been discovered, include 649 billion barrels of oil, 4,669 trillion cubic feet (130 trillion cubic meters) of gas, and 207 billion barrels of natural gas liquids. The USGS used geologic knowledge and theory to determine that these volumes exist, although it did not identify the specific locations, amounts, or quality of the resources. The potential reserve growth, or volumes that would be found and be extractable from discovered fields, was 612 billion barrels of oil, 3,305 trillion cubic feet (92 trillion cubic meters) of gas, and 42 billion barrels of natural gas liquids. These volumes have been identified and can be extracted for a profit at current prices using existing technologies.

The total amount of petroleum that potentially can be recovered from the earth is 2,120 billion barrels. Over the past century, the world has used around 20 percent of the supply of petroleum, or 539 billion barrels, and around 7 percent of the known supply of natural gas. The report indicates that 75 percent of global petroleum has been discovered, including cumulative production, remaining reserves, field growth, and undiscovered resources; 66 percent of natural gas has also been identified.

The estimates were based on geological and statistical analyses made by a team of more than forty geoscientists and support staff. Overall global assessments of petroleum were larger than those made for the last study in 1994 but with some regional variation. Specifically, the Middle East, the northeast Greenland shelf, the Niger and Congo delta areas, and western Siberia and the Caspian Sea region were estimated to have the largest supplies of undiscovered conventional oil. The 2000 estimates, in comparison with those made in 1994, lowered the volumes of resources in offshore areas of China and Mexico and increased those in the Middle East and offshore areas of South America and Africa. Some areas with no production history were highlighted for their oil resources potential, including northeast Greenland and offshore Suriname.

The estimates for natural gas declined worldwide with the Arctic fringes of the former Soviet Union and parts of China and Alberta Canada being particularly lower. The areas having the largest amounts of conventional natural gas included the western Siberia basin, shelves of the Barents and Kara Seas, offshore western Norway, and the Middle East. Natural gas liquids estimates increased dramatically, largely because of improved data analysis techniques.

Better Imaging, Drilling, and Recovery

New diagnostic and imaging technologies, better drilling, and enhanced recovery have all improved the amount of petroleum that can be extracted worldwide. These have increased production of petroleum and generally lessened environmental impact.

New two- and three-dimensional seismic technologies have led to the discovery of more new fields, frequently off shore. Reflected sound waves reveal features of the structure of the crust under the sea floor. Modern computer applications then use millions of bits of data to create three-dimensional images or maps that visualize the geologic structures, facilitating the identification of possible petroleum reserves. Sometimes, the petroleum found may be from a completely new source. Other times, drilling in existing wells may reveal new, previously unknown pools and reservoirs.

Technological change has allowed more efficient extraction of oil resources, enabling producers to tap previously unusable sources, decrease costs, and reduce environmental impact. For example, improved oil field technology enables producers to use fewer wells to extract oil. One domestic well can extract as much oil as four did in 1985. In addition, new drilling techniques, including slimhole drilling and modular drilling rigs, reduce the area used for drilling. As a result, less drilling waste is created and less land is disturbed, leading to a decrease in environmental degradation. Microdrilling systems are being developed and tested by the Los Alamos laboratory of the Department of Energy along with several major oil companies. These well bits are 1 inch (2.5 centimeters) in diameter compared with the current 6 inches (15 centimeters) or more and reduce the space used for drilling by 80 percent and the cost by 90 percent.

Oil can be extracted from deeper locations not only on land but also off shore. In the 1970's, extracting oil found in 200 or 300 feet (60 to 90 meters) of water was thought to be too difficult and dangerous. By 2000, the average depth of a well in the Gulf of Mexico was 475 feet (148 meters). A well of more than 7,000 feet (2,134 meters) has been commercially successful; 10,000 feet (3,048 meters) is considered within the realm of possibility in the near future.

In May, 2000, several major petrochemical corporations entered a joint project designed to combine enhanced oil recovery with the reduction of greenhouse gases.

Less oil is being left behind in the crust as enhanced recovery techniques have been improved. When oil is first pumped out, with primary recovery, a good proportion is left behind. To extract the remaining oil, producers implement secondary recovery techniques, in which water is injected into the well to force out more of the heavy oil. Since significant amounts of heavy oil are still left behind, frequently tertiary or enhanced recovery techniques are employed. Steam or carbon dioxide is injected into the well to facilitate the extraction of remaining petroleum. Steam flooding reduces the viscosity of the oil by increasing its temperature, and the oil can flow more easily out of the reservoir and the well. The increased oil that is extracted more than makes up for the extra costs of the process. Carbon dioxide is used to enhance recovery in around seventy oil fields around the world.

In May, 2000, several major petrochemical corporations entered a joint project designed to combine enhanced recovery with the reduction of greenhouse gases. Carbon dioxide will be captured from power generation plants and industrial sources and then stored in geologic formations below the surface of the earth. The companies plan to use their experience in using carbon dioxide for enhanced recovery in developing the appropriate capture and storage techniques. This approach is expected to increase the amount of oil removed while at the same time lessening the impact of human-produced carbon dioxide.

Oscillating Oceans and Changing Climates

➤ *The Jet Propulsion Laboratory and the National Aeronautics and Space Administration report a two-year cooling of the eastern tropical Pacific Ocean that might indicate a new phase in sea-surface temperature oscillations.*

Predicting weather and climate patterns involves understanding not only atmospheric processes but also their interactions with the oceans on a global scale. These relationships are called teleconnections. Although El Niño and La Niña are two of the most well-known teleconnections, other longer term and larger area associations exist. In 2000, a clearer understanding of the interactions of the atmosphere with the tropical Pacific Ocean and the North Atlantic Ocean emerged through new studies and continued data collection and analysis by atmospheric and ocean scientists associated with a variety of institutions. Much of the recent research at the Jet Propulsion Laboratory, the National Climate Prediction Center, and Lamont Doherty Observatory of Columbia University concentrated on influences on weather in the United States and Europe.

Pacific Ocean

Changes in the sea-surface temperatures of the tropical Pacific Ocean appear to be long term, lasting two to three decades and then reversing for another two to three decades. Over these multiple-decade periods, the eastern and western sectors of the Pacific Oceans alternate as the warmest parts.

The Jet Propulsion Laboratory and National Aeronautics and Space Administration (NASA) reported in January, 2000, that measurements of ocean temperatures by the Topex/Poseidon earth satellite showed that the eastern part of the Pacific Ocean has been cooler than the rest of the ocean for the last two years. This distribution is similar to the period from the mid-1940's to the mid-1970's. From the mid-1970's to the end of the 1990's, the eastern tropical Pacific had been warmer than the rest of the ocean. However, it is not clear whether the change has really occurred; data will need to be examined for five to ten more years.

Whether or not a new phase has begun, the Pacific Decadal Oscillation of temperatures appears to be related to El Niño and La Niña and to weather conditions in the

United States and around the world. El Niños are stronger with a warmer eastern Pacific, as occurred from the mid-1970's to the end of the 1990's. The emerging pattern is associated with La Niña and, therefore, causing new winter atmospheric conditions in North America. Specifically, the eastern United States in early 2000 had a relatively warm winter, with occasional cold snaps. The Northwest was especially stormy, and the South was drier than normal. However, predictions of weather in the South are complicated further by an expected increase in the numbers of hurricanes.

The impact of the Pacific Decadal Oscillation appears to extend beyond the weather of the United States in that it seems to have caused global temperatures to cool. Global temperatures, which had been rising through much of the early twentieth century, leveled off during the previous phase of the oscillation, from the mid-1940's to the mid-1970's. Since the mid-1970's, the Pacific Decadal Oscillation has demonstrated a warmer western Pacific, possibly contributing to warmer global temperatures. Because the Pacific Decadal Oscillation influences weather worldwide, it cannot be consid-

A three-year drought lowered the level of this South Carolina lake, making it impossible for Al Jones to run his landing and campground facility. Scientists hope to understand the effects of the North Atlantic Oscillation and the Pacific Decadal Oscillation on the weather so as to mitigate any negative economic consequences. (AP/ Wide World Photos)

ered in isolation. One possibility is that the Pacific Decadal Oscillation and North Atlantic Oscillation are interconnected.

North Atlantic Oscillation

The North Atlantic Oscillation positively correlates variations in atmospheric pressure in the North Atlantic with temperature and precipitation in the eastern United States, Europe, Greenland, and the Middle East. Its impact is impressive. The North Atlantic Oscillation refers specifically to changes in atmospheric pressure over Iceland and the area over the Azores Islands near Africa. Iceland consistently has low pressure, while the Azores consistently have high pressure. The intensity of these two pressure centers varies over the year but are inversely related to each other. If the pressure is higher than average near the Azores, then it is usually lower than average near Iceland; this situation is called the positive phase of the North Atlantic Oscillation. In its negative phase, the pressure is higher than average near Iceland and lower than average near the Azores. The shift in pressure causes changes in the movements and location of the polar front jet stream, westerly winds, and winter storms in Europe and over the ocean. Usually one phase dominates for years, with only brief interruptions by the opposite phrase.

Although the North Atlantic Oscillation exists throughout the year, it is strongest during the winter months. When the winter positive phase is strong, the eastern United States and northern Europe have above-normal temperatures; Greenland and frequently southern Europe and the Middle East and North Africa have below-normal temperatures. Northern Europe has above-average winter precipitation, associated with greater numbers and more intense storms in the northeastern Atlantic and Southern Europe, and the Middle East has below average precipitation. During the negative phase, the opposite conditions exist.

From the mid-1950's through the winter of 1979-1980, the North Atlantic Oscillation negative phase was dominant; then, the North Atlantic Oscillation abruptly switched to a mostly positive phase (with some intervening negative phases for a few months) from the winter of 1979-1980 for about fifteen years to the winter of 1994-1995. Since then, no clear-cut pattern has emerged as neither phase has been dominant for more than one winter. However, some atmospheric scientists and oceanographers think that the North Atlantic Oscillation positive phase may now be becoming more pronounced. Several groups including the Atlantic Climate Variability Experiment (ACVE) and Climate Variability of the Atlantic Sector (CLIVAR), are intensifying their study.

The Human Connection

Atmospheric scientists and climatologists are trying to identify the interconnections of the North Atlantic Oscillation with long-term and short-term weather conditions. Being able to identify the specific trend is necessary because the positive and negative North Atlantic Oscillation affect not only the weather but also human activities and economies. The impact is wide-ranging. The negative phase could cause such local effects as lower levels of snowfall in the Alps, which damage the ski and tourist industry. The positive phase may be the cause behind the growth of glaciers in Scandinavia, a sharp contrast to the melting of glaciers in most of the rest of the Northern Hemisphere. Fluctuations in the North Atlantic Oscillation also appear to affect natural fisheries and the production of zooplankton.

The North Atlantic Oscillation refers specifically to changes in atmospheric pressure over Iceland and the area over the Azores Islands near Africa.

The North Atlantic Oscillation is arguably largely a natural phenomenon. However, human activities may greatly influence it, and a better understanding of its pro-

cesses will be useful. Because the North Atlantic Oscillation parallels changes in surface temperatures in the Northern Hemisphere, the apparent global warming might be more a result of fluctuations in the North Atlantic Oscillation than of the increase in greenhouse gases. No one knows the true influence of anthropogenic (human-caused) activities on natural fluctuations in the atmosphere and oceans, so the variability in the North Atlantic Oscillation might be caused by human activities, and the change in North Atlantic Oscillation in turn might influence global temperature change.

Global Connections and the Future

Although the teleconnections of El Niño and La Niña are strongly suspected to be wide-ranging, influencing weather far beyond their origin in the eastern Pacific, those of the North Atlantic Oscillation and the Pacific Decadal Oscillation are proving to be even larger in scale and persisting over longer periods of time. Atmospheric and ocean scientists generally agree that the North Atlantic Oscillation and Pacific Decadal Oscillation should not be considered in isolation as they probably reinforce each other as well as teleconnections associated with the other oceans. Thus, ocean and atmospheric scientists continue to strive to understand these oscillations in order to improve predictions of weather and climate patterns worldwide.

Summer 2000
From Too Much to Not Enough Heat

➤ *In the United States, the Southeast, the Plains, and parts of the West experienced drought and extreme heat while the Northeast was cool and wet.*

Summer 2000 in the United States was one of extremes. Large parts of the Southeast, the Plains, and the intermountain West suffered severe drought and, in some cases, record high temperatures, while the Northeast experienced one of the coolest and wettest summers on record. Crops withered in the drought-stricken areas, and beachfront businesses in the East suffered losses as rainy days chased sun worshipers away.

Drought, defined by the National Oceanic and Atmospheric Administration (NOAA) as a "persistent and abnormal moisture deficiency having adverse impacts on vegetation, animals, or people," is not unusual; most climates have periods of little rainfall. However, the consequences can be severe, killing people, wildlife, and livestock and damaging crops. Drought is not as dramatic as floods, but it does cause millions of dollars worth of damage each year worldwide. For large parts of the United States, the damage was significant in the summer of 2000. Similarly, excessive rainfall is not unusual, but heavy rains in the summer of 2000 caused crop damage and devastating floods in parts of the Northeast. For much of the Northeast, the excessive rainfall was more a nuisance than a danger, creating a summer dominated by cold dreary days.

Drought is defined as a persistent and abnormal moisture deficiency having adverse impacts on vegetation, animals, or people.

The drought of 2000 was more severe than the flooding in its economic and environmental impacts, which included devastating wildfires in July and August. In some parts of the country, the drought started in 1999, in others in 1998. Its specific causes are not definitively known, but La Niña is one probable contributing factor. La Niña is the name given to climate conditions characterized by colder than average water temperatures in the eastern South Pacific; it is the counterpart to the better known El Niño. With La Niña, large-scale atmospheric-oceanic interconnections frequently appear to result in colder and wetter winters in the Northwest, warmer winters in the Southeast, and dry weather in the Southwest. In 2000, these predicted characteristics were closely but not precisely correlated with the actual weather conditions.

What Happened Where

Rain occurs with low atmospheric pressure. Air rises and cools, reaching its dew point temperature, at which the water vapor is converted to water droplets, producing clouds and frequently rain. In contrast, with high atmospheric pressure, air sinks or subsides, and warms. As it warms, it can hold more water vapor, which means that the likelihood of clouds, and thus rain, is small.

During the summer of 2000, defined by NOAA as June through August, the locations of large-scale pressure systems varied a bit but in general resulted in dry and warm conditions over the intermountain West, much of the Great Plains, and the Southeast. Variations occurred, at times intensifying drought and contributing to hotter temperatures, and other times leading to precipitation and cooler temperatures. Specifically, in June, the Southeast was dominated by the Bermuda High, which normally intensifies in the Atlantic Ocean, pumping moist and warm weather to the East Coast. However, this year, the Bermuda High extended into the Southeast, contributing to warmer than normal temperatures and drier conditions. At the same time, the Southwest was dominated by low pressure. Therefore, moist air masses from the Gulf of Mexico were able to penetrate deeper inland, resulting in more rainfall than usual in the Southwest as well as in the mid-Mississippi and Ohio Valleys and the mid-Atlantic states.

By July, high pressure, manifested as an upper-level ridge, dominated the western two-thirds of the country. At the same time, low pressure, manifested as an upper-level trough, existed from the mid-Mississippi Valley into the Northeast. The ridge and the trough were related to motions of the polar front jet stream (a band of high-altitude winds separating cold polar air masses and warm tropical air masses). One might imagine the jet stream meandering across the continent moving north and south, with one northward bend and one southward bend during this summer. The northward bend (the ridge) brought warm air from the south, and the southward bend (the trough) brought cold air from the north. Thus, much of the western two-thirds had warm dry air, while much of the northeast had moist cool air. The August pattern was similar but with a new trough over the Pacific coast. Thus, the upper-level ridge moved a bit to the east, dominating the central Rockies to the Mississippi River, and the trough was still over the eastern third of the country.

Measurement Indices

The United States Geological Survey uses several indices to measure drought severity. The Palmer Drought Severity Index, developed by Wayne Palmer in the 1960's, is considered the most effective indicator of long-term drought. It combines temperature and precipitation to indicate comparative dryness. It considers water supply (precipitation), water loss (runoff, or water flowing on the surface and through the soil), and water use (evapotranspiration, or the evaporation of water from land and plants). However, it does not take into account water potentially available from snowmelt.

The Palmer Drought Severity Index, developed by Wayne Palmer in the 1960's, is considered the most effective indicator of long-term drought.

The index indicated that by the end of summer 2000, severe to long-term drought existed almost continuously from the Southeast to the Southern Plains, through the Rocky Mountains into the Far West, as well as in the Central Plains. In contrast, the Northern Plains, parts of the Great Lakes, and into the Northeast, covering about 5 percent of the country, were unusually wet.

The Crop Moisture Index, another invention of Wayne Palmer, measures short-term changes in moisture content more effectively than the Palmer Index. It is especially relevant to farmers because it measures crop need versus available water in a 5-foot soil profile. By the end of August, it shadowed the long-term drought index, indicating excessively or severely dry conditions for south-central states extending into Oklahoma, parts of Nebraska and Kansas, and into Montana.

The Palmer Z Index indicates short-term drought and wetness. The Southern Plains and southeastern states during July and August experienced severe short-term drought, while much of the western United States was persistently dry.

The Standardized Precipitation Index (SPI) is a probability index that considers only precipitation and not water demand and loss as the Palmer Z Index does. The spatial patterns of the SPI for August, 2000, indicated locations, intensities, and durations of the several drought regions.

Droughtlike conditions in Montana in June dried and cracked the surface of Deadman's Basin, a holding reservoir, near Harlowton. (AP/Wide World Photos)

The Southern Plains drought was short term (July to August); the Southeast and northern Rockies drought was longer by at least several months. The Southwest and Central Plains regions, despite intermittent rains, exhibited severe drought conditions for six to twelve months.

The Impact

The Deep South suffered its driest twelve-month period (from September, 1999, to August, 2000) since 1895, when records were first kept. As a result, millions of trees died in Florida, and tens of cities in Georgia were short of water. Losses to agriculture were estimated at around $700 million by the University of Georgia.

Agricultural losses in the southern Great Plains were around $600 million according to the Texas Agricultural Extension Service. The months of July and August were the driest on record in Texas. No rain fell in the Dallas-Fort Worth area for the entire two months and into September. Although a drenching rain of around 2 inches (5 centimeters) fell in Dallas in early September, it did not fall at the airport where the rain gauge is located and data are collected. Therefore, for the

record at least, it did not end the drought. Throughout Texas and the southern Great Plains, new records were set for the number of days the temperature surpassed 100 degrees Fahrenheit (37.8 degrees Celsius).

The heat and drought resulted in dozens of deaths and billions of dollars in crop and livestock damage. Corn, soybeans, sorghum, and other autumn crops failed in many areas extending from Nebraska to Texas. Farmers lost millions of dollars, and states asked the federal government for assistance. In some parts of the Plains, production of grains grown as cattle feed was much lower than usual at the same time that pasture grass was drying up. Scores of people in Texas and the Deep South were killed by the triple-digit heat.

Those whose health, homes, and livelihoods were most affected by the drought and dreariness will not forget the summer of 2000.

In contrast, the Northeast had one of the coolest and wettest summers ever. New York City's July temperatures were the coolest since 1914. Rain was plentiful all summer, with exceptional deluges. About 10 inches (25 centimeters) of rain fell on Sparta, New Jersey, in a half-day on August 12, causing street closures, evacuations, damaged residences, and mudslides and sending the side of a small mountain sliding onto a major road.

Despite the drought and heat, the Department of Agriculture reported that record crops of soybeans and corn were to be harvested in 2000, largely resulting from timely rains and excellent growing conditions in the Midwest. Those whose health, homes, and livelihoods were most affected by the drought and dreariness will not forget the summer, and scientists and government officials are likely to re-examine the events as they try to lessen the impact of future floods and drought. However, most people suffered no negative consequences and so will probably gradually forget the summer's fires, weather-related deaths, and broken temperature records. This is a peculiar characteristic of the weather: People talk about it but frequently remember only bits and pieces.

Forest Fires Devastate the West

➤ *More than eighty thousand wildfires burn nearly twice as large an area as normal in the western United States.*

Millions of acres of forests were burned, thousands of homes and buildings damaged or destroyed, countless businesses disrupted, and untold lives and vacations disturbed as the worst forest fires since 1910 swept through the West in the summer of 2000. An unusually severe drought combined with high temperatures provided a major prerequisite, as logging practices may have. Strained human firefighting resources combined with strong winds contributed to the fires' ability to spread, often uncontrollably, once they had started. The fires, spectacular in themselves, were more of a problem than they might have been decades ago because of the many new homes that have been built over the years in or near forested areas.

Wildland fires are not unusual in the United States. On average, more than 66,000 fires occur each year across the United States, burning more than 3.1 million acres. Therefore, fires are not a surprise, and federal, state, and local governments have plans to deal with them. However, in 2000, the numbers were overwhelming, with 84,960 fires by October 21, burning 6.9 million acres, more than double the average number of acres. At times fifty to sixty wildfires were burning simultaneously. In addition, the fires started with greater intensity early in the season and lasted much longer than usual.

Multiplying Fires, Limited Resources

The ultimate cause of the drought and high temperatures that contributed to the fires was probably La Niña, the name given to weather conditions that produce colder than normal temperatures in the eastern Pacific Ocean, affecting temperatures across the globe. In the United States, significant areas of the West and South had been experiencing drought for more than a year, resulting in vegetation that was drier than usual. The potential for fire was high, and the fires started in mid-February. Within a month, thirteen states reported fires, with Florida being especially hard hit. In May, the fire in Los Alamos, New

Mexico, which started from a controlled burn, eventually destroyed more than 230 homes, attracted national attention, and raised concern, later aroused again and again by widespread fires across the West.

The tremendous number of fires strained firefighting resources. In July, more than 1,000 U.S. Marines and U.S. Army troops came to the assistance of the more than 20,000 civilian firefighters. However, even more assistance was needed, and firefighters came from Canada and, for the first time, from Mexico. Inmates from prisons were also enlisted. By early August, more than 30,000 firefighters were involved.

In August, the fires were so overwhelming that some people perceived government officials to be nearly powerless against the forces of nature. Lightning strikes started fires, and strong winds carried them for miles. According to *The New York Times*, on August 4, 2000, about 75,000 lighting strikes sparked more than 400 new forest fires in Montana. The winds not only spread the fires but also consolidated them in some cases, making them even more dangerous. In Bitterroot Valley in Montana, the fires formed a 20- to 40-foot wall of fire. The federal government

A plane drops slurry and water on a hot spot in Mesa Verde National Park, Colorado, in July. The fire burned for nine days. (AP/Wide World Photos)

was criticized for not deploying even more firefighters, but Bruce Babbitt, secretary of the interior, said more firefighters might prove counterproductive, as they would have to be trained properly before they could start.

The Mesa Verde Fires

Among the year's many fires, those in the Mesa Verde National Park stand out because they provide insights as to how and why the fires of 2000 were so devastating. Mesa Verde National Park in Colorado is not a stranger to fire. However, the fires of 2000 were the most destructive in the history of the park. The first major fire started on July 20 from a lightning strike on private land, the Bircher farm, near the entrance to the park. It burned 1,000 acres the first day and 5,000 acres the second day before it became out of control. Fed by dry forests of Utah juniper, piñon, and Gambel oak, the flames were so high, up to 300 feet, that they hampered firefighters' efforts to get near enough to control them. In addition, the flames were so hot that fire retardants did not work. Thunderstorms, instead of stopping the fire with rains, actually helped spread the fire, as the strong downdrafts of air carried the fire to cliffs, rimrock, and inaccessible slopes. After five days, the fire was slowed when it moved to an area of sparse vegetation caused by earlier fires, and firefighters were able to attack it from the air. The fire was contained on July 29, nine days after it had started and after it had burned an area eight miles long and four miles wide.

Considering the intensity and size of the fire, damage to structures was relatively minimal. However, areas never surveyed for archaeological sites but probably containing them were within the burn area. Because the large areas of evergreen woodlands that were burned might take centuries to return, erosion is a concern, and with it, the possibility of the exposure and destruction of archaeological sites.

On August 4, Mesa Verde was struck by another devastating fire. This one, called the Pony fire, was caused by a lightning strike two days earlier in Pony Canyon on the Ute Mountain Reservation. Despite early thoughts to the contrary, the fire could not be contained when the wind shifted, moving the fire toward the park. This fire burned in steep, rugged inaccessible canyons and dry mesas, which combined with the unpredictable winds and low humidity allowed the flames to flourish. The fire was fought by air tankers and helicopters dropping water and fire retardant as well as by removing dry vegetation that could serve as

fuel in advance of the fire. The fire was contained on August 11.

Compared with the Bircher fire, the Pony fire caused more damage to park buildings, including day-use facilities and shelters that protected archaeological pueblos and pithouses. Although one popular site, Chapin Mesa, was closed after the fire to ensure visitor safety, more famous structures, such as Long House and Step House, suffered little damage. However, the fire destroyed historic Ute structures in the Ute Mountain Reservation and probably structures and archaeological remains that had not yet been discovered.

The impact of both these fires was substantial. The park had to be closed and people evacuated, which disrupted local economies and upset vacation plans. The Department of the Interior's Burned Area Emergency Rehabilitation (BAER) team is working to lessen the damages, including implementing emergency measures to minimize soil erosion and to protect rare species from the damage done to their habitat. Overgrown vegetation will be reduced to minimize future fire damage. Efforts will be made to prevent or inhibit invasion by nonnative plant species.

Controversies Exposed by the Fires

Reactions to the fires included cries of government mismanagement, debates about the amount of logging necessary to avoid large fires, and concerns about problems arising from population growth in wilderness and forested areas. Reduced logging, resulting from environmental protection, was cited as one reason for the great wildfires. Because fewer trees had been cut, more fuel was available to burn. However, in September, a congressionally appointed bipartisan research group reported that heavy logging occurring over many years may have actually contributed to the fires because logging removes large trees and leaves smaller trees, heavy brush, twigs, and needles. The report analyzed fires over the past two decades and found that states with the greatest logging had the most fires. Because the report was released a few weeks before the presidential elections, it was used to some extent politically to defend or attack the forest management policy of the current administration, depending on the party affiliation of the commentator.

The intense fires were not a surprise to some experts, including Wally Covington, director of the Ecological Restoration Institute at Northern Arizona University, who spoke before the House of Representatives Subcommittee on Forests and Forest Health. The problems included excessive amounts of forest fuels combined with increased numbers of people living in or near forested areas. Forest stands that have been protected from fires for decades are less diverse ecologically. They have less plant, animal, and insect diversity and are less resistant to insect attack. He recommended using prescribed fires to lessen forest floor fuels, combined with thinning trees and breaking up interconnected canopy fuels to lessen the potential of crown fires.

In September, President Bill Clinton and six governors from Western states devastated by the fires agreed to a new forest management policy. The policy called for substantially more federal funds for fire management. When the fires raged in the summer, the federal government was harshly criticized as a contributor to the fires because it had not provided sufficient funds for prevention and protection. The $2.8 billion in the agreement is more than twice the $1.2 billion that had been designated for use in wildland fire programs, but around $800 million will be used to pay back emergency firefighting funds used in 2000. The agreement also called for more cooperation among federal, state, and local governments. The policy does not advocate commercial logging or the building of new roads to reduce fire risk, and the secretary of the interior stated that the policy would not be used as part of the debate on logging and other forest economic issues. Instead, the policy was designed to improve forest restoration and make firefighting programs less reactive by focusing on fire prevention measures. Suggested actions include more controlled burns and greater thinning of forest materials. The policy report is not a directive from the federal government. It calls for cooperation among all levels of government and emphasizes the importance of local input. Although the policy report appeared within a few months of a national election, it gained support from both Republican governors and a Democratic president.

The federal government was harshly criticized as a contributor to the fires because it had not provided sufficient funds for prevention and protection.

There is no question but that the fires of 2000 were disastrous. No one knows for sure how much damage could have been prevented. Most experts agree that fire is part of the natural cycle of forests and grasslands, and human beings must learn to live within that natural cycle. The year 2000 may be the beginning of new understanding and appropriate actions.

Wildfires and Los Alamos

➤ *Wildfires threaten government research labs and structures containing radioactive materials at Los Alamos and scorch millions of acres in Florida, Colorado, New Mexico, Texas, Washington, Montana, and Northern California.*

Wildfires are a normal occurrence over much of the United States, with hundreds of thousands of acres burned by tens of thousands of fires each year. The year 2000 was anomalous, however, with millions of acres burned across the country. Florida, Colorado, New Mexico, Texas, Washington, Montana, and Northern California were especially hard hit. One fire in particular captured the attention of the country and the world in May of 2000 because it resulted largely from human error and burned the area around Los Alamos, the laboratory that developed the atomic bomb in the 1940's and continues to be the major weapons-producing and research laboratory in the United States.

A Controlled Burn Goes Wild

The fire started as a controlled burn, not in Los Alamos but in the Bandelier National Monument in New Mexico. The object was to rid an area of around 1,000 acres (about 400 hectares) of excess dry wood, thereby lowering the likelihood of a disastrous fire. Instead, the fire got out of control and moved toward Los Alamos National Laboratory and the city of Los Alamos. A combination of human errors and mismanagement together with weather conditions led to the evacuation of 18,000 residents, the destruction of 235 homes, and the burning of tens of thousands of acres. As the fire spread, fears rose about the possible consequences of the destruction of the structures housing the laboratory and radioactive wastes, but no releases of radioactivity as a result of the fire were reported. The National Park Service accepted blame, and the United States government agreed to compensate the homeowners for their monetary losses. The controlled burn policy was temporarily suspended for all federal agencies in the western United States.

Setting prescribed fires is a procedure accepted and used throughout the United States as a way to burn off excess fuel and thus avoid naturally large fires. Such fires must be set in the absence of strong winds or dry air, which

A wildfire rages half a mile from Los Alamos National Laboratory (foreground and to the left). Laboratory spokespeople assured the public that dangerous materials were adequately protected from the fire. (AP/Wide World Photos)

could spread the fire beyond the prescribed area. In addition, the burn starters must have sufficient knowledge of the terrain and surrounding vegetation to be aware of the potential of spread and should work cooperatively with other agencies.

In this case, in the evening of May 4, 2000, the burn boss, a Park Service employee, ordered around 1,000 acres of brush to be burned in what was called the Upper Frijoles Prescribed Fire. The burn boss was relatively inexperienced. The fire was planned and procedures were followed, but errors were made. Before such fires are set, the Park Service uses a complexity rating system to evaluate the potential for danger. According to the Park Service, Bandelier staff used incorrect values that led to a numeric rating of 87, a low-moderate complexity; when the correct numbers were used, the rating was 137, a moderate-high complexity. Bandelier, however, had acted on the lower rating.

The forecasted weather conditions were not unfavorable for the containment. Specifically, the Haines Index, which measures change in temperature with altitude and moisture content, indicated that the fire would be safe to set and would be controllable once burning. However, a wind-speed forecast was not available or used for the three

to five days following the setting of the fire. Although winds in the region are not too strong during much of the year, they can be strong and sustained during the spring. Therefore, the fire starters should have been more cognizant of the importance of the winds, which contributed significantly to the spread of this fire.

Rapid Changes, Inadequate Responses

Early on May 5 at 1:00 A.M., as a result of wind changes, the fire spread through "slopover," when the fire edge crossed the contained fire line, and "spotting," when sparks and embers were carried by the wind, starting new fires beyond the boundaries of the main fire. Although the burn boss informed Santa Fe Zone Dispatch, which coordinates state and federal firefighting resources, about the spread and asked for additional firefighters and equipment, no additional firefighting resources were sent for hours. Discussions over funding slowed the use of firefighters and equipment. By 10:30 A.M., the funding problem was resolved, and more equipment and people were sent. The fire was officially termed a wildland fire on May 5, 2000, at 1:00 P.M.

However, the National Park Service indicated that based on the review by the Investigation Team's fire behavior specialists, if the appropriate firefighting resources had been applied early on May 5, the slopover might have been contained. The prescribed burn probably would not have been converted to a wildfire and would have burned its course through forest fuels, thereby either slowing or burning out completely. Instead the forest fuels were preheated and dried by the heat of the fire, making them an ignition source and allowing the fire to gain considerable intensity two days later. The fire was contained on May 6, but in the late morning of May 7, as the wind speed increased to gusts as high as 50 miles per hour (80 kilometers per hour) from the west, the fire was out of control. It had become a crown fire, with the fire leaping from the top of one tree to another. By May 10, the fire was at its most intense and moved into Los Alamos Canyon, threatening the towns of White Rock and Los Alamos. It was then renamed the Cerro Grande Prescribed Fire. Firefighters worked for many hours and at considerable personal peril to limit destruction as much as possible. Despite their efforts, later that day the fire destroyed hundreds of homes and other buildings and burned more than 18,000 acres. It spread toward the Los Alamos National Laboratory, and although no radiation was released, several structures were damaged. The fire continued to spread and burned parts of the San Ildefonso Pueblo and Santa Clara Pueblo Reservations as well as private lands. In all, more than 1,200 fire fighters battled the fire, which was declared contained on June 6 with the total land area burned estimated at 47,650 acres (19,283 hectares).

Evaluation and Aftermath

After the Los Alamos fire, the national press covered wildfires more than it had in previous years. Therefore, evacuations and destruction of homes in areas around Denver, in New Mexico, Texas, California, and Florida, which normally would have received primarily local and regional coverage, became national stories. Wildfires became national news. The ban on controlled burns by federal agencies other than the Park Service expired on June 12; the Park Service ban remained in effect indefinitely, except for a few exemptions.

More than 18,000 residents of New Mexico were displaced, including many who lost their homes and all their possessions. The National Park Service was quick to accept blame, and the government began an active program to compensate residents. A Burned Area Emergency Rehabilitation (BAER) team was formed to assess the damage and to implement a plan to rehabilitate the area. Experts from many disciplines will work together with the latest technologies to prevent loss of life and property and minimize natural resource damage in the future.

The Interagency Fire Investigation Team, formed by Secretary of the Interior Bruce Babbit, submitted a report, the Cerro Grande Prescribed Fire Investigation Report, which listed as errors the lack of adequate planning beforehand and mishandling of events that developed as the fire was burning. Necessary weather information was not used, especially predictions of winds. The secretary of the interior also appointed an independent review board to review the interagency report. That independent review board emphasized that the existing federal wildland fire management policy is sound and that problems develop when agencies do not "live the plan."

If the appropriate firefighting resources had been applied early on May 5, the prescribed burn probably would not have been converted to a wildfire and would have burned its course through forest fuels.

Superstorms Flood Southeast Africa

➤ *Extreme weather conditions produce damaging floods.*

Devastating floods resulting from torrential rains during February and March, 2000, killed or displaced thousands of people in the countries of Mozambique, Botswana, and Zimbabwe. A respite lasted a few weeks, but by April, more rains and more flooding caused even more damage.

Extremes in weather are considered extraordinary. However, atmospheric conditions are rarely at their average levels, typically reaching above or falling below average. Therefore, one could expect that in any given year, tremendous amounts of rainfall or snowfall or very high winds or very hot or very cold temperatures might occur in various places around the world. In 2000, extreme weather conditions occurred in many locations, including Mozambique, where many died and the economy suffered. Although the meteorological events were remarkable, they were not unusual from the perspective of normal atmospheric activities. What was noteworthy was the timing and scope of the response of the world community, which revealed much about social and political attitudes.

The Rains and Their Immediate Impact

The climate of Mozambique, called tropical savanna or wet-dry tropical, is characterized by warm temperatures all year long with a few months of rain and longer periods of dry weather. Although the four seasons familiar to people of the midlatitudes do not exist, some variation in temperature does occur. The rainy months occur when temperatures are relatively warm with a low-pressure system, characterized by rising air, dominating the area. Because rising air cools and cooler air cannot hold as much water vapor as warmer air, the water vapor will condense to form clouds and frequently rain. Therefore, it is not unusual for February through April to be rainy. More rains also accompany cyclones that migrate into the area.

The timing and scope of the response of the world community revealed much about social and political attitudes.

Starting in early February, 2000, torrential rains lasting over a week resulted in devastating floods as rivers overflowed their channels. The flooding exceeded any that had occurred in the past fifty years. On February 22, the flooding became much worse as Cyclone Eline struck Mozambique and made landfall near the mouth of the Limpopo River in the southern part of the country. Cyclones in the southern hemisphere rotate in a clockwise direction, and Eline's winds and rain produced a storm surge that inhibited the normal draining of waters from the Limpopo River into the Indian Ocean. Water that would normally have flowed out to sea instead flooded the most densely populated area of Mozambique. The number of homeless and dead increased; fortunately, early estimates of the dead were higher than those confirmed a few months later.

Then, in early March, tropical Cyclone Gloria struck. The central part of Mozambique was most seriously flooded, with the Limpopo and Save Rivers reaching extremely high levels. The waters affecting Mozambique did not come from only the skies above but also from rivers of and flooding in neighboring countries.

A Mozambican woman carries her things as she wades across a flooded field near Palmeira in March. (AP/Wide World Photos)

These countries, including South Africa, Botswana, Madagascar, Zimbabwe, and Zambia, suffered devastating losses of their own. Their losses should not be minimized by any means. However, the physical geography of Mozambique intensified the effects of the floods. More than half of Mozambique is coastal lowlands, thus hastening flooding by rain and overflowing rivers. In addition, rivers from neighboring countries run through and drain in Mozambique, sending even more water.

Rescue Efforts

The first floods in February damaged or destroyed bridges, roads, and rails; thousands were homeless and feared dead. The United Nations had difficulty providing flood victims with food and water. The inundated roads forced food and supplies to be transported by helicopters, but their capacity was only about one-third that of trucks. After the rains stopped in the middle of March and the floodwaters receded, supplies could be transported by trucks. Unfortunately, the truck traffic was severely limited because roads had been washed out and bridges destroyed.

The receding floodwaters revealed hundreds of corpses. About seven hundred people were killed by the flooding, and hundreds of thousands were displaced. The Mozambican government and United Nations aid workers cautioned some displaced villagers about returning too early to their homes, but they were concerned about the crops that had been destroyed.

In an effort to alleviate the loss, aid workers searched for seeds to replace the 10,000 hectares (24,000 acres) of rice, corn, and bean crops that had been lost from farms in central and southern sectors of the country.

By the end of March to the beginning of April, the country had some success in recovery. Food and other aid were distributed, and residents returned to their homes to start the process of rebuilding.

In mid-April, another Cyclone Hudah threatened Mozambique. This storm, however, lost much of its intensity before it reached Mozambique because it traveled more over land than had Eline and Gloria and therefore was not able to gain energy from the waters as had its predecessors.

Political Ramifications and Concerns

Land mines, remnants of the sixteen-year civil war that ended in 1994, were loosened from their sites by floodwaters. Despite six years of efforts by the United Nations to remove them, thousands remained before the floods. The waters swept the mines away from generally known locations to wherever the floodwaters flowed or even beyond. Replanting crops, repairing electrical connections, and other recovery operations were all severely hampered by the possibility of being maimed or killed by the mines.

The United States did not immediately offer help to Mozambique. The distance may have been seen as too formidable for transport of rescue equipment; perhaps the intensity of damage and death was not totally evident, or the location in Africa may have lessened the importance of the plight of thousands. Only after millions around the world saw television images of dramatic rescues on television were significant amounts of aid sent. The media provided less coverage of the Mozambicans in the refugee camps. These camps were unsanitary, and outbreaks of cholera, diarrhea, and malaria contributed to the misery of the thousands of displaced people.

The United Nations provided consistent aid. It rescued people and brought in food and other supplies. Workers also recognized that to ensure that the people could feed themselves after the devastating losses, crops had to be planted. The workers set out to provide the appropriate assistance, seeds, and equipment. The food program continued to provide aid.

The Legacy of Mount St. Helens
An Explosion of Knowledge

➤ *Data gained from Washington eruption leads to changes in the theory of ecological succession and enables scientists to better determine when Mount Mayon in the Philippines and Mount Usu and Oyama in Japan will erupt, allowing for timely evacuations.*

On May 18, 1980, Mount St. Helens exploded, killing at least fifty-seven people, reshaping the surrounding landscape, affecting weather, and destroying much vegetation and wildlife. The explosion was expected but not predicted precisely. Mount St. Helens was the first volcano to explode in the contiguous United States since 1921, when Mount Lassen in California erupted. The people of the United States are used to natural disasters. Hurricanes, tornadoes, floods, blizzards, and even earthquakes are frequent occurrences. However, although Hawaii and Alaska have volcanoes, volcanoes are rare in the lower forty-eight states, and eruptions infrequent. Mount St. Helens provided a spectacular show, blowing off one of its sides, emitting hot, pressurized gases into the atmosphere, and turning day into night with ash and other matter.

Research and Protection

The eruption of Mount St. Helens fueled a tremendous growth in the knowledge of volcanoes, and during the twenty years since its eruption, geologists, volcanologists, and other scientists have expanded that knowledge. Mount St. Helens produced pyroclastic flows and emissions of superheated gasses and rocks. The heated materials melted the glaciers on the mountain, producing lahars, or mudflows. The mudflows picked up material on their way downhill, knocking down trees, destroying bridges, causing floods, and changing stream-channel patterns.

Because the events of Mount St. Helens were observed, the resulting landscape features could be used to explain the origins of landscape features in other locations. For example, the north flank of the volcano collapsed, unleashing a lateral blast. This flank collapse led to a debris avalanche deposit that formed a terrain called hummocky, described as rolling to flat topography interspersed with rounded rocks. Therefore, the hummocky terrain around Mount Shasta could be attributed to a similar flank collapse around 300,000 years earlier.

Trees toppled when Mount St. Helens erupted on May 18, 1980, frame the north side of the mountain. (AP/Wide World Photos)

After the 1980 eruption, a number of governments developed cooperative disaster plans and monitoring agencies. The United States created the U.S. Geological Survey Volcano Hazards Program and the Volcano Disaster Assistance Program (VDAP). VDAP was formed by the U.S. Geological Survey and the U.S. Office of Foreign Disaster Assistance in order to respond to volcanic crises around the world to save lives and preserve property. It was formed a year after the November, 1985, eruption of Nevado del Ruiz in Colombia resulted in 23,000 deaths, largely from lahars. Around the world, sophisticated equipment monitors volcanoes, and volcanologists are better able, although not perfectly able, to predict the timing and types of eruption.

Impressive Eruptions Without Loss of Human Life

Largely because of increased governmental efforts, although volcanoes erupted in numerous places around the world in 2000, producing mudflows and pyroclastic explosions, evacuations minimized loss of life. For example, in February, Mount Mayon in the Philippines spouted streams of lava hundreds of feet high; it emitted ash plumes and car-sized boulders. Tremors shook Legazpi, the provincial capital, located around 7 miles (11 kilometers) from the volcano. Thousands of people evacuated their homes through government assistance, although some farmers stayed behind to care for their animals.

In Japan, in April, the National Coordination Commit-

tee of Volcanic Eruption Prediction predicted that Mount Usu in Hokkaido would erupt. Earthquakes (numbering around 1,629) shook the volcano, and new faults were discovered. About 10,000 people living in the city and neighboring towns were evacuated. During the next few weeks, mudflows destroyed homes; underground magma flows cracked the surface, destroying roads and buildings; and boulders flying from the cracks damaged buildings. Yet, because of the government's quick action, no one was killed. Similarly, many lives were saved by the evacuation of residents before Oyama on Miyake Island erupted in July, spewing ash and rocks 3,200 feet (975 meters) into the air. This eruption was followed by larger eruptions in August and continued volcanic activity.

Lessons on Ecological Succession

The changes caused by Mount St. Helens in the landscape led to changes in thinking about ecological succession. The accepted theory said that when a new land mass, such as a sandy beach, a lava flow, or ash-covered area, is formed, plant life changes in a logical pattern, called a sere. First come the hardy plants, able to adjust to harsh environments including bright sunlight and rapid changes in soil temperature. These so-called pioneer plants usually reproduce by wind-dispersion of their seeds. Over time, they change their environment, making it more tolerable for other plants. After perhaps hundreds of years, the vegetation becomes more diverse and complex, resulting in forests in many locales.

In the twenty years since the eruption, however, this sequence has not occurred in some locations on the mountain but has occurred in others, a surprise to many. In the large parts of the blowdown zone, where trees were knocked down or broken and ash, to depths of 3 feet (nearly 1 meter), covered the lower-level species, the succession pattern occurred as expected. Herbaceous plants, which depend on the wind to disperse their seeds, grew back the year after the eruption, and other plants have appeared since then. In higher elevations of the blowdown zone, however, the species that came back were combinations of pioneers and later-successional plants.

In the pyroclastic flow zone, the traditional theory would call for the classical succession pattern on the sterile land, devoid of any pre-eruption plants. Although large sections of the pumice plain have very little vegetation, other parts now have combinations of pioneers and late-successional plants, including conifers. Because the establishment of these plants depends on various species of soil fungi and bacteria previously thought not to be present in the pumice plain, scientists are now rethinking their theories of plant growth.

Twenty Years Later

In 2000, the impact of the May 18, 1980, explosion remains strong. The plant and animal life on the mountain itself is recovering at differing rates. Knowledge of volcanoes has expanded tremendously, facilitated by new monitoring equipment placed on volcanoes around the world. Governments are able to issue evacuation orders in advance of dramatic volcanic explosions.

However, complacency is dangerous. Evacuation orders may not be sufficient to save human lives or to prevent much property destruction. Ash flows, pyroclastic explosions, and mudflows are real possibilities. Many potentially active volcanoes exist perilously close to large populated areas. For example, Mount Hood is near Portland, Oregon, and Mount Vesuvius is near Naples, Italy.

Shaking Things Up
New Approaches to Analyzing Earthquakes

➤ *Seismologists present a new theory of fault rupturing and predict a high probability of a major earthquake in Istanbul in the next thirty years, based on data from 1999 earthquakes. The discovery of the Hector Mine fault in the Mojave Desert causes scientists to rethink some ideas about earthquake predictability.*

Tens of thousands of earthquakes occur each year around the globe. The vast majority (55,000) are light or minor (4.9 or less on the Richter scale). Only about 120 are strong (6 to 6.9), about 18 major (7 to 7.9), and maybe one great (8 or higher). The year 2000 was typical in that distribution, and from a human point of view, it could be considered lucky as even the strong and major earthquakes resulted in relatively few deaths. The most disastrous occurred in Indonesia in June, killing 97 people. The earthquake was centered in the Indian Ocean around 70 miles (113 kilometers) south southwest of the city of Bengkulu on the island of Sumatra. Measuring 7.9 on the Richter scale and followed by hundreds of aftershocks, the earthquake caused extensive damage, trapping and killing people inside the many collapsed buildings.

In 1999, even more disastrous earthquakes occurred, including one in Turkey in August, measuring 7.8 on the Richter scale, killing 17,000, injuring 50,000, and leaving 600,000 homeless, and one in Taiwan in September, 7.6 on the Richter scale, killing around 2,400, injuring around 8,700, and leaving around 600,000 homeless. Another earthquake occurred in the Mojave Desert in southern California in October, measuring 7.1 on the Richter scale; it killed no one because it was in an unpopulated area. Many other significant earthquakes occurred in 1999, killing thousands of people. Geologists typically study earthquakes and their impacts with the goal of learning as much as possible about seismic events and then applying that knowledge to develop new theories to help predict earthquakes and warn the populace. What makes these particular 1999 earthquakes important is that geologists in 2000 used their data to fuel new research and to develop significant advances in the understanding of earthquakes.

A fault is said to rupture like a piece of paper with the speed of rupture related to the velocity of the movement of the tearing point.

Taiwan Earthquake Leads to a New Theory

The 1999 Chi-Chi Taiwan earthquakes and aftershocks provided data that seismologists, including Lean Teng and James Dolan from the University of Southern California and researchers from the Taiwan Central Weather Bureau and the Institute of Earth Sciences, Academia Sinica of Taiwan, used to develop a new theory of fault rupturing, which they presented at the Seismological Meeting in San Diego in April, 2000.

A fault is said to rupture like a piece of paper with the speed of rupture related to the velocity of the movement of the tearing point. The accepted theory said that the tearing of the rupture occurs in a regular pattern in space and time. However, the Chi-Chi earthquakes did not follow this pattern. The tearing points varied over time and in their location, making predictions of the rupture velocity difficult if not impossible to calculate.

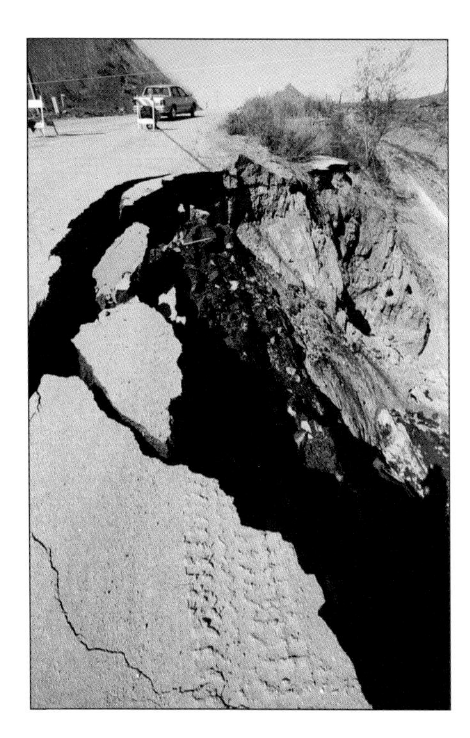

The Chi-Chi earthquakes and aftershocks were impressive for their impact not only on people and fabricated structures but also on natural features. The earthquake occurred as a fault rupture along the 100-kilometer (60-mile) Chelungpu fault and created a 6-meter (18-foot) waterfall on the Ta-Chia River. Seismologists had been aware of the geological significance of the Chelungpu fault and had placed along the fault a dense network of devices to record wave motion outward from an epicenter. These devices were placed very close to each other, about 3 to 5 kilometers (nearly 2 to 3 miles) apart. Therefore, much data were collected, providing detailed information about ground movements and their effects on buildings.

The new rupture theory makes it more difficult to evaluate hazards from earthquakes and complicates efforts to minimize or prevent death and destruction. However, if engineers and seismologists incorporate the new theory and recognize that earthquakes are not going to follow regularly predicted patterns, they can pursue other, more suitable avenues of research. The new data and proposed theory may require development of new solutions to what was thought to be a straightforward theory. The large amounts of data about earthquake wave motions should enable engineers to improve building design by accounting for the patterns of wave frequencies.

Similar Faults: North Anatolian and San Andreas

The earthquake of August, 1999, in Izmit in western Turkey occurred on the North Anatolian fault system, a

The Shihkang water reservoir in Tunchi, Taiwan, gave way in the September 24, 1999, earthquake. Data from this event helped scientists gain extensive new knowledge. (AP/Wide World Photos)

strike-slip fault, which means that the motion of the opposing plates is horizontal. The San Andreas fault system is also a strike-slip fault and is very similar in movement, shape, and length to the North Anatolian fault. The fault slip rate for San Andreas varies from around 20 to 34 millimeters (0.8 to 1.3 inches) per year, and that for the North Anatolian is around 24 millimeters (0.9 inch) with a range of plus-or-minus 4 millimeters (0.16 inch) per year. The two faults are also of similar length and are straight. Because of the similarity of the two faults, Turkish and United States geologists are working together to gather and analyze data from the Izmit earthquake to improve earthquake understanding and prediction in both countries.

The two fault zones also share a similar pattern of faulting. The San Andreas and San Gregorio faults near the Golden Gate Bridge intersect, forming a broad zone of deformation. In Turkey, the Druzce fault near Akyazi is also a zone of deformation. Their component faults share similar patterns; both the Hayward and Rodger's Creek faults under San Pablo Bay in the eastern San Francisco Bay and the fault at Sapanca Lake in Turkey are stepover faults (characterized by displacement) around 5 kilometers (3 miles) in length. Geologists are trying to see how the movements of the Izmit earthquake shifted the faults and how those movements can be applied to predicting fault segment movements near San Francisco.

In April, 2000, a team of scientists led by Thomas Parsons, a geophysicist with the United States Geological Survey, published a report using the results from the Izmit earthquake to predict that Istanbul had a high probability of a major earthquake in the next thirty years. Although some parts of the North Anatolian fault system have had recent earthquakes, including the one in Izmit in 1999, others have not. The research team reasons that the stresses created by earthquakes over the years have built up and increased the stress level farther to the west. Because two major faults, Prince's Islands and Marmara Sea, have not had earthquakes since 1509 and 1776, respectively, the researchers predict a 62 percent probability of a major earthquake in the Istanbul area in the next thirty years. The researchers suggest that their prediction can be used to prepare for the future, through the development of a safety plan to minimize hazards. They are creating a risk analysis for the San Francisco Bay Area.

A Surprise in the Mojave Desert

On October 16, 1999, a 7.1 earthquake occurred on the Lavic Lake and Bullion faults, within the San Andreas fault system. It ran right through Twentynine Palms Marines base, but no one was killed. It was a significant earthquake, and surprising, as it occurred along a known but unnamed fault, later called the Hector Mine fault. If it had occurred in Los Angeles, it probably would have been disastrous.

A team of paleoseismologists, led by Tom Rockwell, in July, 2000, analyzed the layers of rock and soil in the Mojave area to identify past earthquakes, their intensity, and their movements. In the past, earthquakes were thought to release tension on a fault, and then, when more strain built up, to occur again in a predictable pattern. However, earthquake prediction and analysis appears not to be that simple. A large area may not have any earthquakes for centuries and then have many within short periods. From their analysis, scientists have determined that the Mojave area is now an active earthquake cluster.

Major Earthquakes as Research Tools

The Chi-Chi, Izmit, and Hector Mine earthquakes were all major earthquakes, according to the technical definitions of the Richter scale. From a research perspective, they were also major. In 2000, geologists used the data collected and movements analyzed to develop new theories of earthquake activity. As earthquakes are inevitable, this work should prove invaluable in saving human lives and minimizing property destruction.

Resources for Students and Teachers

Books

Bowler, Peter J. *The Earth Encompassed: A History of the Environmental Sciences.* New York: W. W. Norton, 2000. Bowler examines discoveries in geography, geology, and evolutionary biology in describing the history of environmental sciences. He also touches on ethical issues.

Brown, Lester, Linda Starke, and Brian Halweil, eds. *Vital Signs 2000: The Environment Trends That Are Shaping Our Future.* New York: W. W. Norton, 2000. In its annual volume, the Worldwatch Institute provides a graphical look at the environmental trends that affect the future of the planet. Charts, tables, and graphs are provided on topics such as population growth, Internet use, fish catch, and global temperature.

Douglas, Bruce, Michael Kearney, and Stephen Leatherman, eds. *Sea Level Rise: History and Consequences.* San Diego, Calif.: Academic Press, 2000. Traces the history of the changes in the sea level (a consequence of global warming) since the last deglaciation began 20,000 years ago. Topics include measurement techniques, beach erosion, and the effect on the coast. Contains CD-ROM.

Flippen, J. Brooks. *Nixon and the Environment.* Albuquerque: University of New Mexico Press, 2000. Brooks traces U.S. president Richard M. Nixon's early commitment to regulating and protecting the environment as well as his subsequent rejection, then reembracing, of environmentalism.

Hays, Samuel P. *A History of Environmental Politics Since 1945.* Pittsburgh: University of Pittsburgh Press, 2000. Historian Hays traces the emergence of environmentalism in the United States after World War II and the accompanying opposition to "green" causes.

Isenberg, Andrew. *The Destruction of the Bison: An Environmental History, 1750-1920.* Studies in Environment and History. New York: Cambridge University Press, 2000. An examination of the forces that led to the near extinction of the North American bison.

United Nations Environment Programme. *Global Environmen-*

tal Outlook 2000. London: Earthscan Publications, 2000. Provides a look at the current global environment, including trends, policy developments, and future predications. Makes recommendations for new policy.

CD-ROMs and Videos

Conserving Earth's Biodiversity with E. O. Wilson. CD-ROM. Island Press, 2000. This interactive CD-ROM on conservation and environmental science is the product of a collaboration between evolutionary biologist E. O. Wilson and photographer Dan L. Perlman. For high school students and undergraduates.

Earth Science in Action. Video. 14 vols. Schlessinger Media, 2000. The topics covered in this set of educational videos for juveniles include earthquakes, fossil fuels, fossils, geological history, land formations, minerals, natural resources, oceans, rocks, soil, topography, volcanoes, the water cycle and weathering and erosion. Comes with teacher's guide, which is also available online.

Savage Planet. Video. 4 vols. MPI Home Video, 2000. Granada Television, Thirteen, and WNET created this video series on volcanic killers, storms of the century, deadly skies, and weather extremes.

Stormchasers. Video. MacGillivray Freeman Films, 2000. Hal Holbrook narrates this documentary about storm chasers, scientists who pursue storms in order to study them.

Weather: A First Look. Video. Rainbow Education Video, 2000. Peter Cochran produces this educational video on the weather for young people. Contains twelve activity pages and solutions.

Weather and Climate. Video. Earth Science Video series. Prentice Hall, 2000. This examination of the weather and climate covers topics such as hot air balloons, changes in the climate, the greenhouse effect, heat, violent storms, and climate in the United States. For elementary and secondary school students.

What's Up with the Weather? Video. WGBH Boston Video, 2000. This video, originally a NOVA/Frontline special report, was produced and directed by Jon Palfreman. Subjects covered include global warming and the greenhouse effect, climatology, long-range weather forecasting, and meteorology.

Web Sites

The American Geological Institute
http://www.agiweb.org

A professional organization whose membership consists of a variety of private business and academic and research organizations. It produces a monthly journal, *Geotimes: Newsmagazine of the Earth Sciences,* and offers this Web site to provide timely information about earth sciences and the environment.

The Energy Literacy Project (ELP)
http://www.energy-literacy.org

This educational and informational organization analyzes links among energy, the economy, and the environment. It provides projects appropriate for school use and a forum for exchanging ideas about cultural views on energy. Its goal is to change society's views about energy to better balance the interconnections among energy, the economy, and the environment.

National Aeronautics and Space Administration (NASA)
http://www.nasa.gov
Features satellite imagery of earth and space phenomena.

National Oceanic and Atmospheric Administration (NOAA)
http://www.noaa.gov
Explanations and visual imagery of the oceans and the weather, as well as links to the Climate Prediction Center, National Climatic Data Center, and El Niño site.

The National Weather Service
http://www.nws.noaa.gov
The site for the weather service, which is part of the National Oceanic and Atmospheric Administration, provides weather analysis, forecasts, and ancillary information across the nation and worldwide.

Screenscope Films
http://www.screenscopefilms.com/journey.html
Screenscope films, in cooperation with the Public Broadcasting System (PBS), has produced three films in a series entitled *Journey to Planet Earth*. These films, which emphasize the interconnections among the land, oceans, and the atmosphere, are "Rivers of Destiny," "The Urban Experience." and "Land of Plenty, Land of Want."

United States Department of Agriculture
http://www.usda.gov
United States government
Contains explanations of policies and discussions of issues related to agriculture and the environment.

United States Department of the Interior
http://www.doi.gov
United States government
Features assessment of management policies and (especially pertinent in 2000) national fire news.

United States Geological Survey
http://www.usgs.gov
United States government
Provides connections with earthquake and volcanic Web sites worldwide, including real-time monitoring.

United States National Park Service
http://www.nps.gov
United States government
Contains individual park histories and descriptions.

4 · Life Sciences and Genetics

The Year in Review

➤ *John T. Burns*
Department of Biology
Bethany College

➤ *Cheryl M. Speer*
Bethany College

➤ *Michael J. Rafa*
Brooke High School

The year 2000 witnessed an interesting mix of blockbuster events, notable achievements, and the further unfolding of ongoing sagas in the fields of biology and genetics. Developments in stem cell research and cloning made the news, as did the more controversial report that bacteria had survived 250 million years in a salt crystal. Publications in evolutionary psychology and regarding the relationship between science and religion stirred scientific and public debate. However, unquestionably the most spectacular event was the midyear announcement that the human genome had been sequenced.

The Sequencing of the Human Genome

Rival research groups headed by J. Craig Venter of Celera Genomics and Francis Collins of the National Genome Research Institute pooled their results to accomplish the daunting task of sequencing the human genome. Commentators heralded it as the greatest achievement in genetics of the twentieth century. U.S. president Bill Clinton and British prime minister Tony Blair immediately pushed for the free availability of all human genomic information. Beyond its importance for understanding fundamental biochemical processes, the sequencing of the human genome promises to enhance the ability to diagnose and treat diseases ranging from Alzheimer's, various forms of cancer and diabetes, to heart disease, hypertension, and some types of mental illness. Pharmaceutical companies are rushing to develop new drugs that will act on the newly identified genes or on the downstream metabolic processes regulated by them.

The sequencing of the human genome promises to enhance the ability to diagnose and treat diseases ranging from Alzheimer's, cancer, and diabetes to heart disease and some mental disorders.

As with many blockbuster events in human biology, the good news is accompanied by a number of ethical and legal dilemmas. Questions arose as to what extent and under what circumstances the genetics of a future child

Breakthroughs in cloning, which started in 1997 with Dolly, the first mammal to be cloned using a nucleus from an adult animal, continued in 2000 with the cloning of monkeys and pigs. (AP/Wide World Photos)

should be altered, what legal safeguards should be in place to ensure that access to health and life insurance coverage is not denied to individuals with a known genetic predisposition for disease, who will have access to the uniquely private and valuable information in a person's genomic sequence, and what the penalty will be for the misuse of such crucial information.

Stem Cells and Cloning

Closely rivaling the human genome sequencing story for importance are advances in working with stem cells and in cloning. Stem cells obtained from Adam Nash, the world's first designer baby born on August 29, 2000, may well save the life of his sister Molly, who suffers from the inherited blood disease called Fanconi anemia. While yet an embryo, before he was implanted into his mother's uterus, Adam was selected for his stem cell traits from a group of eight-cell-stage siblings. After he was born, healthy stem cells were taken from Adam's placenta and transfused into his six-year-old sister to alleviate her anemia.

Cloning breakthroughs during 2000 have been equally dramatic. Although researchers used different approaches to fashion the clones, a cloned monkey was born in January of 2000, and cloned pigs were born later in the spring. In the case of the monkey, an eight-celled rhesus monkey was divided into quarters, providing four identical embryos. However, only one

survived. Of course, any success in cloning primates has potential implications for the faster development of human cloning techniques. Producing cloned pigs is a blockbuster event because, after some requisite genetic engineering, such pigs can become a source of compatible organ transplants for humans. Ongoing efforts to clone guars and other endangered species continue as one way to salvage the earth's declining biodiversity. Speculation is ongoing as to whether rare preserved tissues will enable woolly mammoths or other extinct species to someday be cloned.

Notable events in 2000 include the announcement of 250-million-year-old bacteria surviving in salt crystals. The controversial claim hinges on the efficacy of techniques employed to isolate the bacteria from ancient salt deposits from New Mexico. Also, the discovery of what may be the oldest fossil vertebrates with feathers may provide new insights into the ancestry of birds. This early reptile-bird predates the dinosaur candidates that heretofore have been considered the most likely ancestors to the birds.

From Bacteria to Science and Religion

Other year 2000 developments have taken place in diverse fields. These include several new roles for the clock genes in both animals and people, the threat of invading giant cloned algae to marine ecosystems, the efforts to reintroduce Canadian lynx to the wilds of Colorado, and the four-year-early arrival of seventeen-year cicadas. Publications and news stories dealt with the so-called new science of love, which tries to find evolutionary reasons for people's innermost desires and motives, and the ongoing science-religion debate in public schools and elsewhere. The administration of President George W. Bush, with its promised concern for science education, may have to grapple with curricula standards and other aspects of the teaching of organic evolution.

The year 2000 brought a number of other developments in life science. Creatures and their environment were the focus of news items on the invasion of Florida by pernicious Asian swamp eels that threaten native species, the release of a long-captive California condor into the wild, the reported extinction of the rare Miss Waldron's red colobus monkey (caused by African hunters seeking bush meat, sometimes for upscale restaurants), and plans to reintroduce the imperial moth into Massachusetts from remnants surviving on Martha's Vineyard. Other "creatures" being studied by scientists included microbes and worms. In 2000, scientists examined tiny microbes recovered from sandstone 3 miles (4.8 kilometers) deep into the earth's crust, smaller than any bacterium ever found and measuring just 20 to 150 nanometers in length. They have been found to contain deoxyribonucleic acid (DNA) and to have other characteristics typical of living things. Other accounts describe evidence that the giant tube worm, Lamellibrachia, growing to more than 10 feet (3 meters) in length and living in the Gulf of Mexico, may live for as long as 170 to 250 years. If so, these mouthless and gutless creatures would be the longest-lived of the known invertebrates.

Producing cloned pigs is a blockbuster event because, after some requisite genetic engineering, such pigs can become a source of compatible organ transplants for humans.

The year 2000 encompassed a rich variety of advances and discoveries. Some, such as sequencing the human genome or cloning, have immediate applications to human medicine or people's well-being, and others, such as bacteria possibly surviving 250 million years or cicadas emerging early, largely just make the world a more interesting place.

The Human Genome Project
Seeing the Big Picture

➤ *The entire human genome is mapped for the first time in history.*

The cover of the July 3, 2000, issue of *Time* magazine featured photos of J. Craig Venter and Francis Collins, accompanied by the banner "Cracking the Code! The inside story of how these bitter rivals mapped human deoxyribonucleic acid (DNA), the historic feat that changes medicine forever." This landmark accomplishment heralded a more comprehensive understanding of human physiology, an improved medical diagnosis and treatment of patients, and, probably least foreseeable, a tangle of new legal and ethical issues for society to sort out.

The two men, Venter of Celera Genomics, a private enterprise, and Collins, director of the National Institutes of Health's National Human Genome Research Institute, approached the task of sequencing human DNA with somewhat different motivation. Venter sought to make his company profitable by patenting the sequencing information before marketing it to the pharmaceutical industry, and therefore his operation was interested in moving quickly. Collins, however, was committed to making the decoding endeavor a worldwide, noncommercial project. For more than a decade, the Human Genome Project has placed the DNA code information, gleaned by more than 1,100 scientists, onto a free Internet site (www.ncbi.nlm.nih.gov). The coming together by Venter and Collins to make the joint announcement of the human genome decoding was carefully negotiated and even involved pressure from the White House to make it happen.

Venter sought to make his company profitable by patenting the sequencing information before marketing it to the pharmaceutical industry; Collins was committed to making the endeavor a worldwide, noncommercial project.

The outcome of the two contrasting approaches shows the power of the entrepreneurial spirit. Celera Genomics first developed many of the innovations that led to rapid increases in the decoding process. In its procedure, numerous chromosomes are chopped into small fragments, the DNA sequence of each fragment is deter-

mined, and then massive computer algorithms analyze the bits of overlapping sequences to reconstruct the original sequence on a chromosome. This conceptual breakthrough and the private money provided for the newest laser-based sequencing equipment speeded the decoding of the human genome. A driving force was to make money by decoding parts of the genome and getting patents on that information, which was vital to the development of new drugs and diagnostic procedures. Access by Venter and his colleagues to the Internet data bank of the Human Genome Project, however, was also invaluable. This database enabled scientists at Celera Genomics to direct their efforts to unstudied regions of the genome and to check some of their sequences against the work of others.

By mid-2000, Celera Genomics had sequenced 97 percent of the human genome. The slower Human Genome Project had completed a smaller portion but perhaps with greater accuracy. The human genome has about 50,000 genes that are composed of a total of about 3.1 billion chemical "letters." Only four chemical "letters," or nucleotides, are involved in the deceptively simple code. The code is based on the sequence

British prime minister Tony Blair (on the monitor) speaks as U.S. president Bill Clinton (left) and Craig Venter, president of Celera Genomics, listen during the joint announcement by Celera Genomics and the International Human Genome Project that they had decoded the entire human genome. (AP/Wide World Photos)

of the nucleotides adenine, thymine, cytosine, and guanine—A, T, C, and G—which occur in pairs to make up the so-called steps in the spiral double helix structure of DNA. Of course, each individual, other than an identical twin, has a unique combination of alternative forms of genes. Therefore, additional examples of human genomes, from a variety of genetic backgrounds, need to be decoded to gain a better overview of human diversity. Although a representative genome has been decoded, at least in terms of the sequence of the nucleotides, how all of this folds into actually making a human is yet to be understood.

From Genes to Proteins

Fundamentally, it is known that genes provide the information for the synthesis of the body's proteins from amino acids. The all-important instructions for the assembly of the various kinds of amino acids into long chains and the elaborate folding of those chains to form proteins is encoded in the genes. These proteins provide much of the structure of human cells and tissues, and often more important, many proteins serve as enzymes that catalyze the innumerable biochemical reactions of human metabolism. There are tens of thousands of different proteins within each person, each with its own unique shape and biochemical properties. The varieties of structural proteins account for much of our individuality, ranging from blood types to the far more numerous, and difficult to match, tissue types. The inability to synthesize suitable proteins can cause type I diabetes (defective synthesis of the protein insulin), hemophilia (the lack of essential blood-clotting proteins), or myriad other genetic-based disorders.

Although a representative genome has been decoded, at least in terms of the nucleotide sequence, how all of this folds into actually making a human is yet to be understood.

The emerging field of proteomics aims to characterize the complex makeup and shape of protein molecules and the subtleties of protein functioning. The decoding of the human genome and that of several other organisms such as the geneticist's fruit fly *Drosophila melanogaster* has to an extent shifted the spotlight from decoding genomes to understanding what the genes control: the synthesis of proteins. In the past, many pharmaceuticals that modify protein synthesis and biochemistry were typically developed by trial and error. Now, drugs will be designed for their ability to influence the activity of genes within the human genome or to alter the process by which a gene's code is copied and transferred to the cell's "protein factories," or ribosomes, where proteins are assembled from amino acids. So-called designer drugs developed for an individual's particular genetic makeup so as to be more effective may become commonplace.

Legal and Ethical Concerns

In March of 2000, President Bill Clinton and Prime Minister Tony Blair declared that all genomic information should be freely available, a pronouncement that caused a temporary sag in the share price of Celera Genomics. However, it was later stated that patent protection should continue to be granted to stimulate genome-based drug research. Frankly, no one could foresee all the implications of the decoding achievement. Even James Watson, Nobel prizewinner with Francis Crick for their discovery of the double helix structure of DNA in 1953, suggested that patenting parts of the human genome was "sheer lunacy." He feared that legal tangles would hamstring scientific progress that humanity desperately needs to treat diseases with a genetic predisposition, such as Alzheimer's disease, Huntington's chorea, breast and colon cancer, diabetes, heart disease, hypertension, and some forms of mental illness.

Predictably, privacy issues loom large. One unanswered question is who will have access to an individual's DNA code once it is determined.

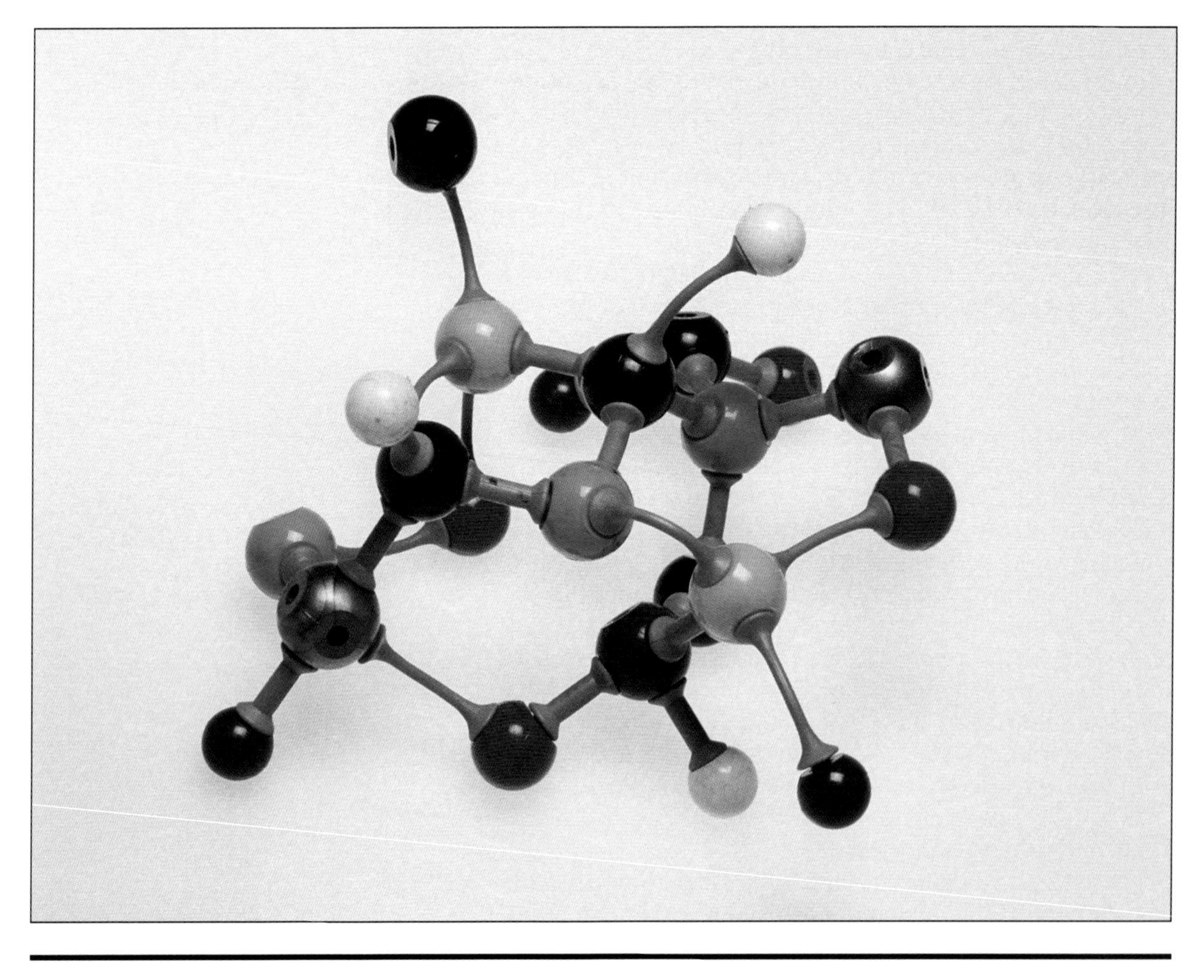

Although there is much interest in finding the genes within the human genome that underlie a predisposition for disease—even identical twins with the same genome often do not develop the same diseases—there remain concerns. Frequently, people are not sure they want to know that they have bad genes that may lead to serious disease. Such knowledge may detract from the enjoyment of their healthy years. However, being aware of a genetic predisposition may facilitate the monitoring of one's health so that an anticipated disease onset will be diagnosed at an early and perhaps more treatable stage.

Predictably, privacy issues also loom large. One unanswered question is who will have access to an individual's DNA code once it is determined. Health and life insurance companies or governmental agencies could use genomic data to limit or deny services provided to an individual. Schools or employers might use the information to discriminate on the basis of genetic traits. Also, prospective parents who know their genomic makeup might want to employ new technologies to make either deletions or additions of traits in their children. Some of these alterations would be to remedy serious genetic disorders but others might be to produce children with the preferred hair or eye color or equip them with greater than average athletic or intellectual skills. The rapidly occurring technological advances are likely to make these scenarios possible before society has found satisfactory legal and ethical answers.

J. T. B.

Cloning 2000
From Monkeys to Prehistoric Animals

➤ *Successful cloning of monkeys and pigs gives researchers hope of saving endangered species and possibly resurrecting extinct animals.*

On October 25, 2000, pioneer cloning researcher Thomas King died at the age of seventy-nine. His experiments in the 1950's with leopard frogs first highlighted many of the technical questions with which the science of cloning has had to grapple. King knew that a fertilized egg's nucleus contains all the genetic information needed for the development of the myriad tissues in an adult. What was unknown was whether, as a zygote underwent cleavages to become a multicellular embryo, the nuclei gradually lost their ability to sustain early development. Another question was whether a nucleus removed from an adult's differentiated muscle or skin cells, if transferred to an enucleated zygote, could still direct a zygote through normal cleavages and further development.

Frogs were ideal for such experiments because zygotes and embryos of frogs can be kept alive in the laboratory more easily than those of higher vertebrates. King and colleague Robert W. Briggs succeeded in transferring nuclei from progressively older leopard frog embryos to zygotes that had their original nuclei destroyed. Their successes, however, were limited to transfers with nuclei taken from rather early embryonic stages. During successive divisions of the fertilized egg, or zygote, the overall size of the early embryo stays constant. Consequently, the embryo's cells become smaller and smaller as the number increases. The smaller cell size in the later embryo makes nuclear transfers involving this donor source more difficult. Then in 1962, John B. Gurdon unexpectedly reported taking the tiny nucleus from a gut cell of an adult South African clawed frog and transferring it into an enucleated zygote that then developed normally. Although Gurdon's claim was at first questioned, he had in fact demonstrated that a nucleus harvested from differentiated tissues of an adult could be used to form a clone. Zygotes and embryos do not reject foreign transplanted material such as nuclei because they as yet have no immune system.

These early studies show that the one-celled zygote,

or fertilized egg, that marks the beginning of most individuals has two essential parts: the nutrient-rich cytoplasm and the gene-packed nucleus. All cloning efforts depend on somehow combining these two essential components, even if they are often from disparate sources. Additional obstacles often must be overcome to allow for the embryonic and, for mammals, the fetal development of the clone. One of these is the "activation" of the zygote that must occur for the zygote to be stimulated to undergo cleavage. This naturally occurs when the sperm enters the egg, but it can be done artificially by the prick of a needle for frog eggs, or more commonly in the case of cloned mammal zygotes, with an electric shock. The techniques for cloning vary considerably with the species involved, and sometimes the donor nucleus must be transferred more than once from egg to egg before it is ready to direct a zygote's development. So far, cloning is never a sure thing, and hundreds of attempts may be required to obtain a clone that survives until birth. Of course, the exact type of clone that may be constructed depends on the motives of the scientist. A particular cloning project may be simply an effort to solve a technical problem in cloning procedure, or the project can have a more spectacular goal that reminds one of something out of the motion picture *Jurassic Park* (1993).

So far, cloning is never a sure thing, and hundreds of attempts may be required to obtain a clone that survives until birth.

Monkeys, Pigs, and Mammals

The year 2000 saw several advances in the field of cloning, including the cloning of a monkey and several pigs. The rhesus macaque monkey, named Tetra, was formed by simply splitting apart an early embryo and did not involve nuclear transfer. Many embryos have the ability, when split apart, to undergo "embryonic regulation," or form as many new whole individuals as there are parts. This happens naturally when identical twins are formed. Tetra was the only survivor from a process that involved in vitro fertilization and the splitting of 107 embryos into two or four pieces to obtain 368 new embryos. Out of thirteen tries at embryo transfer to impregnate surrogate mother monkeys, four pregnancies occurred, with only Tetra surviving to term. Tetra developed from two cells that were split off of an eight-cell embryo. If this cloning technique is perfected, several individuals, or clones, can be produced from a single embryo. Also early in 2000, a Scottish com-

Piglets Millie, Christa, Alexis, Carrel, and Dotcom, born on March 5, 2000, were cloned by PPL Therapeutics. (AP/ Wide World Photos)

pany, PPL Therapeutics, announced that it had cloned pigs, and in July, a Japanese laboratory produced another cloned piglet named Xena. Of course, details of newly developed cloning procedures are shared reluctantly by for-profit companies.

Between 1996 and 1998, scientists cloned a ewe named Dolly, cattle, and a mouse named Cumulina. News of Dolly's existence was released in 1997. Dolly was the first mammal to be cloned using a nucleus from an adult animal. In Dolly's case, the genetic material from an egg was removed and replaced with a nucleus from a cell of an adult ewe, and an electric shock was given to activate the zygote to divide. Later, the resulting embryo was implanted into the uterus of a surrogate mother ewe. However, Dolly was the single successful product of 434 attempts at nuclear transfer.

Sheep, goats, and cattle are desirable subjects for cloning, in part because they are good producers of milk. After being subjected to genetic engineering, these clones can produce large amounts of milk that contain valuable proteins. A ewe named Polly, born in 1997, was genetically engineered to produce the human blood-clotting protein factor IX. A source of this protein is essential for treating the inherited disease hemophilia B. Most hemophiliacs do not want to receive protein factor IX supplements from human blood transfusions because of possible contamination with the AIDS virus. Fortunately, cloning allows a whole herd of genetically identical animals to be produced so large amounts of such proteins or other products can be obtained. The cloning of the mouse Cumulina mainly has implications for physiology and genetics and for biomedical research.

Xena and other pig clones, however, offer more readily understood benefits for humankind. Pigs, like humans, are omnivores, and their organs and metabolism are similar to those of people. Pigs are also the right size to be a practical source of organs, such as the pancreas, liver, and kidney. Because of drastic shortages of compatible human organs, many patients are unable to receive new organs when they need them. Someday, prospective transplant recipients may be able to donate a few cells so that some of their genes can be incorporated into a pig zygote. Later, a genetically altered, designer pig organ would be harvested that matches the patient's unique human tissue type, a practice that would minimize rejections. However, all the problems in cloning

pigs have not yet been solved. In the case of Xena, more than a hundred embryos were implanted into four surrogate mother sows before one live birth was obtained.

Prize, Rare, and Extinct Animals

The applications for cloning include perpetuating genetic lines of beloved domestic pets and other prized animals, salvaging exotic animals that are on the verge of extinction, and, possibly, using remnants of genetic material from extinct animals to recreate living individuals. Sometimes the surrogate mother need not be of the same species. Surrogate mothers, of course, can be implanted with embryos, perhaps from eggs fertilized in vitro, so that embryo transfer is involved, but not cloning. For example, a domestic cat has served as a surrogate mother for an African wild cat, and an eland has given birth to a bongo antelope. Success with such interspecies embryo transfer is crucial to some of the more ambitious cloning plans. Cloning, of course, depends mainly on having available suitable nuclei, and these can be kept frozen or as actively dividing cell lines held in vitro. Maintaining such a source of nuclei is much easier than finding a ripe egg and the sperm to fertilize it. Immediate plans exist for cloning rare species such as cheetahs and giant pandas, which breed so poorly in captivity. The importance of maintaining genetic diversity in small populations of animals is an ongoing problem for endangered species, and cloning would not help in that regard. Additionally, although the nucleus contains most of the deoxyribonucleic acid (DNA) of a cell, there is also some DNA in the mitochondria of the cytoplasm, and clones made with enucleated zygotes of other species will have both kinds of DNA.

A recently extinct Spanish mountain goat called a bucardo lives on as a frozen tissue sample, and it may be possible to clone it someday.

Advanced Cell Technology in Worcester, Massachusetts, a company experienced in cloning, is heavily involved in efforts to clone endangered species. One of its ongoing projects involves fusing skin cells from a male gaur, a rare oxlike mammal native to India, with enucleated cow eggs to obtain more gaurs. Several surrogate mother cows have failed to carry their cloned gaur fetuses through the nine-month pregnancy, but success is expected soon.

Predictably, the cloning of an extinct species remains a distant dream. A frozen Siberian mammoth unearthed in 1999 was found to have partially destroyed DNA in its cell

nuclei. Whether repair techniques can be developed to recreate mammoth chromosomes that can be used in cloning remains to be seen. A recently extinct Spanish mountain goat called a bucardo lives on as a frozen tissue sample, and it may be possible to clone it someday. The Audubon Institute Center for Research of Endangered Species in New Orleans and the Center for Reproduction of Endangered Species at the San Diego Zoo both maintain banks of frozen cells from many species. These precious samples may well be the last hope for resurrecting species that will not survive the current wave of worldwide habitat destruction.

J. T. B.

The New Science of Love
The Evolution of Human Sexual Attraction

➤ *Do men and women possess different psychological traits that favor reproduction? David Buss's* The Dangerous Passion, *a study of the evolution of jealousy, presents the evidence.*

What makes people fall in love? Is it romance? Has each person's one true love been predestined since the beginning of time? Or, is it in the genes? During 2000, numerous magazine articles, research reports, and books attempted to answer this question that sparks such interest in so many people. For example, a *U.S. News and World Report* cover story asked, "Why do we fall in love?" In February, 2000, *The Dangerous Passion: Why Jealousy Is as Necessary as Love and Sex*, by David Buss of the University of Texas at Austin, presented an examination of jealousy and love and attraction through the perspective of evolutionary psychology.

Those who study the evolution of sexual attraction usually base their ideas on Charles Darwin's theory of natural selection. This theory states in part that in nature, typically more members of a species exist than the environment can support. Only the fittest of these are likely to survive to reproduce. Also, because of the nature of inheritance and of mutations that occur in the genetic code of living things, each member of a species is different from the others in many ways; there is, in other words, wide genetic variation in a species.

Some researchers contend that natural selection favors psychological mechanisms that guide people to produce the maximum number of children. These mechanisms have developed very differently in men and women.

Sometimes these mutations are harmful to an animal; a peacock with a heart that does not work properly will probably not live long enough to reproduce. On the other hand, a mutation may help the animal; a peacock with larger and brighter feathers than normal may attract more females and thereby pass on this trait to more peacocks. Over time, the animals with these helpful, or adaptive, characteristics will survive at the expense of the others until what was once a mutation becomes an integral part of the genetic makeup of the species. Researchers such as Buss and his colleagues have applied these principles of natural selection to the psychological mechanisms that rule human mating relationships.

A Reproductive Urge

Reproductive biology has implications for each person. There is a basic inequality in the amount of time men and women must minimally invest in childbearing. Theoretically, men can invest no longer than a matter of minutes to have intercourse and conceive a child. Women, on the other hand, must carry a child for nine months, during which time they do not have the opportunity to conceive more children. This means that a man with an unlimited number of sex partners could father many children in a year, whereas a woman takes months to complete a successful pregnancy. Some researchers contend that natural selection favors psychological mechanisms that guide people to produce the maximum number of children. These mechanisms have developed very differently in men and women.

What do supermodel Kate Moss and the Venus de Milo (opposite) have in common? They both have waist-to-hip ratios within 0.04 of the 0.7 ideal. (AP/Wide World Photos)

In his book, Buss provides much scientific evidence to support this view. For example, men desire to have more sexual partners throughout their lives than do women, and they also fantasize about having sex with a greater variety of partners. Buss explains a classic experiment performed by Russell Clark and Elaine Hatfield in which an unknown woman or man walked up to a member of the opposite sex and asked him or her to have sex. Of the women approached by a man, 0 percent agreed to have sex; of the men approached by women, 75 percent of those propositioned by a strange woman agreed to have sex with her. This drastic difference shows that seeking frequent casual sex must have a much greater adaptive value for men than women.

In contrast, women, who typically invest a considerable portion of their lives in child rearing, would evolve a

reproductive strategy that maximizes their ability to choose the mate who will be the best provider for their children; thus they would tend to refrain from sex until such a mate is found. Buss cites this waiting for the optimal mate as the reason that women typically become very angry when men are aggressively sexual: Having a limited potential for bearing children, a woman impregnated by an untrustworthy man has been deprived of an irreplaceable resource.

If all this is true, then why do women have one-night stands or brief affairs? Buss suggests several answers, including the theory of the "sexy sons." Researchers could not understand why a woman would choose to have a one-night stand with a famous or powerful man such as a professional basketball player or a rock star who had no intention of forming a long-term relationship with her until they considered the matter from an evolutionary perspective. The theory of the "sexy sons" explains that a woman may choose to have a brief fling with a man based on his looks and ability to attract members of the opposite sex. Sons fathered by these men should inherit these qualities, which will in turn enable them to attract a large number of women. This provides the original woman with a maximum amount of descendants and thus fulfills an adaptive value.

Flirtatious Fertility

Beyond the strategies men and women use to transmit their genes to the next generation, recent studies reveal much about the types of characteristics that people find attractive in the opposite sex. Evolutionary psychologists theorize that natural selection may have caused men to seek women who would be likely to be able to conceive and bear a child. The easiest way to find such women is to look at them. A woman's waist-to-hip ratio (the circumference of the waist divided by the circumference of the hips, which is a dependable measure of the distribution of body fat) has received much attention recently as a signal of reproductive capability. Devendra Singh, also of the University of Texas at Austin, finds that both women and men rate female figures with lower waist-to-hip ratios as being healthier. There seems to be a reasonable evolutionary explanation for this. During early human history, when food was often scarce, it would be adaptive for men to find a way to be sure the women they inseminated would be capable of effectively carrying a pregnancy to term. Then and even now, women who carry fat on the hip, rear, and thigh are more likely to have successful pregnancies. The optimal waist-to-hip ratio is 0.7, and the attraction to this figure can be documented by considering the women said to be the most sexually desirable throughout the ages. For example, the February 7, 2000, issue of *U.S. News & World Report* reported that the figures of Venus de Milo, Marilyn Monroe, and Kate Moss all have waist-to-hip ratios within 0.04 of the 0.7 ideal.

In addition to waist-to-hip ratios, youth is another indicator of reproductive capability. Victor Johnston of New Mexico State University found that faces of women near the age of twenty-two—when a women's fertility reaches its peak—were judged to be the most beautiful. In addition, Buss reports that both sexes find good body symmetry very attractive. This probably stems from the fact that bilateral symmetry is a sign that a person had a normal development and was able to fight off environmental dangers such as toxins. Interestingly, over the millennia, women have developed a way to detect symmetry—they smell it. In an experiment, women were given T-shirts worn by symmetrical or asymmetrical men, and during the ovulation phase of their menstrual cycle when they are most likely to conceive (and only at this time), they claimed the shirts worn by the symmetrical men smelled better.

Women have also evolved an inventory of characteristics they look for in a partner. Buss reports that one set of these characteristics involves a man's ability to protect his mate; this list includes being athletic, protective, strong, muscular, and physically fit. The other includes a man's ability to provide food, shelter, protection, education, and status for his children. In the modern era, as women's salaries are beginning to equal those of men, and women are often quite able to provide for themselves, it might be assumed that these male characteristics would no longer be as important to women. Buss maintains, however, that this genetically dictated desire does not just "magically disappear." His studies have found that women with higher incomes will usually seek men who can provide even greater assets.

Buss reports that both sexes find good body symmetry very attractive. Interestingly, over the millennia, women have developed a way to detect symmetry—they smell it.

C. M. S.

Stem Cells and Healing

A Designer Baby Saves His Sister's Life

➤ *Placental stem cells from a baby boy cure his sister's inherited blood disease.*

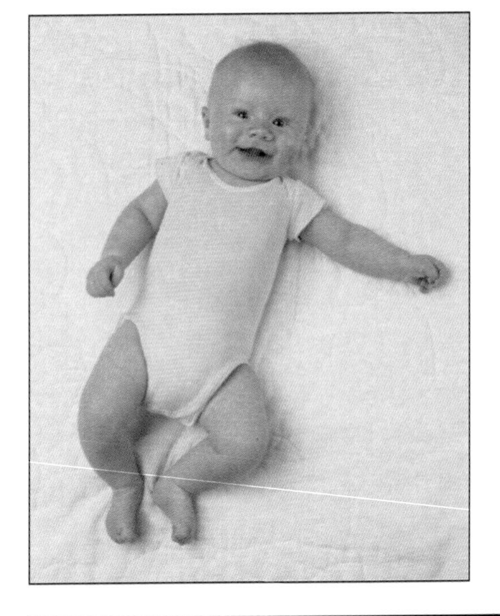

On August 29, 2000, in Englewood, Colorado, Adam Nash was born. He is the first designer baby selected for his stem cell traits. Before Adam, scientists and doctors had made simpler selective attempts such as choosing the sex of a child. In both artificial insemination and in vitro fertilization, special techniques can partially isolate sperm with either X or Y chromosomes to combine with an egg (having an X chromosome) to increase the chances for either a male (XY) or a female (XX) child. However, 2000 saw the birth of a baby who was conceived with the express hope that it would have specific stem cell traits. The unique stem cells were needed to save the life of Adam's older sister Molly.

In late 1999, Jack and Lisa Nash, a Colorado couple, provided eggs and sperm for in vitro fertilizations; subsequently, a number of viable embryos were produced. A cell or two was removed from each eight-cell-stage embryo and tested with molecular probes to determine each embryo's stem cell traits in a process called preimplantation genetic diagnosis (PGD). The embryo that would become Adam was then implanted into Lisa's womb. After a normal pregnancy, Adam was born. Healthy stem cells were taken from Adam's placenta and transfused into six-year-old Molly. Molly's inherited blood disease, Fanconi anemia, was expected to kill her by her mid-teens. The early follow-up tests indicate that the transfusion of cells to reseed Molly's bone marrow was successful.

As is apparent from Molly's case, the stakes are high for the millions of patients who suffer from a host of inherited and other diseases that could be treated, if not outright cured, with stem cell technology. The diseases include various types of leukemia, Hodgkin's disease, sickle-cell anemia, and thalassemia. Stem cells are expected to provide a solution for at least some of these disorders.

It All Starts with a Stem Cell

Often, all cells in an early embryo can be considered of the stem cell type. However, the degree to which this is true depends on the specific type of organism. For example, mollusk embryos that lose a few cells may later simply lack part of their body, such as skeletal muscle, as they continue to develop. In contrast, the early embryonic cells of humans and other mammals are totipotent; that is, the cells left behind can take over the role of the cells that were lost. A complete embryo may then form and develop into an adult with full differentiation of its tissues. This explains how identical twins can develop from the splitting of a single embryo. How a partial embryo "recognizes" that it is incomplete and then reorganizes (a process called embryonic regulation) to form a new whole is still a mystery. As little understood as embryonic regulation may be, it was essential to success in the Nash family case. After a cell or two had to be removed from early embryos for PGD testing, it was essential that the remaining cells were able to form a whole embryo.

The early embryonic cells of humans and other mammals are totipotent; that is, the cells left behind can take over the role of the cells that were lost.

Of course, humans all start out as a single-celled individual called a zygote, which could be considered the ultimate stem cell. As the zygote undergoes mitotic divisions to form an embryo, the descendant cells are normally identical in their genetic makeup with the ancestral zygote. One of the great mysteries of developmental biology is what accounts for the fact that all these cells, with the same genetic heritage in their deoxyribonucleic acid (DNA), do not become the same type of tissue. Thus, although genetically identical, human skeletal muscle cells are highly specialized morphologically and physiologically to contract, whereas nerve cells (neurons) are otherwise highly specialized to carry nerve impulses and secrete neurotransmitters. The process by which these differences arise is called tissue differentiation.

Tissue differentiation is thought to depend on which of the tens of thousands of genes within the given cell express themselves. However, the cell's environment and the presence of various hormones, growth factors, embryonic inductors, and other agents can all contribute to what happens. In the simplest schemes, it is said that the body has only four types of tissues: epithelial, connective, muscular, and nervous. When all the subtypes are considered, however, there are dozens upon dozens of kinds of tissues. The plasticity of cells to develop into any number of these many kinds of tissues is remarkable. The term "stem cell" refers to such plastic

Popular actor Michael J. Fox, chairman of a foundation for Parkinson's disease, was among those testifying before a Senate Appropriations subcommittee hearing on stem cell research. (AP/Wide World Photos)

cells, whether they are cells in the embryo or are other immature cells in the human body that can still seed the development of more differentiated cell types. This plasticity is normally lost as a cell differentiates.

The variety of stem cell locations is revealed in the terms "embryonic stem cell," "tissue stem cell," "blood stem cell," "placenta stem cell," and so forth. Unexpectedly, new evidence reveals that stem cells can give rise to new neurons in the mammalian brain. Previously it was thought that adult mammals could not produce new neurons. In the case of Parkinson's disease, the patient's own brain cells slow in their production of the neurotransmitter dopamine. Taking L-dopa tablets, a precursor to dopamine, can augment the ability of the declining population of neurons that produce dopamine and thereby temporarily ameliorate the symptoms. The actor Michael J. Fox, since his diagnosis as suffering from Parkinson's disease, has urged the U.S. Congress and others to vigorously pursue stem cell research.

Ethics and Hope

A plethora of ethical concerns have been raised over the use of embryonic and fetal tissues, either in the clinic or research laboratory. Fortunately, the discovery within the last couple of years of the more widespread occurrence of stem cells throughout the adult body has opened up new avenues for research. The study of such nonfetal cells does not engender as much debate. However, stem cells from the adult body seem more restricted in their developmental potential than those from fetuses.

No doubt the Nash family has already influenced how most people will view stem cells. They wanted more children, but they did not want to pass on a deadly inherited disease to another child. Without the PGD screening, there would have been a one-in-four chance of any new child having Fanconi anemia. They also wanted to help their daughter Molly live past her mid-teens. They have said that their decision was based on what would be best for their family's future. The use of PGD and the birth of their designer baby Adam has opened the door to their brave new world.

J. T. B.

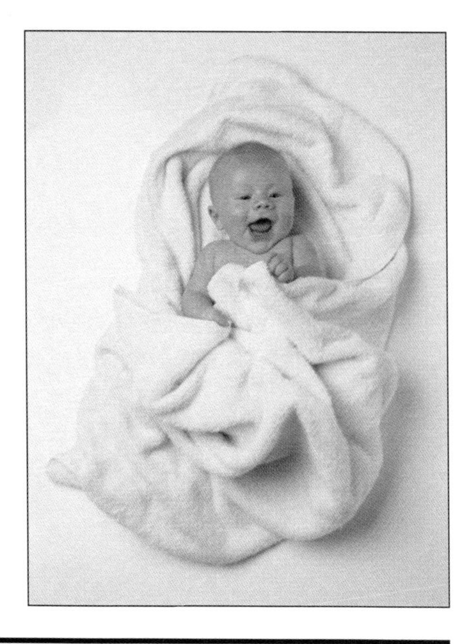

Clock Genes
Of Honeybees and Zebrafish, Introverts and Extroverts

➤ *Researchers report evidence that circadian clocks are active earlier than previously thought in vertebrate development and that the inherited clocks may affect human behavior and personality.*

Studies of the genetic basis of circadian rhythms are yielding profound insights into the complex physiology and behavior of animals and humans. Circadian rhythms are generally considered to be endogenous fluctuations that have a period of approximately twenty-four hours (circadian means "about a day"). Typically, they are synchronized with the twenty-four-hour light-dark cycle. However, under unchanging environmental conditions, the rhythms become desynchronized and "free run" so that the true period lengths, typically not exactly twenty-four hours, are expressed. In the 1960's, cave explorer Michel Siffre documented free-running circadian rhythms in his own body when he spent months in caves away from time cues. At least some of the circadian rhythms serve as "biological clocks" that temporally coordinate diverse physiological and behavioral processes within living things. Rather than being phenomena of interest only to specialists, research on circadian rhythms and biological clocks in everything from cyanobacteria to humans is increasingly in the news.

Winding the Clock

For example, in the July 14, 2000, issue of *Science*, a team of French researchers report new evidence that circadian clocks are operational earlier in vertebrate development than was previously thought. Specifically, their study on zebrafish demonstrates that even in the egg, before fertilization, at least one clock gene is active. The active gene is a form of the *Period* gene called *Per3*, which is a fundamental part of the cell's internal timing mechanism. Because the eggs under study are as yet unfertilized, the gene is in the haploid set of chromosomes inherited only from the mother zebrafish. The eggs essentially inherit a "ticking clock" from their mother. Expression of the gene involving the synthesis of particular proteins occurs rhyth-

Just what adaptive advantage there may be for the unfertilized egg and early embryos of the zebrafish to have an operational biological clock is unknown.

mically, with a circadian period, in the developing zebrafish's central nervous system and retina. A few other examples of "maternal inheritance" are known to control processes in the unfertilized egg and even later in the early development of the embryo as it begins to undergo cleavage. Interestingly, the zebrafish embryos can synchronize their circadian rhythm in *Per3* activity to the environmental twenty-four-hour light-dark cycle. Just what adaptive advantage there may be for the unfertilized egg and early embryos of the zebrafish to have an operational biological clock is unknown. Perhaps it temporally organizes essential biochemical processes during early zebrafish development.

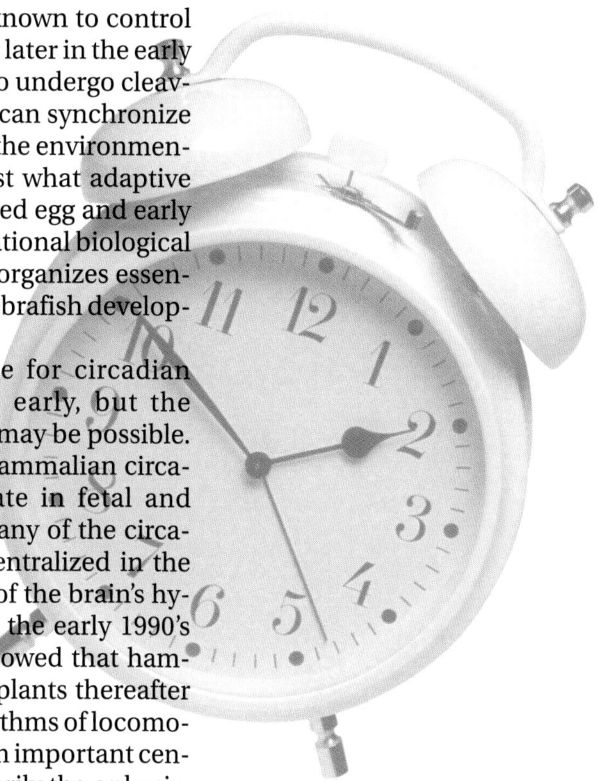

In mammals, there is no evidence for circadian rhythms that are operational so early, but the zebrafish researchers believe that it may be possible. So far it is generally thought that mammalian circadian clocks may not function until late in fetal and postnatal life. Evidence indicates that many of the circadian clock functions in mammals are centralized in the bilobed suprachiasmatic nucleus (SCN) of the brain's hypothalamus. Convincing experiments in the early 1990's by M. R. Ralph and Michael Menaker showed that hamsters receiving period mutant SCN transplants thereafter have altered periods in their circadian rhythms of locomotion. Therefore, in mammals, the SCN is an important central circadian clock, although not necessarily the only circadian clock. In some mammals, such as rats, there is evidence that the mother rat can synchronize the circadian rhythms of her pups. This occurs by way of an unusual means of communication between mother and pup. The mother's pineal gland produces melatonin in high amounts during the nighttime, and this hormone is passed in the mother's milk to the pups during nursing. Melatonin thereby synchronizes the pups' circadian rhythms with those of the mother rat. However, as any parent can attest, human babies take weeks to develop predictable circadian rhythms, at least in their all-important sleep-wake cycle. Therefore, the degree to which animal studies may apply to humans will be answered with further research, but the insights from animal experiments are invaluable.

Clocks and Behavior

Another advance reported in the June 6, 2000, *Proceedings of the National Academy of Sciences* details new functions for the *Period* gene in honeybees. Gene E. Robinson and colleagues discovered that the gene is more active

in the brains of the older forager honeybees than in younger honeybees. Ever since Karl von Frisch's research in the early half of the twentieth century, it has been known that forager honeybees depend on a precise circadian clock for timing their daily activities. Foragers employ circadian clocks for navigation during nectar and pollen gathering and in the subsequent communication of the location of such finds to other foragers by so-called round and waggle dances. Those activities would be simply impossible without the forager having an extremely accurate circadian clock. It remains to be determined just how the higher activity of the *Period* gene in older honeybees may contribute to the developmental shift from youthful hive duties to later outside foraging.

Finally, there are some suggestions that circadian clock genes in humans may have far-reaching effects on behavior and personalities. Standardized methods such as the Horne-Ostberg questionnaire for morningness and eveningness can be used to evaluate an individual's behavior. A tendency toward morningness or eveningness is reflected in the timing of myriad daily activities—that is, when during a twenty-four-hour period a person prefers to awaken, eat, perform exercise, and sleep. In 1998, a research team that included Emmanuel Mignot at the Stanford University Sleep Disorders Center reported that there are slight differences in the clock gene alleles present in persons scoring higher for morningness as compared with those scoring higher for eveningness. This finding could indicate that people inherit their pattern of daily life.

Moreover, preliminary research at Bethany College by John T. Burns and students indicates the existence of a link between morningness and eveningness and introversion and extroversion. The researchers used activity monitors worn on the wrist to provide data that correlate with morningness and eveningness and determined a subject's degree of introversion or extroversion through a psychological inventory test. Psychologist Carl Jung developed the theoretical basis for the personality types of introvert and extrovert and believed that an individual could possess traits of both types. The Bethany College study was published in the journal *Chronobiology International,* and correlations were reported between early daily peaks in wrist activity, morningness, and introversion. By extrapolation, this means that the particular alleles of the clock gene that people inherit may contribute to the molding of their personalities.

J. T. B.

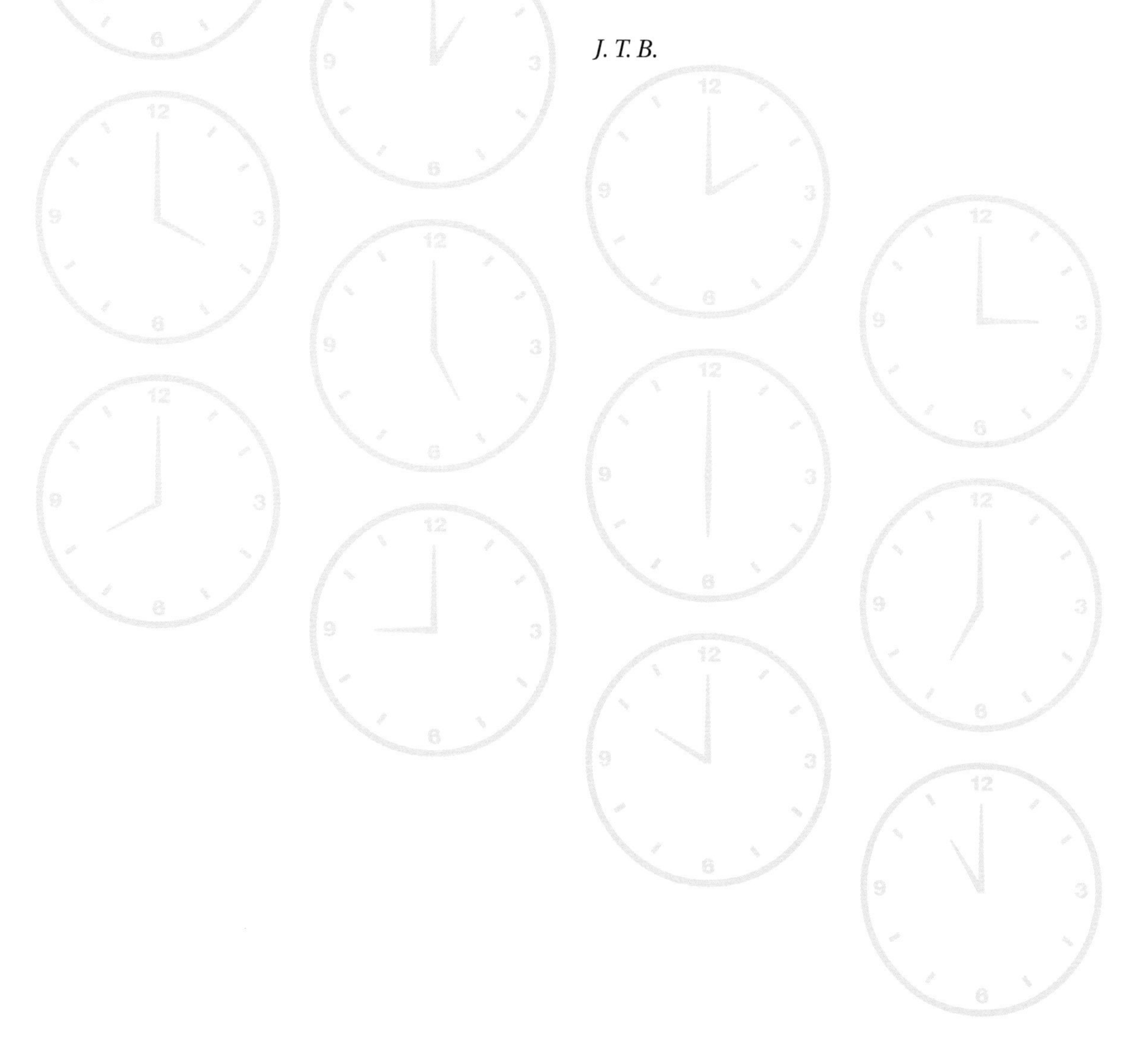

220-Million-Year-Old Feathers Discount Dinosaur Ancestry for Birds

➤ *Scientists cite featherlike appendages on fossil as evidence that birds are cousins, not descendants, of dinosaurs.*

For paleontologists John Rubin and Terry Jones, a visit to a local Kansas City shopping mall became the beginning of an exciting scientific adventure. When they arrived at the mall, where Rubin was to give a talk on dinosaur biology, they noticed a strange specimen in a touring display of Russian fossils. It turned out to be a fossil recovered from an ancient lake bed in Kyrgyzstan in Central Asia. For decades, it had languished in a drawer at the Russian Academy of Sciences' Paleontological Museum in Moscow. Rubin and Jones quickly concluded that the ancient fossil was an extremely early feather-bearing vertebrate. This conclusion, published with Alan Feduccia and other colleagues in the June 23 issue of *Science*, quickly rekindled an ongoing scientific debate on just how birds evolved, and whether the earliest reptiles or the later dinosaurs were the immediate ancestors of birds.

Feather or Not?

Russian paleontologist Alexander Sharov described the rare fossil in 1970, considered it a reptile, and named it *Longisquama insignis*. The generic name *Longisquama*, meaning "long scale," refers to the curious, long, thin appendages covering its back (but, significantly, not on its forelimbs). The question is whether these appendages can be considered feathers and thus whether this approximately 220-million-year-old fossil provides new insights into the evolutionary origin of birds.

***Longisquama*, if it had true feathers, fits the theories of scientists who think that dinosaurs were "distant cousins" but not ancestors to birds.**

The thin appendages do have a central shaft, or midrib, and finer narrow ribs fanning out to the edges, much as feathers do. The appendages seem to have, depending on the bias of the observer, a possible calamuslike basal structure that could have developed in a follicle, similar to the growth of modern bird feathers. Some scientists have suggested that the appendages are merely ribbed membranes

What appear to be feathers are apparent in the 220-million-year-old fossil Longisquama insignis. (AP/Wide World Photos)

or other structures not homologous with feathers. If the latter is true, *Longisquama* remains an interesting fossil but most likely sheds no light on the origin of birds. The fossil antedates the so-called ornithischian dinosaurs that have frequently been proposed as giving rise to birds. Thus, *Longisquama*, if it had true feathers, fits the theories of scientists such as Feduccia who think that dinosaurs were "distant cousins" but not ancestors to birds. In their view, dinosaurs and birds share only a remote, common reptilian ancestor. The second edition of Feduccia's book, *The Origin and Evolution of Birds*, published in 1999, reviews the controversy before the new information on *Longisquama* "feathers."

Evolutionary Links

Birds have always caught the imagination of humans, not only because of their impressive ability to fly, but also because they maintain a warm body temperature (endothermy) as mammals do, sing and show other courtship behaviors, build intricate nests, display maternal instinct, and migrate long distances. Avian adaptations for flight include feathers; lack of a heavy urinary bladder; hollow, air-filled bones; fusion of their skull, synsacrum, and wing bones; lack of heavy teeth; powerful flight muscles anchored to a keel, or carina, on their sternum; highly efficient lungs and air sacs; and a high rate of metabolism. Ornithology was a popular avocation for generations of wealthy dilettantes of the eighteenth and nineteenth centuries.

Charles Darwin, who along with Alfred Russel Wallace proposed the theory of natural selection, prominently featured birds—including varieties of domesticated pigeons and finches in the Galapagos archipelago—as examples in his 1859 classic, *On the Origin of Species by Means of Natural Selection*. It was Darwin's arch defender, Thomas H. Huxley, who in 1868 noted similarities between two approximately 150-million-year-old fossils collected from the Solnhofen limestone quarries in Germany: the small theopod dinosaur *Compsognathus* and the fossil feathered bird *Archeopteryx*. Huxley stated that unique skeletal characteristics, such as the birdlike pelvis and hind limb in *Compsognathus*, suggested that dinosaurs were the ancestors of birds. Huxley overlooked the teeth in the *Archeopteryx* fossil but could hardly miss the prominent imprint of feathers and the fragile, birdlike bones. Such a "missing link" was exactly what Huxley needed for his effort to prove that all organisms, including humans and other vertebrates, had evolved from a long line of more primitive ancestors.

In the fossil reptile-birds that have been recovered, what has been preserved is mainly bones and only rarely scales, feathers, or other softer tissues.

The evolutionary linking of these two amazingly diverse groups of vertebrates, dinosaurs and birds, is perhaps just too appealing to cast aside easily. However, the Victorian anatomist Robert Owen, who coined the term "dinosauria," vehemently opposed Huxley's avian evolutionary scenario. Owen thought that *Archeopteryx* was more closely related to the extinct flying reptiles called pterosaurs. Now, scholars believe that pterosaurs developed flight independently of birds. Some scholars have suggested that Owen may have even considered *Archeopteryx* to be a fake. In any case, the general scarcity of early fossils in the reptile-to-bird transition has made it difficult to resolve the dinosaur-bird lineage question.

Certainly, embryological similarities between reptiles (class Reptilia) and birds (class Aves) leave little doubt that these two vertebrate classes are closely related. The cleidoic egg of reptiles is quite similar to that of birds and to that of primitive egg-laying mammals such as the duckbill platypus and the spiny echidna. Keratinized reptile scales, bird feathers, and mammalian hair all have developmental similarities, no doubt because of their common reptilian origin. However, in the fossil reptile-birds that have been recovered, what has been preserved is mainly bones and only rarely scales, feathers, or other softer tissues. Furthermore, it is partially a matter of definition as to what minimal characteristics are required for a particular animal, extinct or extant, to be classified as a member of a given taxonomic group.

Longisquama is just one of several fossil finds that have joined the seven known examples of fossils like *Archeopteryx* for paleontologists to study. A number of recent finds in China, such as *Confuciusornis*, similar in age to *Archeopteryx*, are particularly noted for their preservation of soft tissues and, in some cases, feathers, along with the bones. Much earlier in 1986, two crowlike fossils christened *Protoavis* were found in a mudstone quarry in Texas. These 225-million-year-old fossils did not provide any evidence of feathers and a debate, based on skeletal details, has raged as to whether they were indeed birds.

J. T. B.

Evolution of the Science-Religion Debate

➤A Kansas primary election opens the door to a more tolerant attitude toward the teaching of the theory of evolution in the state's schools, as a physicist and theologian publishes a book arguing for the integration of science and religion.

In August, 1999, the theory of evolution was dropped from all science curricula standards for public schools in the state of Kansas. The change meant Kansas students would no longer be tested on their knowledge of evolution. Several members of the school board, despite having a limited understanding of the natural sciences, decided that organic evolution was not an appropriate part of a public school curriculum. However, in August of 2000, the antievolution school board members lost in a Republican primary election. Their defeat was expected to result in a greater tolerance toward the teaching of the theory of evolution in Kansas classrooms. Nonetheless, other ongoing efforts to either suppress the teaching of evolution or promote the teaching of creationist theory are under way in several other states.

Some critics view the theory of evolution as simply another unproven scientific theory. Thus categorized, it is no more worthy of special attention than dozens of other unproven theories that exist in the marketplace of ideas. Often, the critics of the theory of evolution are Christians who believe in the literal word of the Bible and look to the book of Genesis to understand when and how the earth and its life originated. For them, it is particularly galling to be required to pay taxes to support public schools that teach doctrines incompatible with their religious beliefs.

Public Schools and Religion

Proponents of religious agendas for the public schools often cite the fact that the founders of the United States of America incorporated many aspects of the Judeo-Christian tradition into its Constitution and its laws. Opponents point out that the changing population of the United States has come to include people with a wide variety of cultural and ethnic heritages. Therefore, it is inappropriate, if not unconstitutional, for a "government of the people" to endorse or promote religious beliefs or agendas held by only part of the citizenry. As a case in point, in 1981, the Arkansas legislature required that "creationist theory" be given equal consideration with evolution theory in Arkansas schools. The following year, the U.S. District Court overturned the law, in part because it violated the constitutional separation of church and state.

A number of religious groups have attacked the separation of church and state on another front. In recent years, public schools have come under pressure to reverse the trend toward increased violence in the classroom and teenage pregnancy and alcohol and drug use. Members of these religious groups believe that public schools should teach good character, allow open prayer in the classroom, and emphasize "traditional values," including the Golden Rule and the Ten Commandments. Additionally, the general public is demanding higher academic standards and new accountability for both teacher effectiveness and student achievement. As a consequence of doubts regarding public schools, home schooling, previously largely a vehicle for religious indoctrination, is gaining popularity among families. In many cases, these families want both to avoid a potentially harmful public school environment and to enhance learning. However, religion-based home schooling can negatively affect students' performance on standard college entrance exams and their ability to compete for scholarships.

Creationism, of course, is primarily based on religious convictions. As such, it draws on different sources of validity than does a scientific finding. People's cultural background and early life experiences are often key in determining their lifelong beliefs. Most psychologists claim that the formation of basic emotional structure begins in early childhood. The rejection of such early childhood teachings is usually not easy or painless. Thus, unfamiliar scientific explanations such as the theory of evolution may have to compete with religious convictions already well entrenched in an individual's belief system.

A Variety of Views

The year 2000 saw the publication of *When Science Meets Religion* by Ian G. Barbour, professor emeritus of physics and religion at Carleton College. The book summarizes many of the arguments in the science-religion debate from the point of view of a believer. Barbour won the 1999 Templeton Prize for Progress in Religion for his efforts to advance the study of religion and science. In the book, Barbour states that science and religion can be viewed as having four possible relationships: conflict (the same per-

In the 1925 Scopes trial, Clarence Darrow argued the right of his client, John Scopes, to teach the theory of evolution in Tennessee public schools but lost the case. To this day, the science-religion debate continues. (Library of Congress)

son cannot believe in both evolution and God), independence (science and religion can both exist if they maintain independent paths), dialogue (the two sides talk to each other), and integration (in which science and religion work to inform each other). He emphasizes integration as the most productive relationship.

The theory of organic evolution was made more feasible by the elaboration of the process of natural selection, independently discovered by Charles Darwin and Alfred Russel Wallace. A review of Darwin's letters shows that early in life, he professed a belief in God and even studied for the clergy. Later, he became less certain of his convictions and eventually concluded that it was beyond the capacity of the human mind to resolve whether God exists. Darwin's arch defender, Thomas H. Huxley, invented the term "agnostic" for those who are simply unsure whether a god does exist. Wallace, Darwin's colleague, was more mystical and devoted the last years of his life to a study of spiritualism. After the famous Scopes trial in Tennessee in 1925 regarding the teaching of evolution in public schools, many church leaders and government officials have deferred to scientists on the question of the validity of the theory of evolution.

Numerous modern writers such as Edward O. Wilson,

Richard Dawkins, Paul Kurtz, and the late comedian Steve Allen have elaborated the rationalist's view of nature that does not require religion. Stephen Jay Gould, a writer of many popular books on paleontology and evolution, has claimed that science and religion deal with independent realms, or "nonoverlapping magisteria." Inadvertently, Gould's somewhat pretentious scheme has tended to please proponents of supernatural forces. Many of these writers may be considered secular humanists, individuals who draw their moral convictions from a general sympathy for humanity, rather than from a specific religious tradition. They are content with a world that does not include, in their view, supernatural forces or events such as religious miracles.

Of course, in the hundreds of thousands of scientific papers published each year in technical journals, the religious convictions of the authors are almost never apparent. Although religion may be an integral part of a researcher's life, the scientific analysis being reported generally stands separate from the private life of the individual scientist. For example, the decomposition of water yields one hydrogen and two oxygen atoms, no matter whether the chemist is a Christian, Buddhist, Hindu, Muslim, agnostic, or atheist. Only a minuscule number of scientific studies deal with topics that would be considered religious in nature, such as determining the true age of the shroud of Turin (shroud alleged to belong to Jesus of Nazareth) or gathering geological and historical evidence for the biblical Great Deluge.

Implications for Students

The increasing pervasiveness of science and technology in people's daily lives is a trend that is likely to continue into the foreseeable future. If a student has an aversion to major premises of modern science, that may lead to problems. Occasionally, living with such a conflict between the views of society at large and one's own convictions may be intellectually stimulating. In practice, however, it more often causes frustration and failure on the student's part. That is, strongly held but poorly examined religious convictions may hinder the student's ability to understand and absorb new information and to develop critical thinking skills. Unfortunately, a young person who is not knowledgeable in science will most likely not be able to compete in the evermore technological society.

J. T. B.

250-Million-Year-Old Bacteria Surviving in Salt Crystals?

➤ *Scientists report growing bacteria captured inside ancient salt crystals from New Mexico.*

Stories from the nineteenth century report that living toads, salamanders, or horned toads upon occasion have unceremoniously emerged when cornerstones of old buildings were opened. Other tales recount the freeing of creatures from natural geologic cavities, presumably after eons of imprisonment. Furthermore, old archives give dubious longevity records for plants or animals far beyond what now is accepted as possible. Claims of extreme longevity for living things are always of interest and often evoke in people a sense of awe at the tenacity of organisms in their struggle to survive. A recent article, in the October 19, 2000, issue of the respected journal *Nature*, contends that some bacteria trapped within ancient salt crystals can survive for up to 250 million years.

Since its beginning in 1869, *Nature* has published, along with its rather staid laboratory research articles, an occasional highly controversial claim. Often, the unusual proposition or suggested discovery does not easily lend itself to testing by the experimental method. For example, after the journal in 1988 published a paper on the efficacy of homeopathic dilutions from Jacques Benveniste's laboratory, editor John Maddox was drawn into a heated controversy on the elusive properties of such dilutions. Eventually, Maddox, popular magician James Randi, and Walter Stewart visited Benveniste's laboratory to review the research notes and witness firsthand the experimental techniques. The techniques under scrutiny involved testing high dilutions of solutions for the lingering presence of anti-immunoglobin E. The staining response of white blood cells, or basophils, was the basis for the assay for such traces. According to mainstream science but not to homeopathic theory, the dilutions were too high for any trace of the solute to remain. One member of the review team, Randi, has become especially well-known

Claims of extreme longevity for living things are always of interest and often evoke in people a sense of awe at the tenacity of organisms in their struggle to survive.

through his popular books that recount his numerous investigations to expose scientific frauds and delusions. The team concluded that Benveniste's years of research supporting the effectiveness of homeopathic dilutions, well intended as they may have been, had indeed produced only a delusion. However, this finding was not accepted by Benveniste, or by the homeopathic medical community. Such forays by *Nature* into what some would call pseudoscience have not escaped the notice of skeptics. Nonetheless, *Nature*'s liberal publication guidelines continue to support a forum where brash new ideas or even old controversies can be freely discussed.

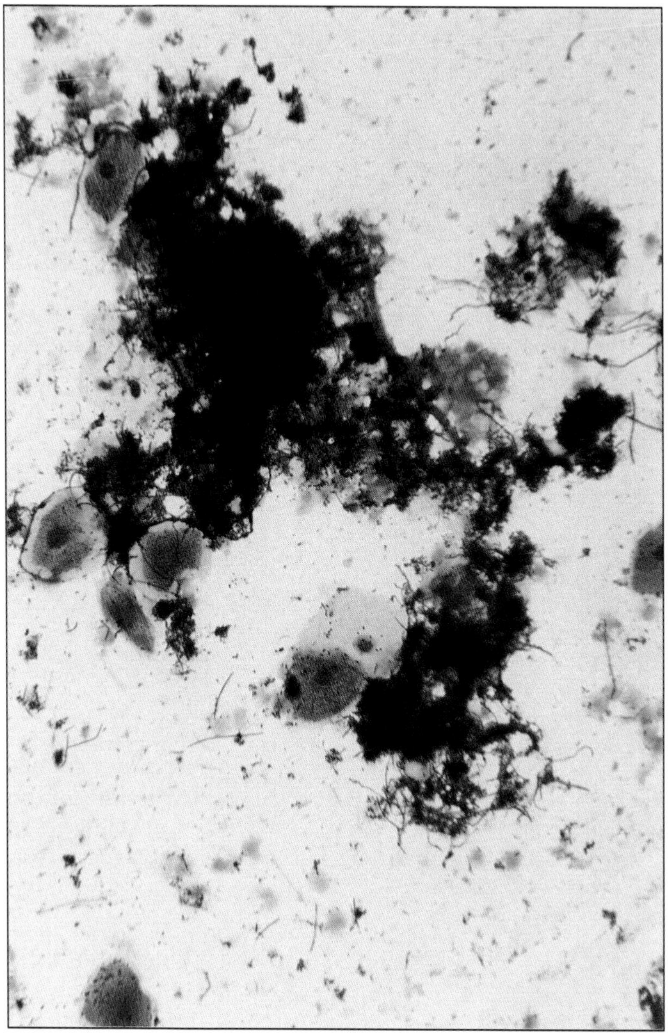

The Bacteria in the Brine

The current claim, sure to evoke an outcry from skeptics, is that individual bacteria, probably in the form of spores, can remain alive for as long as 250 million years. The organisms were recovered from ancient salt crystals collected at a depth of 569 meters (622 yards) from the Permian Salado Formation in Carlsbad, New Mexico. During the work, elaborate safeguards were taken to ensure that contamination of the salt crystals with modern bacteria was highly unlikely. The living material was reportedly found in a few of the many droplets of brine entrapped within the walnut-sized salt crystals. The bacterium is a hitherto unknown *Bacillus* species, designated 2-9-3. Analysis of its 16S ribosomal deoxyribonucleic acid (DNA) has revealed that it is closely related to other known, present-day species of bacteria.

Sterile techniques were used to remove the brine samples from the salt crystals. The recovered brine samples were used to inoculate plates that then grew colonies of bacteria, but it is not known whether the original brine contained whole bacteria or the more resistant spore form. In addition, water seeping through the geologic formation over the ages could have

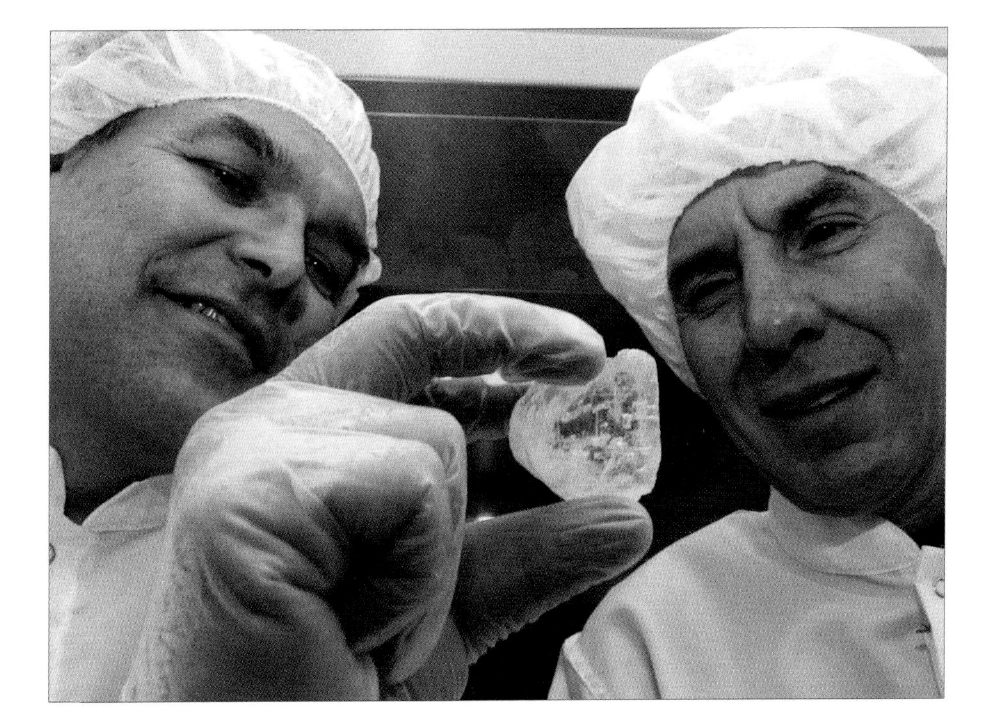

Russell H. Vreeland (left) and William D. Rosenzweig of the Department of Biology of West Chester University in Pennsylvania examine the salt crystal in which they found a 250-million-year-old bacteria. (AP/Wide World Photos)

brought in fresh bacteria. The whole report hinges on the confidence that other researchers will have in the laboratory skills of the authors of the paper, Russell H. Vreeland and William D. Rosenzweig of the Department of Biology of West Chester University in Pennsylvania and Dennis W. Powers, a consulting geologist from Anthony, Texas.

Questions and Considerations

Throughout its life, an organism must repair biopolymers that tend to degrade, an ongoing energy-requiring process that is part of normal metabolism. Over hundreds of millions of years, many such repairs might be required. On the other hand, bacterial spores are viewed as almost inert structures that have extraordinarily minimal metabolic needs. Of course, it is difficult to imagine any living thing surviving so long.

An odd comment in the report is "Once the organism was fully encased, evolutionary pressures would have been relieved." No doubt, natural selection, the accepted mechanism for organic evolution, continues wherever an organism happens to live. Significant evolutionary change usually requires thousands of generations, with natural selection acting over eons on millions of genetically varied and mutating individuals. Perhaps it is more accurate to say that the evolutionary process would have had a minimal effect on a population of organisms that could live individually for hundreds of millions of years. It is worth remembering that such ancient bacteria would predate the age of the dinosaurs.

In the mid-1990's, it was reported that bacterial spores were recovered from a bee encased in amber for 25 million to 40 million years. In that case, the bacterium was a species found living only within bees. That unique circumstance reduced the likelihood of the recovered bacteria being the result of contamination with modern bacteria. In the 1960's, other scientists first began reporting bacteria surviving after millions of years in salt deposits, but the recovery techniques were heavily criticized for their lack of rigor. Even the current findings, which are based on better techniques, need substantiation. For example, it would help to be able to visualize the bacteria or spores in the brine droplets before they are inoculated onto the nutrient plates where they grow into colonies that can be seen.

If such long-term survival is confirmed, it could have far-reaching implications. Among others, British astronomers Fred Hoyle and N. C. Wickramasinghe have theorized that some sort of rudimentary spores may have seeded the universe with life. Such spores would have been resistant enough to survive dispersal within cosmic dust, and any evidence that bacterial spores can live hundreds of millions of years would support this theory. According to Hoyle and others, life would not necessarily have first originated on the earth. Rather, scientists should probably look to the heavens to find where life on earth began.

The evolutionary process would have had a minimal effect on a population of organisms that could live individually for hundreds of millions of years.

J. T. B.

Giant Cloned Algae in California Waters

➤ *Fast-growing clone invades lagoon in Southern California, smothering native fauna and flora.*

On June 12, 2000, an ecological catastrophe was averted with the chance discovery by biologists of an invading giant, cloned alga in a lagoon near San Diego, California. Despite frantic efforts by officials to prevent the notorious seaweed from spreading to U.S. waters by adding the alga to the U.S. federal noxious weed list, the feared alga had arrived. When it moves into a new area, the vigorously growing mutant smothers indigenous fauna and flora on the sea floor under a blanket of green fronds. At the California invasion site, a mat of the alga measuring one-third of a meter (more than a foot) thick had already smothered several hundred square meters of eelgrass, a vital species in the lagoon ecosystem.

The alga is derived from the far more docile, wild alga *Caulerpa taxifolia*, which is found in both Caribbean and Mediterranean waters. The mutant, a cloned variety that grows rapidly and reaches a much larger size than its ancestral form, was developed in the 1980's for ornamental displays at the Stuttgart Aquarium in Germany. Subsequently, curators at aquariums in France and Monaco also employed the luxuriant plant to make their fish tanks more attractive. However, in the middle 1980's, the giant cloned alga escaped from the famed seaside Oceanographic Museum of Monaco into the Mediterranean. At first, the alga occupied only a few square meters of ocean bottom in front of the museum building, and experts believed that within a few years, a chance cold winter would no doubt kill the tropical invader. However, the alga, which can grow to almost 10 feet (3 meters) in length, steadily spread in the Mediterranean so that it now covers more than 6,000 hectares (nearly 15,000 acres) of sea floor.

Most likely, the giant cloned alga in California waters came from a home aquarium that someone dumped into a storm drain.

Despite the seaweed's potential to destroy ecosystems, it continues to be used by some hobbyists to decorate their fish tanks. Small fragments of the plant are inad-

vertently passed from one aquarium to another with the routine transfer of fish or gravel. These fragments quickly grow into whole new plants. Most likely, the giant cloned alga in California waters came from a home aquarium that someone dumped into a storm drain.

Unfortunately, once the alga gets a start, it crowds out all other plants so that instead of a rich diversity of organisms, all that remains is a monoculture of mutant *C. taxifolia*. No longer is there room for the indigenous ecosystem that evolved in the area over millions of years. Furthermore, the fast-growing clone has strong characteristics not seen in its ancestor. For example, it withstands harsh temperature extremes and survives after being out of the water for as long as ten days. In addition, the alga produces toxins that can inhibit the development of other marine organisms. Thus, the invading alga presents a threat against which other organisms have few defenses. The development of any "defenses" by vulnerable members of the disturbed biotic community would depend on their own evolution and natural selection, inherently slow processes requiring eons. Airlines and other means of rapid transportation are

Breeding Like Rabbits

In 1859, Thomas Austin brought twenty-four domestic rabbits from England and released them at his home on the Australian mainland. They multiplied, and just seven years later, sport hunters killed 14,253 rabbits on Austin's property. The rabbits rapidly spread across Australia to the delight of the sportsmen and the dismay of the farmers. The government attempted to eradicate the rabbits by employing trappers and shooters, who had a financial incentive to prolong the problem, and later by constructing fences, which harmed many native species. The soaring rabbit population caused a number of problems, economical and financial. The rabbits changed the composition of the ecosystem, displaced small to medium-sized marsupial natives such as the greater bilby and the burrowing bettong (leading to regional extinction), and caused an increase in the number of feral cats and foxes and thereby a decrease in the number of small native mammals. The efforts to poison and shoot rabbits resulted in the deaths—inadvertent and otherwise—of wildlife such as rat-kangaroos and wombats. In 1950, the disease myxomatosis was deliberately released in Australia, and two years later, the rabbit population had fallen from 600 million to less than 100 million. Myxomatosis is believed to have reduced the population to about 5 percent of its pre-1950 numbers in wetter areas and 25 percent in the drier areas. As time progressed, some rabbits developed resistance to the disease. In 1995, rabbit calcivirus disease, a viral haemorrhagic disease that was still in testing, escaped and spread among the rabbit population. It has substantially reduced the rabbit population, and in 2000, Brian Cooke of CSIRO Wildlife and Ecology received the POL Eureka Environmental Research award for his work with calcivirus.

bringing together species that never before were in competition. Laboratory cloning and other alterations of the genetics of organisms are substantially increasing the rate of mutation or change usually seen in natural species. Never before has the natural world been so threatened. Therefore, prompt human intervention to fight invaders is often essential to protect increasingly vulnerable local ecosystems.

Concern and Intervention

As was reported in the August 3, 2000, issue of *Nature*, the coordination and organization of the required human intervention, as is occurring in California, is challenging. Scientists recruited by government agencies often have technical, scientific concerns that seem irrelevant to bureaucrats or the general public. Susan Williams, the director of the Bodega Marine Laboratory in California, is quoted as complaining that as a technical member of an advisory team, she felt as if she were "shouting into a hurricane" when she tried to make suggestions. The California Water Quality Board is heading up the eradication effort, but it receives input from many individuals, committees, and agencies. Therefore, the federal Invasive Species Advisory Committee, the Aquatic Nuisance Species Task Force, the Institute for Biological Invasions at the University of Tennessee at Knoxville, and the U.S. Department of Agriculture, among other agencies, are all involved.

The book *Killer Algae* (1999) by biologist Alexandre Meinesz of the University of Nice-Sophia Antipolis describes the devastation that could ensue from the unchecked spread of the giant cloned alga. Meinesz shares the concern of those who think that this particular invader will cause worldwide havoc with delicate marine ecosystems. His book recounts the insurmountable problems he faced when he tried to alert various authorities and governments of the potential damage that the invading alga could cause in the Mediterranean.

Upsetting the Ecological Balance

Certainly, there have been many examples of similar ecological disasters, ranging from the introduction of rabbits into Australia to the bringing of house sparrows to the United States. Such immigrant species easily become pests when they no longer are held in check by the natural predators or competitors of their former homelands. A wide variety of foreign plants such as water hyacinths and various shade trees or vines also have proved unsuitable when transplanted to new ecosystems. As many as 4,500 exotic species are believed to have successfully invaded the United States. The foreign animal or plant species invariably has a negative impact on the unique local fauna and flora and, in the worst cases, may cause the extinction of indigenous species and the degradation of the environment. Commercial fisheries and sport fishing have been early victims of the spread of devastating blankets of giant cloned algae in the Mediterranean. Fishing nets have helped spread the alga to new areas as widely traveling commercial fishermen deploy nets that have been contaminated with fragments of the alga.

Although it may be easy to ignore the early stages of such invasions, once the foreign species gains a foothold, it may soon outstrip any human means of countering the onslaught. The intricate interdependence of species, including humans, within an ecological community makes it impossible to predict the ultimate extent of the damage.

J. T. B.

U.S. Grants Threatened Species Status to the Canadian Lynx

➤ *The Fish and Wildlife Service's listing of the Canadian lynx as a threatened species in the lower forty-eight states provides protection to the animal.*

In the spring of 2000, the U.S. government granted threatened species status to the Canadian lynx, *Lynx canadensis*, marking a significant milestone in a complex struggle between various interest groups. The U.S. threatened species status applies only to Canadian lynx in the lower forty-eight states. For now, the Canadian lynx is not considered to be in sufficient peril to be classified as an endangered species.

The U.S. Fish and Wildlife Service's listing of the lynx as a threatened species followed a six-year legal battle that raged over just what place the lynx should have in a increasingly complex world with ever decreasing wilderness. Much of the controversy has centered on the states of Washington, with its natural Okanogan Meadows lynx population, and Colorado, where efforts have been made to reintroduce lynx into areas of interest to the winter sport, ranching, and timber industries. Animal rights activists have been concerned about the suffering and high mortality of lynx trapped in Canada, shipped to the United States, and released into situations in which their survival may be unlikely.

The Lynx in Nature

Few would dispute the beauty of the Canadian lynx, which weighs about twenty pounds, has a short, black-tipped tail, characteristic tufts of hair on the tips of its ears, and long legs with large padded feet that can function as snowshoes. The luxurious silver-white fur of the lynx is long and thick, with a frosted appearance on the back, dark stripes and spots on the lateral surfaces of the body, and a light belly. Individual lynx range over territories encompassing many square miles. They usually hunt at night, making the sighting of a lynx an extremely rare experience, even for trappers and naturalists who each year spend many months in the back country. Preferring secluded old-growth forests, lynx spend their days in crude nests concealed under rocky ledges or logs and prowl during the night under the cover of forest branches and through

A Canadian lynx, named a threatened species in 2000, heads for the woods in the first step of a reintroduction program in February, 1999. (AP/Wide World Photos)

dense thickets, avoiding clearings. Occasionally, a severe winter ice storm may force a lynx to become active in the daylight hours, increasing the chance that it will be seen or, perhaps, be photographed.

Typically, the lynx patiently stalks its prey or waits for hours in ambush rather than trying to run down its prey in an open chase. In the wild, it may eat only every two or three days and feeds primarily on snowshoe hares, rabbits, squirrels, rodents, grouse, and small birds. The lynx favors the snowshoe hare, or *Lepus americanus*, to the point that the survival of a lynx population in a given locality is largely dependent on the availability of an adequate number of these hares. In regions where the lynx must share its habitat with farms and ranches, it may kill and eat small domestic animals such as lambs. This behavior has angered sheep owners, whose way of life is also increasingly threatened in the modern world.

Reintroduction Efforts

The lynx is the only wild cat whose distribution spans the continents of the Northern Hemisphere. Extensive efforts have been made in Europe to preserve the large Eurasian lynx and to reintroduce them into parts of their former range where they have become extinct, including some mountainous regions of Austria, Germany, France, and Italy.

The twenty-pound Canadian lynx's similarity to the domestic cat and the kittenlike appearance of its young were bound to stir the emotions of the general public. This similarity may in part explain why animal rights activists have criticized the care with which the reintroduction programs have been conducted in Colorado. The released lynx were fitted with radio transmitter collars that allowed them to be tracked. The tracking devices revealed that several lynx either starved to death or were killed on highways, resulting in repeated newspaper headlines during 2000. The Colorado lynx reintroduction program is centered in the San Juan Mountains in the southwestern part of the state. Of the forty-one lynx released in 1999, twenty-two had died as of September, 2000. Additionally, fifty-five lynx were released between Interstate 70 and the San Juan Mountains during 2000. It is thought that they may have a better chance of survival because of extra feedings given to fatten them up before release. However, the preferred prey of the Canadian lynx, the snowshoe hare, apparently is not abundant in the release area. The Colorado Division of Wildlife, which has carried out the reintroduction program, has been criticized for not estimating the local snowshoe hare populations with sufficient accuracy to know where to release the lynx.

A further complication is that the snowshoe hare and the Canadian lynx both display about a ten-year cycle in their population densities. The Hudson Bay Company and Statistics Canada have records extending back more than two hundred years documenting the yearly number of pelts offered for sale by trappers. The numbers can vary around one-hundredfold for the snowshoe hares and about twentyfold for the much rarer lynx over the course of the cycle. The ultimate cause of such cycles remains controversial, but their existence means that some years—corresponding to peaks in the population

Captured in British Columbia and transported to Colorado, this Canadian Lynx is being reintroduced into the wild by the Colorado Division of Wildlife. (AP/Wide World Photos)

densities of snowshoe hares—may be better for reintroducing lynx than others. The reintroduction program in Colorado was reportedly timed to match an expected end-of-the- century peak in the local snowshoe hare population.

The Forest Service will have to balance the interests of many different groups who want access to wilderness areas for recreation or to earn their livelihood against its legal mandate to ensure the survival of what is now recognized as a threatened species. Like the American alligator and the bald eagle, in the years ahead, the now partially protected Canadian lynx may be restored to its place in the ecosystem in a few of the lower forty-eight states.

J. T. B.

Cicadas Emerge Four Years Early

➤ *In Ohio, thousands of periodical cicadas emerge four years early.*

In May, 2000, all around Cincinnati, Ohio, hordes of seventeen-year cicadas of the Brood X group emerged four years before their expected year of emergence. Also, during the summer that followed, outbreaks of many billions of seventeen-year periodical cicadas occurred in Georgia, North Carolina, and South Carolina. In 2001, similar outbreaks of seventeen-year periodical cicadas are anticipated in New York, centered in the Finger Lakes region, and outbreaks of thirteen-year cicadas are expected in Louisiana and Mississippi. In 1999, outbreaks of seventeen-year cicadas took place in Ohio, Pennsylvania, Virginia, and West Virginia.

Even though cicadas do not bite, most people find the sudden emergence of thousands of these noisy 1- to 2-inch (2.5- to 5-centimeter) long insects a nuisance during hot summer days. However, the deluge of emerging insects every thirteen or seventeen years and their dominance, at least for a while, of people's yards and woods reminds them of a wilder, more exuberant side of nature not often seen in the urbanized world. Periodical cicadas are winged insects with sucking mouthparts used in feeding on plants. They are members of the insect order Homoptera, along with aphids and leafhoppers. Unlike annual cicada populations, which have some members emerging every year because their two-year life cycles are unsynchronized, the thirteen-year and seventeen-year periodical cicadas have almost completely synchronized life cycles so that huge outbreaks of these insects occur, and then the cicadas are not seen again for many years. However, there are broods of cicadas within each species, and the broods each have their own schedule of emergence. Consequently, during a given year, thirteen-year or seventeen-year periodical cicadas are likely to be emerging somewhere.

A "predator satiation" principle has been invoked to explain the adaptive advantage to periodical cicadas having evolved synchronized, many-year-long life cycles.

The phenomenon seen in Cincinnati of seventeen-year cicadas emerging four years early in 2000 rather than

A seventeen-year cicada emerges from its nymphal skin at a sanctuary in Pennsylvania. (AP/ Wide World Photos)

in 2004 is little understood. These insects seem to have skipped a four-year period of relative inactivity that normally occurs in seventeen-year periodical cicadas just before the year of emergence. The relatively long life cycles of periodical cicadas no doubt hampers cicada study by impatient scientists who generally seek short-term research projects.

An Adaptive Advantage

Six species of periodical cicadas, three with thirteen-year life cycles and three with seventeen-year life cycles, are native to eastern North America and are members of the genus *Magicicada.* (Although periodical cicadas are often referred to as "locusts," true locusts are different insects related to grasshoppers and crickets.) It is thought that the synchronized emergence of periodical cicadas gives them an advantage over their possible enemies. Predators such as birds, foxes, dogs, and cats tend to maintain relatively stable populations that cannot increase in numbers to respond to the sudden bounty of thousands of cicadas to eat during a few weeks every thirteen or seventeen years.

Thus, a "predator satiation" principle has been invoked to explain the adaptive advantage to periodical cicadas having evolved synchronized, many-year-long life cycles. Cicadas spend almost all of their life as nymphs living underground. There they feed on the sap of small roots and go through five juvenile stages before emerging in the spring of their thirteenth or seventeenth year. Typically the nymphs tunnel to the surface and then leave their burrows after sunset. They climb up trees and bushes where they undergo a final molt to become winged adults. The tan exoskeletons of the nymph stage are left behind clinging to the bark of the tree as the newly formed adult moves on a few inches. There it waits a few days for its exoskeleton to harden before flying away. The adults first are a ghostly white color but soon become dark green with red eyes and red wing veins. They live only thirty to forty days after their emergence, a period during which the males relentlessly produce a buzzing noise that is part of the cicada's mating ritual.

The life-cycle length in years of periodical cicadas may have evolved to be a relatively large prime number because this reduces the likelihood of a joint emergence of the two types.

Following mating, the female inserts her ovipositor into the tips of twigs to lay her eggs. After six to ten weeks, the eggs hatch and the tiny first-instar nymphs drop from the trees and then burrow into the soil where they will feed off the sap in roots. There the nymphs remain and go through their molts until their emergence from the subterranean habitat in thirteen or seventeen years.

Generally, seventeen-year cicadas are more northern in their distribution than the Southern and Midwestern thirteen-year cicadas. However, both types can be found in the same locality, and in such locations every 221 years (13 times 17), both types will emerge the same year. Such a joint emergence of thirteen-year and seventeen-year cicadas tried the patience of inhabitants of western Missouri in 1998. In fact, one theory is that the life-cycle length in years of periodical cicadas has evolved to be a relatively large prime number because this reduces the likelihood of such a joint emergence of the two types. A joint emergence would be expected to reduce the reproductive success of both species involved. Perhaps some periodical cicada species of eons ago had shorter or more often conflicting life-cycle lengths, but these may have given way to the present forms.

J. T. B.

Resources for Students and Teachers

Books

Ehrlich, Paul R. *Human Natures: Genes, Cultures, and the Human Prospect.* Washington, D.C.: Island Press, 2000. Distinguished biologist and environmentalist Paul R. Ehrlich makes the case that biological determinism simply cannot be justified from the available scientific data. That is, human behavior cannot be substantially explained by genetic inheritance, and a complex interplay of environmental and cultural forces must be added to whatever part genes may play. Erhlich's opinions stem not from a religious conviction but from an evaluation of the scientific evidence. He suggests that genes "whisper suggestions" rather than dictate human behavior.

Fiffer, Steve. *Tyrannosaurus Sue: The Extraordinary Saga of the Largest, Most Fought Over T-Rex Ever Found.* New York: Freeman, 2000. *Tyrannosaurus rex* fossils can be spectacular, and the specimen described, named Sue, is exceptional in both size and completeness. Sue, excavated on an Indian reservation by private fossil hunters, was eventually sold at auction by Sotheby's for $8.36 million and is on exhibit at Chicago's Field Museum of Natural History. Sue has been a marketing and publicity boom for exhibit sponsors McDonald's and Disney.

Fortey, Richard. *Trilobite! Eyewitness to Evolution.* New York: Alfred A. Knopf/HarperCollins, 2000. This book recounts the career of paleontologist Richard Fortey, whose lifelong passion for collecting trilobites began at age fourteen and led to discoveries of trilobites in geological deposits in China, Thailand, and Australia. The study of these extinct marine invertebrates has contributed to various evolutionary theories, including the "punctuated equilibria" theory of paleontologists Niles Eldredge and Stephen Jay Gould. According to this idea, the rate of organic evolution may occur discontinuously, rather than in a steady and gradual fashion.

Keller, Evelyn Fox. *The Century of the Gene.* Cambridge, Mass.: Harvard University Press, 2000. Provides an overview of the important developments in genetics in the 1900's. These include the rediscovery of Mendel's laws in 1900, Thomas Hunt Morgan's pioneering work with fruit flies in the early decades of the century, James Watson and Francis Crick's elucidation of the double helix structure for deoxyribonucleic acid (DNA) in 1953, and the ultimate success of the Human Genome Project in 2000. This book presents the historic perspective and basic information needed by teachers and students in biology and genetics.

Keynes, Richard Darwin, ed. *Charles Darwin's Zoology Notes and Specimen Lists from HMS Beagle*. Cambridge, England: Cambridge University Press, 2000. Naturalist Charles Darwin's work continues to be central to modern biology and genetics. These particular zoological notes fall somewhere between the full-fledged journal and the pocket-sized field notebooks that Darwin kept. He mined these informal reports for details to be worked into his more polished writings; however, many gems remain in these heretofore unpublished pages. These notes also reveal the more spontaneous and human side of the adventurous young explorer and naturalist.

Laurance, William. *Stinging Trees and Wait-a-Whiles: Confessions of a Rainforest Biologist*. Chicago: University of Chicago Press, 2000. Biologist William Laurance deals with the issue of rainforest fragmentation in Queensland, Australia. He spent eighteen months on location trying to convince authorities to salvage the remnants of this remarkable ecosystem. By describing the exotic flora and fauna of the region, Laurance makes a passionate case for its preservation.

Turner, J. Scott. *The Extended Organism: The Physiology of Animal-Built Structures*. Cambridge, Mass.: Harvard University Press, 2000. In this long-neglected subfield of biology, the challenge is to explain mechanistically, in terms of physiology and genetics, just how animals achieve their marvelous feats of construction. The author claims that no sharp line of demarcation exists between what constitutes the organism itself and what it may construct in the environment. He argues that such distinctions are more a convenience for humans' compartmentalized thinking process rather than an objective truth. The discussion is highly recommended for students interested in yet another unsolved biological mystery.

CD-ROMs and Videos

Australia's Kangaroos. Video. National Geographic, 2000. Traces the adventures of a red kangaroo mother and child in the Australian desert as the child grows to an adult.

Great White Shark. Video. National Geographic, 2000. *Jaws* author Peter Benchley and undersea photographer David Doubilet examine the great white shark in its natural habitat.

Raising the Mammoth. Video. Discovery Channel Video, 2000. Describes the excavating of an entire woolly mammoth from the frozen tundra near the Arctic Circle, where it died about 23,000 years ago.

The Swarm: India's Killer Bees. Video. National Geographic, 2000. Follows a large group of "killer" bees on a one-hundred-mile trip from part of India to the foothills of the Himalaya Mountains.

Web Sites

About Animals/Wildlife
http://animals.about.com/science/animals/
This site features articles on a wide range of animals, from alligators to zebras. Also provides links to wildlife organizations, zoos, and zoological museums.

About Biology
http://biology.about.com/science/biology/
This site looks at current topics in biology and covers many basic concepts. Contains explanations of biology prefixes and suffixes and features a number of simple how-to projects.

About Genetics
http://genetics.about.com/science/genetics/
This site covers the latest topics in genetics as well as presenting background information. Contains links to many related topics in science.

Biological Timing Center
http://www.cbt.virginia.edu
University of Virginia, Brandeis University, Northwestern University, Rockefeller University, and Scripps Research Institute
This university-based center's outreach program is intended to promote a greater understanding of the biological clocks of plants, animals, and humans. This Web site has a teacher's manual and an on-line science experiment that helps explain why in terms of biological clock theory adolescents are sleepier in the morning than adults.

Biology4all
http://www.biology4all.com
Department of Biological Sciences, University of Central Lancashire, United Kingdom
Designed for students, teachers, and the general public. Contains numerous texts and diagrams on biological topics, as well as several hundred links to other Web sites for museums and research institutes, professional organizations as well as those that provide on-line courses and opportunities for undergraduate and graduate training.

Carolina Biological Supply Company
http://www.carolina.com
A tremendous resource for both teachers and students. Articles from Carolina Tips cover topics ranging from deoxyribonucleic acid (DNA) experiments, to wild silk moths, to the contents of owl pellets. A Teacher's World section gives tips for grant writing, procedures for laboratory safety, and other teaching information.

Department of Botany, National Museum of Natural History
http://www.nmnh.si.edu/departments/botany.html
Smithsonian Institution
This site describes the museum's collections. It provides information on the museum's research and allows users to search through the department's photographs and illustrations.

Endangered Species Program
http://endangered.fws.gov/
U.S. Fish and Wildlife Service
The official site of this governmental agency describes the efforts being made to conserve and restore endangered species. Provides information on current programs, legal issues, and a Kid's Corner, for young students and their teachers.

High School Hub
http://www.highschoolhub.org
The biology section offers many intriguing experiments and newsy items sure to interest young scholars.

Human Genome Project Information
http://www.ornl.gov/TechResources/Human_Genome/home.html
Human Genome Program
This site describes the Human Genome Program and the research that has been done. It contains numerous teaching aids, including presentations and videos.

Internet Resource Guide for Zoology
http://www.york.biosis.org/zrdocs/zoolinfo/zooinfo.htm
Contains detailed descriptions of thousands of animals, organized by taxonomic group.

National Association of Biology Teachers
http://www.nabt.org
The association's site offers a wide variety of membership benefits such as conventions, a database of job openings for teachers, a professional journal, and special publications. Its award winning laboratory guides enable students at all levels to learn about biology and genetics through hands-on experiments and demonstrations.

Science Kit and Boreal Laboratories
http://www.sciencekit.com
Features biology laboratory experiments for students in middle school and above.

UCMP Exhibit Halls: Evolution Wing
http://www.ucmp.berkeley.edu/history/evolution.html
University of California, Berkeley
Features an explanation of the theory of evolution and links to a site designed to help teachers educate students about evolution. The Evolution Forum contains activities, resources, teaching tips, and documentation.

World Wildlife Fund's Global Network
http://www.wwf.org/
World Wildlife Fund
Provides information on the organization and its ongoing efforts as well as current environmental news. Also contains a Kids and Teachers area with fact sheets, teachers guides, and other educational tools.

M. J. R. and J. T. B.

5 · Mathematics and Economics

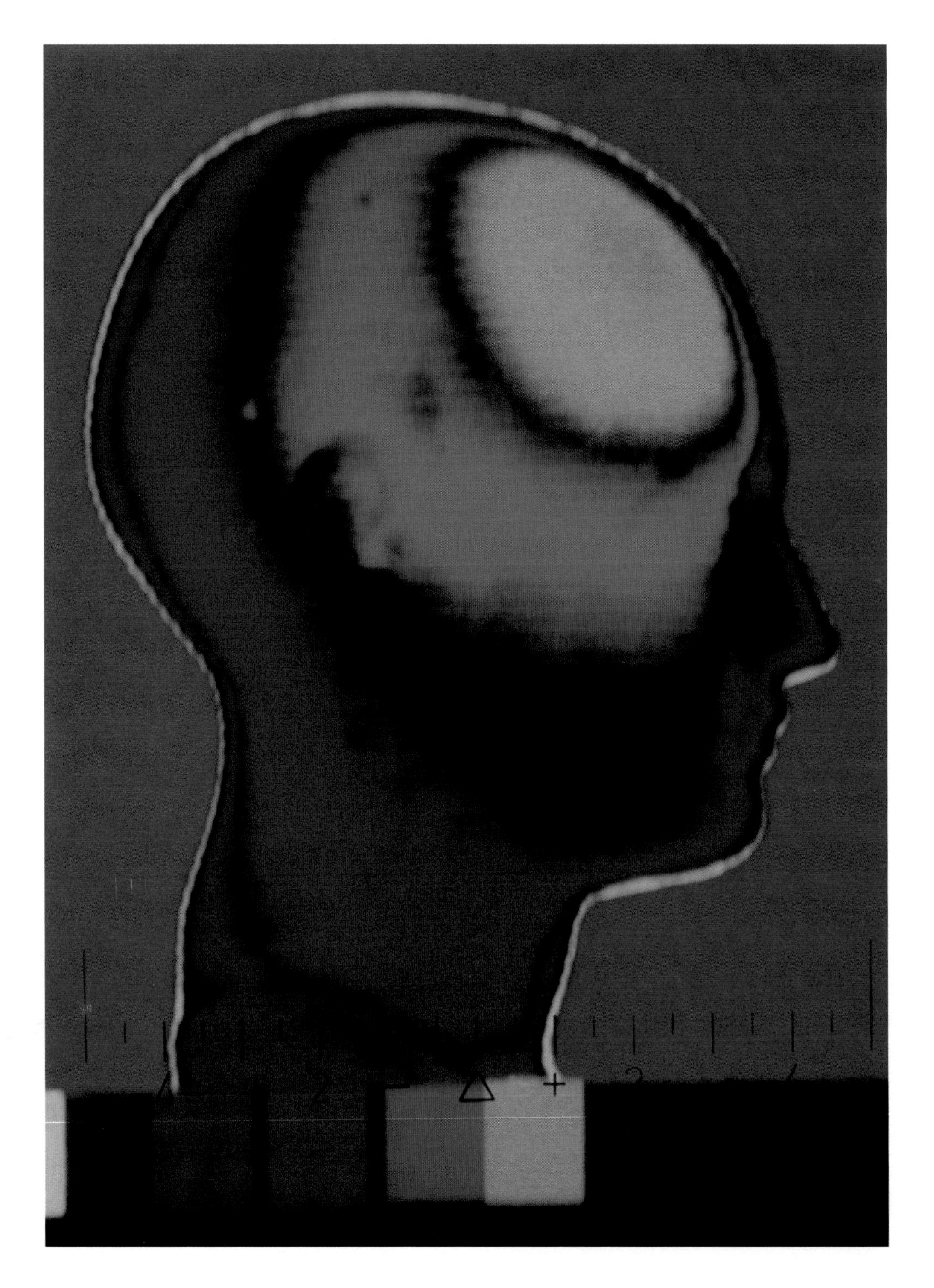

$$\frac{1}{\Gamma(z)} = ze^{\gamma z}\prod_{n=1}^{\infty}\left[\left(1+\frac{z}{n}\right)e^{-z/n}\right] \quad (|z|<\infty)$$

$$\sqrt{}$$

$$\Gamma(z) = \lim_{n\to\infty}\frac{n!\,n^{z}}{z(z+1)(z+2)\ \ldots\ (z+n)} \quad (z \neq 0,-1,-2,\ldots)$$

$$\left(\sum(a_i b_i)\right)^2 \leq \left(\sum a_i^2\right)\left(\sum b_i^2\right)$$

The Year in Review

➤ *Zachary Franco*
Community School of Naples, Florida

From the expensive analysis of the Y2K bug to the counting of popular and electoral votes during the presidential election, mathematics appeared to affect many aspects of people's lives in 2000. E-commerce continued to grow, and an Advanced Encryption Standard was selected for use in the twenty-first century. The Clay Mathematics Institute offered seven $1 million prizes for solutions to important mathematics problems such as the Riemann hypothesis.

E-Commerce and Encryption

E-commerce, once limited to business-to-business information exchange, grew from the occasional sale of an item posted by an individual on a computer bulletin board

Amazon.com chief executive officer Jeff Bezos and electronics vice president Christopher Payne demonstrated the company's new wireless phone Web site in Seattle right before the Christmas season. (AP/Wide World Photos)

to a projected $61 billion in sales in the year 2000 by a seemingly endless array of electronic shops. Another area of growth was in consumer-to-consumer auctions, where sales were expected to reach $3.1 billion for the year. However, although some companies have flourished by selling books, flowers, financial and travel services, and subscription services for adult entertainment over the Internet, other dot-coms have failed. One of the concerns expressed by customers of online firms is security. In order to safely process a credit card transaction over the Internet, card numbers are encrypted using an algorithm. Other sensitive material, such as government documents and data, is also encrypted. As technology advanced, it became obvious that the current standard, which was developed in 1977 and uses a 56-bit key, would soon be obsolete, so in 1997, the National Institute of Standards and Technology announced a competition for the development of a better encryption system, to be known as the Advanced Encryption Standard.

On October 2, 2000, the National Institute of Standards and Technology announced the winner of its competition. Two Belgian cryptographers, Vincent Rijmen and Joan Daemen, used algebra, combinatorics, and number theory to create a very secure cryptosystem called Rijndael, using 128-, 192-, and 256-bit keys. Their method will be used starting in 2001 to encrypt passwords, credit card numbers, financial transactions, and other sensitive data.

The Mathematics of Choice

The 2000 Nobel Prize in Economics went to James Heckman of the University of Chicago and Daniel McFadden of the University of California, Berkeley, for their work in microeconometrics. Both developed statistical techniques of analyzing microdata (economic information about individual and household behavior) related to selection processes and are interested in their applications to social problems. The greater availability of microdata and powerful computers has led to studies of individual choice among a finite set of decision alternatives, called discrete choice analysis.

James J. Heckman, Nobel Prize in Economics. (AP/Wide World Photos)

By means of discrete choice analysis, researchers are able to determine which alternative, from a set of possible alternatives, a person is likely to choose to accomplish a task or reach a goal. For example, discrete choice analysis can determine whether people will choose to use a particular mode of transportation such as walking, driving, or

Daniel L. McFadden receives the Nobel Prize in Economics from Carl XVI Gustaf, king of Sweden. (AP/Wide World Photos)

using a train, given known data attributes regarding those individuals' preferences. Some attributes that affect people's decisions remain unobservable and difficult to factor out for each individual. For example, the earning power of college graduates can be observed only for those who have graduated college. We cannot observe what a person's salary would have been had he or she chosen a different alternative.

Bus, Train or Car?

An example of discrete choice analysis is the analysis of the demand for urban travel. Let A represent the set of data attributes one has collected for a society of individuals and I = {walk, bus, car, train . . . } represent the set of travel alternatives. Let $P(i \mid a, I)$ be the probability that an individual will choose alternative i from set I, given observed attributes $a \in A$.

The conditional logit model devised by McFadden gives an estimate of this probability under the following assumption: An individual's preference (utility) for the alternative i can be represented as the sum of two components: society's preference for alternative i, $v(i, a)$ and the individual's preference, $\varepsilon(i, a)$. Then,

$$P(i \mid a, I) = \int_{-\infty}^{+\infty} \frac{\partial F}{\partial x_i} [x + v(i,a) - v(1,a), \ldots, x + v(i,a) - v(I,a)] \, dx$$

where F is the (joint cumulative) distribution function (d.f.) for the individual's preference $\varepsilon(i, a)$

Also, $P(i \mid a, I) = \exp(i,a)/\Sigma_{j \in I} \exp(j,a)$ if the $\varepsilon(i,a)$ are independently distributed.

Mathematics in Entertainment

Although mathematics seems an unlikely topic for the stage or a popular work of fiction, several mathematical plays opened in 2000 and a mathematical novel was published with great fanfare. On April 27 and May 24, *The New York Times* reviewed the plays *The Five Hysterical Girls' Theorem* by Rinne Groff and *Proof* by David Auburn. An off-Broadway musical called *Fermat's Last Tango* dealt with seventeenth century mathematician Pierre de Fermat's last theorem. A popular mathematics novel was Apostolos Doxiadis's *Uncle Petros and Goldbach's Conjecture* (2000). Faber and Faber and Bloomsbury, the British and U.S. publishers of the English translation of the 1992 Greek novel, offered a million-dollar prize to the first person to solve Goldbach's conjecture.

Fermat's Last Tango was inspired by Princeton University professor Andrew Wiles's efforts to solve Fermat's last theorem. In 1630, Fermat wrote a marginal note in his copy of Diophantus's *Arithmetica* regarding the statement that there are no positive integers x, y, z and $n \geq 3$ for which $x^n + y^n = z^n$:

> I have discovered a truly remarkable proof which this margin is too small to contain.

However, he wrote no more on the topic. Wiles secretly worked in his attic for seven years on this problem, the most famous unsolved problem in mathematics. Then, in

1993 at a conference in Cambridge, England, he asked if he could speak for three hours instead of the usual one hour. It was not until the end of his third talk that Wiles announced that he had proved Fermat's last theorem. The details would fill more than two hundred pages, and after lengthy review by the few who could understand it, the proof was found to be incomplete. With the help of Richard Taylor, however, this gap was filled in 1995.

The methods leading to the solution, the study of elliptic curves, led to many more interesting problems. Perhaps the best known of these is the ABC conjecture. It states that if three integers a, b, c have no common divisor and $a + b = c$ then c cannot be much larger than the product of the prime numbers dividing abc. More precisely, if $P(a, b, c)$ denotes the product of the primes dividing abc and $\epsilon > 0$, there is a constant K depending only on ϵ for which

$$c < K[P(a, b, c)]^{1+\epsilon}$$

Many scientists believe that a proof of the ABC conjecture would simplify proofs of theorem such as Fermat's last theorem because the ABC conjecture implies Fermat's last theorem (that is, Fermat's last theorem is a special case of the ABC conjecture). Let $a = x^n$, $b = y^n$, and $c = z^n$. Then, the ABC conjecture says that

$$z^n < K(xyz)^{1+\epsilon} < K(z \times z \times z)^{1+1/4} = Kz^{3.75}$$

This implies that $n = 3$, a case known to be true in the 1700's.

Solutions and Education

On May 24, 2000, the Clay Mathematics Institute offered a million dollars each for solutions to the following problems: the Riemann hypothesis, P versus NP, the Poincaré conjecture, Yang-Mills existence and mass gap, the Navier-Stokes existence and smoothness equations, the Birch and Swinnerton-Dyer conjecture, and the Hodge conjecture. The institute decided to offer these prizes as a way of emphasizing the importance of mathematics in everyday life. Unlike many contests, the winning of any of these prizes is a multiyear process, requiring publication of the solution and peer review and approval.

Two Belgian cryptographers used algebra, combinatorics, and number theory to create a very secure cryptosystem called Rijndael.

Although only a few people may be capable of solving these complex problems, the National Council of Teachers of Mathematics hopes to raise the level of mathematical ability among the general populace. In an effort to update U.S. mathematics standards, the institute issued the *Principles and Standards for School Mathematics* (2000). The book, based on various groups' studies of mathematics education in primary and secondary schools, makes recommendations for improving the teaching of mathematics in the United States. Mathematics education has come under fire in recent years as a result of the poor performance of U.S. students compared with Asian students in international math tests. Critics of the council's effort claim, however, that very little has been altered and that the standards still place too much emphasis on calculators and discourage memorization.

High-Volume E-Trading
Using Math to Make Money

➤ *Investors go online, trying to turn stocks' volatility into profit.*

The stock market of 2000 has been compared to a roller-coaster ride. Although the Dow Jones Industrial Average stayed within one thousand points of its average for most of the year, it moved more than that amount every month of the year. An investment in the popular market index for an entire year is subject to many gains and losses, but they cancel one another out to reduce the net gain or loss. In fact, the price of a stock or market index will be reported "unchanged" for the day even if it takes a roller-coaster ride that closes at its opening price. Like the coastline of a country, the closer one looks, the more detail one finds. If enough people were to trade a stock, it could take a roller-coaster ride every hour. This variation is called volatility.

Market analysts describe a stock's volatility by its beta value. Thus if the market (index) loses 1 percent of its value, a stock with beta of 1.5 would expect to lose 1.5 percent of its value. For example, the betas for McDonald's (MCD), IBM, and America Online (AOL) are 0.61, 0.99, and 1.69, respectively. Volatility makes most investors nervous because in a volatile market, stocks can lose their value very quickly. In 2000, it was not uncommon for a stock to lose 20 percent or more of its value in a single day.

A small number of investors thrive and try to take advantage of volatility. Perhaps gambling is a more accurate description of what they are doing, because they never own shares in a company long enough to be considered investors. Indeed, there is more opportunity to gain (and lose) in a volatile market. Furthermore, the more predictable the swings are, the easier it is to profit. For example, if a stock price continually oscillated between \$10 and \$20, one would simply buy it each time it reached \$10 and sell it each time it reached \$20.

Maximum Total Variation

The maximum total variation of the price of a stock can be determined mathematically, providing an aid to investors. Let $f(t)$ represent the price of a certain stock at time t: Let $t_1, t_3, t_5, \ldots$ denote the times when the stock price is at a peak (the price rises immediately beforehand and falls immediately afterward) and $a = t_0, t_2, t_4, \ldots, t_n = b$ denote the low points (a = beginning of day or month, b = end of day or month). The net gain of the stock on the interval $a \le t \le b$ is $G = f(b) - f(a)$, but the total variation on that interval is

$$V(f) = [f(t_1) - f(t_0)] + [f(t_1) - f(t_2)] + [f(t_3) - f(t_2)] + \cdots [f(t_{2n+1}) - f(t_{2n})] = 2\sum_{k=0}^{n} f(t_{2k+1}) - f(t_{2k+2}) + G$$

$V(f)$ represents the combined amount by which f increases plus the amount by which it decreases. In a volatile market, the little rises and dips made by a stock (variation) add up to a lot of movement (total variation). It is possible to make money when the market goes down. Betting against the market is called taking a short position.

Note that when making several consecutive investments, the returns are multiplicative, not additive. Consider the investor who buys a stock at \$10 a share, sells it at \$15, and uses the proceeds to buy a second stock at \$20. If he or she sells the second stock at \$30, the net effect is that his investment grows by a factor of

$$\frac{15}{10}\frac{30}{20} = \frac{9}{4} = 2.25$$

for a 125 percent return. The actual dollar gain depends on the initial investment. The length of time required to achieve the return is also important. If the foregoing example takes place in a period of one year, the annual rate of return is 125 percent. If it took six months, then the 9/4 growth factor could occur twice in a year, yielding a factor of

$$\frac{9}{4} \times \frac{9}{4} = \frac{81}{16} = 5.0625$$

for an annual rate of return of 406.25 percent. Allowing for taking short positions, the maximum that a \$1 investment can become is the product of the maximum returns, or

$$M(f) = \frac{f(t_1)}{f(t_0)}\frac{f(t_1)}{f(t_2)}\frac{f(t_3)}{f(t_2)}\frac{f(t_3)}{f(t_4)} = \left(\prod_{k=0}^{n}\frac{f(t_{2k+1})}{f(t_{2k+2})}\right)^2 \frac{f(b)}{f(a)}$$

E-trade in Palo Alto, California, is one of the largest online brokerages. (AP/ Wide World Photos)

In other words, $\log M(f) = V(\log f)$. For large intervals $[a, b]$, the net return $f(b)/f(a)$ is usually very small compared to $M(f)$. Although market indices and their daily changes were higher in 2000 than they have been in past years, their relative variation $V(f)/f$ has also risen, especially in the Nasdaq issues.

Individual Investors Online

The number of investors using cheap online brokerage accounts and their data skyrocketed in 2000. The fifteen largest firms make up 96 percent of the online brokerage market. The three largest are Charles Schwab, with 4.1 million online accounts; E-Trade, with 3 million; and Ameritrade, with 1.3 million. With electronic trading easily available, new breeds of investors have emerged. Many investors now focus on short-term opportunities in the market. They may trade dozens of times a month, profiting from small share price movements.

Among these online investors, the number of traders called day traders rose dramatically in 2000, estimated at fifty thousand individuals. Day traders try to make a living by buying and selling large quantities of stock, sometimes within minutes, to capitalize on a small price increase.

They must contend with the possibility that a stock will not go up, but day traders are quick to cut their losses and generally trade in stocks that are known to be volatile. It is not unheard of for day traders to make one thousand trades in a day, knowing little about the stocks they traded. They often buy after a stock reacts to bad news, gambling that the reaction was an overreaction and that the stock will soon bounce back. Day traders also subscribe to additional trading information called level two quoting services set up by the Nasdaq exchange. Normal (level one) quotes list only the best bid and ask prices. A level two quote lists bids and asks for the stock at all levels, who ordered how many shares, when, which orders were filled, among other information.

The maximum total variation of the price of a stock can be determined mathematically, providing an aid to investors.

There is a dark side to the stressful lifestyle of a day trader. The success rate is low, and day traders often gamble with borrowed money (called trading on margin). Even if stocks continue to be volatile, they cannot support an unlimited number of successful day traders. Eventually, as their numbers increase, day traders will have to compete with each other for smaller gains, and those without sufficient capital to ride out a flat market will fail. However, individual investors continue to seriously consider doing this full time and many master traders are writing books about their secret strategies. *How to Get Started in Electronic Day Trading: Everything You Need to Know to Play Wall Street's Hottest Game* (2000), by David Nassar, and *The Electronic Day Trader: Successful Strategies for On-line Trading* (2000), by Marc Friedfertig and George West, which topped the electronic bookstore Amazon.com's best-seller list, were very popular in 2000.

More People Go Shopping with a Mouse
The Future of E-Commerce

➤ *E-commerce continues to grow amid concerns about its profitability and sales-tax-free status.*

Just as the Internet was born decades ago to service the military and scientific communities, e-commerce was born in the 1970's when banks started transferring funds electronically and corporations began using Electronic Data Interchange (EDI) to share information with partners and suppliers. Now, the term "e-commerce" refers to much more than these business-to-business (B2B) transactions. Only a few years ago, shopping by computer was limited to a few classified ads on electronic bulletin boards. Today, most major corporations have Web sites where customers can access information about the company and its products, and numerous retailers—both purely online companies and those that also have brick-and-mortar shops—offer a wide array of products online. With the help of 128-bit encryption, secure sites allow for safe credit card transactions and help millions of customers feel safe when purchasing items online.

Internet retail revenues have been approximately doubling each year, from $600 million in 1996 to a projected $200 billion by 2004. Some analysts' estimates are even higher, placing 2000 sales at $61 billion. The Internet commerce research firm Jupiter Communications projects e-commerce revenues for just the 2000 holiday season (November and December) to total about $12 billion compared with $7 billion in 1999 and $3.1 billion in 1998. Another area of note is the consumer-to-consumer online auction sector, where sales are expected to grow from $3 billion in 1999 to $15 billion in 2004. However, business-to-business transactions will continue to represent the largest revenues. The *eMarketer* newsletter projects revenues of $268 billion by 2002, up from $5.6 billion in 1997. Another aspect of e-commerce that drives profits is advertising. When measured by revenue, online advertisements account for 6 percent of all U.S. advertisements.

Many brick-and-mortar merchants have been reluctant to create Web sites because they are not confident that the online market is large enough to justify their expenses.

Tax Moratorium

In the mid-1990's, the growth of e-commerce outpaced the ability of the U.S. government to decide how to tax these staggering revenues, especially given the great uncertainty regarding the future shopping habits of the consumer. To make a fair assessment and to encourage the growth of e-commerce, in 1997 the government approved a tax moratorium on Internet commerce, the Internet Tax Freedom Act, that would be in effect from October, 1998, to October, 2001. Three years later, legislation was proposed that would extend this moratorium until October, 2006. The bill passed the House of Representatives 352-175 in May, 2000. Because of opposition from some state governments, a two-year extension is more likely to pass the Senate and obtain the president's signature. Although the bill was sent to the Senate, no action was taken on it before the year's end.

As it was in 1997, the Internet Tax Freedom Act is unpopular with the nation's governors. Local governments claim they are losing 10 percent of sales tax revenue because of the Internet and are campaigning hard against the extension. Not all states opposed the Internet's tax-free status, however. California governor Gray Davis made the following statement to the California Assembly:

> In order for the Internet to reach its full potential as a marketing medium and job creator it must be given time to mature . . . at present, it is less than ten years old. Imposing sales taxes on Internet transactions at this point in its young life would send the wrong signal about California's international role as the incubator of the dot-com community.

Profitability

In general, although Internet sales are soaring, profits are not. Most notable is the example of Amazon.com, the online seller of books and other products, which continues to post losses despite huge sales. Many businesses have failed to meet revenue expectations and incur large expenses setting up and maintaining their Web sites. Pets.com, an online company selling pet food and related products, went out of business in November, 2000, and its Web address was purchased by Petsmart.com, the online version of the brick-and-mortar pet superstore Petsmart. Some online merchants were not able to keep up with the large number of orders that arrived during the 1999 holiday season, producing a backlog that resulted in late

In February, North Dakotan antique dealers Shane Balkowitsch and Sharon Balkowitsch were selling more than four hundred items a month on eBay, a popular Internet auction site. (AP/Wide World Photos)

Christmas presents and high levels of customer dissatisfaction. Many brick-and-mortar merchants have been reluctant to create Web sites because they are not confident that the online market is large enough to justify their expenses. Others worry about the process of fulfilling online orders and the level of security the Web offers, especially for credit card transactions.

Shoptok, an e-commerce consulting firm, lists six kinds of e-shoppers and offers marketing recommendations that target each group. The new-to-the-Net shopper typically uses the Web to purchase small and safe items and responds to product pictures and validation signs from organizations such as the Better Business Bureau. The reluctant shopper uses the Web to research purchases, planning to buy offline, and will respond to clearly stated security and privacy policies. The bargain shopper uses comparison shopping tools to find the best price and responds to sections featuring discounted or sale-priced items. The surgical shopper knows exactly what he or she wants before going online and responds to product reviews, configurators with many options, and informative customer service representatives. The enthusiastic shopper purchases frequently and adventurously and responds to dynamic merchandise demos and personalized product recommendations. The power shopper shops out of necessity and responds to convenient navigation tools, quick access to information, and customer support.

Perhaps a seventh kind of e-shopper should be added, the criminal shopper. Many larger Internet businesses are finding too many credit card orders being charged back, many the result of stolen credit cards. Without a signature on a credit card order, the law has generally been on the consumer's side. Federal law caps a consumer's liability to fifty dollars, so Internet businesses that are not careful can end up with heavy losses. According to First Data Corporation, the United States' largest credit card processor, 1.25 percent of all Internet transactions are charged back, compared with 0.33 percent of mail orders and 0.14 percent of storefront retail transactions. In a survey of 156 retailers with median revenue of $250 million, the firm Gartner found that 2.64 percent of their Internet transactions are charged back, compared with about 1.24 percent of their offline transactions. Many of the merchants said they believed that more than one-third of the Internet charge backs involved stolen credit cards. These businesses are caught in the middle: If they publicize the fact that credit cards are being stolen online, they will lose business, but if they continue to ignore the problem of credit card theft, they will go out of business. In 2000, the U.S. attorney general and the Federal Bureau of Investigation cyber crime unit were working on a solution to this serious problem.

The Future of E-Commerce

Young people are generally more at ease with computers and are more likely to shop online. However, this group has relatively small purchasing power. Over the next few years, they will become larger consumers, helping to drive up e-commerce sales. In fact, many of them will be employed in fields related to e-commerce. In its listing of the highest rated jobs, the 1999 *Jobs Rated Almanac* included several that are essential to e-commerce. The top ten listed, from one to ten, were Web site manager, actuary, computer systems analyst, software engineer, mathematician, computer programmer, accountant, industrial designer, hospital administrator, and Web developer.

There is no doubt that e-commerce is here to stay. It may take a while to iron out the numerous problems, but its convenience is unparalleled. In many areas, such as travel, entertainment, computers, and flowers, it is already profitable.

Factoring Large Numbers
Internet Security

➤ *Internet security is enhanced by the selection of a new algorithm as the encryption standard.*

In 2000, after a three-year search, the National Institute of Standards and Technology selected an algorithm called Rijndael, devised by two Belgian cryptographers, to be the Advanced Encryption Standard (AES). Although the algorithm was not to become the standard until the summer of 2001, the October 2 announcement marked the end of IBM's reign as standard-bearer, which began in 1977, when its algorithm was adopted as the standard. In related news, two weeks earlier, on September 20, the seventeen-year-old patent for the RSA public-key algorithm officially expired, and the algorithm was released to the public domain.

IBM's algorithm, the Data Encryption Standard (DES), was the most widely used cryptosystem in the world. It was used to encrypt credit card numbers conveyed through the Internet, computer passwords, television broadcasts, and many other electronic exchanges. DES was created by IBM in 1977 in response to a public solicitation for a cryptographic algorithm. It used a 56-bit key, which remained secure for much longer than expected. However, faster computers, better factoring methods, and the ability of thousands of PCs to work together via the Internet have combined to make DES no longer secure. In 1997, the National Institute of Standards and Technology announced a competition for a replacement for DES, to be called the Advanced Encryption Standard (AES). Out of twenty-one submissions from cryptographers and companies all over the world, including IBM, the institute selected Rijndael (pronounced "rain-doll") for the Advanced Encryption Standard. Rijndael, written by Belgian cryptographers Joan Daemen and Vincent Rijmen, was selected for its simple and elegant design, speed, and compactness in hardware and on smartcards. It allows for 128-, 192-, and 256-bit keys and makes use of the finite field of 256 elements generated by the irreducible polynomial $x^8 + x^4 + x^3 + x + 1$.

The best current encryption algorithms use basic facts of number theory known to mathematicians hun-

President Bill Clinton (far right) discusses ways to improve security on the World Wide Web with members of the computer and Internet industry. Clockwise from Clinton are Vinton Cerf, MCI Worldcom; Christos Cotsakos, E-Trade; Douglas Busch, Intel; Michael McConnell, Booz-Allen; Attorney General Janet Reno; and Peter Solvik, Cisco Systems. (AP/ Wide World Photos)

dreds of years ago. They rely on the assumption that factoring a number that is the product of two large prime numbers is difficult. It is not known if this assumption is valid. No fast factoring method is currently known, and a proof that none exists has yet to be exhibited. A factoring algorithm is considered fast if it takes polynomial time. That is, for $n > 1$, there is an exponent p for which it takes less than n^p seconds to factor an n digit number. The method one learns in school for factoring a number N involves up to $\sqrt{N}$ trial divisions. If N has n digits, $N \approx 10^n$ so the number of divisions is about $10^{n/2}$, which is not polynomial time. Deciding if a polynomial time factoring algorithm exists is one of the most important questions of computer science.

In 1977, Massachusetts Institute of Technology professors Ronald Rivest, Adi Shamir, and Leonard Adleman (RSA) developed an encryption algorithm (known as the RSA algorithm) to help Internet security. Surprisingly simple, it also allows the encoding algorithm to be public so that posting one's public key enables one to receive secure messages from anybody. Public-key algorithms use two separate keys, a public key and a private key. The data are encoded using the public key, and decoding is done with the private key. Because only the public key is transmitted, it is possible to use an ordinary channel rather than a secure channel. The algorithm can be described with just a small background in number theory.

A number n is called prime if its only divisors are n and 1. Euclid proved that there are infinitely many prime numbers. In fact, there are about $x/\log x$ prime numbers less than x. The fundamental theorem of arithmetic states that every whole number is the product of prime numbers and the representation is unique apart from the order in which they are multiplied. For example, $72 = 2 \times 2 \times 2 \times 3 \times 3 = 2^3 \times 3^2$.

Euclid's Algorithm

Euclid's algorithm shows how successive division of remainders of two numbers will produce their greatest common divisor (gcd) without factoring them. It also shows how to express the gcd as a combination of the two numbers: For example, the gcd of 48 and 72 is 24, and $24 = (-1)48 + (1)72$. The gcd of 17 and 29 is 1 because they have no common factor other than 1. To see how Euclid's algorithm works with $m = 17$ and $n = 29$; first divide 17 into 29. Continue by dividing each remainder into the previous divisor and write the results so the same numbers (like the 12) line up diagonally from top right to bottom left. Then solve each equation for the remainder as shown in the third column of the table. The number just above the 0 in column three is the gcd: In this case, it is 1 as seen before.

EUCLID'S ALGORITHM EXAMPLE

Step Number	Division Result	Remainder Equals
1	$29 = 1 \times 17 + \mathbf{12}$	$12 = 29 - 1 \times 17$
2	$17 = 1 \times \mathbf{12} + 5$	$5 = 17 - 1 \times 12$
3	$\mathbf{12} = 2 \times 5 + 2$	$2 = 12 - 2 \times 5$
4	$5 = 2 \times 2 + 1$	$\boxed{1} = 5 - 2 \times 2$
5	$2 = 2 \times 1 + 0$	$0 = 2 - 2 \times 1$

In the gcd equation $1 = 5 - 2 \times 2$, substitute for 2 the expression on the line above it $(2 = 12 - 2 \times 5)$ to find

$$1 = \mathbf{5} - 2 \times (\mathbf{12} - 2 \times \mathbf{5}) = (1 + 2 \times 2) \times \mathbf{5} - 2 \times \mathbf{12} = 5 \times \mathbf{5} - 2 \times \mathbf{12}$$

Now, in the equation $1 = 5 \times \mathbf{5} - 2 \times \mathbf{12}$, substitute for 5 the expression on the line above it $(5 = 17 - 1 \times 12)$ to find

$$1 = 5 \times (\mathbf{17} - 1 \times \mathbf{12}) - 2 \times \mathbf{12} = 5 \times \mathbf{17} + (5 \times -1 + -2) \times 12$$
$$= 5 \times \mathbf{17} - 7 \times \mathbf{12}$$

Finally, in the equation $1 = 5 \times \mathbf{17} - 7 \times \mathbf{12}$, substitute for 12 the expression on the line above it $(12 = 29 - 1 \times 17)$ to find

$$1 = 5 \times \mathbf{17} - 7 \times (\mathbf{29} - 1 \times \mathbf{17}) = 12 \times \mathbf{17} - 7 \times \mathbf{29}$$

This expresses 1 as a combination of 17 and 29.

Finally, use a result of Euler that states that if p and q are prime numbers that do not divide T; then $T^{(p-1)(q-1)}$ leaves a remainder of 1 when divided by pq. If $p = 17$ and $q = 29$, this says that if T is not divisible by 17 or 29 then $T^{16 \times 28} = T^{448}$ leaves a remainder of 1 when divided by $17 \times 29 = 493$.

The Public-Key Algorithm

The public-key algorithm has a trap-door function that is easy to compute, but its inverse is very difficult to compute (as it requires factoring a large number). This makes it relatively simple for senders to encrypt their messages and hard for people to decipher them. The algorithm begins with two large primes p and q for which $p \pm 1$ and $q \pm 1$ have large prime factors. Then a number d is chosen that has no factors in common with $p - 1$ or $q - 1$: The gcd of d and $(p - 1)(q - 1)$ is therefore 1. By Euclid's algorithm, it is possible to express 1 as a combination of d and $(p - 1)(q - 1)$. This means it is possible to find numbers e and f for which

$$1 = de + (p - 1)(q - 1)f \qquad (1)$$

If T is the message to be encoded and $T < pq$, raise T to both sides of equation (1)

$$T^1 = T^{de+(p-1)(q-1)f} = T^{de} \times [T^{(p-1)(q-1)}]^f \qquad (2)$$

Divide both sides of equation (2) by pq and use the fact that the remainder of a product is the product of the remainders. Then since $T^{(p-1)(q-1)}$ leaves a remainder of 1, T equals the remainder when T^{de} is divided by pq. So to send a message T to someone, you need to know only e and the product pq. Simply compute the remainder when T^e is divided by pq and send it to the other person. The receiver can decode the message by raising it to the dth power and dividing it by pq. The remainder will be T because $(T^e)^d = T^{de}$ leaves a remainder of T when divided by pq.

Return to the example in which $p = 17$ and $q = 29$ (for illustration purposes only—this clearly would not lead to a secure encryption). Choose $d = 11$ and use Euclid's algorithm to find f so that $1 = 11e + 16 \times 28 \times f$.

$$448 = 40 \times 11 + 8$$
$$11 = 1 \times 8 + 3$$
$$8 = 2 \times 3 + 2$$
$$3 = 1 \times 2 + 1$$

Therefore, $1 = 3 - 1 \times 2 = 3 - 1 \times (8 - 2 \times 3) = 3 \times 3 - 1 \times 8 = 3 \times (11 - 1 \times 8) - 1 \times 8 = 3 \times 11 - 4 \times 8 = 3 \times 11 - 4 \times (448 - 40 \times 11) = 163 \times 11 - 4 \times 448$, so $e = 163$ and $f = -4$. Suppose the message you want to send is 241. Using the encode key $e = 163$, compute the remainder when 241^{163} is divided by 493, which is 180. The decoder then uses the decode key $d = 11$ to compute the remainder when 180^{11} is divided by 493, which is 241, the original message.

Rijndael (pronounced "rain-doll") was selected for its simple and elegant design, speed, and compactness in hardware and on smartcards.

Note that when p and q are larger, the messages can be fairly large numbers but in practice will still require splitting into smaller parts and encoding separately. For example, to encrypt a long message, it may be necessary to encrypt each paragraph separately. Also, computing the remainder of such large powers is made easy by the fact that the remainder of a product is the product of the remainders. For example since $163 = 128 + 32 + 2 + 1 = 10100011_2$ in base 2, $T^{163} = T^{128} \times T^{32} \times T^2 \times T^1$. This means that after seven successive squarings of T and computing remainders when divided by pq, one need only multiply the four factors together, computing the remainder when divided by pq. In our example, $pq = 493$, and if the message we wish to send is 241, successive squarings give $241^2 = 58{,}081$ which leaves a remainder of 400, 400^2 which leaves 268, 268^2 leaves 339, 339^2 leaves 52, 52^2 leaves 239,

239^2 leaves 426, and 426^2 leaves 52. Thus, we need only find the remainder when $52 \times 239 \times 400 \times 241$ is divided by 493, which is 180.

Usage

The RSA public-key cryptosystem is the de facto standard in industrial cryptography and used by a number of international organizations. Because its copyright has expired, it is likely to be incorporated into more works. Public-key systems solve the primary problems involving Internet security: confidentiality and authentication. Only the person who holds the decryption key can read the message, and because a private key was used, the recipient can be sure of the sender's identity. These systems can be attacked through factoring and algorithms. The factoring attack attempts to determine a secret key from its public key, either by factoring a number associated with the public key or solving mathematical problems. Algorithm attacks try to find a flaw or weakness in the mathematical problem used in developing the encryption system.

If a code cracker machine could determine a DES key in one second, which would take 2^{55} tries, that machine would require 149 trillion years to crack a 128-bit AES key.

The Advanced Encryption Standard will be available free to all users. One use will be to encrypt sensitive but unclassified electronic documents for the U.S. government. Its use is likely to spread into Internet security, bank cards, and ATMs. The standard is expected to become an official standard in late spring, 2001. Although in the late 1990's, "DES cracker" machines were built that could determine a DES key after a few hours, experts believe the creation of an "AES cracker" to be less likely. If a code cracker machine could determine a DES key in one second, which would take 2^{55} tries, that machine would require 149 trillion years to crack a 128-bit AES key.

The Riemann Hypothesis:
Are We Any Closer to a Solution?

➤ *An institute offers $1 million to anyone who can solve the Riemann hypothesis or one of six other unsolved mathematical problems.*

The Riemann hypothesis has been one of the outstanding problems of mathematics ever since Georg Friedrich Bernhard Riemann posed the problem in an eight-page paper published in 1859. In 1900, mathematician David Hilbert gave a famous talk in Paris in which he stated that proving or disproving this hypothesis is one of the most important unsolved problems confronting modern mathematics. He proposed twenty-three problems that set the course for twentieth century mathematics. The Riemann hypothesis is one of three that remain unsolved but is widely believed to be true and is supported by much numerical evidence. Furthermore, many mathematical results now assume the Riemann hypothesis or a generalized version of it in their proof, so an ever- increasing part of mathematics depends on its solution. With the solution of Fermat's last theorem in 1995, the Riemann hypothesis is now considered by many to be the most famous unsolved problem in mathematics.

In its simplest form, the Riemann hypothesis states that if $\zeta(s) = 0$ for a complex number $s = x + iy$, then $x = 1/2$. The infinite sum

$$\zeta(s) = \frac{1}{1^s} + \frac{1}{2^s} + \frac{1}{3^s} + \cdots \quad ,$$

when extended to the complex plane, is known as the Riemann zeta function but was studied as early as 1748 by mathematician Leonhard Euler for complex numbers s with real part $x > 1$. For example, Euler showed that

$$\zeta(2) = 1 + \frac{1}{4} + \frac{1}{9} + \frac{1}{16} + \cdots = \frac{\pi^2}{6}$$

When studying the distribution of prime numbers, Riemann extended Euler's zeta function to the complex plane with simple pole at $s = 1$. Riemann noted that his zeta function had so-called trivial zeros at –2, –4, –6, . . . and that

Georg Friedrich Bernhard Riemann. (Library of Congress)

all nontrivial zeros would have to be symmetric about the critical line $x = 1/2$. The Riemann hypothesis states that the nontrivial zeros are all on this critical line. The first such zero is approximately $s = 0.5 + 14.134725i$. The Riemann hypothesis is related to the distribution of prime numbers, which can be seen by Euler's product representation

$$\zeta(s) = \prod_{p \text{ prime}} \left(1 - p^{-s}\right)^{-1}$$

where the product (Π) is over the prime numbers 2,3,5,7,11, . . .

Who Wants to Be a Millionaire?

On May 24, 2000, the Clay Mathematics Institute announced that it had designated the Riemann hypothesis as one of its seven millennium problems. For each of these problems, a prize of $1 million will be awarded to whoever comes up with the first correct solution. The announcement was made at the Collège de France in Paris during the centennial celebration of Hilbert's 1900 lecture. The six other millennium problems are P versus NP, the Hodge conjecture, the Poincaré conjecture, Yang-Mills existence and mass gap, the Navier-Stokes existence and smoothness equations, and the Birch and Swinnerton-Dyer conjecture. Boston businessman Landon Clay founded the institute, which has been supporting mathematics since 1998.

Another million-dollar prize has been offered by British and U.S. publishers Faber and Faber and Bloomsbury Press in connection with the publication of the English translation of the novel *Uncle Petros and Goldbach's Conjecture* (2000) by Apostolos Doxiadis. Goldbach's conjecture states that every even number beginning with four can be written as the sum of two prime numbers. For example, 4 = 2 + 2, 6 = 3 + 3, 8 = 3 + 5, 10 = 3 + 7 . . . . Unlike the Clay prizes, however, this offer has a deadline of March, 2002, for a solution. Having remained unsolved since 1742, it is unlikely to be solved by then.

The Numerical Evidence

In 1859, Riemann verified his hypothesis for the first 3 zeros. Over the years, other mathematicians made progress toward proving his hypothesis. In 1903, Jorgen Gram verified the Riemann hypothesis for the first 10 zeros. In 1915, G. H. Hardy proved that an infinite number of the zeros do occur on the critical line. In 1925, J. I. Hutchinson verified the Riemann hypothesis for the first 173 zeros. In 1935, Edward Titchmarsh and Leslie Comrie verified the Riemann hypothesis for the first 1,041 zeros. In 1953, Alan Mathison Turing used an electronic computer to verify the Riemann hypothesis for the first 1,104 zeros. In 1955, Derrick Lehmer used Turing's method to verify the Riemann hypothesis for the first 25,000 zeros. In 1966, Sherman Lehman verified the Riemann hypothesis for the first 250,000 zeros. In 1968, J. Barkley Rosser, James Yohe, and Lowell Schoenfeld verified the Riemann hypothesis for the first 3.5 million zeros. In 1986, Richard Brent, J. van de Lune, Herman te Riele, and D. T. Winter verified the Riemann hypothesis for the first 1.5 billion zeros. In 1989, Brian Conrey proved that over 40 percent of the zeros in the critical strip are on the critical line. However, despite this overwhelming evidence supporting the Riemann hypothesis, it could still be false, as no proof has been discovered.

Connections to Other Areas

A proof of the Riemann hypothesis could revolutionize encryption, which is used to secure information such as credit card numbers sent over the Internet. Both are based on prime numbers, the building blocks of arithmetic. In fact, it was the calculation of a sum related to the distribution of prime numbers that revealed the bug in the first Pentium chip. Mathematics professor Tomas Nicely was estimating the sum

$$\sum_{p,p+2\text{ both prime}} \frac{1}{p}+\frac{1}{p+2}$$

when he discovered an error in the tenth decimal place in the computer's computation of 824633702441^{-1}. After running the same calculation on a 486 (non-Pentium) computer, the error disappeared. Further investigation showed that it was the Pentium chip that was responsible for the error, and its manufacturer, Intel, corrected the error.

Another surprising connection is that of the Riemann hypothesis to quantum mechanics. Hilbert stated that the zeros of the Riemann zeta function are distributed like the eigenvalues of a random matrix. It has been shown that the zeros are statistically distributed like the energy levels of a time-irreversible and chaotic quantum mechanical system. The discovery of such a system would prove the Riemann hypothesis.

The Mathematics of Molecular Biology and DNA

➤ *Technological advances allow mathematical fields such as combinatorics and topology to play important roles in biological research.*

In June, 2000, the completion of the first draft of the Human Genome Project was announced. Mathematics played an important role in mapping the human genome, partly through Eric Lander, one of its project directors, who was a mathematician turned biologist. His involvement underscores the importance mathematics is gaining in biological research, which until recently has not involved mathematics as much as some other sciences have. Technological advances have allowed scientists to study life at the cellular and even molecular levels. The high level of structural detail and complexity observed has sparked mathematical interest. New applications have been found for such mathematical fields as combinatorics, the mathematical study of elements in sets or in geometric configurations, and topology, the mathematical study of shapes. The subfield of topology called knot theory has been most useful in studying deoxyribonucleic acid (DNA).

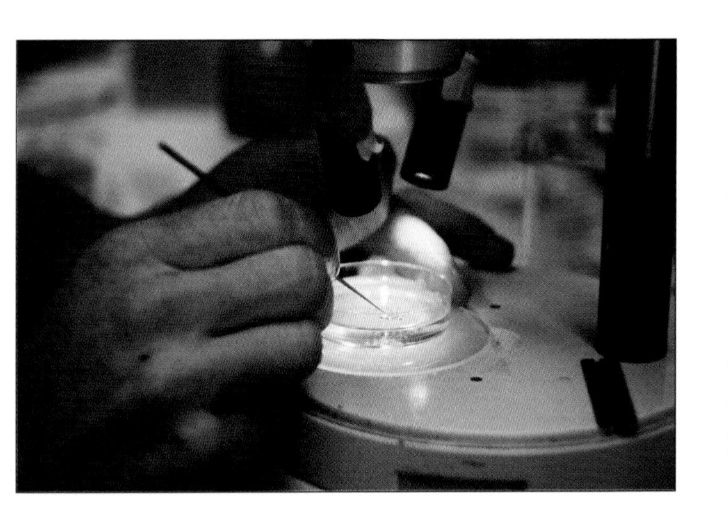

So far, the achievements in the laboratory have been spectacular. Experiments formerly thought of as material for science fiction are carried out in laboratories around the world and commonly reported in newspapers and magazines. From cloning to mapping human genes, these advances depend on technology that allows scientists to "edit" living cells on a microscopic level. Electron microscopes can be used to "cut and paste" parts of molecules. This technology is part of a larger industrial revolution called nanotechnology.

One example of the use of nanotechnology is the construction of smaller circuits to make faster computers. The manipulation and construction of materials on the molecular level presents other possibilities. For example, when the creation of material on the molecular level becomes a reality, manufacturers will be able to use strong, lightweight materials such as diamonds in the construction of airplanes, allowing them to be much lighter and cheaper. The National Aeronautics and Space Administration (NASA) and the defense industry are particularly interested in nanotechnology's potential in this area, and President Bill Clinton doubled the nanotechnology research budget for fiscal year 2001.

Topology

Cloning is essentially the removal of genetic makeup (DNA) from a cell, and inserting it in a live embryo. In the 1950's, James Watson and Francis Crick discovered the structure of DNA to be two long strands wrapped around each other in the shape of a double helix. Each strand contains the cell's genetic makeup, a sequence of bases labeled A, T, C, or G (adenine, thymine, cytosine, guanine). The DNA inside a cell can be visualized as one hundred miles of dental floss coiled up inside a basketball. As the DNA moves and reacts, it can become tangled. Because it needs to replicate and recombine and has limited space in which to do so, it is useful to study the ways it can be tangled and untangled. The mathematical study of these shapes and their transformations is called topology. Cells have enzymes called topoisomerases that perform topological manipulations on DNA. By studying the resulting shapes when these enzymes act on circular DNA, scientists have gained a better understanding of what the topoisomerases are capable of doing.

Experiments formerly thought of as material for science fiction are now being carried out in laboratories around the world.

Topology studies those properties of objects that remain unchanged when the object is continuously transformed by movements such as stretching, twisting, and unraveling. For example, a coffee cup and a doughnut are topologically equivalent to a single-holed surface called a torus, and a pair of scissors is topologically equivalent to a two-holed torus. These are two-dimensional objects in three dimensions. In three dimensions, it is also useful to study one-dimensional objects (strings) called knots, links, and braids.

Knot Theory

Mathematician William Thomas, the first baron of Kelvin, introduced knot theory in the late nineteenth century along with his conjecture that atoms are knots in the ether (the medium then believed to occupy outer space). Although his conjecture is incorrect, the theory is useful in studying polymers, which are knots. Unlike a knot that one ties with a piece of rope, a mathematical knot is a single string that connects back onto itself to form a loop. A simple loop like a rubber band is called the zero knot or an unknot. Circular DNA is also an example of a zero knot. Aside from the zero knot, no knot can be "flattened" into two dimensions. In fact, in dimensions higher than three, the unknot is the only knot because all loops can be unraveled.

The simplest non-zero knot is the trefoil knot. To visualize a trefoil knot, one can take a piece of string and wrap one side over the other and back under and through the "hole," then connect the two ends to form a loop. By moving the string around, one sees that the trefoil knot has three parts that are symmetric about the center. There are actually two different trefoil knots: One is formed by wrapping one end of a string under the other and back over through the resultant hole, then connecting the two ends to form a loop. This forms the mirror image of first trefoil knot and cannot be continuously transformed to the first. There is another way to create a trefoil knot that is more enlightening—by taking a narrow strip of paper and giving one end a half-twist, then taping the two ends together. This is called a Möbius strip, the first known surface with only one side. A trefoil knot can be created by giving the narrow strip of paper three half-twists (instead of the single half-twist), then cutting the strip down the middle. Giving the strip two half-twists and cutting the strip down the middle, produces two linked zero knots. This is called a link because it involves more than one strand. Links occur commonly in chains and even in the logo of the Olympics.

Like integers, knots can be factored into simpler knots. The trefoil knot is an example of a prime knot, one that cannot be decomposed into simpler knots, just as prime numbers cannot be factored into smaller numbers. A loop with a trefoil knot followed (before closing the loop) by another trefoil knot is called a granny knot. A loop with a trefoil knot followed by its mirror image is called a square knot. Thus, in order to classify all knots, it is necessary to classify only the prime knots. One often classifies knots by their crossing number, which is the least number of times a knot crosses over or under itself. The number, $f(n)$, of prime knots with crossing number equal to n, forms an interesting sequence for which the exact pattern is still unknown. The first few values of the sequence are given in the following table:

NUMBER F(N) OF PRIME KNOTS WITH CROSSING NUMBER N

n	1	2	3	4	5	6	7	8	9	10	11	12	13	14
$f(n)$	0	0	1	1	2	3	7	21	49	165	552	2,176	9,988	46,972

It is very difficult to determine whether two knots are different or same knot tangled somewhat differently. It is sometimes easier to study the complement of a knot rather than the knot itself. To visualize the complement of a knot, one can consider a worm eating through an apple. When it forms a loop, the wormhole is the knot and the part of the apple remaining is its complement. It is known that different knots must have complements that are not topologically equivalent. Numerous attempts to decide whether two knots are different make use of polynomials. The first such attempt assigned to each knot is an Alexander polynomial. By showing that two knots had different Alexander polynomials, one could prove that two knots were different, but if the Alexander polynomials were the same, one could not conclude that the knots were the same. It turned out, unfortunately, that both the unknot and the pretzel knot had Alexander polynomials equal to 1. Some improvements have been made in the Jones and HOMFLY polynomials, but similar problems arise. Finding a fast algorithm that computers can use to decide whether two knots are the same remains an important problem.

Another important question for scientists to answer is to determine the probability of whether a specific strand of DNA has formed a loop or knot somewhere. One solution to this question was to discretize it by assuming the strand is made up of n line segments of length one that pass through lattice points in the x-y-z grid. It has been shown mathematically that for large values of n, a strand of length n will almost always close with a non-zero knot. However, the proof does not reveal how large n has to be to ensure that most strands of length n have knots. For $n = 1{,}600$, only 1 percent of strands have knots. Another unanswered question is how many knots one should expect on a strand of length n.

By answering questions such as those posed by knot theory and using it and other mathematical devices to examine structure, scientists can determine the shape and form of objects that cannot readily be perceived, helping them learn more about biology at the molecular level.

HOME
CALC
REPORT
R
T
Y
G
H
ç
B
N
SPACE

Resources for Students and Teachers

Books

Bunch, Bryan. *The Kingdom of Infinite Number: A Field Guide.* New York: W. H. Freeman, 2000. The author takes a field-guide approach to looking at numbers and their properties and thus presents a new way of viewing mathematical patterns and relationships.

Casti, John L. *Five More Golden Rules: Knots, Codes, Chaos, and Other Great Theories of Twentieth Century Mathematics.* New York: John Wiley & Sons, 2000. Presents five modern mathematical theories—knot theory, functional analysis, control theory, chaotic systems, and information theory—in an interesting and understandable manner and relates them to everyday life.

Devlin, Keith. *The Math Gene: How Mathematical Thinking Evolved and Why Numbers Are Like Gossip.* New York: Basic Books, 2000. Mathematician Devlin argues that all humans possess an inherent ability for mathematics and suggests that this ability provided an evolutionary advantage.

Drazen, Allan. *Political Economy in Macroeconomics.* Princeton, N.J.: Princeton University Press, 2000. This book examines the effects of politics on economics through an examination of models of political economy.

Katz, Victor, ed. *Using History to Teach Mathematics.* Washington, D.C.: Mathematical Association of America, 2000. Mathematical historians from around the world demonstrate why history is a necessary component of mathematical education.

Nassar, David. *How to Get Started in Electronic Day Trading: Everything You Need to Know to Play Wall Street's Hottest Game.* New York: McGraw-Hill, 2000. Nassar, who runs a trading firm, describes the types of electronic access to markets, the forces that drive stock prices, and the strategies of successful day traders.

Schneider, Leo J. *The Contest Problem Book VI: American High School Mathematics Examinations, 1989-1994.* Washington, D.C.: Mathematical Association of America, 2000. This guide offers practice problems and solutions in algebra, geometry, and number theory designed to help students who will participate in the U.S. Mathematical Olympiad.

CD-ROMs and Videos

Amazon.com. Video. Films for the Humanities & Sciences, 2000. Amazon.com and its leader Jeff Bezos, booksellers, and Internet marketing are the topics covered in this program, originally produced for the *NewsHour with Jim Lehrer.*

Internet Shopping: Interactive or Invasive? Video. Films for the Humanities and Sciences, 2000. Internet companies, electronic commerce, and marketing and advertising on the Web are the subjects of this program, originally broadcast on the *NewsHour with Jim Lehrer.*

Math Advantage 2000. CD-ROM. 3 vols. Encore Software, 2000. Pre-algebra, algebra I, geometry, measurement, statistics, and probability are covered in this interactive learning aid. For students age eleven and up.

The Web World of Business. Video. Television Education Network, 2000. E-commerce, the Internet form of business, is discussed in this video.

Web Sites

Interactive Mathematics Miscellany
http://www.cut-the-knot.com/content.html
This interactive site provides a variety of games, puzzles, and interesting facts about mathematical subjects, including arithmetic, geometry, and probability. Contains quotations and speculations on a variety of subjects.

Mathematics Information Server
http://www.math.psu.edu/MathLists/Contents.html
Provides links to mathematics departments around the world as well as to mathematical societies, institutes, journals, and mathematics software.

The Millennium Problems
http://www.claymath.org/
The Web site of the Clay Mathematics Institute provides information on the millennium problems and on the prize for solving them.

Visual Calculus Exercises and Demonstrations
http://archives.math.utk.edu/visual.calculus/
Teachers and students can use this collection of tutorials on calculus. Topics offered include precalculus, limits and continuity, derivatives, applications of differentials, integration, applications of integration, and sequencing and series.

6 · Medicine and Health

The Year in Review

➤ *Thomas C. Jefferson, M.D.*

Advances in the field of medicine in 2000 have deepened our understanding of human health and illness and have shown the way toward even greater knowledge. None has been as spectacular as, for example, the invention of the first polio vaccine or the discovery of the atomic structure of deoxyribonucleic acid (DNA). However, many of them will be seen as important steps forward when remembered by future researchers. The major medical stories of 2000 can be grouped into four categories: genetics and gene therapy, hormones and endocrine function, infectious diseases, and public health.

Genetics and Gene Therapy

The major medical stories in genetics during the year 2000 centered on the progress being made in practical genetic therapy after decades of research. Unfortunately, the first human death attributed to gene therapy was reported in late 1999. The ensuing scrutiny of gene therapy research led to the discovery of a second death possibly caused by gene transfer that had not been reported to monitors at the National Institutes of Health (NIH). The laboratories involved in the deaths were required to stop their gene transfer trials, and many institutions put their gene research on hold.

It appears feasible to use gene transfer to block the growth of new blood vessels in solid cancers, thus starving the tumor cells to death.

Late in 2000, the NIH issued new regulations that require closer oversight of human gene research. The regulations include requirements for the provision of more complete information to study subjects regarding the risks and possible benefits of the research for them personally and a requirement for stricter adherence to research protocols. Protocols are detailed plans written before an experiment begins that tell how test subjects will be chosen, how the research will be performed, and how patients will be safeguarded against harm from the research itself.

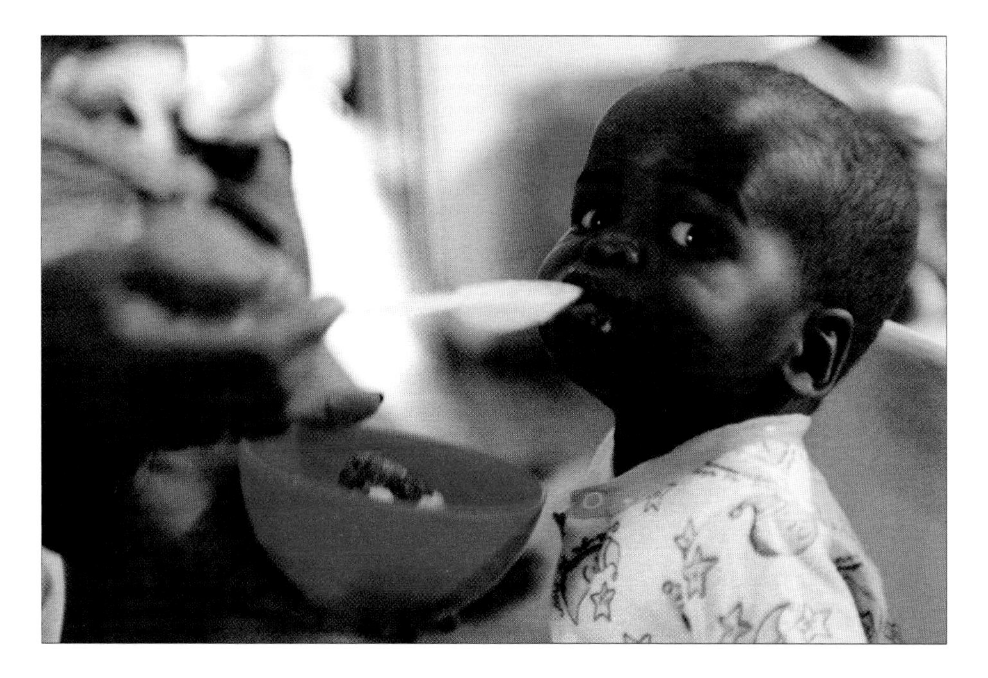

A worker feeds a baby at a home for HIV-positive children in South Africa. Experts believe that in a few years, the population in some African countries will stop growing or even fall because of the large number of AIDS deaths. (AP/ Wide World Photos)

During the year, the furor over problems with gene transfer research in the United States overshadowed a hopeful story from France. For the first time, it appears that scientists have possibly cured a human genetic defect. French researchers inserted a gene into the cells of two babies born with severe combined immunodeficiency (SCID) that corrected their lack of protection from infection. There are a number of forms of the disorder, which causes multiple defects in a person's immune system. The most severely affected children die within the first two years of life. It is not yet possible to know if the corrections are permanent or if all SCID defects in these two infants were corrected.

Many other uses of gene therapy are under investigation. It appears feasible to use gene transfer to block the growth of new blood vessels in solid cancers, thus starving the tumor cells to death. Cancers of the breast and lung are among those that might respond. Also, genes could be placed in healthy cells, causing them to produce antitumor proteins that would be released into the circulation to attack a cancer elsewhere in the body. At the same time, it also appears that the transfer of other genes could stimulate the growth of new arteries that could carry blood around coronary arteries blocked by atherosclerotic plaques. Genes may even help to replace antibiotics in this age of bacterial resistance to the common antimicrobial drugs, including penicillin and erythromycin. Viruses called bacteriophages, which infect bacteria by injecting their own genes, as nucleic acids, into the bacterial cell, can be developed to kill specific bacterial strains. Resistance to antibiotics provides no defense for the germs against these viruses.

Hormones and Endocrine Function

Many of the body's most important regulatory functions are maintained by hormones, messenger molecules that manage the level of activity of various biochemical processes within cells. Insulin, thyroid hormone, and estrogen are only three of the many hormones that circulate in each person's body. One hormone that was only recently described is leptin, which seems to control body weight by altering metabolic rate and hunger. Its function is complex, and it has not yet proved to be the weight-loss boon that many had hoped it would be.

The medical abortion drug mifepristone (RU-486), which blocks the stabilizing effect of progesterone in early

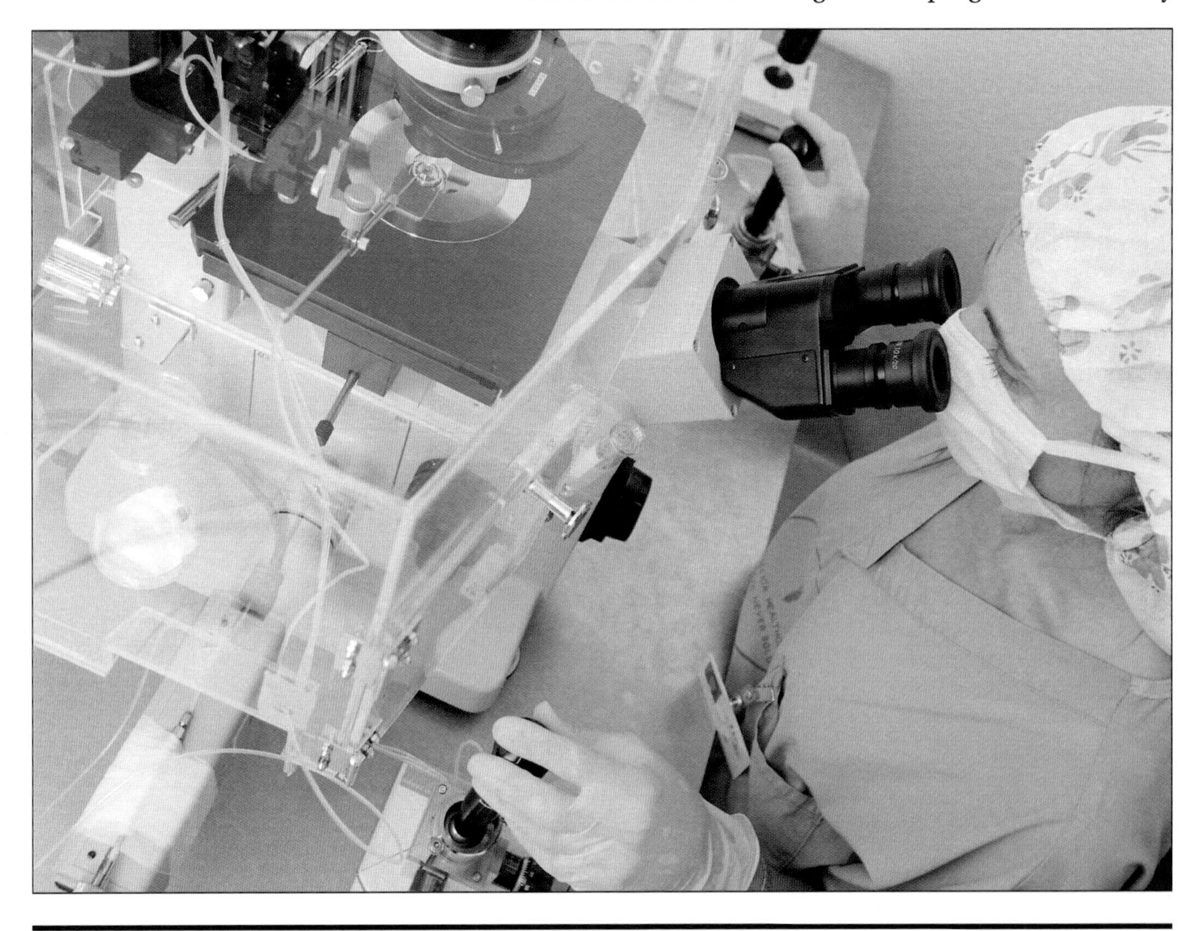

pregnancy, received Food and Drug Administration (FDA) approval for use in the United States more than a decade after its acceptance in Europe. Use of the drug will be under unusually strict controls put in place by the FDA. The drug will not be available by prescription in pharmacies but only directly from physicians who have agreed to abide by the controls.

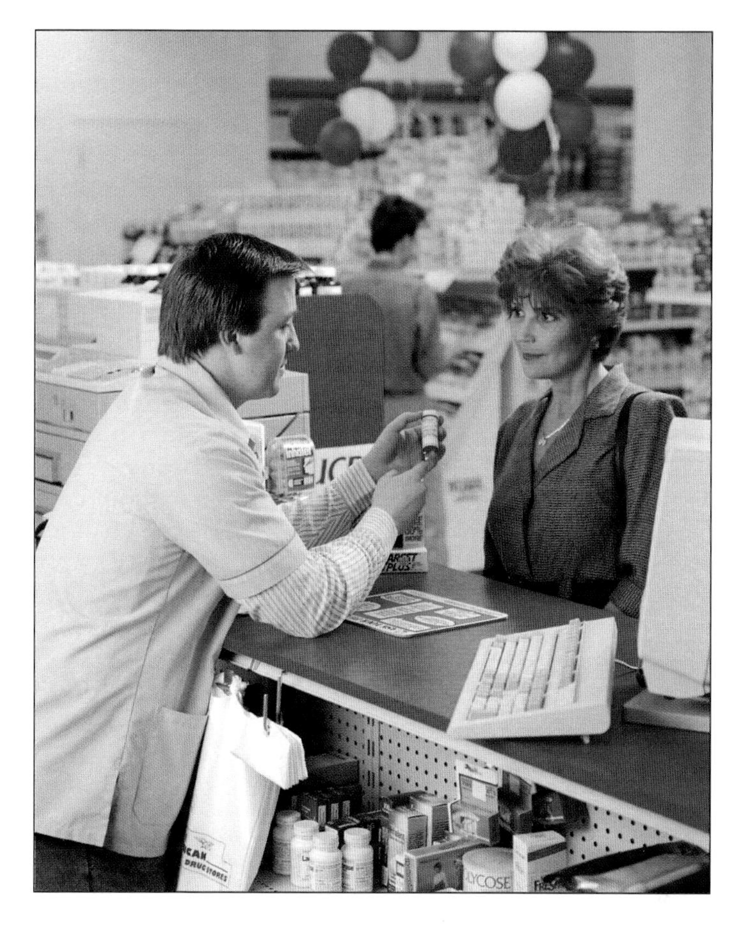

In 2000, much attention was given to estrogen, progesterone, and testosterone, three of the hormones that regulate reproductive function and maintain adult muscle strength and bone mass. They may also be important in maintaining good mental function as people age. The blood levels of all three hormones fall progressively lower as people become older. Whether the normal decrease of these hormones with age has a direct effect on coronary artery disease, either to slow or to speed its development, is hotly debated. Even harder to determine is the possibility that they might have the beneficial effect of slowing the aging process if their levels are kept at normal midlife levels.

There is no question that extra estrogen can be a great help in easing the most uncomfortable experiences of menopause for women. However, estrogen, given without the opposing effects of progesterone, can increase a postmenopausal woman's chance of developing endometrial cancer in the uterine lining and may increase the possibility of developing some forms of breast cancer. For men, although extra testosterone seems to ease some of the changes of aging, particularly muscle strength and sexual function, the presence of high levels of testosterone may encourage rapid growth of any prostatic cancer that might already be present but undetected.

The human body appears to be designed to age and apparently can tolerate only a certain amount of interference with that process. It seems most prudent at this point for men not to routinely take medicinal testosterone as they age. Hormone replacement therapy with estrogen and progesterone in women does alleviate the symptoms of menopause and maintains bone strength, but these may vary greatly between individual women, especially with regard to the possible benefits in terms of heart disease.

Infectious Diseases

One development in the area of infectious diseases in 2000 was the possibility of using bacteriophages to supplement antibiotics. Another was the steadily mounting evidence that chronic infection may be an important part of the chain of events leading to atherosclerosis, the blockage of coronary arteries, and heart attacks. The third big story in infectious diseases was the remarkable and frightening increase in the incidence of the human immunodeficiency virus (HIV) and the acquired immunodeficiency syndrome (AIDS) in sub-Saharan Africa. Some African countries will actually experience zero or negative population growth within only a few years because of the large number of AIDS deaths. Already, life expectancy in years is declining in parts of Africa because of HIV. The development of new antiviral drugs has not kept pace with the ability of the AIDS virus to adjust its defenses. Another problem is that the cost of anti-HIV drugs is so great that many poor governments cannot afford to purchase the medications for their citizens.

Chronic infection may be an important part of the chain of events leading to atherosclerosis, the blockage of coronary arteries, and heart attacks.

In addition, the relative number of adolescents with HIV is increasing, not only in Africa but also in the United States. Sexually transmitted disease has become a teen epidemic, and the presence of one such infection makes the passage of HIV more likely between sexual partners.

Public Health

A number of important medical stories in 2000 were in the arena of public health. They included the growing evidence that a person's race and sex may determine the quality of care he or she receives for a number of conditions, including heart attack and organ transplants. Two studies during the year also showed that racial/ethnic minority patients are less likely than whites to receive adequate pain medication in emergency rooms for their injuries. Poverty has long been recognized as one of the strongest public health factors linked to poor health. Now

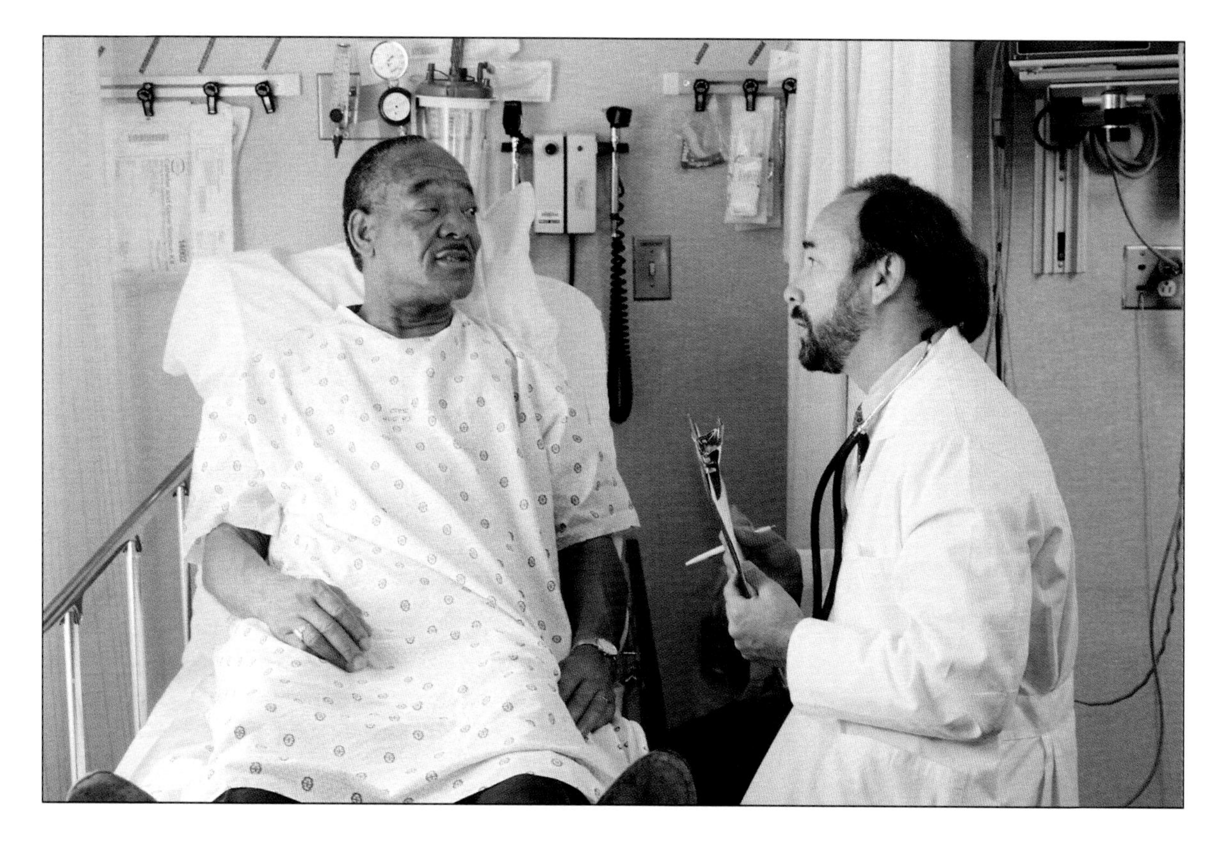

it appears that the degree of difference between the wealth of the most-well-off and those in poverty may be an even more important factor, even in developed nations.

The troubling announcement that as many as 98,000 Americans die each year from medical errors in hospitals was made public late in 1999. Major reports and studies in response to this threat were published in 2000. The Institute of Medicine, the Institute for Healthcare Improvement, and the American Medical Association, among others, are part of the active search for ways to minimize or avoid such incidents.

A number of other developments concerning public health began to become more visible during 2000 and promise to increase in prominence in coming years. These include terrorist use of biological weapons in the United States, concerns about the safety of vaccines for anthrax and Lyme disease, and research to determine whether gun safety education leads to better weapons security in homes with children.

Medical Research and Patient Rights
Whose Genetic Information Is It?

➤*A lawsuit filed in Chicago in autumn, 2000, illuminates remarkable developments in the rights of patients to have a say in the use of their own genetic information.*

According to a November 19, 2000, article in the *Chicago Tribune*, the parents of two children who died from Canavan disease, a rare genetic brain disease, and other parents and concerned organizations filed suit against Miami Children's Hospital. They charge that researchers at the hospital, who discovered the gene responsible for the disorder, have hindered further research into the disease. The hospital is accused of having done this by strictly controlling the licensing of other laboratories to conduct the test that its researchers developed to identify the gene's presence. Scientists used blood from the plaintiffs' children, who suffered from the fatal disorder, to isolate the gene. Once they succeeded in identifying the gene, the researchers patented the nucleic acid sequence of the gene and the test they developed to detect its presence.

The strict control includes a fee of $12.50 (originally $25) payable to Miami Children's Hospital for every test performed. The parents and others who joined them in the suit fear that the license controls are so restrictive that they could prevent couples from learning that they carry the gene for Canavan disease until they have given birth to a baby who suffers from it. The Canavan Foundation, a nonprofit organization that is also a plaintiff, claims that it was no longer able to provide the test for free after being notified that it must comply with all the license requirements.

Scientists, universities, and corporations have been patenting genes from plants, animals, and humans for more than two decades.

On the other hand, Miami Children's Hospital argues that it spent a large amount of money on the research that identified the gene. As a result, it says, the hospital needs to recover some of the costs of the program.

The parents in this suit do not contest the right of Miami Children's Hospital to receive a patent on the gene itself and on blood tests to detect its presence. Scientists, universities, and corporations have been patenting genes from plants, animals, and humans for more than two de-

cades. Many research universities and corporations hold patents on the deoxyribonucleic acid (DNA) sequence of specific genes, and Miami Children's Hospital is not the only institution that charges for the use of its proprietary genes. Myriad Genetics of Salt Lake City, Utah, holds the patent on the DNA sequence of the genes *BrCa1* and *BrCa2* and on the method of detecting their presence in a person's genome. These two genes are associated with a significantly increased risk for breast cancer in the women who carry them. *BrCa2* is also a risk factor for breast cancer in men. The firm charges $2,580 for its commercial *BrCa* testing but has recently agreed to accept $1,200 per person for research use of the test by scientists associated with the National Institutes of Health. Myriad claims that this lower fee yields no profit to the firm.

Not all researchers take a restrictive approach with the patents that they obtain. When scientists at the University of Michigan were the first to identify the gene for cystic fibrosis, a genetic disease of the lungs and digestive system, they obtained a patent. However, they chose not to permit exclusive licenses and charged a royalty of only $2 a test, a negligible fee for detecting a fatal genetic disorder that is far more common than Canavan disease.

Canavan Disease

Canavan disease, also called spongy degeneration of the brain, is a degenerative disorder of the central nervous system that is principally found in families of Ashkenazi Jewish heritage, an eastern and central European heritage common to nearly 90 percent of American Jews. Even within this background, it is strikingly uncommon, occurring in about one out of every sixty-four thousand births. An autosomal recessive disorder, it occurs only when the infant receives a defective gene from both parents.

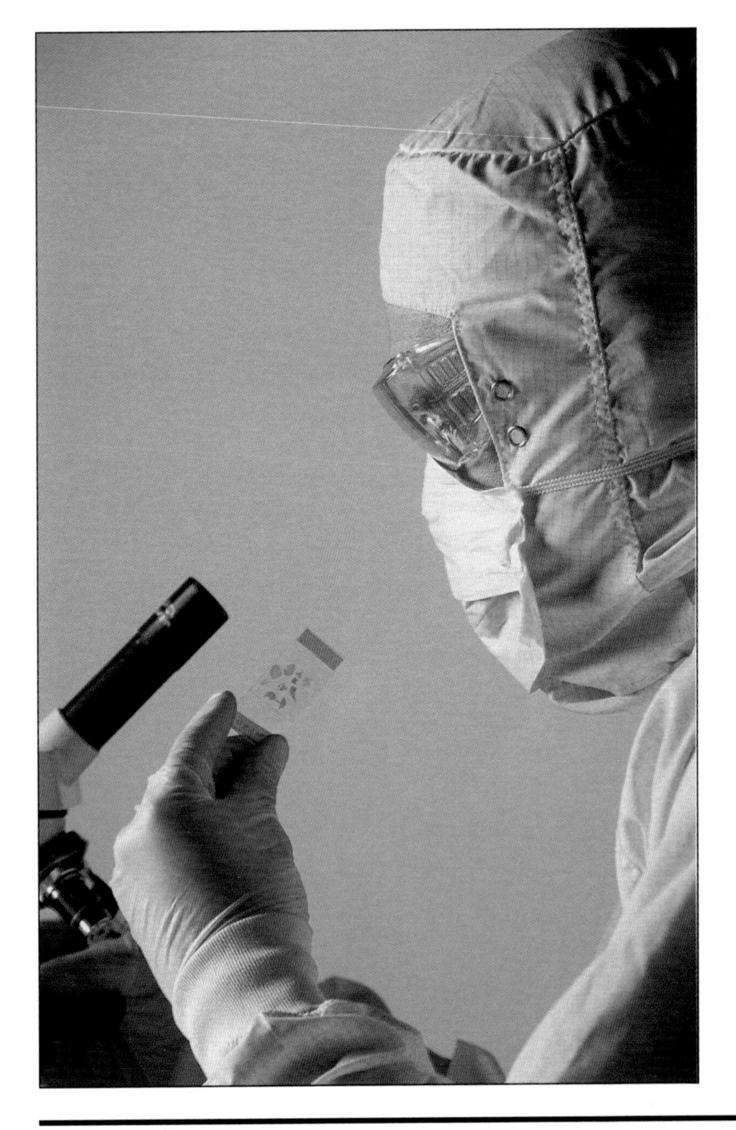

In Canavan disease, nervous system damage first becomes apparent between three and six months of age in previously normal-appearing infants. The infants develop progressive weakness, followed by the loss of any ability to think, express feelings, interact with their parents, or see. As the nervous system continues to deteriorate, affected children develop seizures and such severe loss of muscle control that voluntary movement becomes virtually impossible.

The disease is caused by a defect in a gene on chromosome 17 that controls the production of aspartoacylase, an enzyme that preserves the function of myelin in the brain. Myelin provides critical structure support and insulation within the brain, keeping the impulses of the individual nerves orderly and properly directed. Because of a complete absence of aspartoacylase in these babies' central nervous systems, brain damage begins to accumulate early, becoming evident within the first three to six months of life.

Essentially unable to think, experience, or be consciously aware, affected children die in mid-childhood, usually by ten years of age. The experience can be devastating to a family. Worse, before the gene was identified, some families had to experience the death of more than one child as each of their children would have a one-in-four chance of being affected. This shared experience caused many parents, including most of those in the suit, to press for research into the cause of the disease. That research has now borne fruit, and the families of the children who died do not want a limitation on testing and further research. The DNA in the original study came from their own children, the parents say, and that gives them an interest in how the gene and the test are used.

The Patent Question

Although many patents on genes—such as that granted in August, 2000, to the Burnham Institute for E1A, a tumor inhibitor gene used in cancer research—do not provoke controversy, others do. In February, 2000, the Associated Press reported on the debate that followed the receipt of a patent by the biotechnology firm Human Genome Sciences for a gene that controls how the human immunodeficiency virus (HIV) gains entry to the body. When the firm, which isolated and decoded the gene, applied for a patent in 1995, it had no idea of the gene's function. A group of academics, including University of Maryland's Robert Gallo, learned later that year that the gene is a receptor for HIV and that a defective version of the gene produces a protein that prevents HIV from attaching to the cells, thereby preventing infection. This might explain why some individuals, despite repeated exposure to HIV, do not become infected. The academics argue that they, not Human Genome Sciences, should receive the patent because they understood the gene's function. The biotech firm, however, believes the patent recognizes the work that went into the isolation and decoding process, and it plans to make the gene available for academic research at no cost. AIDS activists were sharply critical of what they termed the biotech company's emphasis on profits over progress in the fight against AIDS.

Gene Patents

The patenting of genes taken from the genetic material of humans and other living organisms raises a number of questions. Does a patent on a gene mean that the holder owns a small part of humanity or of each human? Can an individual patent his or her own genome? Should anyone be permitted to control access to a gene all people hold in common? Is the widespread search for medically important genes in developing societies a new form of subjugation by more advanced nations, a sort of biocolonialism? Will it deprive the citizens of poor countries of even their own genetic material for the profit of already wealthy societies? These are but a few of the questions that physicians, attorneys, religious leaders, ethicists, and leaders of indigenous groups are debating.

Some of the questions have practical answers. Humans the world over differ genetically from each other by less than one-thousandth of the genes that make up their individual genomes. That means that almost all human genes are shared by everyone. For a corporation to own a patent on a common gene or one for a specific disorder certainly does not mean that nobody can have it in their genome without permission. The patent simply gives the patent holder the right to exclusive use of that gene outside the genome, usually for research or for products, for the duration of the patent. Many people feel that being human, in the largest sense, involves much more than just the mechanisms of their DNA. To them, the possibility that somehow geneticists might come to control the source of all life seems extremely unlikely.

This year, poet Donna MacLean applied to the British patent office to receive a patent on her personal genome. However, because her genes came to her from her parents with no work on her own part to invent them, she did not meet the criteria for a patent under the European Patent Convention. For something to be eligible to receive a patent, it must be an invention, be novel, have been actively invented and not simply discovered (such as a new moon of Jupiter), and have some sort of industrial application.

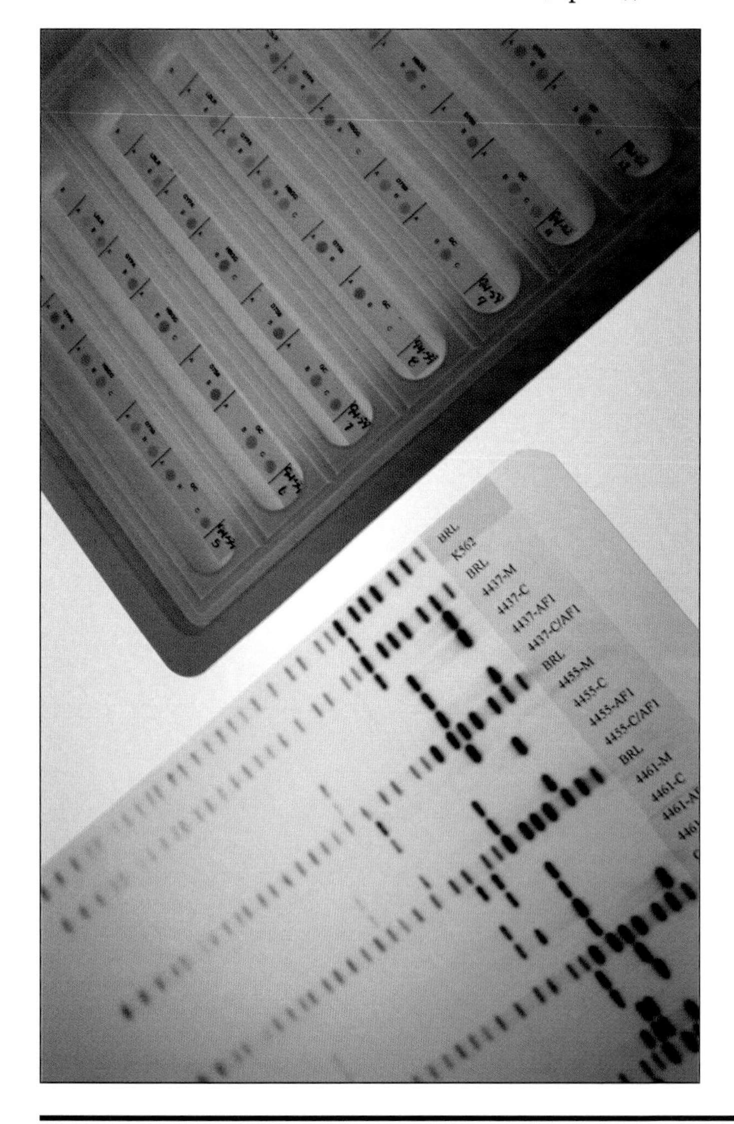

In the United States, a patent application has to meet three requirements. It must be novel, that is, new and original; have some usefulness that can be described; and be a finding that is not obvious to others, one that has required some work to achieve. The U.S. Patent Office has so far ruled that the laboratory work of isolation, description, and production of a gene is sufficient for it to have been nonobvious before its discovery. As a result, most applications for newly isolated genes are approved. On the other hand, a number of those involved in the debate point out that the techniques used to learn the genetic information are widely available and obvious. They argue that a person has only made a discovery, not invented something new, by determining the nucleic acid sequence of a naturally occurring gene.

Questions involving relations between developing and developed countries are much harder to answer. In 1995, the U.S. National Institutes of Health patented a line of cells that are genetically resistant to some forms of leukemia. The genetic material came from blood that had been drawn from one person in a remote New Guinea hill tribe. If the agency profits financially from this genetic information, should the individual and his or her tribe also benefit? Around the world, many people, including many members of indigenous tribes, such as the Maori of New Zealand or First Nation tribes of Canada, believe it is morally wrong to prevent the tribesman from receiving part of the value of his own genetic material. They argue that this is simply another example of exploitation of poor societies by rich ones, not particularly different from the forced removal of native resources by colonialists around the world for centuries.

A person's right to limit access to and to determine the uses of his or her own intimate personal molecular information has never required consideration before now.

For medical science, as the Chicago lawsuit shows, a new sort of conflict is developing. The remarkable new techniques in genetics raise serious academic, business, legal, and moral questions that people have never before had to consider. A person's right to limit access to and to determine the uses of his or her own intimate personal molecular information has never required consideration before now. Determining fair and equitable answers will probably require as many years as have already passed in the pursuit of molecular genetic knowledge.

Gene Therapy Using Adenoviruses
Riskier than We Thought?

➤ *The year brings the revision of federal rules regarding research in human gene transfers in the light of two deaths and the first successful use of the technique to treat a disease in humans.*

The death in the summer of 1999 of a young man who was a subject in a gene transfer experiment led to a major reevaluation of the safety of gene therapy, the adequacy of information provided to subjects before they signed their consent to participate, and the quality of supervision over such studies. It culminated in the publication in October, 2000, of the National Institutes of Health (NIH) *Guidelines for Research Involving Recombinant DNA Molecules,* a 126-page revision of federal rules for the conduct of gene-transfer research.

The man's death was believed to have been the first to occur in ten years of human gene transfer research. He suffered from a mild form of a genetic liver disorder, ornithine transcarbamylase deficiency. While some forms of the disease are fatal to infants, his milder illness had been brought under control through medicine and dietary restrictions. The research was intended not to cure this patient's disorder but to obtain information that would help design the proper treatment for severely affected infants.

For many diseases, the best chance of getting genetic material into enough of the body's cells to improve the disease requires viral transport of the gene.

In the study, researchers were using an adenovirus to carry genes into the subject's cells. Adenoviruses cause only mild infections, such as colds, in humans. The patient had experienced mild reactions to earlier injections of the virus that carried the corrected gene to his cells. When he received the maximum dose of gene-containing viruses that was permitted under the study's guidelines, more than 250 billion for every pound he weighed, his body responded with a massive immune reaction that led to the failure of multiple organs. He died a few days later when his father agreed to remove him from life-support machinery.

The death of another subject of gene transfer studies was discovered in March, 2000, at a different research cen-

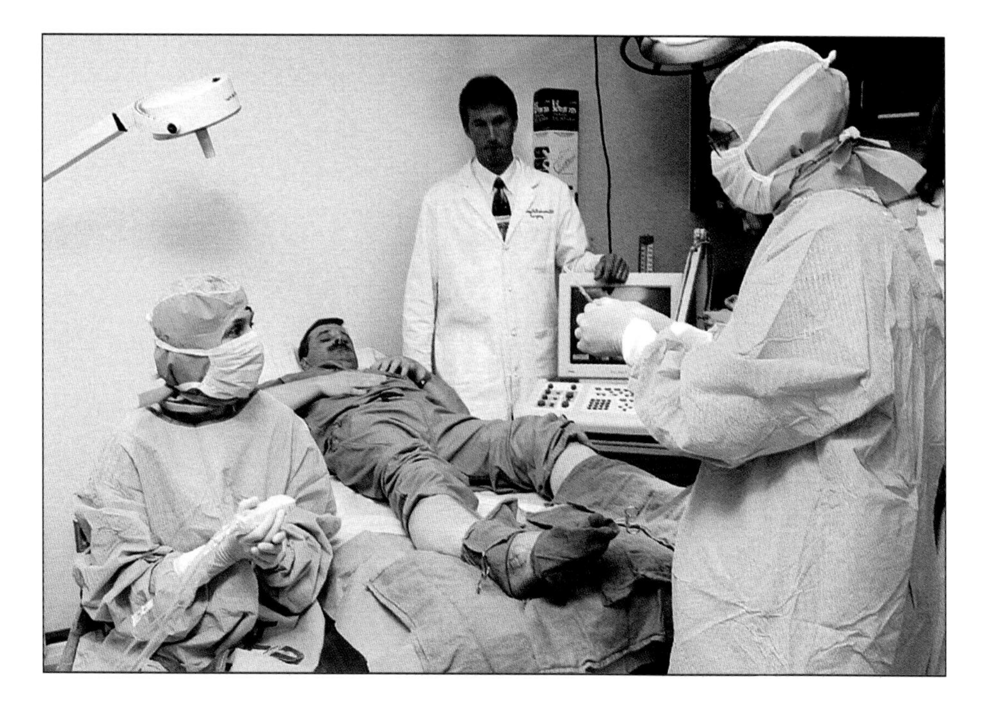

Patient Donovan Decker waits for doctors to inject him with gene-carrying viruses as part of a six-week gene therapy treatment for muscular dystrophy. (AP/Wide World Photos)

ter. This death had never been reported to Food and Drug Administration (FDA) officials, who uncovered it during a routine review of the center's research files.

How Gene Transfer Therapy Works

For years, medical scientists have known that defects in the genetic code for certain proteins are at the heart of many serious illnesses. They have also gained insight into how a virus infects an animal by entering its cells and forcing their internal machinery to respond to the viral genes to make many copies of the virus. These newly created viral particles can then move on to infect other cells. Scientists reasoned that viruses could be modified to carry copies of the corrected gene into the cells of people with illness-producing defects in their genetic code. Each cell infected with the virus would then incorporate the new gene into its own protein production. The defect would be corrected, and the patient would no longer suffer from the effects of the disease. Other ways of sneaking a new gene into living cells have been developed as well. However, for many diseases, the best chance of getting genetic material into enough of the body's cells to improve the disease requires viral transport of the gene.

Several classes of virus are employed in gene transfer studies. All undergo laboratory modification before they are injected into humans. The gene to be transferred has to

GENE THERAPY VECTORS AND TARGETED DISEASES	
Retrovirus	cancer
	hematopoietic disorders
	AIDS
Adenovirus	Cystic fibrosis
	Duchenne muscular dystrophy
	glioblastoma multiforme (brain tumor)
	anemia associated with chronic renal disease
	hemophilia A
	viral disease
	cancer
	liver diseases
Adeno-associated Virus	Cystic fibrosis
	Fanconi anemia
	HIV infection

SOURCE: *University of Toronto Gene Therapy Web site*

be incorporated into the virus's own genetic material. Also, the viral genes that would stimulate the infected cells to produce more viral copies must be inactivated so that the virus cannot replicate and spread as an infection.

Retroviruses can carry a gene into the nucleus of a cell and attach it to the animal's own deoxyribonucleic acid (DNA). Then, during the natural course of cellular metabolism, the new gene rather than the defective one will be used to produce correct proteins, thereby curing the disease. Much of what is known about retroviruses comes from the study of the human immunodeficiency virus (HIV) that causes acquired immunodeficiency syndrome (AIDS). In addition to HIV, there are many other retroviruses, many of which do not cause disease in humans and can successfully transport genes into cells. These viruses are small, though, and this limits the size of the gene that can be carried. A more serious concern is the possibility that the new gene could be spliced into the host's genetic material at a location at which it would damage another important gene or set off a cancerous growth. So far, this event, called insertional mutagenesis, has never occurred, and it remains only theoretical.

Adenoviruses, a family of viruses that can cause mild respiratory and other infections, are also important gene carriers, or vectors for genetic material. These viruses have the advantage that each has room to transport hundreds of copies of the desired gene. Also, they are easily grown, so billions of them can be injected at one time. However, they stimulate the host's body to produce protective antibodies against them, just as occurs in a common cold. The host's immune system can then attack and destroy large numbers of the viruses in each succeeding injection. This decreases the ability of the virus to enter enough cells to make a difference in the genetic disease.

Another group of viruses that can be employed to carry the gene are the adeno-associated viruses, which can replicate within a host's cells only with the help of an adenovirus. This characteristic makes them advantageous because viral replication in host cells is not desired in gene transfer.

Other families of viruses can also be used. In addition, in some cases, it is possible to place the raw genetic material directly into the body or to attach it to fat molecules, allowing host cells to take it in. The nonviral delivery systems are relatively inexpensive, but they are not particularly efficient at penetrating large numbers of cells.

Safety Problems

Over its ten-year history, until the recent deaths, gene transfer research had maintained an excellent safety record. However, it had failed to cure any genetic disorder. Although the deaths of two research subjects over a decade of research could simply be the result of chance, FDA officials were concerned that they had not been notified of these and other serious adverse effects by either research center. An important part of the monitoring of genetic research is the requirement that virtually any adverse health events occurring to a subject during the research must be reported. Yet this clearly was not happening.

A heightened review of procedures at gene therapy research institutes also revealed problems with the way potential participants were being recruited. Many subjects, or their parents if they were children, were not receiving adequate information about the research design, safeguards, and risks, established beforehand in a written protocol, prior to agreeing to become a test subject. In the case

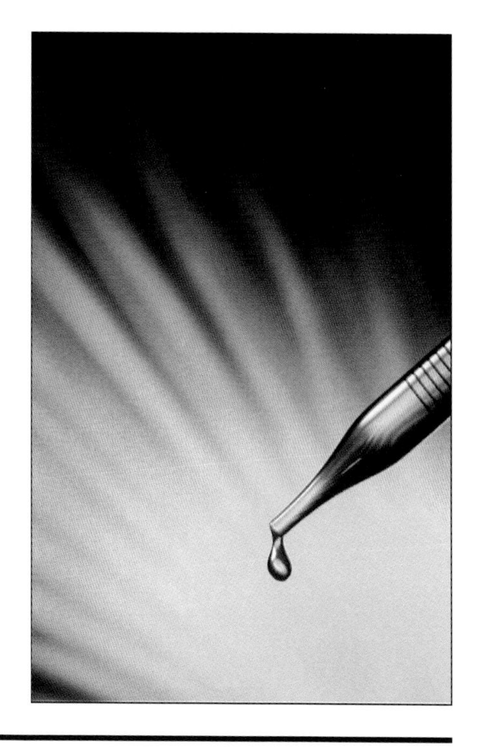

of the first death, the patient's father said that he and his son were never notified of the deaths of monkeys earlier in the trial or of a transient worsening of liver function that some other human subjects had experienced.

Every organization that conducts medical research must convene a committee, an institutional review board (IRB), to approve and monitor all studies in advance. The IRB is particularly concerned about the safety and the level of risk to which subjects are being exposed. A protocol, which describes the exact procedures that researchers will follow to ensure scientific accuracy and safety, must be approved before any study begins. Boards must also make sure that every research subject receives sufficient detailed information about the test and its possible risks before agreeing to take part. It is common for an IRB to oversee more than one hundred research protocols at the same time. The result is that when an occasional lapse occurs in adherence to protocols, the boards have to rely on the researchers to notify them of the mistake.

Scientists are under pressure from many sides to achieve success in their studies. The sponsoring university or institute can expect to receive increased funds for future research if a successful trial is completed. Academic advancement for the researchers usually depends in part on the publication of articles that detail an advance in their field of study. Researchers must also enroll an adequate number of subjects for gene therapy trials, just as in other medical research. There are relatively few patients who suffer from some of the disorders being studied, and subjects can be hard to find.

Another source of pressure comes from the fact that much of modern research is funded by corporations eager to recover their large investment with the discovery of a successful treatment. Personal pressure to complete the study can also exist if researchers will benefit financially from a successful outcome. In the two studies in which deaths occurred, a lead researcher at each institution had ties to the company that sponsored the study.

Scientists and institutions are well aware of the dangers of letting such pressures influence their decision making. They gain IRB approval for the protocol and set up other controls and limits on their research before beginning in order to minimize the chance of a biased decision. However, the careful FDA reviews spurred by these two deaths have uncovered many violations of the research protocols and rules at a number of institutions. Some are simple paperwork oversights, but the pattern suggests that inadequate attention is being paid to the avoidance of protocol violations and biased decisions.

The revised NIH guidelines of October, 2000, require much stricter oversight of gene transfer tests. They also emphasize the requirement that any adverse events noticed in one of these studies must be reported to the FDA immediately.

A 2000 Success Story and the Future

Despite the great amount of negative publicity surrounding gene transfer studies, 2000 was the first year in which a disease was successfully modified in humans. In April, French researchers announced that they had corrected a severe form of immune deficiency in infants. Children with the disease, severe combined immunodeficiency (SCID), cannot fight infections and die within months if they are not kept in a protective "bubble" room. The replacement gene allowed the affected infants to develop antibodies against infectious agents, and they are surviving.

The recent deaths of two gene therapy research subjects have caused scientists to reevaluate this approach to resolving genetic disease. The subsequent reevaluation has led to tightening of the oversight of gene therapy tests. However, despite these serious setbacks, costing patient lives, gene transfer therapy has finally demonstrated its potential to be a remarkably powerful tool for the correction of some human genetic defects.

Obesity
In the Genes?

➤ *In 2000, scientists concentrated on understanding obesity, weaving together earlier research as well as the discovery of a fat-storage gene and new knowledge of the hormone leptin's role in taste.*

More than half of all American adults are overweight. At any given time, as many as two-thirds of adults in the United States are trying to lose weight, a task on which they spend more than $30 billion annually. Sadly, despite recent promising discoveries in the study of obesity, the secret to weight loss remains the same: To lose weight, dieters still have to burn up more calories than they take in.

Nevertheless, scientists are gaining a better understanding of the mechanisms by which the body manages weight. These mechanisms are very complex, and each new finding makes it clearer that obesity is not simply the result of sloth or gluttony. Many of the factors that determine people's appetite and weight are beyond their conscious control.

In 2000, although no dramatic breakthroughs took place, scientists worked at pulling together the results of many strands of recent research. For example, the journal *Medical Clinics of North America* devoted its entire March issue to the topic of obesity. The authors, all active in obesity research or treatment, reviewed current knowledge of the many facets of weight gain or loss and their physiologic control. In addition to the analyses of existing data, researchers reported identifying a fat-storage gene and linking leptin to the sense of taste.

About 97 million adults in the United States are either overweight or obese. The prevalence of obesity in the United States rose 6 percent from 1998 to 1999.

A Chronic Illness

Obesity should be considered a chronic illness that requires long-term treatment to control. One major concern is that as a person's obesity increases, so does that person's risk of early death.

The degree to which a person is overweight is defined by that person's body-mass index (BMI). BMI is determined by dividing a person's weight by the square of his or her height, using metric units: BMI = weight in kilograms/

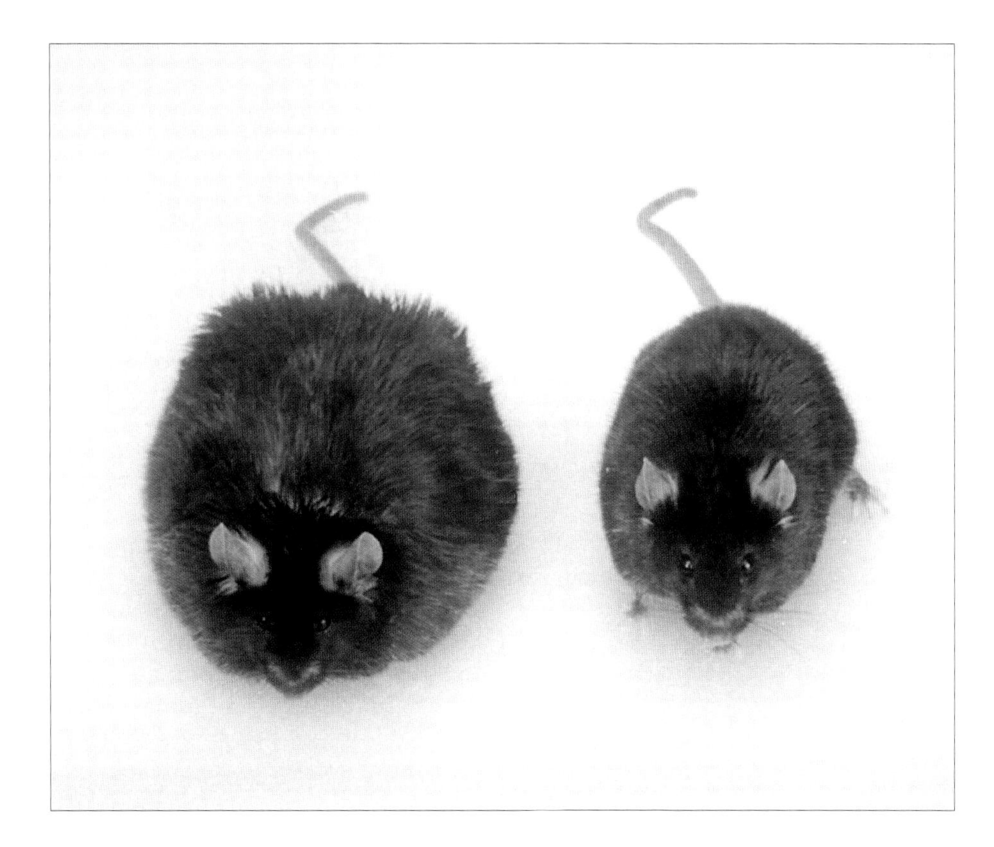

Although both mice have a defective gene that causes them to become fat, the mouse on the right received leptin, which reduced its weight to 35 grams (1.23 ounces), still above the normal 24 grams (0.84 ounces). The mouse that did not receive leptin weighs a hefty 67 grams (2.37 ounces). (AP/Wide World Photos)

square of height in meters. When using English units, a conversion factor of 703 is necessary: BMI = 703 × (weight in pounds/square of height in inches). Since 1997, clinicians have used the World Health Organization's definitions of excess weight and obesity. People with a BMI of 25 to 29.9 are overweight; those with a BMI of 30 or greater are obese.

Based on these definitions, about 97 million adults in the United States are either overweight or obese. The prevalence of obesity in the United States rose 6 percent from 1998 to 1999. Obesity is increasingly common not only in developed nations but also in less developed societies. A 1998 report from the World Health Organization identified obesity as the major worldwide health problem that is receiving the least attention.

The more obese a person is, the greater his or her risk of early death. This is true even if obesity is the person's only chronic illness. Obese people are also at higher risk for developing many other chronic diseases, including type II diabetes mellitus, hypertension, coronary heart disease, and cholesterol and lipid disorders. Increased weight aggravates osteoarthritis, the form of arthritis that develops from frequent overuse of the joints and heavy weight-bearing. The risk of some types of cancer, such as estrogen-sensitive breast cancer and cancer of the colon and prostate, increases with obesity, as well.

Genetics and Environment in Obesity

The interaction between a person's genetics and the environment that leads to obesity is not fully understood, but it is obviously complex. Evidence for a genetic factor is strong, but genetics alone does not explain an individual's tendency to stay at a particular weight. The environment is definitely a factor as well. In fact, it has been pointed out that the American environment seems to maximize the development of obesity.

Evidence for a genetic factor comes from many studies, especially those that compare the adult weight attained by twins. Twins tend to attain similar BMI as adults, with identical twins being more closely matched than fraternal twins. More proof comes from adoption studies. The adult BMI of adoptees is significantly closer to the BMI of their natural parents than it is to the BMI of the adoptive

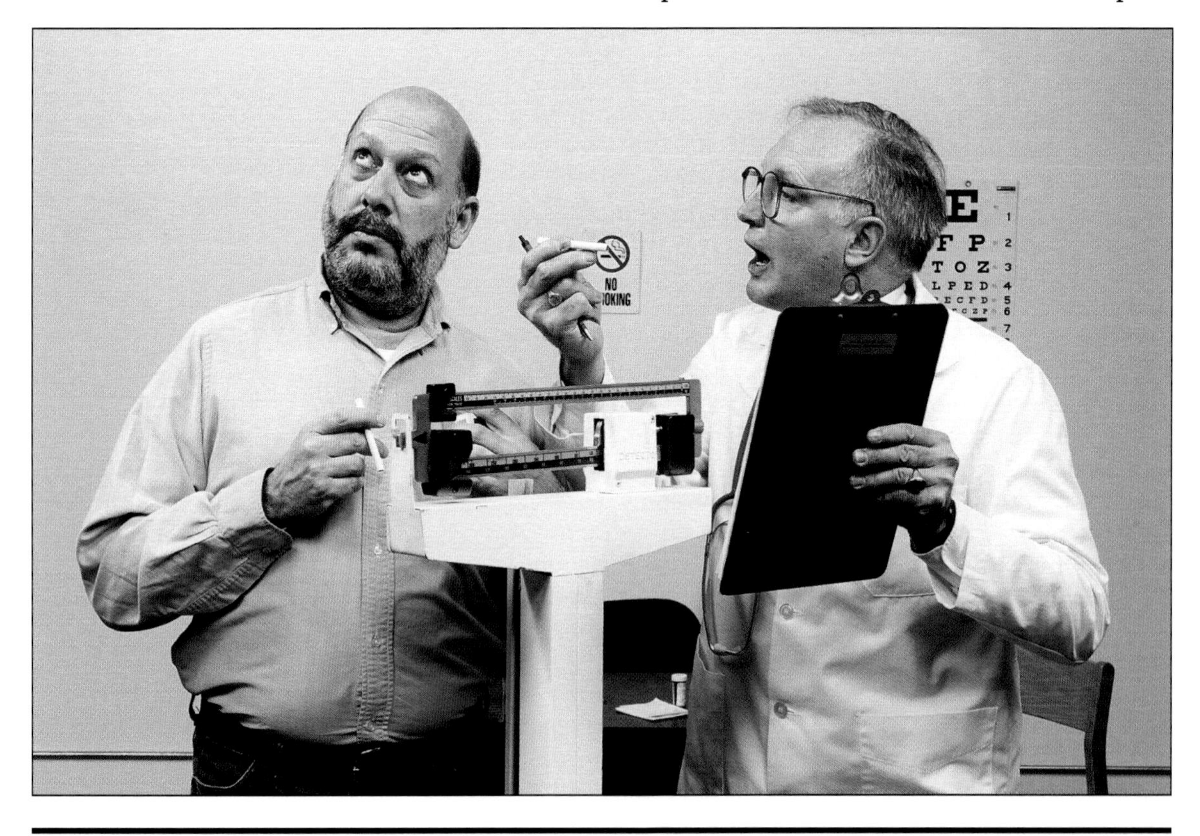

parents who raised them. Genetics may account for up to 70 percent of the variation in weight in a human population.

It is still not clear, however, how genetics affects the variation in body fat between adults in a particular population. Although several obesity-causing genes have been found in mice and rats, not one of these genes has been shown to have a significant effect on human weight.

Clearly, the internal workings of the body adjust to weight change. In a recent study, researchers discovered that people's metabolisms adjust as they either gain or lose weight to keep their weight within a relatively narrow range. They found that human subjects who increased their weight by 10 percent increased their daily energy expenditure by 25 percent, which tended to bring their weight back down. Those who lost 10 percent of their weight experienced an 18 percent decrease in metabolic rate, which caused a tendency to regain weight. This goes a long way toward explaining the difficulty that dieters experience in keeping lost weight from returning. It may be necessary to find ways to reset the body's genetically determined weight to a lower range before people can keep the lost pounds from creeping back.

A role for genetics has been clearly demonstrated in the function of the hormone leptin. This hormone plays an important part in the body's maintenance of a relatively narrow weight range by inhibiting appetite as the body's fat stores increase. In genetic research, mice that are bred to have a defect in the gene that stimulates leptin production become remarkably obese, three times fatter than ordinary mice. When given injections of leptin, their appetites fall and so does their weight.

The remarkable increase in the prevalence of obesity in a very short time, perhaps three decades, cannot be blamed on genetics alone. That is too brief a time period

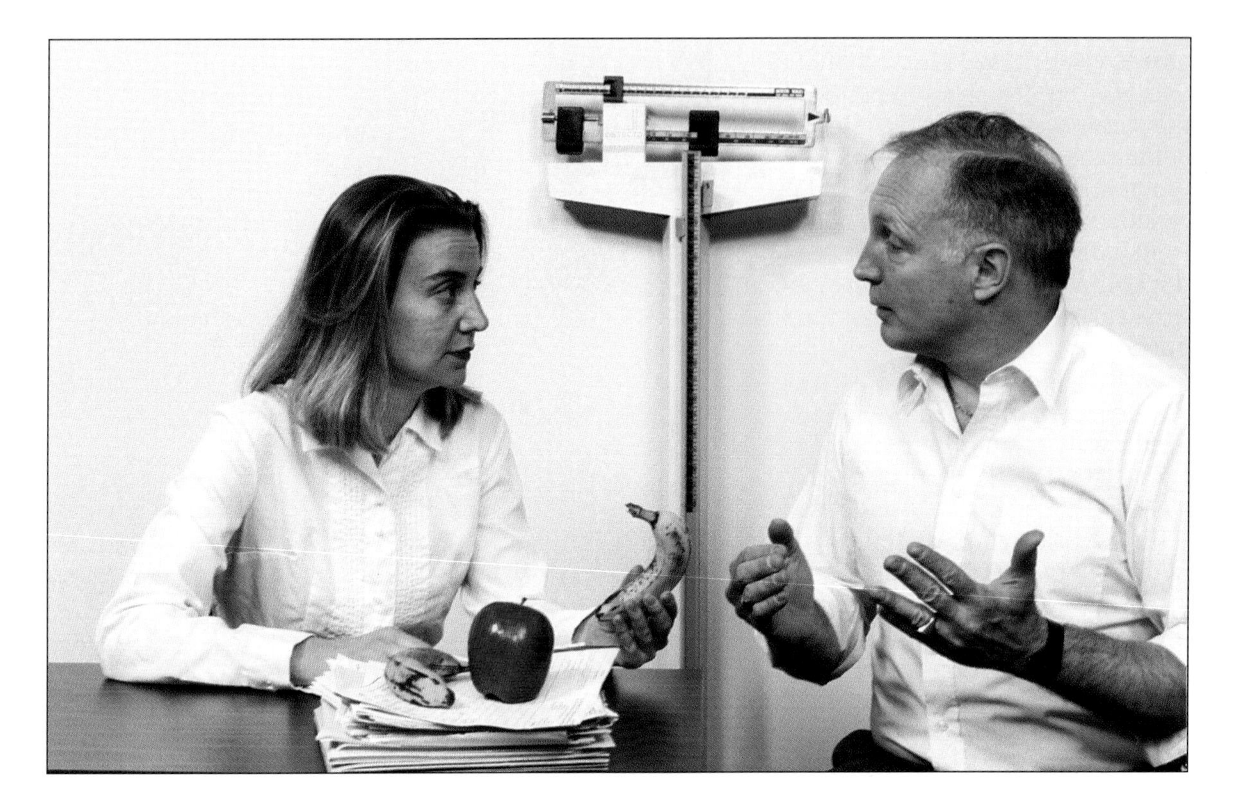

for a population's genetic makeup to be so significantly altered. As a result, much current research involves determining just what changes in the environment have led to such widespread weight gain. At least three environmental factors might be involved. First is an increase in the availability of high-fat foods that are energy-dense. A high-fat diet is generally perceived to be more palatable than a diet lower in fat content and tends to increase the amount of energy eaten at a meal. A very small increase in calories eaten can lead to significant weight gain over time if energy expenditure does not also rise. Even though an adult eats around one million calories per year, an increase of only 100 calories a day, if not accompanied by increased energy use, will cause a ten-pound weight gain each year. Second, food portions have tended to increase in restaurants and fast-food establishments, which often offer the opportunity to purchase a larger portion for only a slightly higher price. In addition, the menu items at fast-food restaurants tend to be high in fat and energy-dense. A third potential factor is the abundance in the environment of palatable, inexpensive food. Very little research has been done to explore the relationship between the cost and availability of food and the energy intake of humans.

Two New Discoveries

Researchers reported two intriguing discoveries involving obesity, at least in mice. First, in April, 2000, a team at the University of Medicine and Dentistry of New Jersey reported that the alteration of a single gene in mice allowed them to eat a high-calorie diet without gaining weight. The gene, named HMGIC, is involved in increasing the body's storage of fat when an animal routinely eats a high-fat diet. Mice that had both a defect in HMGIC production and a genetic absence of the weight-control hormone leptin did not gain weight, even on a fatty diet. It is far too early to tell if an understanding of this gene will yield practical results for human weight control.

In a report from Kyushu University in Japan published in October, 2000, researchers demonstrated the presence of leptin receptors on the nerves involved in the sense of taste in mice. When the nerves are exposed to leptin, the mice's taste for sweetness diminishes, and they decrease their food intake. Genetically obese mice crave sweet food and appear to lack the leptin receptor on their taste nerves. The researchers suggested that one way leptin may control weight gain is by making food taste slightly less palatable as weight goes up, causing the animal to eat less.

The percentage of chronically obese members of the population is rising rapidly. Genetic factors have a significant effect on the weight that people tend to keep as adults, but the rapid increase in obesity suggests that large environmental effects must also be operating. The weight-modulating hormone leptin appears to have a major role in keeping our weight within a certain range. However, research on the genetics of this hormone has not yet identified a use for it in the management of human obesity.

Cancer
New Gene Therapies Attack Tumors' Blood Supply

➤ *Rather than poisoning cancer cells and sickening the body's healthy cells at the same time, antiangiogenic therapy blocks the ability of solid tumors, such as those found in breast and lung cancer, to nourish themselves.*

The year 2000 saw remarkable advances in the science of controlling blood vessel growth. A review by Andrew L. Feldman and Steven K. Libutti, published in the September 15, 2000, issue of the journal *Cancer*, shows how close researchers are to making antiangiogenic therapy a reality for treating solid cancers in humans.

To stay alive, all living cells require a constant supply of nourishment, including oxygen, sugar, fats, and proteins. This nourishing supply line is the network of arteries, capillaries, and veins that carries oxygen and nutrients to virtually every nook and cranny of the body and then carries away the by-products of energy production. Tumor tissue is no different from healthy tissue in its need for nourishment. If anything, tumors may need more nutrients than normal tissues because they are undergoing such rapid growth. A tumor is unable to grow larger than two millimeters (seven-hundredths of an inch) unless new blood vessels grow with it.

The process of stimulating and controlling the growth of new blood vessels, called angiogenesis, is remarkably complex. In the healthy body, this activity is kept in balance by the interactions of a large number of enzymes and other proteins. Scientist have identified more than forty of the body's own substances that can inhibit vascular growth. In addition, a wide variety of manufactured proteins, including some medications, will either promote or disrupt vascular growth. One of these, thalidomide, is well known from the 1960's and 1970's for its effect on the fetuses of pregnant women who took the medication. In addition to its medical uses, such as treating nausea and morning sickness, thalidomide had the completely unintended ability to stop vascular growth in the extremities of the fetus, resulting in a child with absent or poorly developed arms or legs.

A tumor is unable to grow larger than two millimeters (seven-hundredths of an inch) unless new blood vessels grow with it.

Solid cancers produce their own substances that can hijack the body's control of vascular growth. The cancer releases proteins that stimulate blood vessels in the surrounding tissue to send branches into the tumor, keeping it alive and expanding.

For nearly thirty years, researchers have studied angiogenesis, convinced that if they could block the blood supply to a tumor, they could kill it. Until now, though, there has been no reliable way to send the necessary genes and messenger proteins into cells. Proof is rapidly accumulating that gene transfer therapy can successfully change the signals for protein production within a cell, whether healthy or cancerous. The necessary technical skills and techniques have now been developed to make this treatment a real possibility.

Cancer Facts 2000

The American Cancer Society estimated that about 1,220,100 new cases of cancer, plus 1.3 million new basal and squamous cell skin cancers, were expected to be diagnosed in 2000 in the United States. These cancer cases join the 8.4 million Americans (estimated by the National Cancer Institute) who have cancer or are considered cured. The five-year relative survival rates (people who are alive five years after diagnosis, whether in remission, diseasefree, or in treatment) for all cancers is 59 percent. The American Cancer Society estimated that in 2000 about 552,200 Americans, or 1,500 people per day, would die from cancer.

Studies in laboratory animals of drugs that block vascular growth have shown it is possible to shrink tumors by blocking the spread of new blood vessels. Researchers are now moving to study their safety and efficacy against cancer in humans. They plan to flood either the tumor or the surrounding normal tissue with genes that will stimulate the production of proteins that can halt blood vessel proliferation.

Gene Therapy

The chromosomes present in every cell of the body consist of long chains of the nucleic acids that make up deoxyribonucleic acid (DNA). The cell uses its DNA as a memory bank or recipe to guide the production of thousands of proteins that are necessary for sustaining the cell's function. The segment of DNA that serves as the recipe, or template, for production of a single protein is called a gene. Other nucleic acid chains made up of ribonucleic acid (RNA) carry out the replication, or building, of the protein called for by the recipe. Each cell contains the genes for tens of thousands of proteins, although only the ones necessary for that cell's normal function are employed.

Gene therapy permits scientists to add a new "recipe" to the cell's DNA or RNA. The protein that is produced can carry out very specific tasks within the cell, or it can be released from the producing cell to act on other cells of the body. Intracellular proteins, those that remain within the

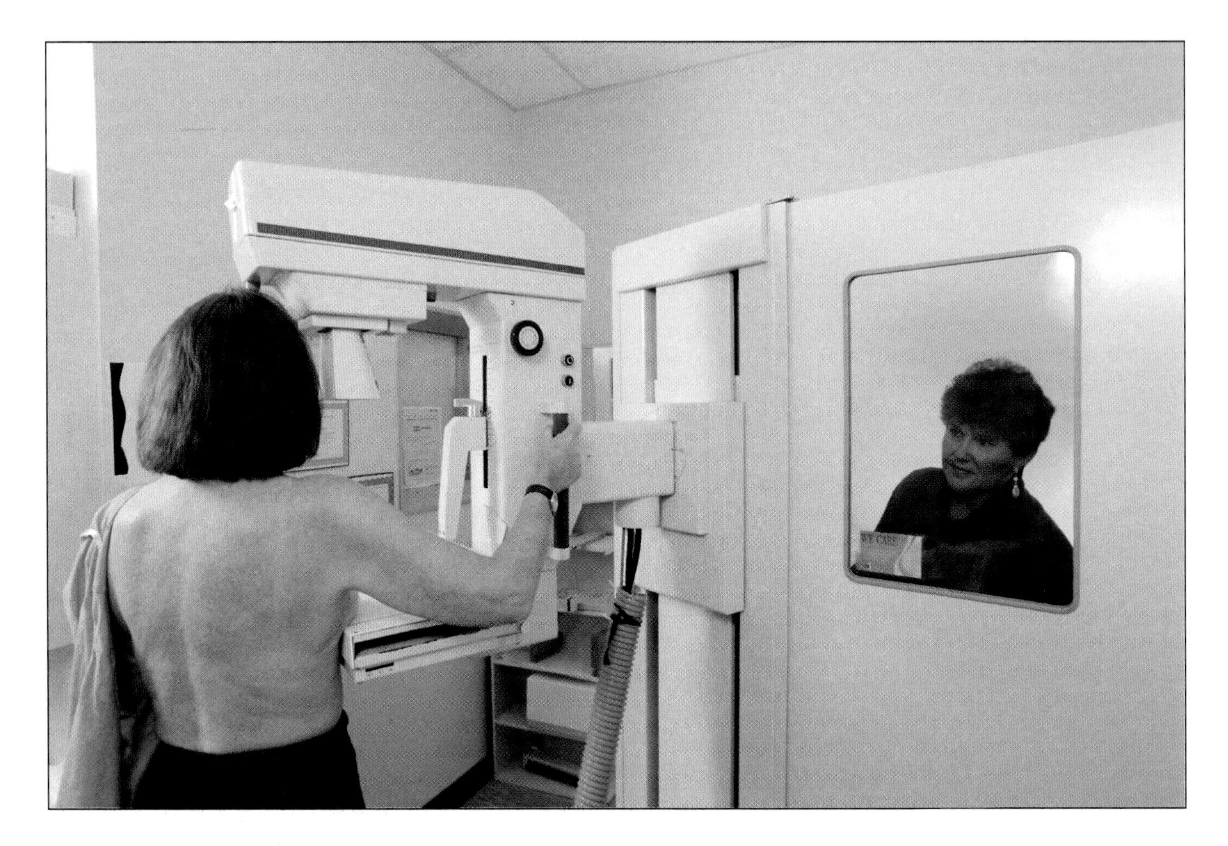

cell, carry out such business as building cellular structures and accumulating energy. Examples of the proteins that are released include hormones such as insulin, which directs the control of energy-producing sugar throughout the body. All living organisms contain genes. Even viruses, hundreds of times smaller than bacteria, contain enough DNA or RNA for the production of proteins that allow them to enter a cell and cause the cell's own resources to create more viruses.

Gene therapy is the term for the use of a group of techniques to sneak new nucleic acid segments, that is, new genes, into a cell or a tissue, such as the liver, to modify the function of the cell or the tissue. To date, most gene therapy research has focused on inserting genes that will correct abnormal cellular function in order to cure a genetic disorder such as cystic fibrosis, a chronic disease that destroys the lungs, and various types of immune deficiency.

A number of means can be used to package the genetic material, making it suitable for entering a person's cells. Genes can be wrapped inside particles of fatty acids or other metabolic products. They can also be placed in a wide variety of viruses that are known to infect the particular type of cell that must be modified. The virus infects the target cell and releases the new gene. To prevent the perpetuation of actual infection, the virus has had its own genes modified so that it cannot force the cell to make more viral copies.

Antiangiogenic therapy uses the same variety of particles and viruses to slip genes either into the tumor cells themselves or into other cells of the body, where they produce one of the proteins known to block the growth of new blood vessels. These proteins are almost all enzymes, organic compounds that accomplish most of the body's work by digesting, rearranging, and modifying other molecules for the body's use. The proteins that digest molecules of carbohydrate, protein, and fat in the intestinal tract, for instance, are enzymes.

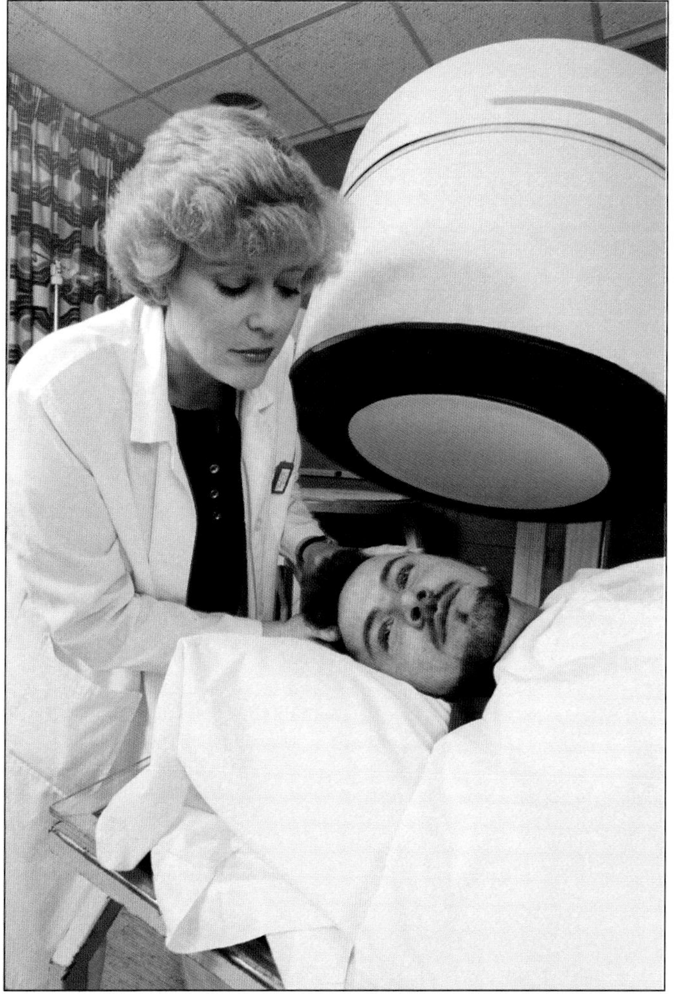

Stopping Vascular Growth

The body's control of blood vessel growth is very complex, and a number of its steps are susceptible to being blocked by genes or the enzymes they produce. Two broad approaches to antiangiogenic cancer treatment have been proposed. The first is gene therapy directed at the tumor itself. This mode of attack sends genes directly to the tumor cells, destroying their ability to influence vascular growth. Unlike standard chemotherapy, the intent is not to kill tumor cells outright but to eventually starve them to death. One limitation of this approach is that a solid tumor may not have developed enough blood supply to carry adequate numbers of genes to its malignant cells. In particular, very early, tiny metastases, clusters of cancer cells that have floated away from the main tumor to establish themselves in distant parts of the body, may not be attacked. Another problem is that as the cancer cells divide, the new genetic material will not necessarily enter each daughter cell, and its effect on the tumor will be diluted. This can be overcome, at least theoretically, by using retroviruses, which attach the new gene directly to the cell's chromosomes.

The second approach, designed to overcome the limitations of gene therapy directed against malignant cells, is to send the genes to healthy cells to stimulate them to produce antiangiogenic proteins. These proteins would circulate throughout the vascular system, blocking new blood vessel formation anywhere it is occurring. With this approach, the gene for the antiangiogenic protein will not be diluted as it could be within the tumor because most of the body's normal cells are not constantly dividing.

Status of Research

More than a dozen antiangiogenic substances have already been studied, either in laboratory cultures of tumor cells or in animals. Several have advanced to preliminary testing in humans. At this early stage, the human tests are solely to ensure the treatment's safety and to determine any side effects. So far, there seem to be few identifiable risks. Tests against actual tumors in people are yet to be conducted.

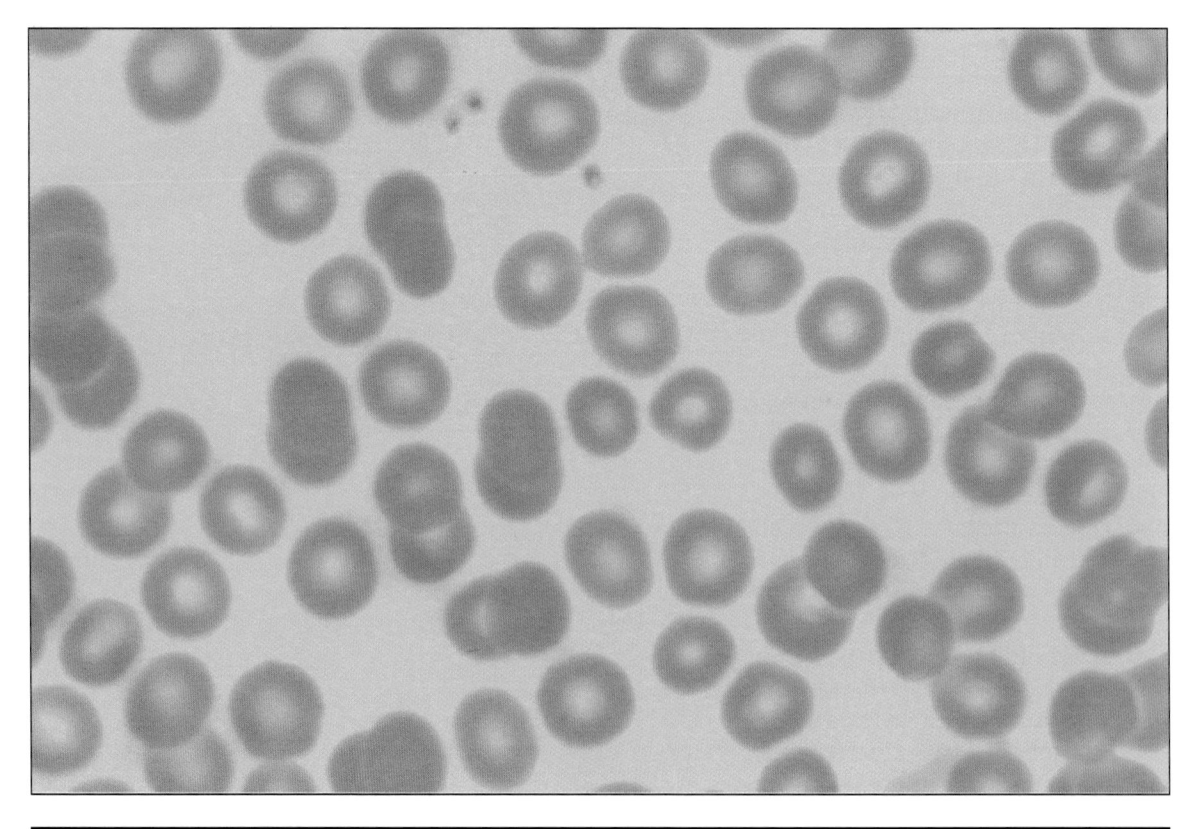

Events in 1999 and 2000 have made researchers more cautious about gene therapy in humans. In 1999, investigation of the first known death of a person serving as a research subject for gene transfer studies revealed what appeared to be inadequate oversight of the research. As a result, some laboratories have been ordered to stop gene therapy trials, and others have come under increased scrutiny. Although the death occurred in a study of a genetic liver disorder, not cancer, the vector virus that carried the new gene into that patient is one that is often used in other gene therapy applications. Because the patient appears to have died from an overwhelming immunologic response to the virus, gene therapy researchers are moving ahead more slowly than before because the risk of such reactions is not known.

Gene therapy is a simple concept that requires extremely complex procedures. However, it holds great hope for cancer patients. Researchers have not identified all the roles antiangiogenic therapy can play in the treatment of malignancies, but it is certain to have a growing role in years to come.

Asthma Treatments
New and Improved?

➤ *Two studies reveal that inhaled corticosteroids taken to treat asthma do not permanently stunt children's growth.*

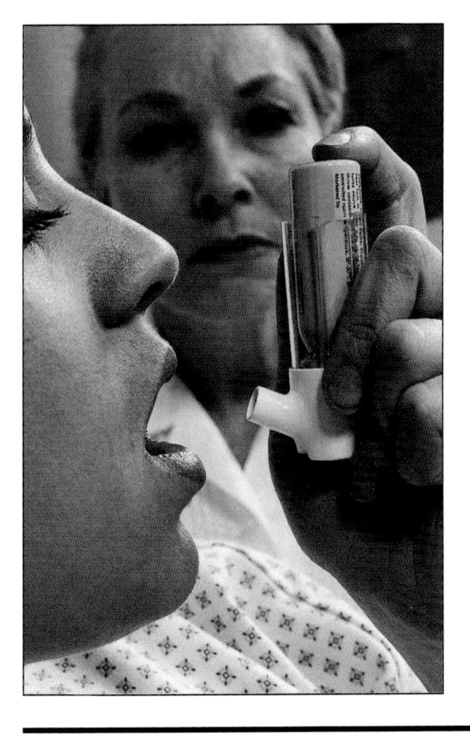

Two reports of major studies concerning pediatric asthma, published in the October 12, 2000, issue of the *New England Journal of Medicine*, have gone a long way toward answering the worrisome question of whether benefits of inhaled corticosteroids outweigh the potential of these medications to slow a child's growth. As a result of these studies, physicians can more confidently prescribe inhaled corticosteroids, a mainstay in asthma treatment, for their pediatric patients.

Many physicians have been reluctant to use this strong medication for asthmatic children unless they were severely affected because of concerns that the drug could cause a decrease in a child's growth rate when taken regularly, as has been proved for orally taken corticosteroids. The two reports show that while there is an initial slowing of growth in children using inhaled steroids, it lasts no more than a year. Equally important, the children eventually make up for this period of slow growth and go on to attain their expected adult height.

The first report, from the Childhood Asthma Management Program Research Group, compared the beneficial effects on mild-to-moderate asthma of two common inhaled asthma medications, the corticosteroid budesonide and another type of anti-inflammatory drug, nedocromil. They found that budesonide was more helpful in decreasing the lungs' reaction to irritants than was nedocromil. However, an additional finding was even more interesting. The children, who took budesonide for up to six years, experienced an initial slowing of their growth. However, it was brief (no more than a year) and was followed by a return to their normal growth rate. No other significant side effects of inhaled budesonide were found.

The second study was conducted by Lone Agertoft and Søren Pedersen in Denmark. They measured the growth of asthmatic children who took inhaled budesonide for up to thirteen years. They followed the growth of these patients until they reached their adult height. As in the first study, these researchers noted a slowing of growth early in the course of treatment. However, the patients eventually grew to their full predicted adult height. This suggested that the negative growth effect was temporary.

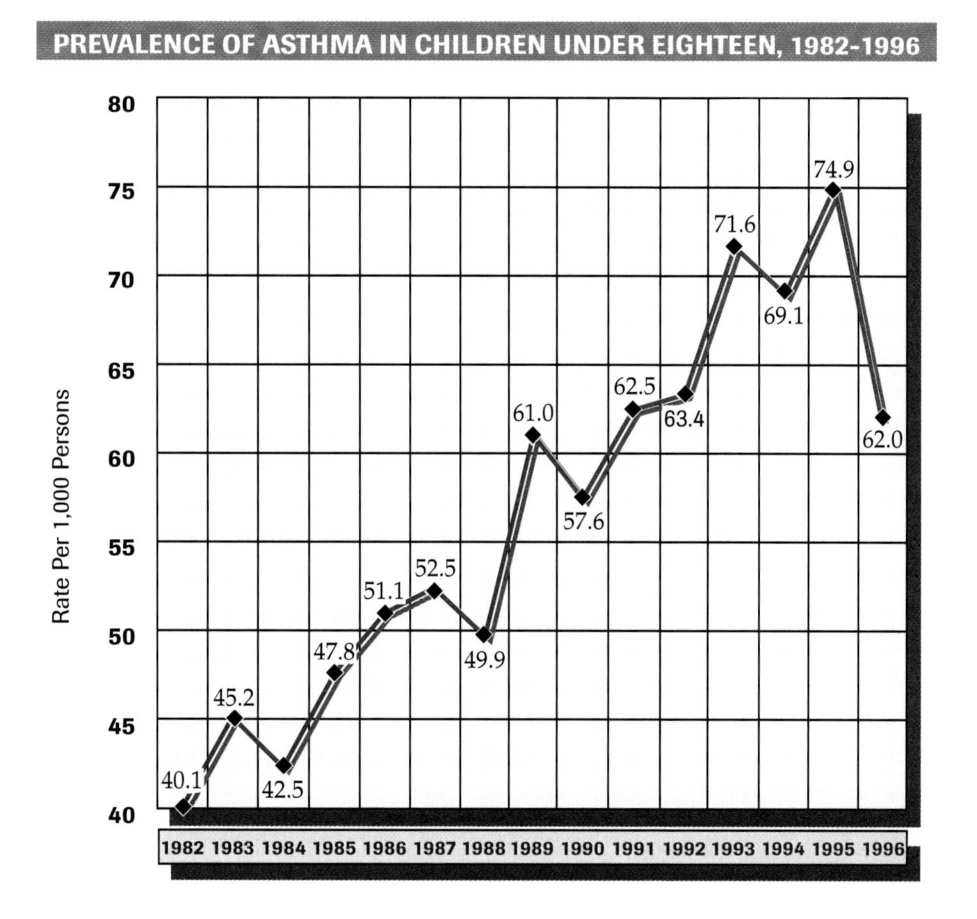

Source: National Center for Health Statistics, National Health Interview Survey, 1982-1996

A Growing Problem

Asthma is the most common serious chronic disease of childhood in the United States, affecting more than 5.3 million children. Its rate of occurrence is climbing steadily. In the United States, the prevalence of childhood asthma has increased more than 50 percent over the past fourteen years. Asthma is the cause of more than 10 million lost schooldays annually. More than 150 children die each year in the United States as a direct result of their asthma. The estimated cost of treating asthma in children under eighteen years old is $3.2 billion every year. Effective control of the disorder will produce significant benefits for the children, their families, and society.

Experts at the World Health Organization are finding a worldwide increase in the incidence of asthma. It is much more a disease of industrialized nations than of those with less developed economies. In fact, studies conducted in less developed countries, such as Albania, showed a remarkable increase in the incidence of asthma as their economies became modernized in recent years.

The Disease and the Drugs

Asthma is primarily a chronic inflammatory disorder. It affects the tubular air passages that branch off from the trachea and enter the lungs. Air passes through these airways—the larger ones called bronchi and the smaller ones called bronchioles—into and out of the lungs as a person breathes. The bronchi are particularly affected in asthma and are abnormally sensitive to irritants and allergens that the person inhales. These irritants include some found in the home, such as dander from pets' fur and the feces from microscopic dust mites, and many environmental pollutants, such as dust and diesel exhaust fumes.

When an asthmatic patient inhales a bronchial irritant, mast cells, a type of cell that line the bronchi, respond by rapidly setting off an intense inflammatory reaction. The bronchial lining swells and produces increased amounts of mucus. Muscles in the walls of the bronchi contract, narrowing the air passage even further. This partial blockage of the bronchi throughout the lungs is responsible for the difficulty breathing, the shortness of breath, and the wheezing that patients experience during an asthma attack.

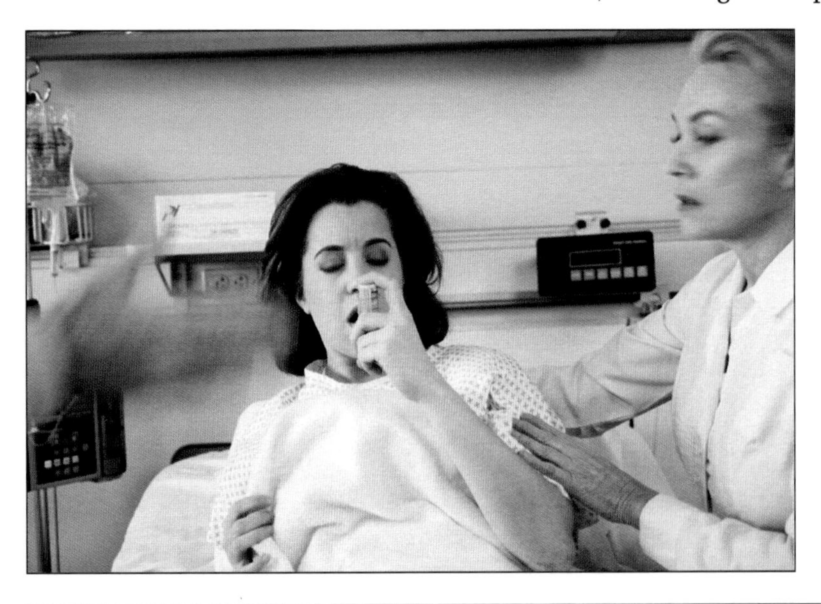

Corticosteroids such as budesonide are powerful anti-inflammatory chemicals. Cortisol is the corticosteroid produced in the human body, where it affects many cellular functions such as energy production and the response to stress. An abnormally high blood level of cortisol, such as occurs in Cushing's disease, can cause the same complications as a high level of medicinal corticosteroid. These include diabetes-like increases in the blood sugar level and fluid retention in the body, plus some complications such as cataracts in the eyes and a decrease in growth rates in children, which occur only after prolonged high cortisol and corticosteroid levels in the blood. Corticosteroids are distinctly different from anabolic steroids, a class of extremely high-risk drugs that some athletes use to increase their muscle mass.

In the United States, the prevalence of childhood asthma has increased more than 50 percent over the past fourteen years.

Inhaled corticosteroids have two important roles in the treatment of asthma. They block the intense inflammatory reaction during an acute attack. On a long-term basis, they also increase the asthmatic's ability to avoid acute attacks by controlling the ability of bronchi to respond to irritants with inflammation. Other medications for asthma include albuterol and other inhaled drugs that cause the bronchial muscles to relax. Like corticosteroids, these drugs can be taken to prevent an attack from occurring or to ease an attack when it happens. However, they do not directly affect the inflammatory response. Drugs that block the onset of bronchial inflammation, such as cromolyn and nedocromil, help prevent an attack. A number of other drugs that block specific biochemical steps in the asthmatic response, such as the group called leukotriene modifiers, are also available or in development.

The best treatment for asthma is for its sufferers to prevent attacks from occurring in the first place by regularly inhaling anti-inflammatory drugs. However, patients with mild asthma may need to inhale a medication such as albuterol only at times when they feel an attack beginning. Those with persistent mild or moderate asthma that does not completely clear between attacks may need to inhale medicine, especially corticosteroids, on a regular basis. For patients with moderate or severe asthma, a combination of regular anti-inflammatory drugs with additional medicine during an attack results in good control of respiratory symptoms.

Alzheimer's Disease
Remembering a Growing Problem

➤ *Research in 2000 produces hope in the shape of a plaque-reducing enzyme, a potentially useful vaccine, and the identification of two destructive enzymes whose action future drugs might stop.*

Sometimes a significant advance in medical research is the discovery of the right questions to ask. That certainly has been the case for Alzheimer's disease research in 2000. At the World Alzheimer Congress 2000 and at other major scientific meetings, researchers shared important new insights into the biochemical and genetic nature of Alzheimer's disease. They also discussed the results of a vaccine that has caused regression of some of the structural brain changes of Alzheimer's, at least in mice. These new insights into Alzheimer's disease have allowed researchers to identify the next level of questions that must be answered as they gain ground on this common neurodegenerative disorder. The questions include whether the brain damage of Alzheimer's results from changes inside the nerve cells or outside them, whether the changes are internal, and which of two biochemical processes is the principal cause.

For more than twenty-five years, it has been clear that two separate types of microscopic change occur in the nervous systems of Alzheimer's patients. The relative importance of each in causing Alzheimer's disease has to be clarified in order to develop a new generation of treatments that could cure or prevent the disease. Each pathway of change has its enthusiasts in the research community. The first type of microscopic change, an accumulation of dense plaques of protein outside the nerve cells (neurons), is seen as the disease progresses. The second type of change, called neurofibrillary tangles, results when the fine branches of neurons, called neurofibrils, that maintain contact with other neurons become tangled together as they lose their internal structure. The structural molecules that become damaged are called tau proteins. Both plaques and neurofibrillary tangles are seen in the brain tissue of Alzheimer's patients when the tissue is examined microscopically.

As the U.S. population ages, the number of people with Alzheimer's disease is expected to increase more than threefold by 2050.

The need for answers to these questions and for the development of new treatments for Alzheimer's disease is obvious. According to the Alzheimer's Association, some 4 million Americans currently suffer from the disease. One in ten adults over age sixty-five, including roughly half of those over age eighty-five, has this devastating disorder. The economic cost in the United States alone is at least $100 billion every year. Most health insurance policies do not cover the long-term treatments and care, either at home or in a nursing home, that most of these patients need. As the U.S. population ages, the number of people with Alzheimer's disease is expected to increase more than threefold by 2050 if no cure or preventive measure can be found.

Protein Clumps

Alois Alzheimer described plaques in the brain of a patient in the 1907 medical article that provided the first clinical and pathological information on the disorder. The plaques are clumps of a protein called amyloid beta-peptide (A-beta). This protein begins as a normal molecule, called amyloid precursor protein (APP), that is a part of the nerve cell's outer membrane. Two enzymes, one that is active inside the cell and another that is active outside the cell, partially digest the original protein, leaving strands of A-beta. These pieces of protein tend to stick tightly to each other, forming large, irregular clumps outside the nerve cells. The bulky, firm clumps are suspected by many researchers of causing damage to nearby neurons, either from the pressure they exert on the cells as they enlarge or through a direct toxic effect on the nerve cells. Recent research has shown that A-beta also accumulates inside neurons, although it may not form plaques there. This has led some scientists to argue that it is the A-beta inside the neurons that actually damages them and that the external plaques are "tombstones," inactive debris from the intracellular destruction. Clearly, further research will

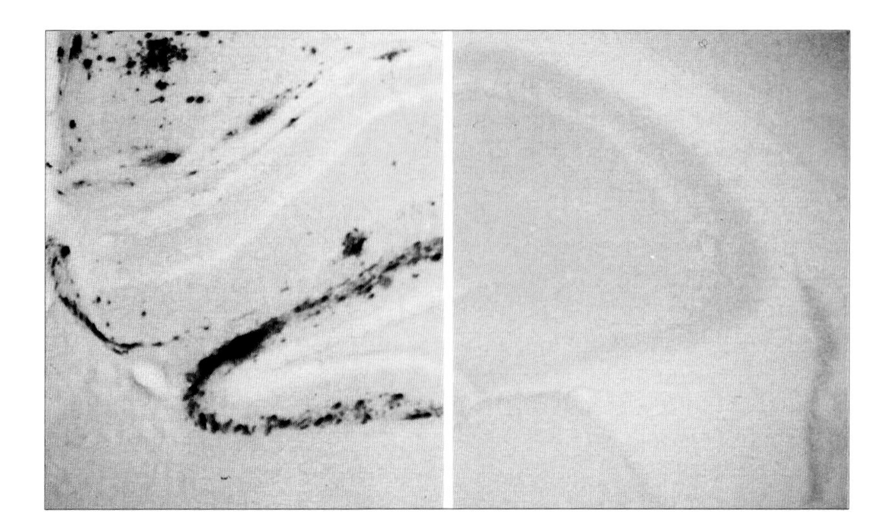

One of the topics at the conference was a vaccine that apparently prevents Alzheimer's disease in mice. Although both mice were genetically engineered to produce the dark protein deposits characteristic of the disease, the brain of a mouse given the vaccine (right) does not show the deposits that appear in the brain of the untreated mouse (left). (AP/ Wide World Photos)

be necessary to answer the question of which location of A-beta actually leads to neuronal damage.

Another major question that has emerged is whether the A-beta clumps form in Alzheimer's patients' brains because the protein is produced at an abnormally high rate or because the body's ability to dispose of the accumulated clumps is damaged. New research, reported in *Nature Medicine* in February, 2000, identified a specific protein-digesting enzyme (protease), called neprilysin, that appears to decrease A-beta plaques in mouse brains. Again, much additional research will be necessary to measure the activity of this protease and to identify situations within the brain in which its activity may be increased or decreased.

Neurofibrillary Tangles

The spacing and structure of the tiny microtubules within neurofibrils is maintained by repeated molecules of tau protein that serve to align the microtubules, acting as a series of external braces. In Alzheimer's disease, tau proteins weaken, break loose of the microtubule surface, and become intertwined in pairs. The neurofibrils lose their shape and become tangled. Eventually, the affected neuron itself dies.

Scientists who believe that tau protein damage is the main cause of Alzheimer's disease point out that the brains of people with some forms of dementia other than Alzheimer's display tangled lengths of tau protein. Plaques of A-beta never form in these patients' brains, yet they develop severe dementia.

A Possible Answer

Two other major recent developments may finally allow researchers to answer the question of which of the two abnormal processes is responsible for the dementia of Alzheimer's disease. Researchers at a pharmaceutical laboratory administered a vaccine made of A-beta fragments to mice that were genetically engineered to develop plaques in their brains. The plaques disappeared in all mice and had not reappeared in all but two animals (seven of nine treated) after one year of observation. The second discovery was a clear identification of the two enzymes that digest APP into A-beta. The identification of these enzymes enables researchers to design drugs that will block their action. The combination of both discoveries will permit scientists to design research studies that soon should clarify whether A-beta or tau protein is the primary cause of Alzheimer's disease.

Of course, neither asking the right questions nor finding their answers will immediately eliminate Alzheimer's disease. Also, it seems unlikely that new treatments will be fully restorative to people who already have developed nerve cell damage. Therefore, for many of the 4 million Americans who currently suffer from Alzheimer's and for their 19 million family members, the answers may still come too late.

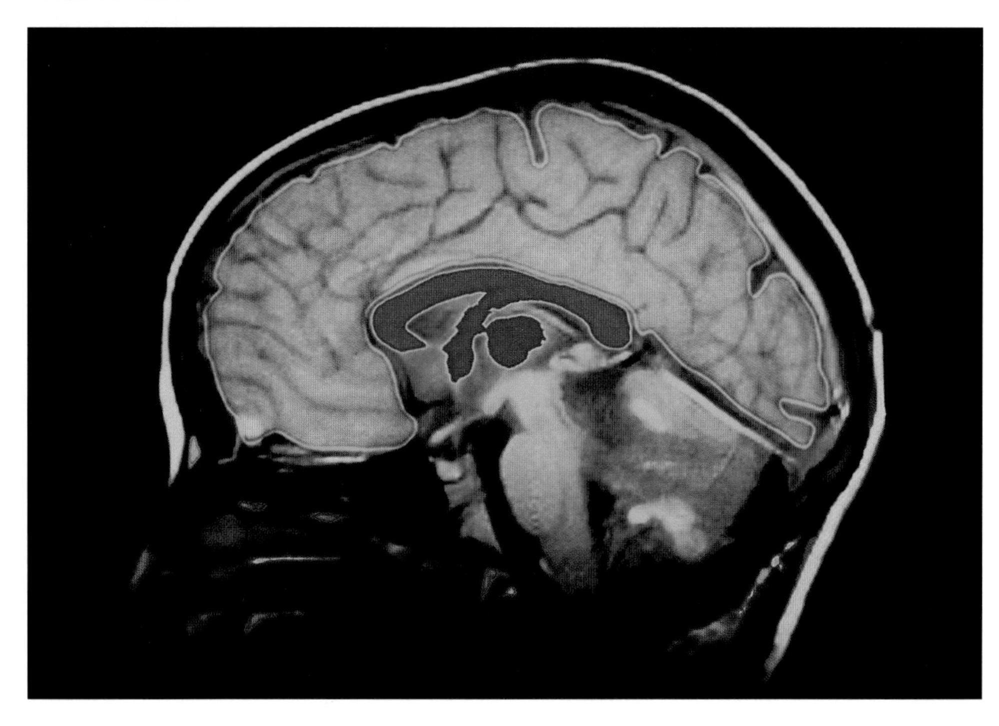

Diabetes Breakthrough
Promise of an End to Insulin Injection

➤ *Researchers reported the successful transplantation of insulin-producing cells, a breakthrough likely to improve the treatment of type I diabetes mellitus.*

In the July 27, 2000, issue of the *New England Journal of Medicine*, A. M. J. Shapiro and colleagues at the University of Alberta in Canada reported that they had successfully transplanted pancreatic islet cells, which produce insulin, into seven diabetic patients using a set of new techniques. The new approach virtually completely overcame the significant problems that had attended earlier transplant procedures. Particularly exciting was the fact that each patient was able to stop taking routine insulin injections. The procedure helps only patients with type I diabetes, which is caused by a deficiency of insulin production in the pancreas, and not patients with other forms of the disease. The majority of child diabetics and 5 percent of adult diabetics have type I diabetes, which is the hardest to treat and is generally the most severe and life-limiting form.

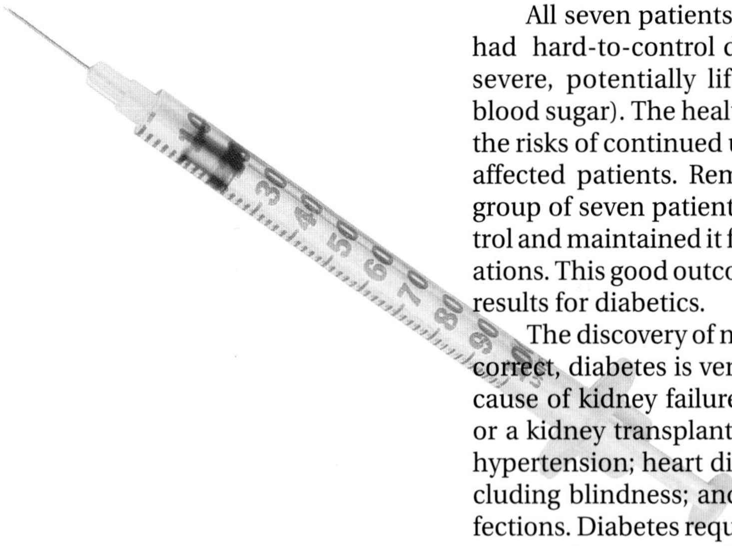

All seven patients who received the pancreatic islets had hard-to-control diabetes, with frequent episodes of severe, potentially life-threatening hypoglycemia (low blood sugar). The health risks of transplant were less than the risks of continued unstable diabetes for these severely affected patients. Remarkably, after the transplant, this group of seven patients all regained normal glucose control and maintained it for at least a year in follow-up evaluations. This good outcome far exceeds previous transplant results for diabetics.

The discovery of new and better ways to treat, or even correct, diabetes is very important. Diabetes is the major cause of kidney failure, which necessitates hemodialysis or a kidney transplant. It is also strongly associated with hypertension; heart disease; nervous system damage, including blindness; and an increased susceptibility to infections. Diabetes requires major life changes for patients. Its complications are difficult and expensive to treat. Most important, diabetes shortens the life expectancy of its victims.

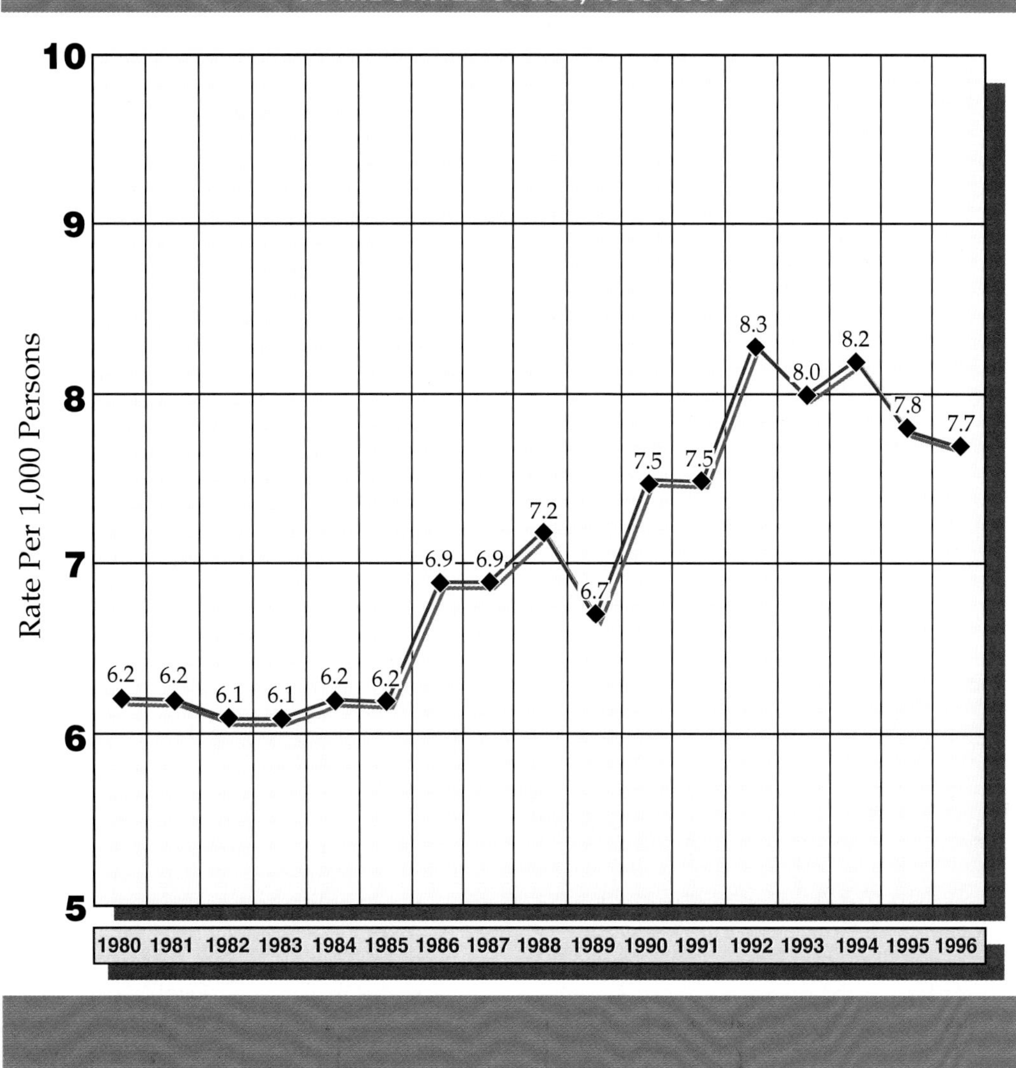

Overcoming Problems

Researchers have shown that diabetic patients experience the fewest complications from their disease when they closely control their blood sugar level. Theoretically, a return to the normal mechanisms of glucose control by transplanting new, functional islet cells should provide the best glucose control possible. One problem with pancreatic transplantation has been the complex nature of the surgical procedure necessary to transplant an entire pancreas into an appropriate site in the abdomen. This can be overcome by simply transplanting islet cells. However, with either technique, the patient must begin to take immunosuppressive drugs to combat rejection of the new tissue. This has always included corticosteroids, such as prednisone, which minimize the formation and action of antibodies against the new cells. Unfortunately, all corticosteroids, even the body's natural corticosteroid, cortisol, exert a major influence on glucose metabolism, worsening the effects of poor insulin function.

To counteract these major difficulties, the research team chose to transplant pancreatic islets, small clusters of insulin-secreting cells, rather than individual cells or an entire pancreas. They then injected the islets into the portal vein, the main source of blood from the intestines to the liver and pancreas. This provided the benefit of exposing the islets to virtually the same blood supply they would have experienced in the pancreas, coming straight from absorbing glucose and other nutrients in the intestine. The team also used a regimen of immunosuppressive drugs that did not include corticosteroids but still successfully combated rejection.

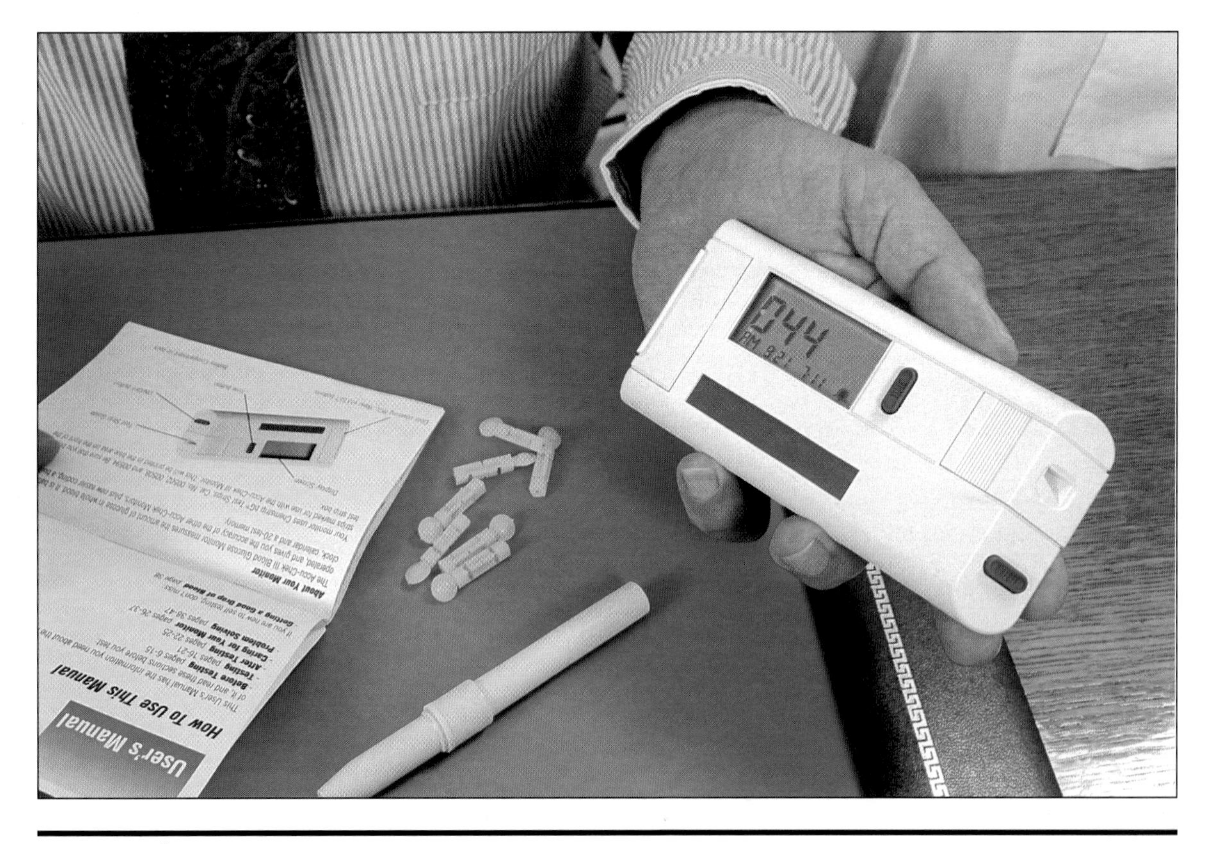

The Mechanics of Diabetes

Diabetes mellitus is a group of disorders in which the main problem is a loss of the body's ability to use insulin to control the level of glucose (the form of sugar most necessary for energy) in the patient's cells and fluids. Lacking the action of insulin for the transport of glucose molecules into the cells, the body experiences a rise in the level of glucose in the blood. However, the cells are literally starving despite an oversupply of glucose in the fluid around them.

In response to cellular warnings of starvation, the body turns to an alternate source of glucose, the digestion of supplies of protein and fat that are already present in tissues. This produces an increased amount of glucose in the blood, but the cells are no more able to use this glucose than the glucose that was already present. Unfortunately, this pathway also leads to the formation of acidic molecules in the body, principally from digestion of fat. It is a combination of the high blood glucose level, the starvation of cells throughout the body, and the formation of toxic levels of acid that leads to widespread tissue damage in the diabetic patient.

The division of diabetes mellitus into two broad categories is based on two different kinds of defects that cause poor insulin function. Type I diabetes occurs as the result of a failure of the pancreatic islet cells to produce enough insulin for the body's needs. Patients with this form of diabetes must receive injections of insulin to meet their body's changing needs for glucose energy, as many as four or five times each day. This type is by far the most common form diagnosed in children and adolescents.

Theoretically, a return to the normal mechanisms of glucose control by transplanting new, functional islet cells should provide the best glucose control possible.

However, the great majority of diabetics, especially adults, suffer from type II diabetes. These patients, rather than producing too little insulin, experience poor insulin function because their cells are very resistant to the actions of insulin to import glucose. In fact, these patients may produce large amounts of insulin at first. Treatment of type II diabetes involves exercise, diet control, weight loss, and medications, all of which increase the body's ability to use insulin. Only in rare instances do type II diabetics require frequent insulin dosages.

Chasing the Fountain of Youth
Hormone Replacement Therapy for Women—and Men

➤ *Studies make a woman's decision whether to use hormone replacement therapy more difficult by showing both benefits and risks associated with the treatment.*

The complex hormonal and bodily changes in middle-aged women, gathered together under the term "menopause," have been recognized for centuries if not millennia. Whether the benefits of hormone replacement therapy (HRT), generally using a combination of the hormones estrogen and progesterone, to counteract these changes are greater than their associated risks has proved to be an extremely difficult question to answer. New evidence presented in 2000 clouded the issue even further.

Much less obviously, men undergo a similar decrease in hormonal function with aging. The term "andropause" has come into use to describe changes that occur in men as they age and are caused by a decrease in male hormones, particularly testosterone. Recent evidence from a major symposium on andropause, compiled in a special supplement to the January, 2000, issue of the journal *Mayo Clinic Proceedings,* shows that scientists have just as far to go to understand the pluses and minuses of testosterone replacement in men.

Hormone Replacement Therapy for Women

Some questions regarding female hormone replacement therapy have been answered. It is clear that this therapy minimizes many of the most bothersome symptoms of menopause, including hot flashes, difficulty sleeping, and altered mood. The therapy also significantly helps keep women's bones and muscles adequately strong by minimizing the progression of osteoporosis and loss of muscle tissue with increasing age.

Hormone replacement therapy is associated with an increased risk of breast cancer, whether a woman takes one hormone or two.

In 2000, research showed that hormone replacement therapy can help prevent the incidence of harmful cardiovascular events such as heart attacks if begun before a woman's first attack. However, in at least one previous study, the incidence of harmful cardiovascular events actually rose for

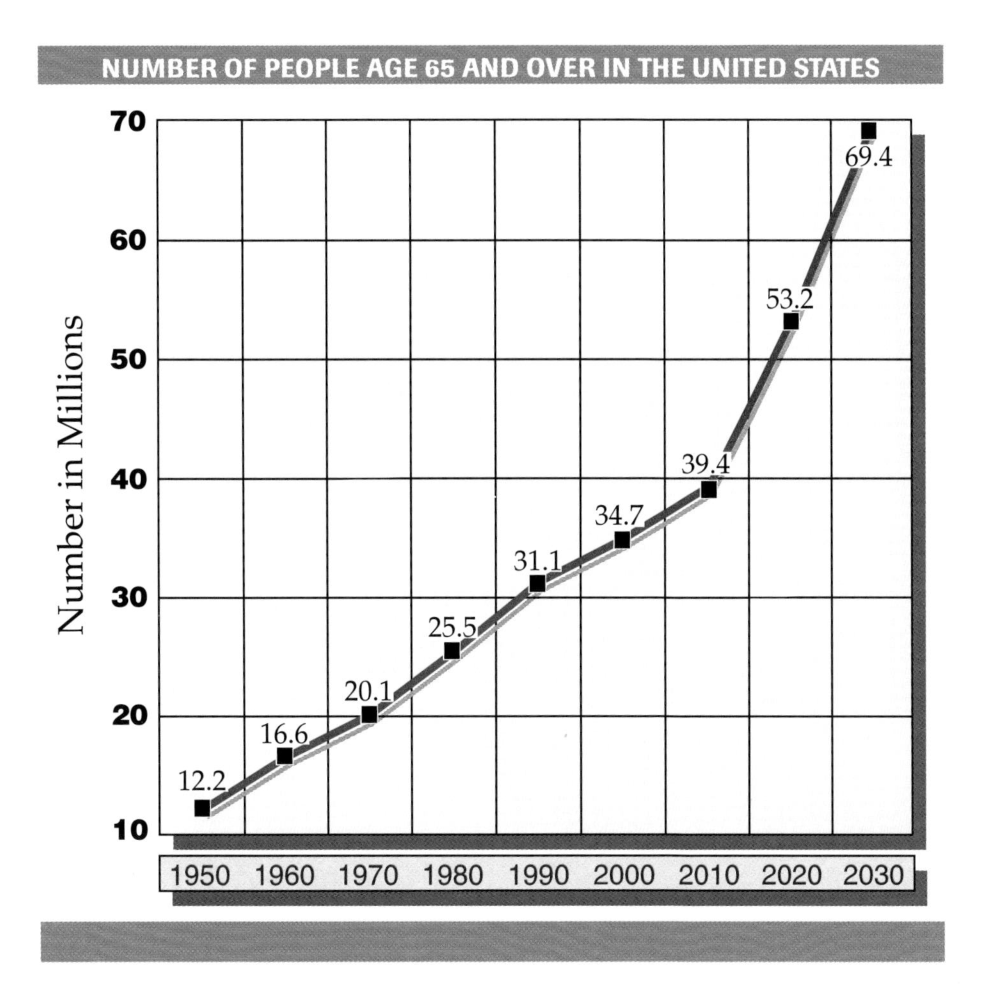

the first two years that women underwent hormone replacement therapy. This is an important concern because coronary heart disease is the most common cause of death for women if measured across all ages. In fact, it is five times more likely to be the cause of death in a woman than is cancer of the breast or lung.

There are other concerns about the safety of hormone replacement therapy. It has been recognized for some time that replacement therapy using estrogen alone is associated with an increased risk of endometrial cancer, which develops in the lining of the uterus. Adding progesterone to the drugs used in hormone replacement therapy significantly decreases this risk by providing a balance for the effects of estrogen.

A more worrisome concern was presented in an article in the January 26, 2000, issue of the *Journal of the American Medical Association* on the results of a remarkably

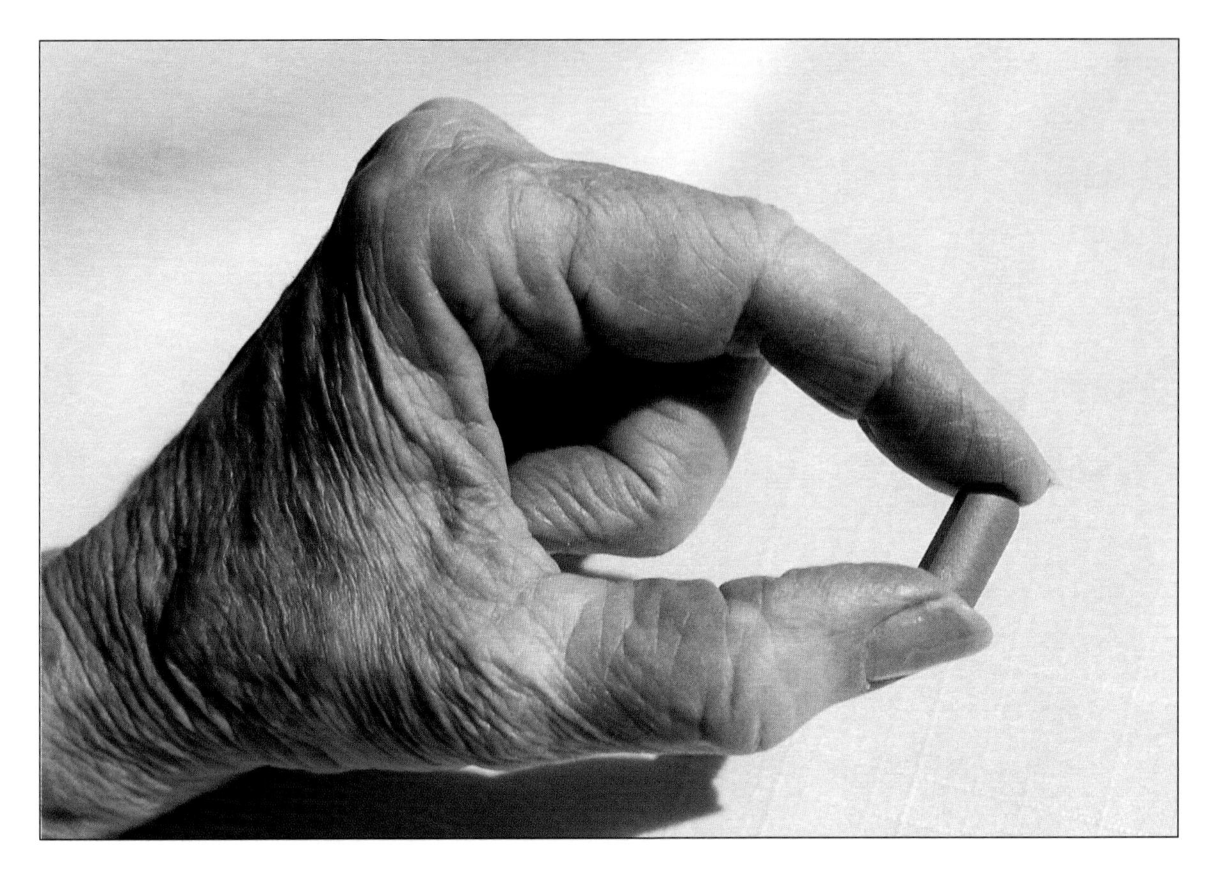

large study that followed more than forty-six thousand women for fifteen years to monitor their rate of developing breast cancer. Women on hormone replacement therapy, whether estrogen alone or estrogen and progesterone, were more likely to develop breast cancer than those who did not take any hormones. In addition, women who took both estrogen and progesterone had a greater likelihood of developing breast cancer than those who took only estrogen. Breast cancer was particularly more likely to develop in those women on two-hormone replacement therapy who were not overweight. The increased risk was relatively small for both groups, but it underscored the fact that hormone replacement therapy is associated with an increased risk of breast cancer, whether a woman takes one hormone or two.

Faced with all the possible combinations of benefits and risks, it seems most reasonable for each menopausal woman and her physician to consider together whether hormone replacement therapy is the right choice for her. The choice is increasingly a very individualized decision.

Hormone Replacement Therapy for Men

The effects of hormone replacement therapy on men have been studied far less than the therapy's effects on women. In men, the therapy involves the replacement of testosterone, which declines gradually as men age. Although men do not experience the relatively sudden change in hormone levels that women do, they undergo many of the same changes. Muscle mass and bone strength decline, accompanied by increased risk of falls and fractures. Mood changes and depression become more common. Sexual interest and function decrease.

The January supplement to *Mayo Clinic Proceedings* makes at least three points clear. First, a man's decision to take hormone replacement therapy should be as individual as it is for a woman. Second, there are both benefits and significant risks to male hormone replacement therapy. Third, research that could measure the balance of benefits to risks with male hormone replacement therapy lags far behind similar research in women.

The benefits of testosterone replacement therapy include the minimization of loss of bone and muscle strength and improvement of moods and sexual function. The therapy also improves cognition, especially in the realm of short-term memory. The effect of hormone replacement therapy on male coronary heart disease has received little attention, especially in comparison to the large studies of its effect on postmenopausal women. The existing evidence shows a modest lowering of the risk of coronary heart disease for older men with adequate levels of natural testosterone. However, it is also true that men, especially under age fifty, have a far higher incidence of cardiovascular events than women. Although estrogen appears to protect against cardiovascular events in women, it is much less clear whether androgenic hormones such as testosterone are involved in the increased risk for men.

In summary, the journal states that it is much too early to recommend hormone replacement therapy for older men. Unlike female hormone replacement therapy, male hormone replacement therapy has not been the subject of long-term studies that can adequately answer whether the benefits of such therapy outweigh the risks for men.

Aging as a Disease

A number of researchers as well as patients have worried that the wide use of hormone replacement therapy will encourage people to think of aging as a disease.

This is a particular concern for women because there is a long history of the "medicalization" of normal female physiologic functions. The tendency to manage pregnancy and childbirth as if this course of natural events were a disease that requires careful monitoring and treatment has received wide comment since obstetrical care was turned over to physicians a century ago. Menstrual and premenstrual changes have until recently received little attention, except as abnormal symptoms to be medicated.

It is important to remember that hormone replacement therapy modifies the normal changes associated with aging. Its role is to provide people with an improved quality of life as they experience the aging process. Both clinicians and patients must keep this clear in their minds as they consider whether to begin hormone replacement therapy. Further exploration of such questions as what constitutes a disease, what counts as treatment rather than assistance with normal change, and what an improved quality of life means must be considered by patients and clinicians alike. Studies of the humanities and philosophy can provide additional insight, but

the answers to such questions will always remain intensely personal.

Will hormone replacement therapy become the ultimate "elixir" of prolonged life? Absolutely not. The human organism is programmed to age steadily, and this inexorable journey into old age will continue. What can be legitimately hoped for is that hormone replacement therapy will improve the quality of life for men and women as they age.

HIV 2000

Growing Rates, Growing Denial

➤ *Both medical journals and consumer publications reported three alarming trends in the character of the HIV epidemic: the increasing prevalence of HIV in adolescents, the likelihood of negative population growth in Africa because of AIDS, and the growth of a social movement that attempts to deny or minimize the causative link between HIV and AIDS.*

Adolescents constitute an increasing portion of those infected with the human immunodeficiency virus (HIV). The highest rate of new infections is among those age fifteen to twenty-four. They count for half of the nearly 6 million new HIV infections identified worldwide in 1999. Of the 40,000 new cases identified annually in the United States, currently 25 percent occur among those age thirteen to twenty-one. This statistic may even underestimate the number of American teenagers who are infected with the virus. Up to 19 percent of new acquired immunodeficiency syndrome (AIDS) cases each year in the United States occur in young adults in their twenties. Because HIV can infect the body for as many as ten years before turning into AIDS, many of these adults undoubtedly were originally infected as teenagers.

Young female Hispanic teenagers and non-Hispanic black teenagers are all strikingly overrepresented among adolescents with HIV, compared with their proportion of the total population. For example, 54 percent of American teenagers with HIV infection are women, 66 percent are non-Hispanic blacks, and only 28 percent are non-Hispanic whites. Among adolescents thirteen to twenty-one years old with AIDS, 58 percent are women, compared with only 23 percent of women over age twenty-five with AIDS. Of all adolescent AIDS cases reported in 1999, 60 percent occurred among blacks, even though they make up only 15 percent of American adolescents. Hispanic teens, 14 percent of the national adolescent population, accounted for 24 percent of all reported adolescent AIDS cases in 1999.

In southern Africa, as many as half of all current fifteen-year-olds will die from AIDS as adults, even if they are not infected now.

The future is even more frightening for teenagers in southern Africa, where as many as half of all current fifteen-year-olds will die from AIDS as adults, even if they are not infected now. For Botswana, the estimate rises to a horrifying two-thirds of all current fifteen-year-olds.

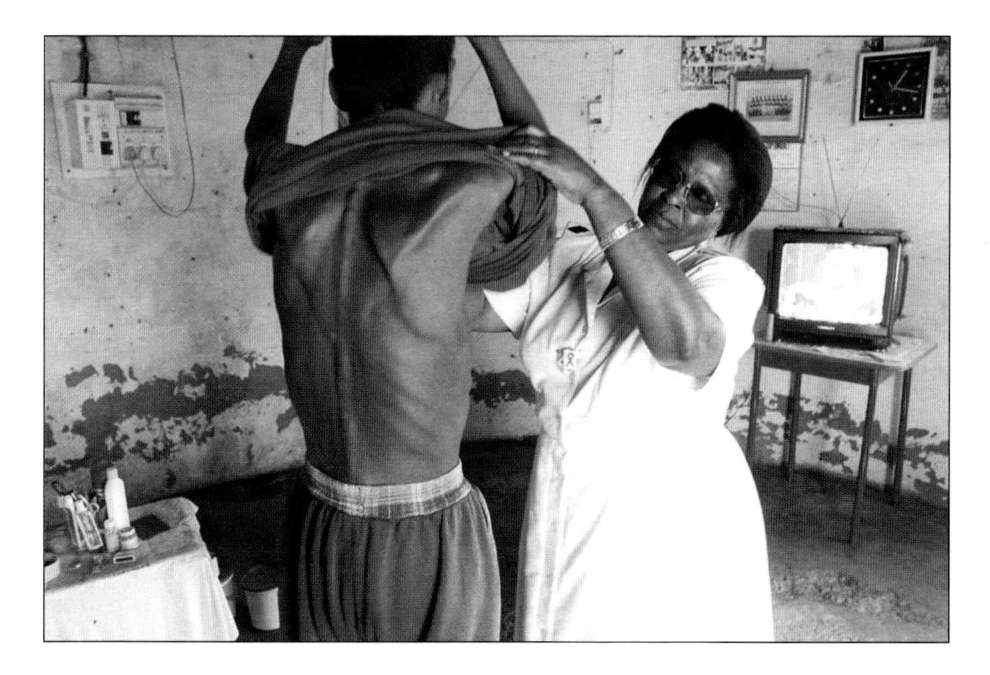

The region of Hlabisa, about 250 miles north of Durban, South Africa, has the highest rate of HIV infection in South Africa. Here, a nun assists a man who has AIDS at his home in Hlabisa. (AP/Wide World Photos)

Adolescents struggle through remarkable changes physically, intellectually, socially, and emotionally. The experiences that go with these changes tend to increase the risk of HIV infection for teenagers. Up to 50 percent of high school students are sexually active, but fewer than half of them use condoms regularly. This fact, plus their tendency to have had more than one sexual partner (16 percent report more than four sexual partners) puts them at increased risk for sexually transmitted diseases (STDs), such as chlamydia, gonorrhea, and genital herpes. In fact, 25 percent of the 12 million STDs reported each year in the United States occur in the adolescent age group.

STDs are particularly dangerous for adolescent women for several reasons. Being infected with an STD increases the risk of HIV infection. This is partly attributed to multiple-partner exposure but also to two other factors. Female genitals provide a larger surface area for possible infection than male genitals. Also, STDs tend to cause fewer symptoms in women than in men. In fact, a woman may experience no infectious symptoms for an extended time. Therefore, they are more likely than men to have sex without realizing that they have an STD.

Teenagers participate in other activities that increase their risk of HIV. Experimentation with drugs and alcohol can impair their judgment and can also interfere with the necessary development of adult social skills and behavior. Teens who are in

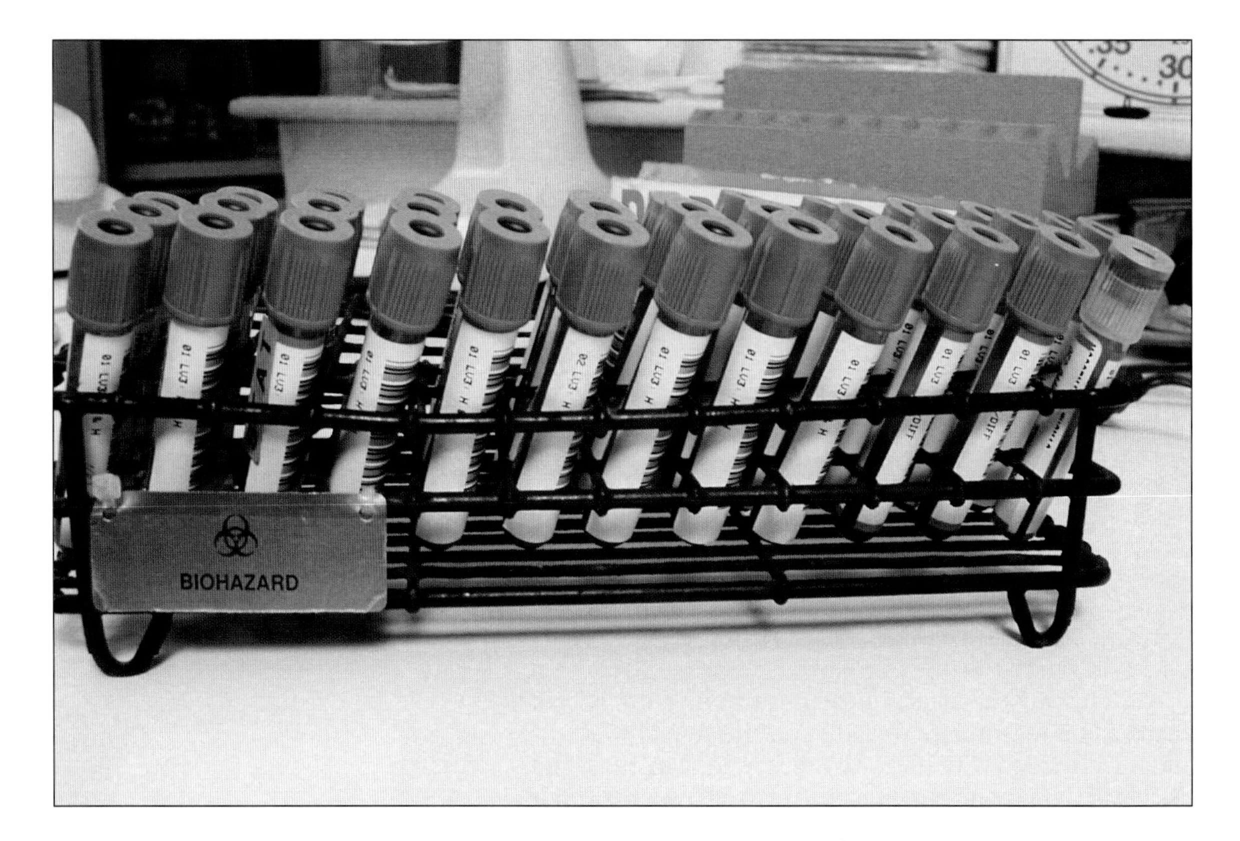

foster care, who are homeless or have run away from home, and who are incarcerated are also at increased risk for STDs and HIV infection. Children and teenagers, as a group, are also the Americans least likely to have health insurance or adequate access to health care. Also, teenagers tend to be extremely concerned about their personal privacy, to fear disclosing too much information to adults, and to lack knowledge of adult bodily functions. At the same time, they tend not to understand how to make the health system work in their favor and are either unaware or suspicious of the privacy safeguards that the health system is required to provide. They tend to receive episodic and disrupted health care, which prevents them from developing a trusting relationship with physicians. Another socioeconomic factor is that teenagers who work tend to have low-paying, entry-level jobs that do not provide much chance for increasing income or for receiving health benefits.

As of 2000, more than eighty thousand American children and teenagers have suffered the loss of one or both parents to AIDS. In addition to the disruption and emotional burden of living with a dying parent, these adolescents face the likelihood of continuing to live in disrupted homes. They are also highly likely to remain in the same socially and medically dangerous neighborhood that first led to their parent's infection and that places them at continuous high risk.

The statistics of HIV infection in the United States have changed considerably since the epidemic began. The citizens with the fewest available resources—young people, women, and ethnic minorities—are now the most severely affected.

AIDS and Africa

Only within the last few years has the severity of the AIDS epidemic in Third World countries begun to be recognized or acknowledged. The Joint United Nations Program on HIV/AIDS (UNAIDS) "Report on the Global HIV/AIDS Epidemic," released in July, 2000, contains HIV-related statistics so bleak that they are hard to comprehend. In sixteen African nations, more than 10 percent of all adults are HIV-infected. In seven nations in the southern cone of Africa, at least one of every five adults is infected. In Botswana, 36 percent of all adults are HIV-infected; in South Africa, the infection rate is 20 percent. South Africa has the world's largest HIV-infected population, 4.2 million persons. One out of four South African women between twenty and twenty-nine years old already is infected with HIV. Because of AIDS deaths, life expectancy in parts of Africa may soon drop below thirty years. At least three African nations—Botswana, South Africa, and Zimbabwe—are likely to experience a decrease in population, or negative population growth.

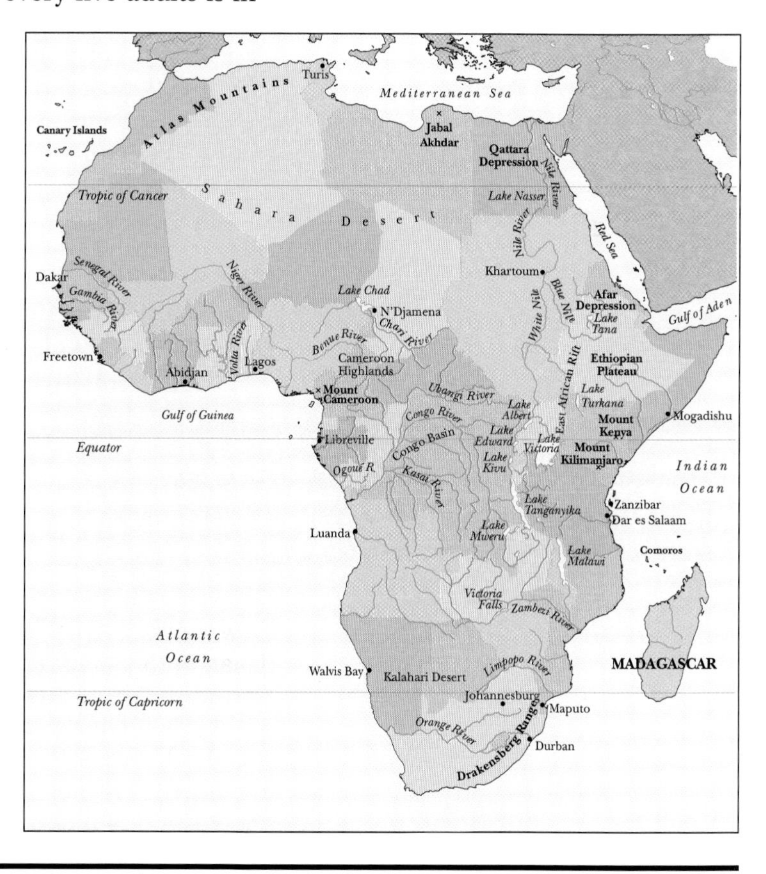

AIDS has already created 13.2 million orphans throughout the world. More than 12 million of those orphans live on the African continent, and their numbers will increase greatly. Half of all fifteen-year-old children currently alive will die of

AIDS in South Africa, Zimbabwe, and other African nations in which the adult infection rate is 20 percent or greater. AIDS will be responsible for nearly 80 percent of all deaths in adults between ages twenty-five and forty-five in any African country with an HIV rate 10 percent or higher.

The emotional and financial effects of such an epidemic are incalculable. Its severity may far exceed the effects of the medieval European plague epidemics, and it will go on for decades. Already, many governments in Africa are unable to afford more than token treatment for their citizens with HIV. Health care institutions are overwhelmed. In some places, HIV patients occupy up to 70 percent of city hospital beds. The cost of providing even the current minimal care for HIV absorbs more than half of the already tiny health care budgets of many African nations.

There are some bright spots on the map. Strong public health programs, primarily aimed at prevention of HIV infection, are having small but significant effects on the prevalence of HIV in Zambia and Uganda and in some specific localities in other nations.

The resources needed to manage the HIV epidemic in Africa far exceed those available. Although nations of every continent are severely affected by HIV, the problems are nowhere so huge as in Africa. The HIV epidemic has become a global phenomenon, and it will take worldwide commitment and effort on the part of governments and individuals to bring it under control.

Denial of HIV-AIDS Link

As widely reported in 2000, a growing number of people are beginning to deny that there is a causal link between HIV and AIDS. They have pointed out inconsistencies in research results, particularly early studies, and broad variations in the course of the disease in those who are infected. The group includes a number of persons infected with HIV. Although they raise some legitimate concerns, the "denialists" generally base their arguments on a faulty understanding or misinterpretation of the scientific literature. The denialists generally claim that there is no solid proof that HIV causes AIDS or that the virus causes a steady decrease in the body's immunity. They tend to blame AIDS on environmental and social factors, such as increasing pollutant levels and overcrowding. Although

The denialists generally claim that there is no solid proof that HIV causes AIDS or that the virus causes a steady decrease in the body's immunity.

they attack HIV science, they have not begun any research of their own to scientifically counter the current explanation for the cause of AIDS.

The clinical and epidemiological evidence for HIV as the cause of AIDS is so overwhelming as to be virtually irrefutable. However, for whatever reasons, the scientific community has been remarkably slow to respond to the criticism. Recently, however, a large number of scientists have begun to actively contest the denialists' claims.

One of the most outspoken denialists is the president of South Africa, Thabo Mbeki. His health minister is also highly critical of the HIV-AIDS relationship. As a result, the direction of South African government interest, resources, and money against the HIV epidemic has been delayed and prevented to the point that only minimal public health efforts are being used to battle the epidemic that engulfs South Africa. Given the viewpoint of these government leaders, it is unlikely that solid advances against HIV will be made in South Africa in the near future.

Cardiovascular Infections?

Chlamydia and Other Coronary Culprits

➤ *Scientists review evidence that points to the Chlamydiaceae family of bacteria as a causative factor in heart attacks.*

Tantalizing hints have surfaced in recent years that coronary artery disease might be caused by infection, at least in part. If this is so, researchers wonder, is there a role for antibiotics in preventing heart attacks? In a supplemental issue of the *Journal of Infectious Diseases*, published in June, 2000, scientists from around the world reviewed the current evidence that favors one particular family of bacteria as a likely participant in causing cardiac ischemia and heart attacks.

The Chlamydiaceae Family

The papers in this issue of the journal focused on *Chlamydia pneumoniae*, a member of the Chlamydiaceae family of bacteria that causes respiratory infections in humans. It is the infectious organism most likely to be associated with the development of atherosclerosis. There are two other species of this bacterial family that infect humans, but neither of them appears to have a role in atherosclerosis. One is *Chlamydia trachomatis*, which is a common cause of sexually transmitted disease in both men and women and which, especially in less developed nations, can cause chronic infection of the cornea and conjunctivas of the eyes. Prolonged infection leads to blindness. In fact, this organism is responsible for as much as 25 percent of all cases of blindness worldwide. *Chlamydia psittaci*, the third species in the family, causes respiratory infections of variable severity in many species, including humans.

Changes in the twentieth century in the annual death rate from coronary artery disease have provided hints that chronic infection might be involved.

Chlamydiae are small bacteria that cannot produce their own energy or reproduce on their own. They must invade another cell and use its metabolism to provide these services. *C. pneumoniae* enters several human cell types, particularly the smooth muscle cells that line arterial walls, and macrophages. Macrophages are circulating cells of the immune system that appear to play an

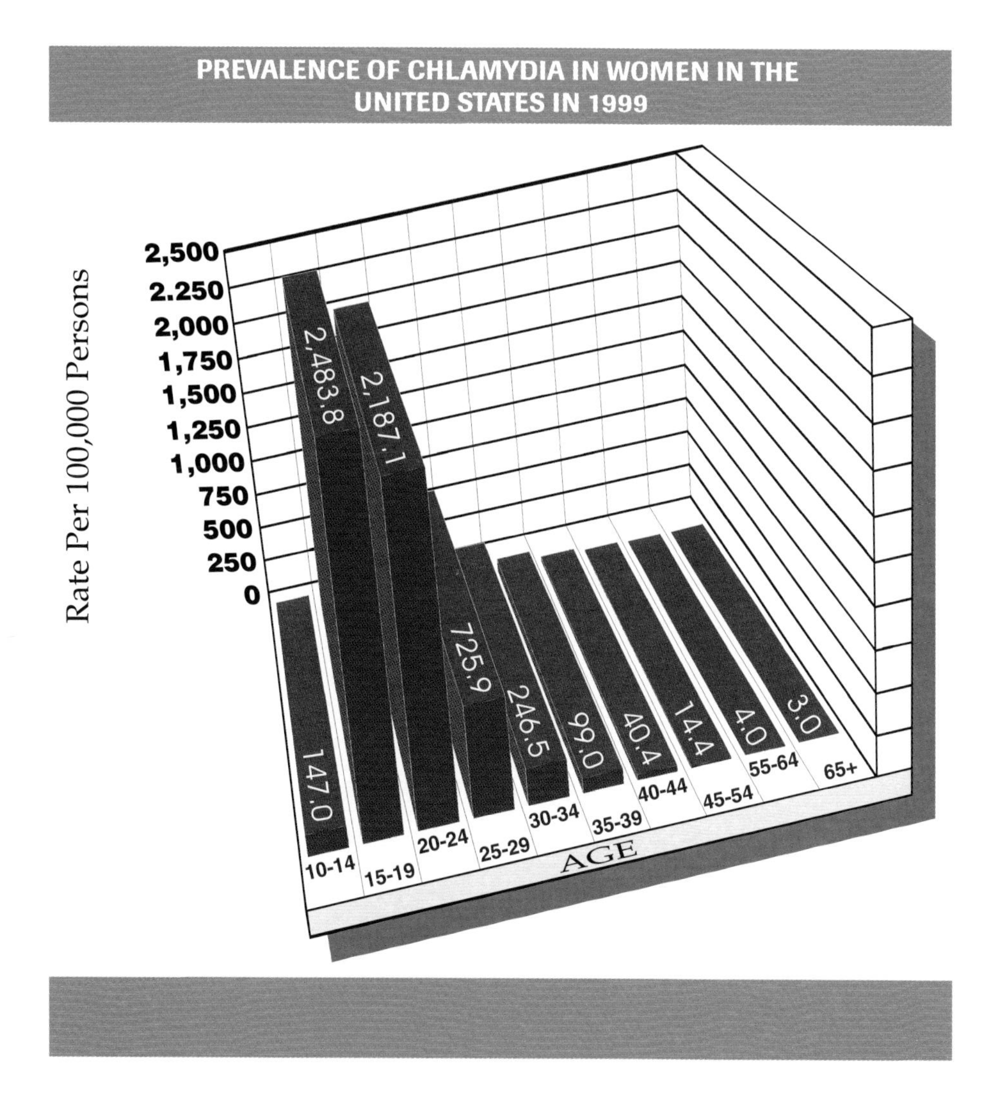

active role in the formation of plaques, the small swellings in the internal walls of blood vessels that cause atherosclerosis.

Although atherosclerotic plaques can occur in blood vessels almost anywhere in the body, their greatest potential for causing serious health problems is when they develop in the arteries that nourish the heart (coronary arteries) and the brain (carotid arteries). Plaques build slowly within the arterial wall and are composed of a mixture of smooth muscle cells, macrophages, dead cells, connective tissue, and lipids such as cholesterol. As a plaque grows, it narrows the opening of the artery by bulging out into the lumen, the hollow center of the vessel, through which blood flows.

Plaques can be in a stable or an unstable state. When stable, they are smooth elevations of the vascular wall, with a thick cap, or covering, and an outer surface made of an even layer of cells that line the inside wall of the blood vessel. A plaque can become unstable when T-cells and other inflammatory cells activate its macrophages. The surface of the plaque becomes irregular and rough, and inflammation causes the cap of the plaque to thin. If the now-damaged plaque ruptures, its loose fragments and its rough, raw edge can cause the sudden formation of clots made of blood cells, proteins, and platelets. These clots block the artery either partially or completely and deprive cells that are beyond the blockage of their critical blood supply. Ischemia, or cellular damage due to lack of oxygen, follows. If the blockage and the ischemia are severe enough, the affected cells die, causing a myocardial infarction, or heart attack, in cardiac muscle or a stroke in the brain.

Evidence for Infection in Atherosclerosis

Several lines of evidence point toward a role for chlamydia in the cause of atherosclerosis. Clinical researchers

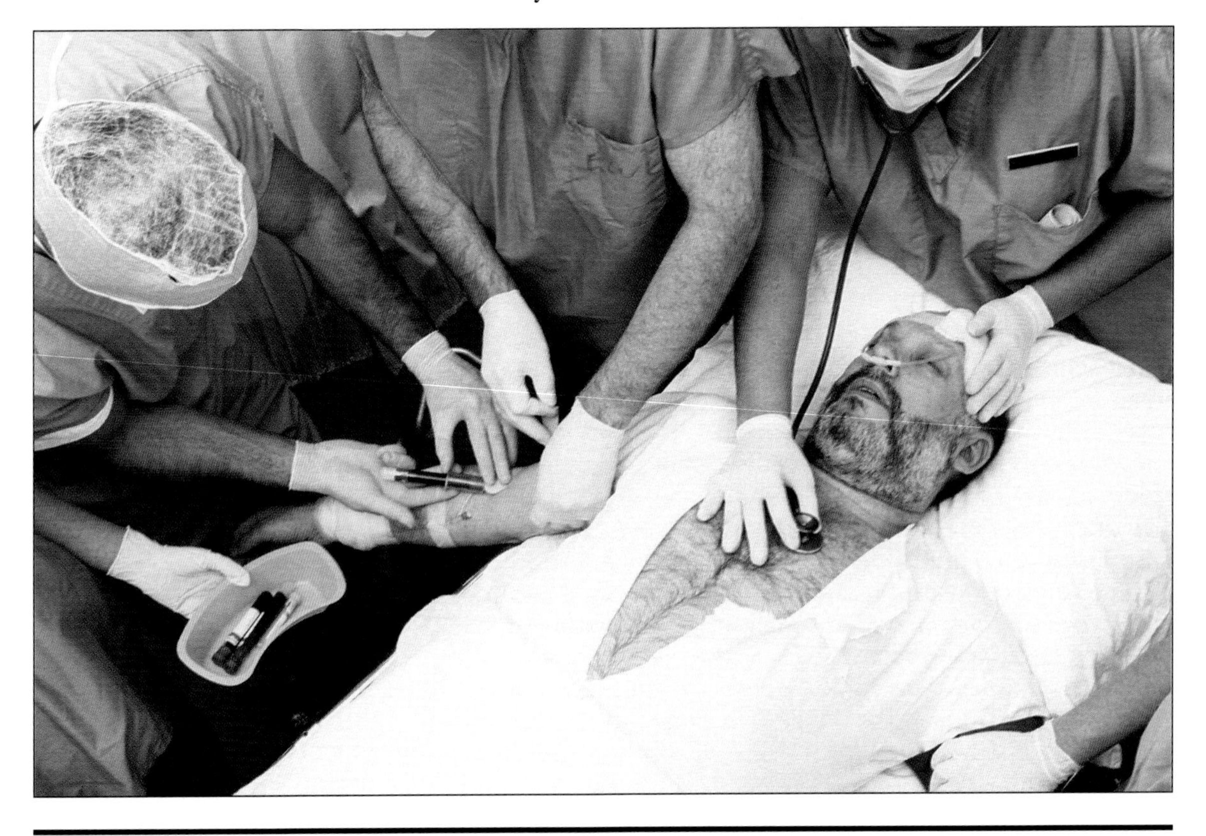

have known for decades that the complex events associated with heart attacks and ischemia include signs of inflammation. Inflammation includes a wide variety of responses by the body to infection and to many other sorts of tissue irritation. Substances in the blood that are part of the body's inflammatory response are present in increased amounts when significant cardiac muscle damage or a heart attack occurs. In fact, increased levels of inflammatory substances in the blood can indicate more than double the risk for ischemia in a person with atherosclerosis, even before any ischemic event has occurred.

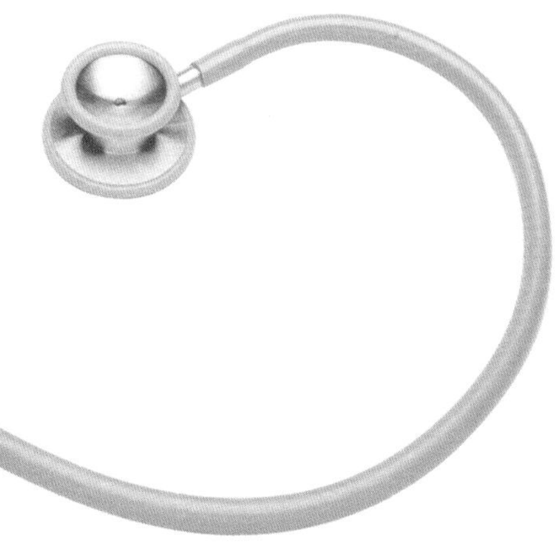

Changes in the twentieth century in the annual death rate from coronary artery disease have also provided hints that chronic infection might be involved. Before the mid-twentieth century, atherosclerosis was relatively uncommon in the United States. The yearly death rate from atherosclerosis began to rise in the early 1950's. The death rate peaked in 1963 and has been slowly dropping ever since. Changes in lifestyle, diet, and exercise levels also changed greatly over the same years, and these are bound to have exerted some effect on the improved statistics. However, to many physicians, these same changes in incidence and mortality could also be explained by a chronic, relatively mild epidemic of infection.

C. pneumoniae is not a new organism, even though it was first identified in 1965. Blood specimens collected long before that year already contained antibodies to the organism when researchers tested them after the 1965 discovery. The tiny organisms have even been identified by electron microscopy in autopsy and surgical specimens collected well before 1965. Both sets of evidence suggest that this organism was already causing human infection much earlier in the century than when it was first discovered.

On the other hand, much of the evidence that points toward chlamydia as the culprit is circumstantial. Chlamydia has been found in many atherosclerotic plaques, but not all. The antibodies that a person's immune system produces to a chlamydial infection are present in more than half of the blood specimens of these patients, but the others have no evidence of a past or current infection with chlamydia. Because chlamydia infection cannot be proven in every patient with atherosclerosis, there must be other causative factors as well. Many, such as cigarette smoking, high blood cholesterol levels, high blood pressure, and diabetes mellitus, have been recognized for a

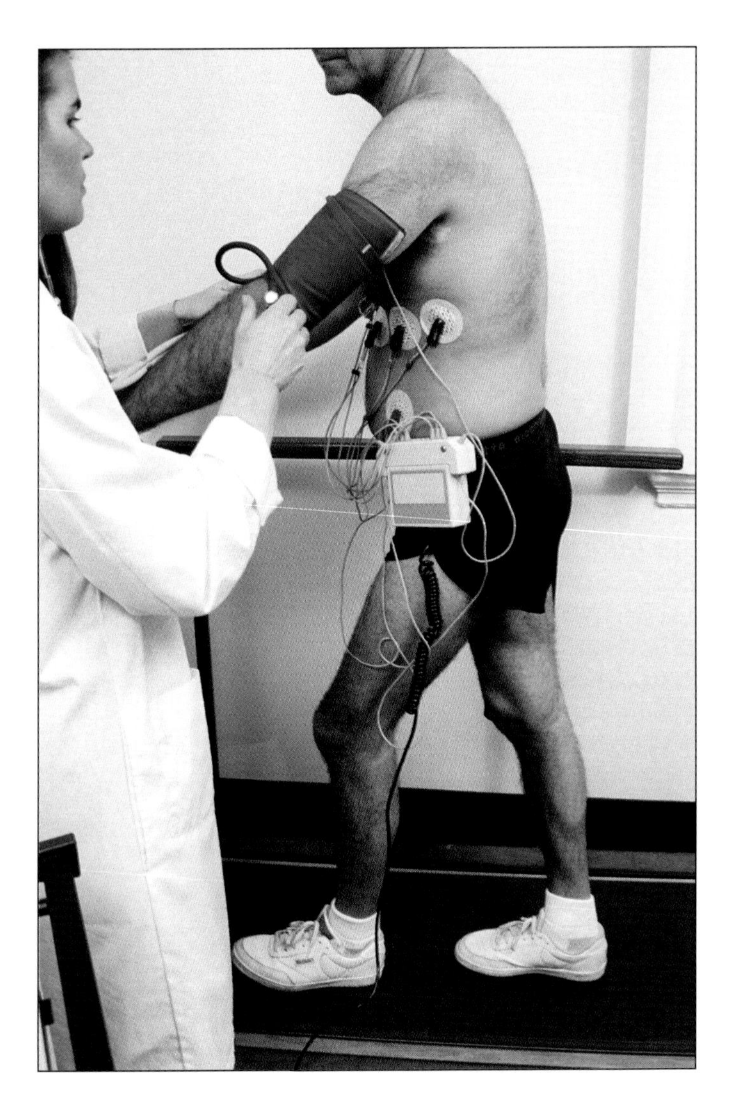

long time to increase a person's risk for developing atherosclerosis. However, when powerful statistical tests are applied to the research results, it is clear that neither solely infection with chlamydia nor solely the presence of one or more known risk factors can completely explain all cases of atherosclerosis. As is the case in so many chronic disorders, probably several combinations of infection and risk factors lead to plaque formation.

Two other microscopic organisms have also been suggested for the inflammatory role in atherosclerosis. One is a virus, cytomegalovirus, which generally causes unnoticed infection in adults. However, it can cause an illness in children that resembles mononucleosis, and it can cause severe birth defects in a fetus exposed to the virus while inside the mother's womb. Two lines of evidence favoring cytomegalovirus are that it is able to persist in the body for an extended time after an acute infection and that a majority of adults have evidence in their blood of past infection with the virus. It is less likely than chlamydia, though, to be present in atherosclerotic patients. The other organism that has raised suspicion is *Helicobacter pylori,* which is now known to be responsible for most cases of gastric and duodenal ulcers in humans. The evidence that favors *H. pylori* is also circumstantial and is not as strong as for chlamydia. More than one infectious agent may be capable of participating in plaque formation. Other bacteria, such as those involved in periodontal inflammation, may turn out to play a causative role as well.

Future Research

Research studies have shown that chlamydia causes accelerated plaque growth in the arteries of laboratory animals, particularly rabbits. Infection with chlamydia increases the speed and extent of plaque formation in rabbit arteries. Interestingly, rabbits that receive azithromycin, an antibiotic, at the time of infection do not show any acceleration in plaque growth. This observation has led researchers to begin several human tests to determine whether the long-term administration of antibiotics lessens the risk of heart attack or stroke.

It will most likely never be possible to perform definitive research in human subjects that can prove, without a doubt, that chlamydia causes atherosclerosis. For one thing, it is not ethical to cause human infection with an agent that may cause a fatal disease. Therefore, human tests similar to the rabbit experiments, in which the subject is actually infected with the agent, will not ever be acceptable. This very appropriate limitation means that it is not currently possible to prove that chlamydia is just an innocent bystander in plaque formation, that it is not simply coincidental that chronic infection and plaque instability coexist in coronary arteries. Another limitation of human research is that the human life span is so much longer than that of most other animals. As a result, it could require decades-long tests of one large group of patients to show whether infection with chlamydia accelerates the development of atherosclerosis in humans.

Therefore, the evidence for an infectious component to plaque formation and for chlamydia's direct participation will continue to be circumstantial for many years. A more definitive answer may require the development of completely new ways to measure the effects of chronic infection on the plaque formation that leads to atherosclerosis.

Substantial evidence is building that inflammation and, more specifically, infection of the coronary arteries may be an important factor in the development of atherosclerosis in humans. Research in this direction is bound to continue to grow. The possibility that simply taking a broad-spectrum antibiotic every day might eliminate atherosclerosis as a fatal human disease is just too attractive an idea to ignore.

Bioterrorism
Surveilling Infectious Secret Agents

➤*A report from the Centers for Disease Control and Prevention (CDC) suggests ways to deal with possible bioterrorist attacks on civilians.*

Since the early 1990's, military and civilian health planners have become increasingly worried that terrorists might unleash biological or chemical weapons on civilian communities. These weapons have become easier to obtain, manufacture, and release covertly, making them more attractive to terrorist organizations, particularly after breakouts of fighting and losses of stable military infrastructure in Eastern Europe and the Middle East. Along with nuclear weapons, these deadly chemical and infectious tools are classified as weapons of mass destruction. They are the most frightening weapons available to terrorist groups that wish to inflict casualties directly on civilian communities. In the last five years, fatal chemical attacks and unsuccessful attempts to release biological agents by Aum Shinrikyo, a Japanese religious cult, have proven the usefulness of these weapons to terrorists. The level of concern has become so high that the U.S. Central Intelligence Agency was reported in late 1999 to have investigated whether that summer's outbreak of West Nile encephalitis in New York might be a terrorist attack. They found no evidence to support the idea.

Several strains or families of bacteria and viruses and a limited number of toxins produced by bacteria and plants have been identified as the most likely terrorist biological weapons. Among infectious bacteria, those that cause anthrax (*Bacillus anthracis*), plague (*Yersinia pestis*), and tularemia (*Francisella tularensis*) top the list, followed by several that cause illnesses less likely to be fatal, such as bacterial dysentery, brucellosis, and cholera. The most worrisome viral agents are the smallpox virus, which now exists only in a few military laboratories; the viruses that cause Ebola and Marburg hemorrhagic fevers; and the virus that causes Lassa fever. Some extremely potent toxins, such as the bacterial toxin that causes botulism and ricin, a toxin made from castor beans that can cause lung and gastrointestinal hemorrhage, are also high on the list of possible agents.

The Response: Military or Civilian?

Initially, planning in the United States emphasized a response to biological and chemical attacks modeled on the military's plans for such defense in warfare. In 2000, however, that approach lost favor as scientists and medical researchers collaborated during a number of planning conferences to address the threat in the civilian community.

Scientists and medical researchers expressed a number of reasons for dissatisfaction with a military-style approach. First, chains of command and cooperation are quite different between the military and elected civic bodies. Second, military planning has emphasized that identifiable military organizations—not anonymous terrorists—would attack using these agents during active combat—not outside the military arena. Third, some of the ways that casualties are cared for vary greatly between combat and civilian settings, particularly in what are called mass-casualty situations, or medical events in which the number of sudden casualties exceeds the ability of the medical system to immediately care for all of them. Fourth, it had become obvious by late 1999 that the U.S. military did not have sufficient resources to establish a nationwide rapid-response system for a major terrorist attack involving possibly thousands of deaths and injuries.

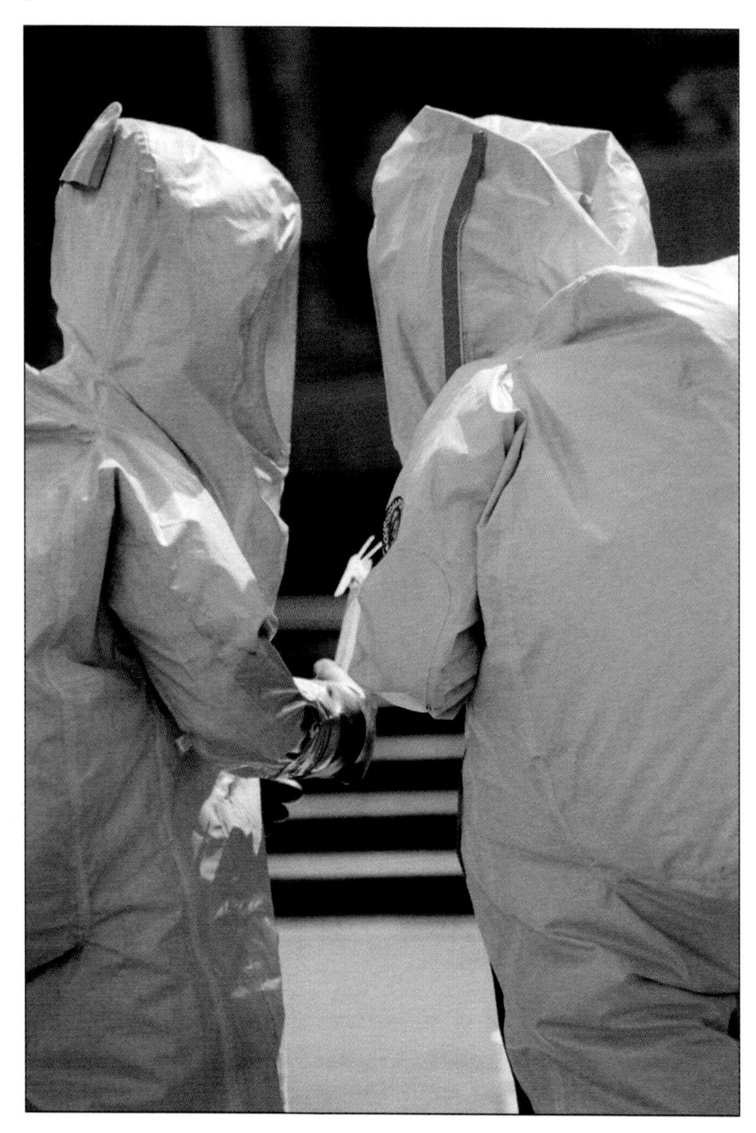

A report on bioterrorism released by the Centers for Disease Control and Prevention (CDC) in April, 2000, identified several crucial differences between a biological attack on unsuspecting civilians and an attack on combat troops. Unlike an overt chemical attack or a focused biological attack on combat troops, a biological attack on civilians would be covert and would at first be very difficult to recognize. An infectious agent's incubation time, which is the number of hours or days that the agent infects a person before signs of illness develop, plus the nonspecific nature of the first symptoms, means that early cases would most likely appear to be simply serious but unexplained infections. Before a biological attack could be recognized, it already would have spread to many other persons. Heightened surveillance by health departments would be necessary to rapidly identify a cluster of illnesses that might represent a biological attack. This level of surveillance is not present in the routine disease-reporting systems of state health departments. Also, the specific microbiological tests to identify these pathogens are routinely available in only a few medical laboratories. As a result, there would most likely be a delay of several days before a reference laboratory could identify the causative agent. During this delay, patients possibly would not receive the proper treatment and would continue to infect other persons. In addition, community physicians, who would most likely be the first to see infected patients, are not trained in the early recognition of these rare illnesses. This would further delay the realization that a biological attack had taken place. Also, physicians might not report these cases to a health department during the days of initial uncertainty regarding the correct diagnosis.

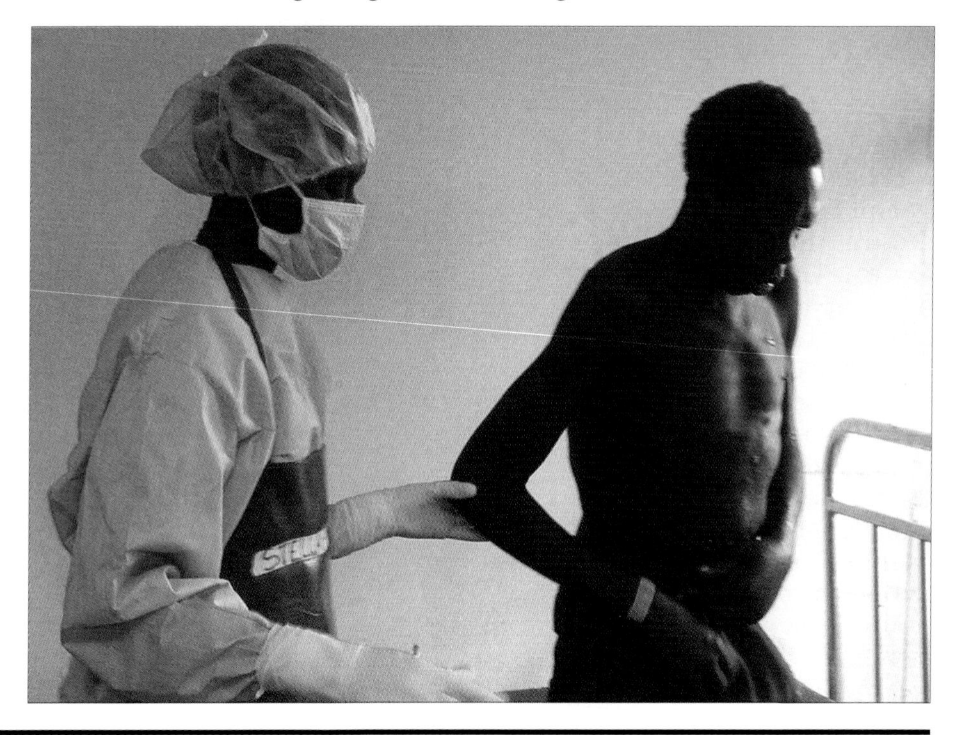

Ebola, a deadly hemorrhagic disease, is one of the viral agents that experts fear bioterrorists may use. Here, a Ugandan nurse helps a man believed to have Ebola at a quarantined hospital during an outbreak in November. (AP/ Wide World Photos)

CDC Recommendations

The CDC report made the following recommendations, which are to be in place by 2004. First, through increased education and awareness of both health department officials and primary care physicians and with advance stockpiling of therapeutic medications, both the public health agencies and care providers must be ready to respond quickly to outbreaks of disease caused by biological terrorism. Second, health department surveillance of infectious diseases must become much more sophisticated in order to provide physicians with the proper tools for reporting unusual infectious events and to allow public health officials to rapidly identify a suspicious cluster of illnesses. Third, the current national laboratory response network for identifying and diagnosing events of bioterrorism must be enlarged to include facilities in every state. Fourth, all health departments, both state and federal, must possess state-of-the-art laboratory and data management systems to allow rapid epidemiological evaluation and control of terrorist biological attacks. In addition, stockpiles of essential medications and equipment for the treatment and prevention of spread of these diseases must be rapidly available. Finally, a group of health care providers and public health workers must be established in every state, ready to respond immediately to such an attack. This preparation will require a significant improvement in communication systems and plans as well.

The CDC report and other medical articles emphasize that these improvements in the public health system will also benefit communities in routine health surveillance. In this sense, planning for bioterrorism will provide a health dividend for all citizens, even if a terrorist biological attack never occurs. Historically, the only known terrorist use of biological agents that resulted in civilian illness occurred in the United States in 1984. Members of a religious commune in Oregon poisoned food items on salad bars in approximately ten restaurants with *Salmonella typhimurium* bacteria. A total of 751 citizens suffered acute salmonella gastroenteritis as a result; all recovered. The causative organism was eventually discovered in a laboratory on the commune grounds.

Planning for bioterrorism will provide a health dividend for all citizens, even if a terrorist biological attack never occurs.

“Killer” Viruses Target Bacteria

➤ *Scientists turn to bacteriophages to combat antibiotic-resistant bacteria.*

This age of antibiotic resistance has generated new interest in a century-old scheme for killing pathogenic (disease-producing) bacteria: the use of “killer” viruses that are tailored to attack a specific strain of bacteria. Lawrence Osborne reported on the medical status of these unusual viruses in the February 6, 2000, issue of *The New York Times Magazine.*

Researchers have studied these viruses, called bacteriophages, since 1896, when a scientist noticed that the dingy water of India’s Ganges River contained an agent that could kill waterborne disease germs such as the ones that cause dysentery. Research into the antibacterial uses of bacteriophages dwindled rapidly after the commercial introduction of penicillin after World War II. In fact, the principal modern use of bacteriophages is for the identification of specific bacterial strains in specialized microbiology laboratories.

In the former Soviet Union, however, decades of research have led to the use of bacteriophages for human infections, alongside or in place of modern antibiotics. Joseph Stalin, the Soviet dictator, was particularly interested in the potential utility of bacteriophages for human disease and implemented large research and production programs. Bacteriophages have continued to be important as treatment for infections in several former Soviet republics and satellite nations, especially in the Georgian Republic and in Poland, both of which have long-standing laboratories for bacteriophage research and production. The bacteriophages are particularly useful in treating or preventing infections in burns and military wounds.

Bacteriophages are viruses that generally possess two sections, a head that contains the genetic material—DNA for these viruses—and a tail that contains the molecules used to latch onto and gain entry into the bacterial cell. A bacteriophage attacks bacteria by attaching to a specific molecule on the bacterial cell’s surface. The virus then injects its own DNA into the bacterium. The viral DNA takes over cellular function, causing the bacterium to manufac-

ture hundreds of viral copies. These new virus particles swell the bacterium until it ruptures, releasing them to seek out, infect, and destroy more bacteria.

Advantages of Bacteriophages

Several characteristics of bacteriophages make them potentially attractive for medical use. The "dose" of viruses rapidly increases once inside the human body. This is distinctly unlike a dose of antibiotic, which immediately begins to decrease after its administration as the body digests or eliminates it. Also, the fact that they target a specific strain of bacteria means that bacteriophages do not kill other bacteria that have a helpful role in the body's function. Antibiotics are indiscriminate in the bacteria that they kill. For example, the yeast rash that often develops on a baby's bottom after antibiotics are administered is a sign that some of the body's normal bacteria have been killed along with the pathogenic ones.

Also, bacteriophages can be administered in a large number of ways, including orally, rectally, intravenously, and as wound dressings. They can also be given as liquid solutions, in capsules, or as a mist to be inhaled. Bacterio-

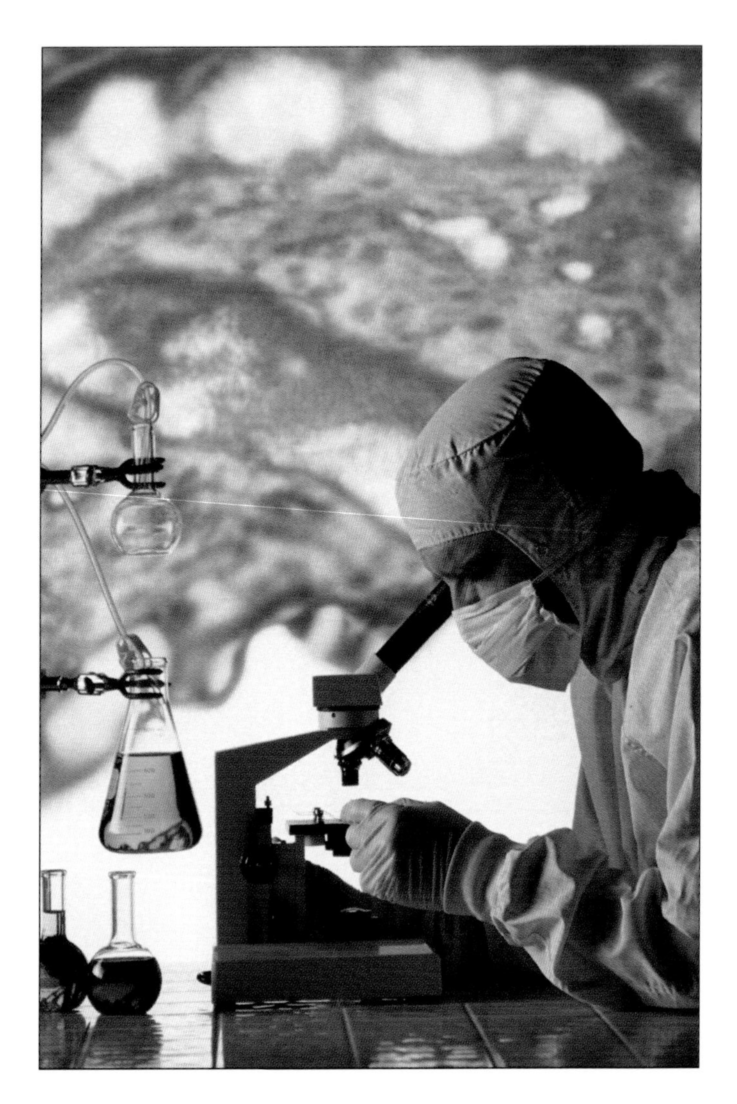

phage mixtures have been used widely as sprays to decontaminate equipment and surfaces in operating rooms and burn wards.

Bacteriophages cease to function inside the body once they have destroyed all bacteria of the specific strain that they were sent to attack. Bacterial cells differ so fundamentally from the cells of more complex organisms that humans are completely immune from infection by these viruses. Antibiotics, on the other hand, continue to affect the patient's tissues and metabolism until they are completely eliminated.

The Problems with Bacteriophages

Bacteriophages also have potential drawbacks. The most basic concern among scientists outside the former Soviet Union, and one that must be addressed before pursuing further development of these viruses, is that the bacteriophages developed to treat illness have never undergone rigorous scientific testing to confirm just how helpful, if at all, they are for human infections. Basically, no clear, unbiased, statistically significant proof exists that bacteriophages work against human infections. Research in the former Soviet Union was limited to reports of single cases or small numbers of patients whose diagnoses were not sufficiently certain by Western standards and who did not receive any long-term observation or follow-up regarding the eventual outcome of their disease. Therefore, it is not known how often they eliminated the infection, how often the same person became reinfected, or how many treated patients developed severe or late complications of bacteriophage treatment.

A particular bacteriophage is active against only one strain of a single species of bacteria; however, virtually all human bacterial pathogens consist of multiple strains. For instance, *Streptococcus pneumoniae*, a common bacterial cause of human respiratory infections, exists as more than twenty infectious strains. One antibiotic, such as penicillin, is sufficient treatment for all of them, as long as they are not penicillin-resistant. However, a bacteriophage treatment would have to contain a mixture of viruses, one to specifically attack each bacterial strain. Eastern European scientists have attempted to overcome this difficulty by developing mixtures of bacteriophages that attack different strains of the same bacteria.

Another problem is that a person may be infected with more than one species of bacteria, especially in wounds and surgical infections, or the patient may suffer from an infection that could be caused by any one of several species, as is often true in cases of bacterial dysentery. For these situations, medical researchers have again depended on mixtures of bacteriophages.

Researchers have been aware from an early point that intravenous mixtures of bacteriophages had to be carefully filtered and cleaned before administration to avoid causing the patient to suddenly worsen because of impurities and bacterial proteins in the solution. However, it is not clear whether bacteriophage mixtures contain other, potentially harmful viruses or whether the bacteriophages are still biologically active once they have been prepared in any form for human consumption.

Besides medical concerns, bacteriophage treatment generates the practical political question of whether physicians and patients outside the former Soviet Union will ever accept a treatment that was developed during Stalin's dictatorship. Many people are likely to harbor a persistent general concern, completely unwarranted in many medical instances, that Soviet science must be viewed with a degree of suspicion. This could be balanced in time by the results of new research, as rigorous tests of bacteriophage therapy take place in Western medical laboratories.

Despite the unknowns and possible drawbacks of bacteriophage therapy, there is no question that the strong Western medical preference for overuse of antibiotics has led to a marked increase in antibiotic-resistant bacteria. A new approach to infection in humans is definitely needed, and bacteriophages just may be the right one.

Medical Abortion

RU-486 Comes to the United States

➤ *A drug that enables women to have nonsurgical abortions receives approval from the Food and Drug Administration.*

On September 28, 2000, the Food and Drug Administration (FDA) approved the use of mifepristone (formerly known as RU-486) for nonsurgical abortion in the first forty-nine days of pregnancy. Approval came some seventeen years after research first began on this use for the drug and twelve years after it was approved for use in medical abortions in Europe.

During the four years before it received FDA approval, mifepristone was being employed an estimated four thousand or more times per year in the United States for medical abortions. These cases involved both subjects of approved research studies and women who obtained the drug outside the United States. In addition, some women were traveling to countries in which the drug was already approved for this use.

Medical Versus Surgical Approach

Before approval of mifepristone, women in the United States had to undergo a surgical procedure in a medical facility to abort a pregnancy. This involved the removal of the uterine contents, including the embryo, with a suction device or by scraping the interior surface of the uterus. Surgical abortions are nearly 100 percent effective. However, both procedures are painful and require anesthesia, and the procedures carry some risk to the woman, although it is small. In the United States, women must also deal with the added stress of the social controversy over abortion. A woman who enters a clinic for an abortion may have to pass by other members of her community who are protesting the procedure with their presence. Physicians in the clinics must often deal with the potential for personal violence against them, their staff, and their patients.

The decision to terminate a pregnancy has generally been viewed as a difficult—often emotionally and spiritually troubling—very personal decision. A pregnancy termination that could be conducted by the woman herself,

French professor Emile-Etienne Beaulieu holds the RU-486 abortion drug that he invented. (AP/Wide World Photos)

in privacy, has been viewed by many women and their physicians to be a major improvement over surgical procedures. In addition, medical abortion is felt to restore the woman's sense of power over her own body

Mifepristone, a progesterone antagonist, blocks the action of progesterone on different tissues in the uterus. Progesterone is the most critical hormone involved in the early establishment and maintenance of a pregnancy. By blocking progesterone from its role as a cellular messenger, mifepristone causes the pregnancy to fail, much in the same way that a spontaneous abortion does. It also acts on the muscles in the uterine wall, making them more sensitive to the effects of drugs that stimulate uterine contractions. When the woman follows the mifepristone with a drug that causes contractions, any remaining tissue related to the pregnancy is expelled from the uterus.

Based on clinical experience in research trials and in widespread use in other countries, medical abortion successfully terminates the pregnancy 95 percent of the time. The risks of medical abortion include blood loss that may require a transfusion in 1 percent of cases and the possibility that a surgical abortion will be necessary if the mifepristone fails to terminate the pregnancy.

If medical abortion fails, it is recommended that the woman undergo a surgical abortion. By blocking so many critical progesterone effects, the use of mifepristone carries a high risk that the embryo, should it survive to delivery, will have serious birth defects. However, risks of surgery are confined only to this much smaller group of patients, unlike in surgical abortion.

The Procedure

Medical abortion using mifepristone is approved only for the first forty-nine days of pregnancy, a very brief span of time following conception. In fact, a woman who chooses this method must begin taking the medication within a month after she misses her first period because of the pregnancy. Although medical abortion can still be performed through the second trimester of pregnancy, roughly six months, the possibility of a failed abortion and the risk of complications increase substantially. It is not routinely administered after forty-nine days of pregnancy in any country that has approved its use.

A medical abortion using mifepristone requires three visits to a physician's office over two weeks. At the first visit, the physician and patient must confirm that the woman is pregnant and establish how far along the pregnancy is. If the woman is within the forty-nine-day period, the physician will give her three tablets of mifepristone, each containing two hundred milligrams of the drug. The medication will not be available through pharmacies. The patient takes all three pills at the same time. She then must return to the physician's office in forty-eight hours for another examination. If an abortion has not already occurred, based on either physical examination or ultrasound, she then receives two tablets of a second medication, misoprostol. This drug is a prostaglandin, a category of medicine that stimulates uterine cramping. Again, she takes both tablets at the same time. This second medication completes the expulsion of any remaining products of the pregnancy. Uterine cramps may be strong enough to require medication for the pain. In addition, the woman often experiences abdominal cramps (96 percent of the time), nausea (61 percent), headache (31 percent), vomiting (26 percent), or diarrhea (20 percent). Less frequently she may experience dizziness, fatigue, or back pain.

The decision to terminate a pregnancy has generally been viewed as a difficult—often emotionally and spiritually troubling—very personal decision.

The third visit to the care provider takes place two weeks after the day the woman took the mifepristone. Even if the woman feels that her pregnancy has been terminated, this return visit is very important. At this time, the physician will confirm, again by physical examination or ultrasound, that a complete termination of the pregnancy has taken place. At this visit, the physician must also ascertain that there is no evidence that this is an ectopic pregnancy, which is an implantation of the embryo in the abdominal cavity rather than the uterus.

Uterine bleeding can persist after the abortion for up to sixteen days routinely and occasionally continues as long as a month. For about 5 percent of women who undergo this procedure, the blood loss may be enough to require close monitoring with blood counts or, very rarely, a blood transfusion.

There have been studies of women's opinions about their abortion method, surgical or medical, and whether they would consider using the other method if another abortion became necessary. Fewer than 9 percent of women who experienced a medical abortion would accept a surgical procedure the second time, but more than 40 percent of those who underwent a surgical abortion would prefer a medical abortion if another were ever necessary.

FDA Restrictions

The FDA placed some unusually rigorous requirements on the use of mifepristone for abortion. This was

(AP/Wide World Photos)

perhaps partially in response to social and political pressure, but restrictions were also felt to be important because of the potential for misuse of the drug by persons outside the health care professions. As a result, mifepristone will not be available in pharmacies. The drug will be shipped directly from the manufacturer to the physician's office, but only after the physician agrees to adhere strictly to the protocol for medical abortion.

The physician is required to sign a detailed agreement before being permitted to dispense mifepristone. The physician must attest that he or she has the proper training and skills to determine accurately the duration of a pregnancy and to diagnose an ectopic pregnancy. In addition, the physician must verify that he or she is able to provide any necessary surgical intervention in case of an incomplete abortion or severe bleeding. If the physician is unable to do this, then a surgeon who can perform the necessary procedures must be identified beforehand. The physician also acknowledges having read and understood the prescribing information regarding this drug.

The physician will dispense the mifepristone to the woman at the first visit, but only after they have thoroughly discussed the procedure and its possible complications. The patient must receive a copy of the detailed patient information that comes with the drug and have an opportunity to read it and ask any further questions. In addition, the woman must sign a patient agreement that covers fourteen points of information, including her agreement to return for the next two visits, before receiving the medication. The physician will give her the misoprostol at her second visit if abortion is not yet complete.

All clinical visits and the prescribing of mifepristone may take place only in a clinic, medical office, or hospital. The health care provider must be a physician who meets the requirements of the physician agreement, or another health professional under that physician's supervision.

It is unclear whether the availability of such a private means of pregnancy termination will lead to an increase in the total number of abortions performed each year in the United States. There is no clear evidence that this has happened in other countries. What is clear is that, with the approval of mifepristone for medical abortion, women and their physicians will have more options when it appears necessary to abort a pregnancy.

Hospital Errors
Who's Minding the Medications?

➤ *The Institute of Medicine reports that more than 40,000 people per year are dying because of errors committed by health care providers.*

Each year, more Americans die from medical errors made in hospitals than from vehicle crashes. Between 44,000 and 98,000 deaths annually may be the result of errors made by health care providers, compared with 43,458 traffic deaths in 1998, according to a report published in early 2000 by the Institute of Medicine. In the report *To Err Is Human: Building a Safer Health System*, the institute found that at least half of the deaths caused by medical error are preventable. These frightening statistics and suggested remedies garnered much media attention during the year.

The institute's report and numerous articles in newspapers and magazines gave individual faces to the people who suffer the sometimes tragic results of medical errors. These include a newspaper health reporter who died from a cancer therapy overdose, an eight-year-old boy who died from a medication error during minor surgery, an elderly woman who received a fatal dose of medicine in her intravenous line, and a man who suffered the surgical amputation of the wrong leg.

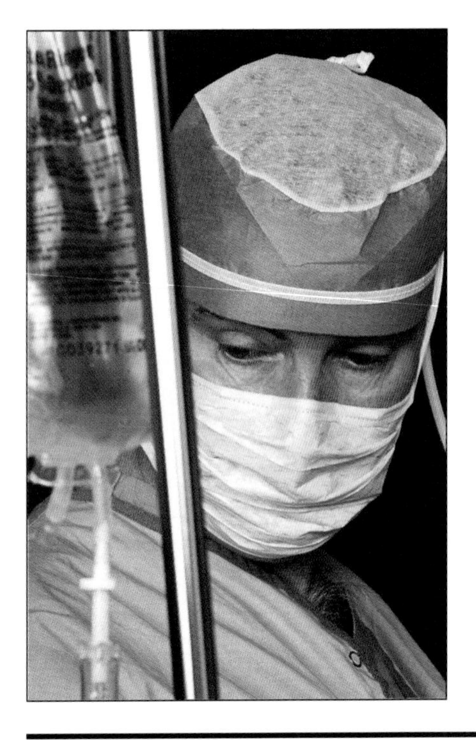

A number of forces in health care have acted to hide the magnitude of the problem. Current error-reporting systems at most hospitals place a greater stress on identifying the person who made the mistake than on discovering ways to prevent a further error. As a result, hospital employees may be reluctant to notify administrators of an error, fearing disciplinary action. Also, the heavy emphasis by health care payers, including managed care organizations, on minimizing the cost and length of hospitalization has created significant pressure on physicians and hospital staffs to pursue a fast-paced approach to diagnosis and treatment. Finally, many doctors have been reluctant to report and to study physician errors, fearing possible lawsuits.

The Institute of Medicine found that because of all these factors, a substantial degree of secrecy has developed around medical errors, making it difficult to study the nature of such errors and to seek solutions. One of the first goals in preventing medical errors, then, must be to change from a punitive approach to a collaborative, learning approach, in which the problem itself, not who caused it, is the center of investigation.

The Cost of Errors

Medical errors vary widely in their severity and complexity. The simplest errors, such as a brief delay in giving a routine dose of medicine to a patient who is temporarily away from the ward, are almost never dangerous. However, when a person who has entrusted his or her life to the medical care of a hospital is injured or dies from a medical mistake, it represents a critical failure of the entire health care system.

A substantial degree of secrecy has developed around medical errors, making it difficult to study the nature of such errors and to seek solutions.

To put this loss of life in perspective statistically, one must consider that 31.8 million people were hospitalized in 1998. Of these, 820,000 died. Medical error is believed to have caused between 5 and 12 percent of all the deaths that occurred during hospitalization. At least half of these error-related deaths could have been prevented, according to the Institute of Medicine. Unfortunately, error-related deaths are probably underestimated because the institute's numbers do not include deaths in other medical settings, such as clinics, outpatient surgical centers, and patients' homes.

Safety-Enhancing Systems

People in every line of work make errors. Most errors, including medical ones, are caught in time or do not cause significant harm. However, when an error could be life-threatening, managers must develop procedures that make it easy for personnel to make a correct decision and avoid a mistake and very hard for them to make the wrong decision. Aviation, especially the airline industry, offers one example of the successful development of safety systems. Airline pilots are not allowed to fly more than a certain number of hours each week to prevent them from becoming fatigued. Also, critical systems aboard modern airliners are designed to help the crew quickly recognize a problem and choose the right response. Important systems may even be present in duplicate, in case one of them fails.

The systems and procedures that enhance safety in the health care field are far less developed than those for aviation and many other industries. Physician orders and

A report on deaths and complications caused by hospital errors set in motion many attempts to fix the problem. Here, Maryland representative Connie Morella announces the introduction of a bill designed to prevent such errors. (AP/Wide World Photos)

patient medication records continue to be handwritten in many hospitals. Poor handwriting can make a perfectly correct order dangerous. Also, patients are sicker, on average, now than they used to be when they were hospitalized.

Modern treatment often necessitates the application of cutting-edge technology and the use of complex equipment. At the same time, the number of registered nurses caring for hospitalized patients has fallen drastically, as hospitals meet financial stresses by placing more patients under the care of fewer nurses.

In its report, the Institute of Medicine recommends a four-level approach to minimizing the occurrence of medical errors. At the first and highest level, there should be a national focus on creating the leadership, information, and means of research to understand and investigate medical error. Second, mandatory reporting of all medical errors should be required. Third, the organizations that most directly influence patient care—health care professional associations, state agencies that oversee care, and the corporate purchasers of health insurance—must insist that all health care providers maintain high safety standards. Fourth, all providers of care must create safety systems that make it extremely difficult for errors to occur in the treatment of any of their patients.

Some physician groups have spoken against the idea of mandatory error reporting, which they fear would be simply another bureaucratic intrusion into their patient-care decisions. However, based on past evidence, it is unlikely that physicians or any other health care providers would rigorously follow a voluntary reporting system.

The prevention of errors in patient care can be a simple task in many instances, once the nature of the problem has been defined. In the case of the man whose left leg was amputated instead of the right, careful review at the hospital revealed that both of the patient's legs were severely diseased. The surgeon did not see the patient that day until his legs had been prepared for surgery and had been covered on the operating table. In addition, a nurse had noted that the operative schedule wrongly indicated the left leg was to be amputated but did not correct all copies of the schedule. The nursing staff decided that no patient would be transferred to the surgical suite until a nurse had physically marked the side that was not to be operated on. There have been no further wrong-side surgery incidents in the five years since.

Herbal Medicines
Safe? Effective?

➤ *Results of study of St. John's wort indicate that additional research on complementary medicine may prove fruitful.*

A research article published in the important *British Medical Journal* in September, 2000, showed that *Hypericum perforatum* (St. John's wort extract) is equivalent to imipramine, a commonly used antidepressant drug, in its ability to improve mild or moderate depression. In addition, St. John's wort was better tolerated, with fewer side effects, than the antidepressant drug.

Although many researchers already thought this to be the case, others questioned whether the previous studies of St. John's wort had been conducted with enough attention to detail. As a result, many clinicians remained skeptical about this herbal treatment for depression. However, this study was so carefully designed and carried out that its results appear to be firmly proven.

In the study, 157 patients received hypericum (the active extract of St. John's wort) and 167 received imipramine. Neither the patients nor their doctors could tell which drug anyone was taking. After six weeks of treatment, the patients and physicians each measured the improvement in the patients' depression. The researchers then determined which patients had received which drug, by breaking the coded identification of each patient's medication. Those who had received the herbal remedy were judged to have had as much improvement in their depression as the patients who took imipramine. Both the patients and their physicians agreed on this. When the patients measured their drug's degree of tolerability (side effects and adverse effects), hypericum was felt to be significantly more tolerable than imipramine.

An Herbal History

The extract of St. John's wort has a long history of use for depression and anxiety in Europe. In ancient Greece, Hippocrates mentioned the plant as a good treatment for demonic possession. Hypericum is available throughout Europe and without prescription in the United Kingdom. The strength and quality of the herbal product are strictly

A bee lands on an echinacea flower on an herb farm near Lyons, Colorado. Echinacea, believed to have immune-strengthening qualities, is one of many herbs used for medicinal purposes. (AP/Wide World Photos)

controlled in Europe. In the United States, however, such regulatory control does not exist for herbal remedies because they are generally considered to be dietary supplements, which are not tightly regulated. As a result, consumers can never be sure either of how strong a dose of hypericum they are taking or if it is the same from batch to batch of the same product. Concerns about the lack of uniformity in potency and purity of herbal preparations has led many to call for the Food and Drug Administration to supervise more closely the dietary supplement market. These are important concerns because hypericum is strong enough to interact adversely with at least one category of prescription psychiatric medications, the monoamine-oxidase inhibitors.

Although the term "complementary medicine" is quite new, herbal and other therapies have been employed for thousands of years. Only in the last two centuries has a distinction developed between those therapies and the scientific treatments used by allopathic physicians. In the United States, herbal and folk remedies are still in wide use. Most primary care physicians can relate stories of patients who gargled with kerosene or used cattle-dehorning medicine on bad cuts. Many social or ethnic groups still recognize the value of their traditional medicines. Within some Native American tribes, ailing people often combine consultations from medicine men and traditional healers with the care provided by modern physicians. Although folk medicines may seem far removed from modern drugs, many of the pharmaceuticals people use today originated as plant materials. Digitalis, widely used for heart failure, comes from the foxglove plant. Willow bark is the original source of salicylate, which is now taken millions of times each day as aspirin.

In the middle of the nineteenth century, Americans began to prefer scientific medicine to other approaches to therapy, such as homeopathy, which came to be regarded as unscientific and potentially dangerous therapies. The American Medical Association, founded in 1847, was organized, in part, to help medical doctors compete for patients against practitioners of other popular theories of health. Despite the demand of modern physicians for rigorous scientific testing of new treatments and drugs, prob-

ably more than half of the currently accepted medical treatments have never received adequate scientific scrutiny.

Effective Medicine

Patients have maintained an interest in alternatives to modern medicine even if physicians have tended to dismiss such approaches as naturopathy and homeopathy as being too unscientific. Within the last five to ten years, researchers have begun to study complementary and alternative treatments rigorously. At the National Institutes of Health, a small Office of Alternative Medicine was established in 1992 to direct research on complementary and alternative treatments of all sorts. In 1998, it was replaced by the National Center for Complementary and Alternative Medicine, which received a budget of $68.7 million for fiscal year 2000, more than triple what its predecessor had received in its first year.

Mainstream researchers are showing increasing interest in evaluating alternative approaches to patient care. The November 11, 1998, issue of the *Journal of the American Medical Association* was given over entirely to reports of the best available research on complementary and alternative medicine. The articles included a study that showed that moxibustion, a method of applying heat to the body (between the toes, in this study), was successful in turning an abnormally positioned fetus in the womb to the correct position for delivery. As then-editor of the journal George Lundberg pointed out, many academic physicians are beginning to think that there should be no such terms as alternative or scientific medicine. Rather, he suggested, whatever treatments are scientifically shown to work should be grouped together as "effective medicine," whether they are traditional, alternative, or modern.

Much of modern scientific medicine had its beginnings in herbal treatments and other therapies that were once what would now be regarded as "alternative." High-quality research, such as the newest report in the *British Medical Journal,* is leading people to reconsider that the ancient roots of human healing may still have much to offer the modern age.

Ethnic and Social Differences

Predictors of Hypertension, Heart Disease, Diabetes, Cancer?

➤ *The U.S. government calls for ending gender, racial, and ethnic inequalities in health care within the next ten years.*

The federal government's latest plan for improving the health care of Americans, *Healthy People 2010*, calls for the elimination of all gender, racial, and ethnic inequalities in health by the year 2010. The call to eliminate these inequities completely is new. Previous health plans have encouraged work to narrow the differences, mainly by improving health care access for minorities and by decreasing the effects of gender, racial, and ethnic bias within the health care system.

Medical research into racial, ethnic, and gender health differences has not been ignored. A great deal of effort has gone into finding possible genetic and physiological explanations for such facts as the increased likelihood for Native Americans to develop diabetes or for African American men to suffer strokes.

An article by Arline T. Geronimus in the June, 2000, issue of the *American Journal of Public Health* provides insights into the possible social causes for ethnic health differences. The author also offers a well-reasoned method for choosing solutions to health care inequity, which is different from the approach being taken by many mainstream medical researchers and policy analysts.

Less health care plus increased risk means that some Americans, particularly black men, experience a significantly decreased life expectancy because of their ethnic identity.

Possible Causes of Inequity

For years, studies have shown that, in the United States, women and members of racial/ethnic minorities receive proportionately lower levels of health care than do men and whites. For various groups, this can include such important procedures as the provision of fewer heart catheterizations and surgeries for coronary artery disease, fewer routine screening tests for cancer, less adequate pain management, and fewer organ transplants.

Worldwide, poverty is the social factor that has the strongest association with poor health. The more dire the poverty of a population, the worse its health. Studies in

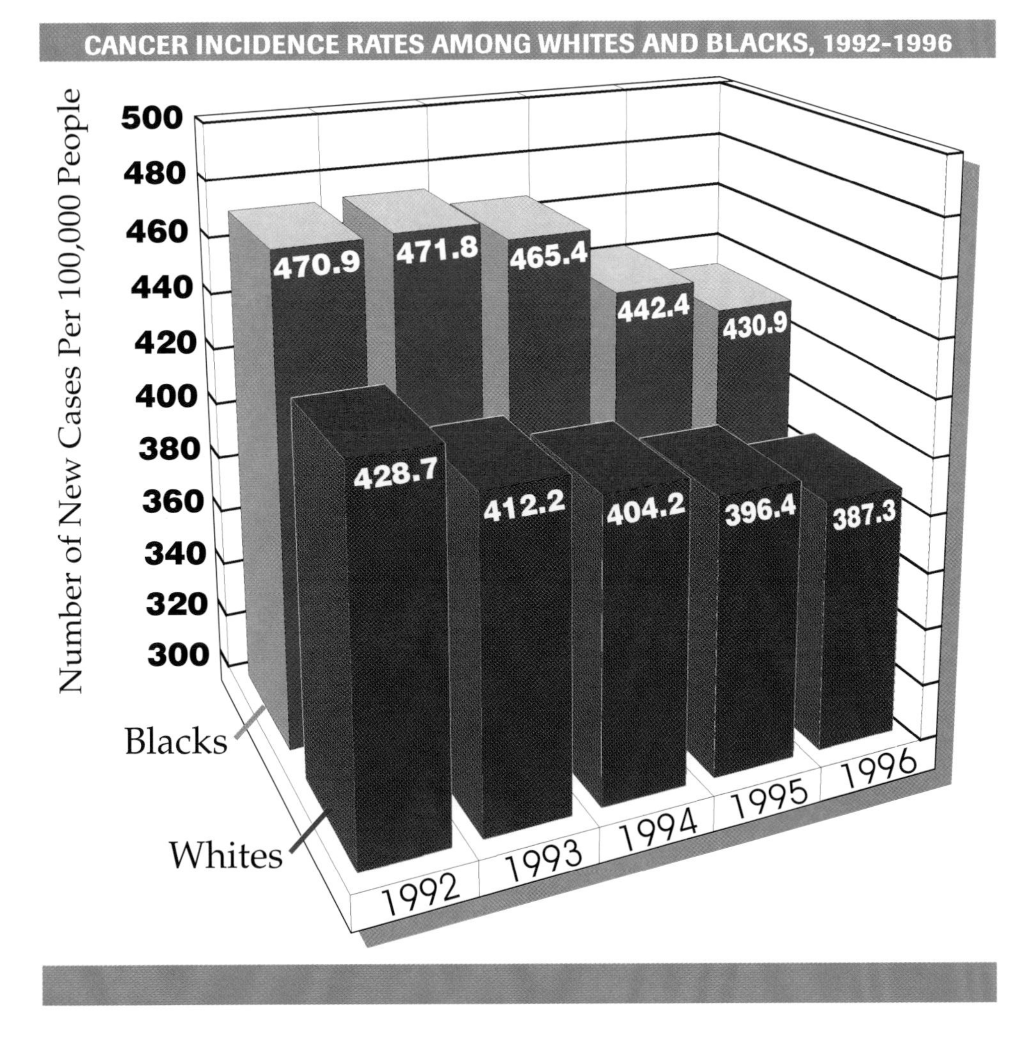

the last few years have shown that an even more critical factor is the income difference between the wealthy and the poor members of a society. The more unequal the distribution of wealth, the greater is its effect on the health of the poor.

Being identified as a member of a racial/ethnic minority also increases a person's risk of such chronic diseases as hypertension, cancer of the prostate and other organs, and diabetes. The result of this "double whammy" of less health care plus increased risk is that some Americans, particularly black men, experience a significantly decreased life expectancy because of their ethnic identity. In the mid-1990's, white men could expect to live, on average, eight years longer than black men, to age seventy-three compared with sixty-five.

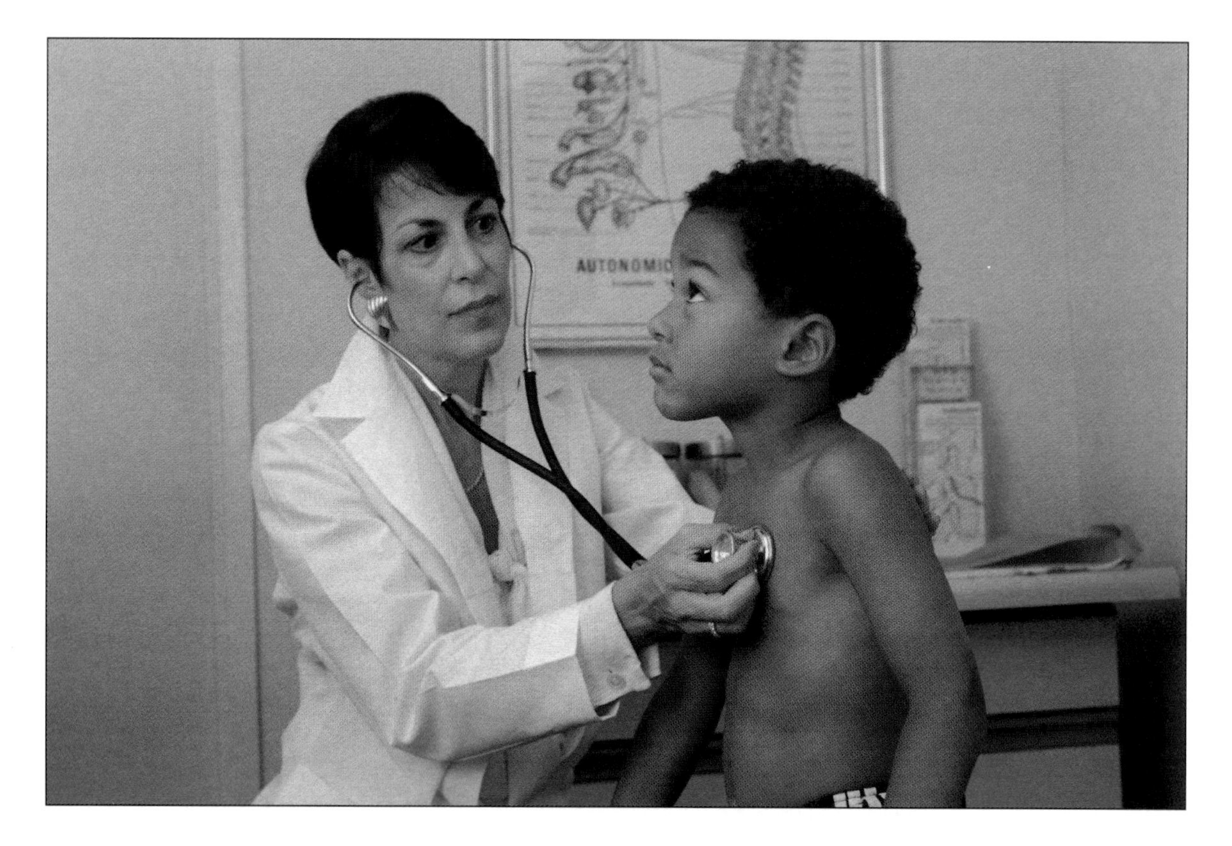

It seems unlikely that these health differences are caused solely by significant genetic variations among races. Of the tens of thousands of genes that make up human deoxyribonucleic acid (DNA), more than 99.99 percent are common to all humans, regardless of our family or racial background. Many researchers suspect that the lifestyles, social biases, stress, and other social factors identified with membership in a minority group may exert as great an influence on a person's health as does any specific genetic composition or outward appearance. In fact, such social disparities may greatly overshadow genetics as a cause. For example, some unhealthy ethnic dietary preferences may be based on a generations-old history of a lack of high-quality produce and protein. As a result of poor food quality, people preparing it over the decades may have routinely added large amounts of salt or fat to improve taste or caloric content. Such food is no longer healthful when diet quality eventually improves but taste preferences remain the same. High levels of fat and salt, for example, may be risk factors for such chronic diseases as hypertension and coronary artery disease, regardless of one's ethnic background.

Geronimus argues that it is impossible to correct the existing health differences simply by making changes in the health care system. It is critically necessary, she says, to recognize the constantly changing nature of social relationships between and within racial/ethnic groups. An important intragroup relationship, according to Geronimus, is the changing role of traditional organizations within ethnic communities. When an identifiable group of people is disproportionately impoverished or restricted in work and living choices, that group creates organizations "to mitigate, resist, or undo" the external effects on their lives from the bias that exists throughout the whole society. These organizations can be social, spiritual, educational, or business structures within the minority society. If such social organizations are damaged by other changes such as massive relocation or pervasive unemployment, the loss of their structure can become an important health factor as well. The author's point, that both the need for special social structures within minority societies and the effects of damage to them may be important health factors, is particularly new.

Impoverished groups, whatever their ethnic background, tend to establish self-protective networks of relatives and friends to share what resources and personal support are available. Yet the limitations that society places on these groups, such as poor housing, inadequate transportation, and lack of jobs with substantial benefits, can weaken or destroy the strength of such informal networks. Some Asian and Hispanic communities, for example, left behind their social and cultural roots within just the last generation as they moved to seek better economic opportunities. Family members who used to be counted on for child care may have stayed behind or must, themselves, now work outside the home. The urban areas into which they move tend to be those least desired by more successful ethnic groups that have already found a way to move out and up.

When changes in social structure are mentioned, many people think first of violence levels in minority communities. However, the number of deaths due to homicide among men in Harlem, for example, has actually dropped steadily since the early 1990's. At the same time, the death rates from heart disease and cancer have doubled for the same group of men. It is clear that very complex relationships between society and health must be involved.

The stress of increased disruption of the group's social fabric for coping, when added to the social limitations that

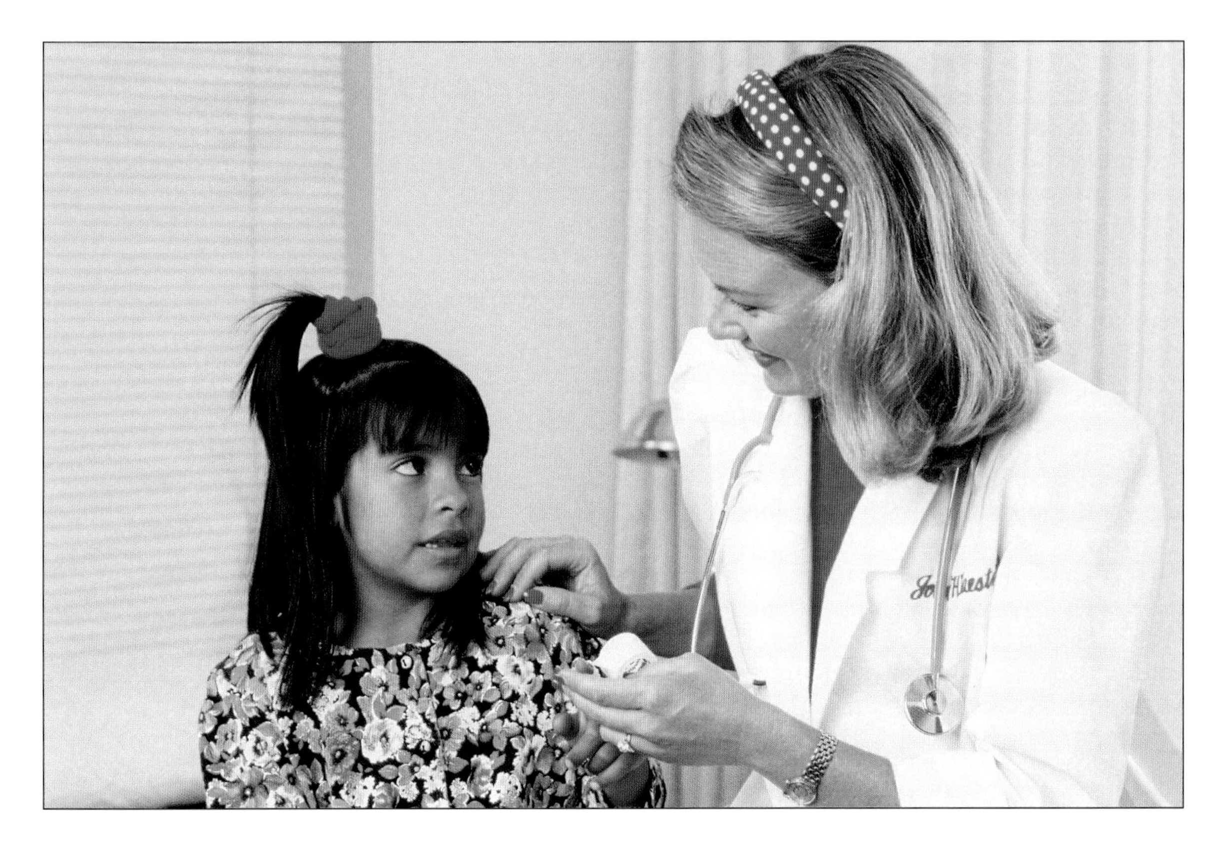

are placed on an ethnic group from outside, may explain most of the increased health problems of ethnic minorities. The point that Geronimus and others are making is that all humans are so essentially identical in their chemistry, physiology, and abilities, that any group placed in a deprived-minority status will suffer worse health than any preferred majority.

Evaluating Proposed Solutions

Geronimus suggests that the approaches to improving minority health care can be divided into two groups. The first group contains the approaches that are currently being emphasized. They attempt to modify the quality of care being provided, the form of the health service system, and the biases of its personnel to make them more responsive to the health needs of the minority. Such programs include the establishment of neighborhood clinics and the improvement of health services to treat specific disorders such as hypertension and diabetes. These changes in the health care system are "ameliorative," according to Geronimus. Such approaches ease the impact of these diseases on an ethnic group but will not eliminate them.

The only way to eliminate health differences between ethnic and social groups in the United States is to emphasize the second group of approaches to inequality, Geronimus suggests. She calls these "fundamental" approaches. They are changes in the whole social system, not just the health care system, that will correct the underlying social problems that face each minority. For example, the availability of affordable, convenient, and reliable child care provides all parents with a wider range of employment and at the same time reduces the stress and worry that less satisfactory care arrangements produce. The provision of affordable and adequate housing improves personal safety and affords an environment that can maintain health, not wear it down. Improved health occurs as a result of these fundamental programs.

The elimination of all health inequity is an important societal goal, Geronimus says. It is unlikely, though, that studying the extremely minor genetic and physiological differences between people will allow us to completely correct racial and ethnic differences in medical care and in personal health. It will require solutions to the fundamental effects that poverty, social bias, and racism have on the lives and, therefore, the health of members of racial and ethnic minorities.

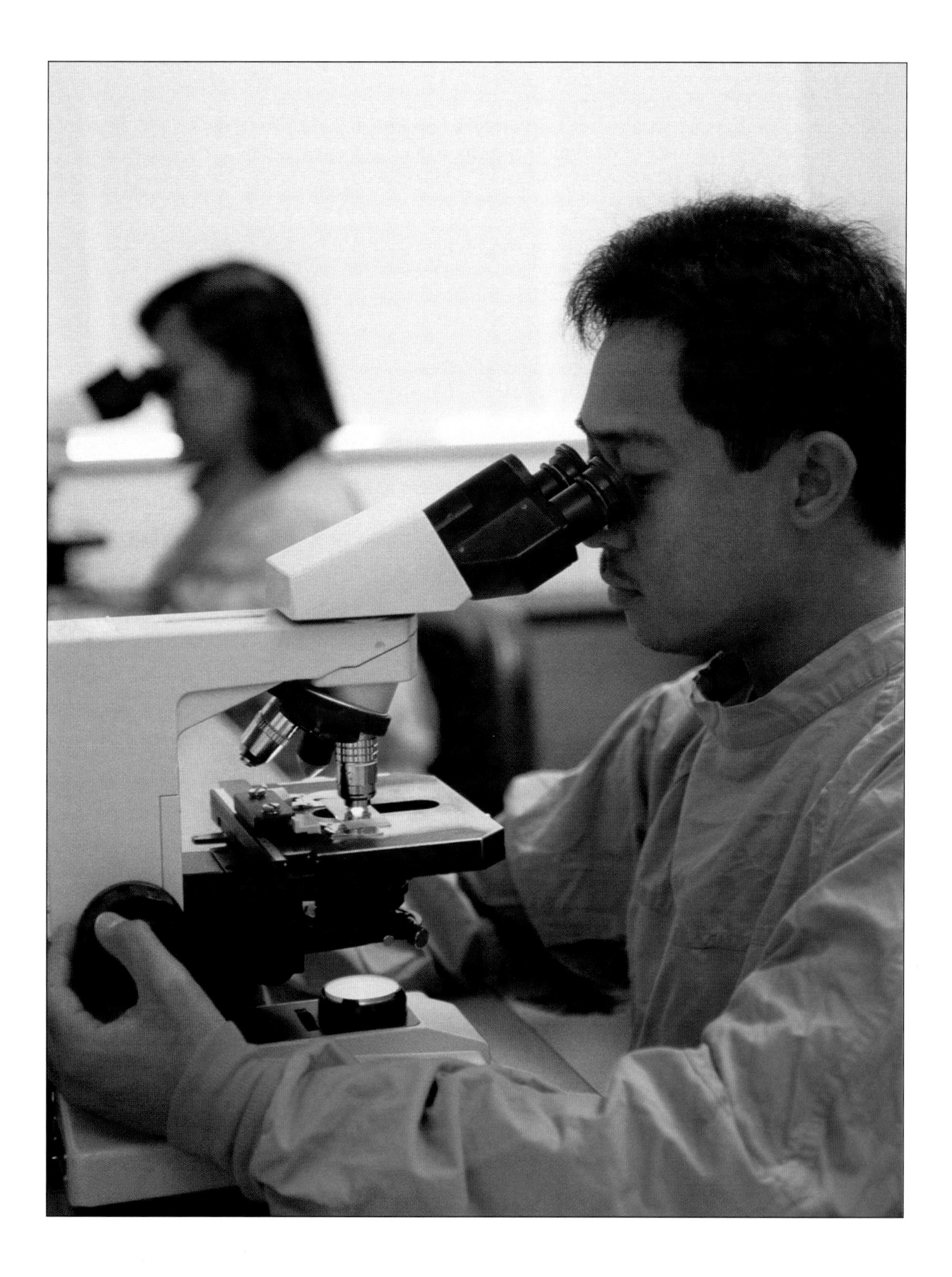

Resources for Students and Teachers

Books

Burnham, Terry, and Jay Phelan. *Mean Genes: From Sex to Money to Food—Taming Our Primal Instincts*. New York: Perseus, 2000. The authors argue that the genetic legacy of humans' hunter-gatherer ancestors helps create problems for modern people, including obesity, violence, addiction, and infidelity.

Ehrlich, Paul R. *Human Natures: Genes, Culture, and the Human Prospect*. Washington, D.C.: Island Press, 2000. Evolutionary biologist Ehrlich takes a controversial approach to the relationship between genes and behavior, claiming that genetics alone cannot account for the behavioral variation.

Ewald, Paul W. *Plague Time: How Stealth Infections Cause Cancers, Heart Disease, and Other Deadly Ailments*. New York: Free Press, 2000. Biologist Ewald presents his controversial theory that germs, which can lurk in the body for decades, play a part in many serious chronic illnesses. He urges conventional medicine to reexamine the role of infectious agents in diseases such as cancer.

Garrett, Laurie. *Betrayal of Trust: The Collapse of Global Public Health*. New York: Hyperion, 2000. The Pulitzer Prize-winning author of this examination of public health conditions around the world concludes that although globalization may create prosperity, it can have adverse effects on the health of the population. She calls for the global monitoring of disease outbreaks and tainted food and water.

Gordon, James S., and Sharon Curtin. *Comprehensive Cancer Care: Integrating Alternative, Complementary, and Conventional Therapies*. New York: Perseus, 2000. At the third annual Comprehensive Cancer Care conference, oncologists met with practitioners and researchers in alternative medicine. They looked at ways to integrate promising complementary therapies with conventional medical treatments.

Klaidman, Stephen. *Saving the Heart: The Battle to Conquer Coronary Disease*. New York: Oxford University Press, 2000.

Journalist and research fellow Klaidman examines the history of the treatment of heart disease, tracking its progress and exploring the ethical issues involved.

Montagnier, Luc. *Virus: The Co-Discoverer of HIV Tracks Its Rampage and Charts the Future.* New York: W. W. Norton, 2000. Montagnier describes his work leading to the discovery of the retrovirus as well as developments in HIV/AIDS research since 1981.

Regush, Nicholas. *The Virus Within: A Coming Epidemic.* New York: Dutton, 2000. Journalist Regush discusses herpesvirus-6 (HHV6), which may play a role in AIDS, multiple sclerosis, and chronic fatigue syndrome. As the author acknowledges in his work, the link between this virus and AIDS is highly controversial.

CD-ROMs and Videos

Critical Condition with Hedrick Smith. Video. Hedrick Smith Productions in asssociation with South Carolina Educational Television, 2000. Examines the state of health care in the United States, including managed health care plans and long-term care.

Eating Well for Optimum Health. DVD. Wellspring Media, 2000. In this lecture, based on his popular book of the same name, Doctor Andrew Weil sets forth his guide to food, diet, and nutrition. He applies his commonsense approach to the subjects of diet fads, the carbohydrate-protein debate, vitamins, and fast food.

Family Medical Reference Library. CD-ROM. 3 vols. Dorling Kindersley, 2000. The three disks in this reference work contain the American Medical Association family medical guide, the ultimate human body, and the ultimate 3-D skeleton, respectively. The medical guide provides information on various symptoms, illnesses, and treatments.

Healthcare Crisis: Who's at Risk? Video. Issues TV, 2000. This documentary looks at the health care system in the United States, focusing on the uninsured and those experiencing problems with the system.

Nutrition Interative: Your Personal Nutrition Tutorial. CD-ROM. Wadsworth, 2000. This interactive disc allows the user to learn about the human body and nutrition.

When Cancer Touches Your Life: Meeting the Needs of Patients and Families. Video. Cleveland Clinic Foundation, 2000. This videotape produced by the Cleveland Clinic presents an overview of cancer treatment and helps patients and their families learn how to cope with cancer, from diagnosis to survivorship.

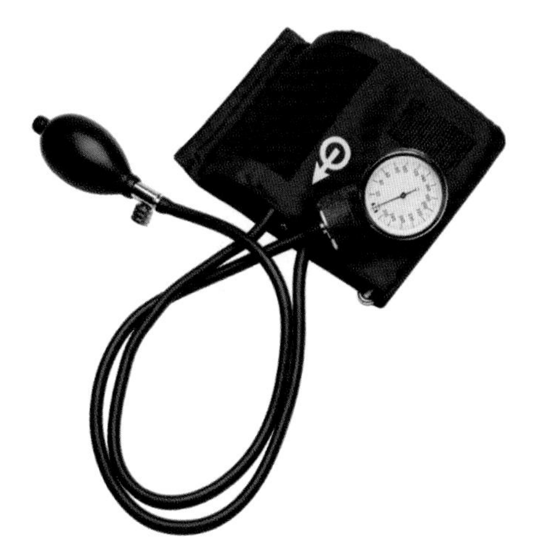

Web Sites

American Diabetes Association
http://www.diabetes.org
The Web site of the American Diabetes Association offers information and other services to people with diabetes, their families, health care professionals, and the public.

American Cancer Society
http://www.cancer.org
The American Cancer Society Web site provides information and services for those with cancer as well as health professionals and researchers.

American Heart Association
http://www.americanheart.org
The Web site for the American Heart Association provides information on heart disease and strokes for consumers and health professionals. In keeping with an emphasis on prevention, the association offers tips on ways to reduce the risk of a heart attack or stroke, including exercise regimes, recipes, and lifestyle changes.

American Medical Association
http://www.ama-assn.org/
The American Medical Association site provides news and information for its members, physicians, health care professionals, and patients. Patient information includes a doctor locator, basic health information, and association news.

Centers for Disease Control and Prevention
http://www.cdc.gov
The agency's site features current health news, information on a variety of health topics (including anthrax, cancer, and zoster), and statistics and data on communicable diseases, cancer, birth defects, and tuberculosis.

How to Search for Medical Information
http://204.17.98.73/midlib/www.htm
Prepared for the Rhode Island Reference Round Table on January 29, 1998, but updated in April, 2000, this site is a superb Internet source for librarians, teachers, and students.

Mayo Clinic Health Oasis
http://www.mayohealth.org
Offers consumer information on a variety of health conditions, from asthma to cancer to women's concerns. A library provides overviews on other health topics.

Merck Manual On-line
http://www.merck.com/
The *Merck Manual of Medical Information* and *Merck Manual of Diagnosis and Therapy* are both available in on-line versions. These well-known resources provide a wealth of medical information.

National Institutes of Health
http://www.nih.gov
Provides a wealth of data for consumers, researchers, and health professionals. Health information includes publications and fact sheets, an A-Z topics index, and MEDLINE, a resource provided by NIH's National Library of Medicine. Also features news and events and a section on scientific research.

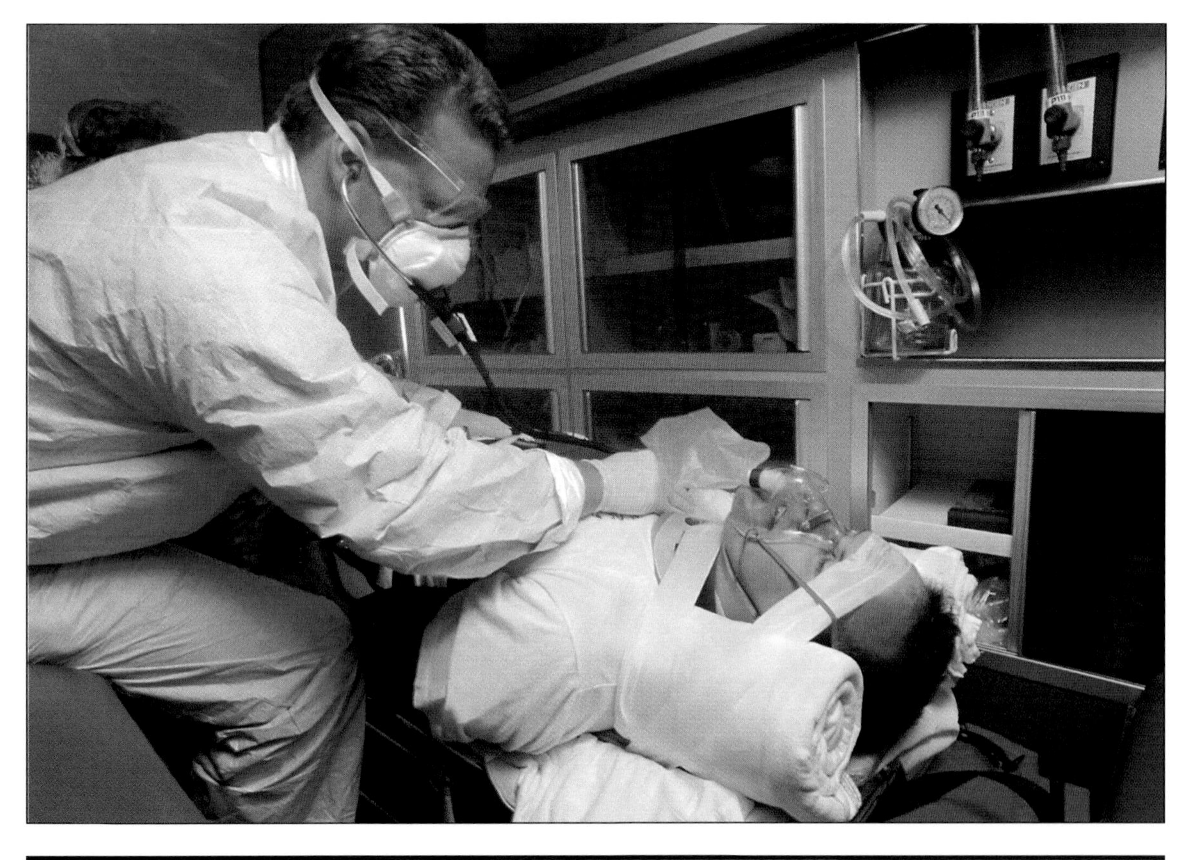

PubMed
http://www.ncbi.nlm.nih.gov/PubMed/
National Library of Medicine
Lists millions of entries for medical articles published since the 1960's in thousands of journals. Information is organized by "MESH" headings, which are standard medical terms. Beside many PubMed article entries will be an icon that will provide a secondary list of rticles that are cross-referenced to the article.

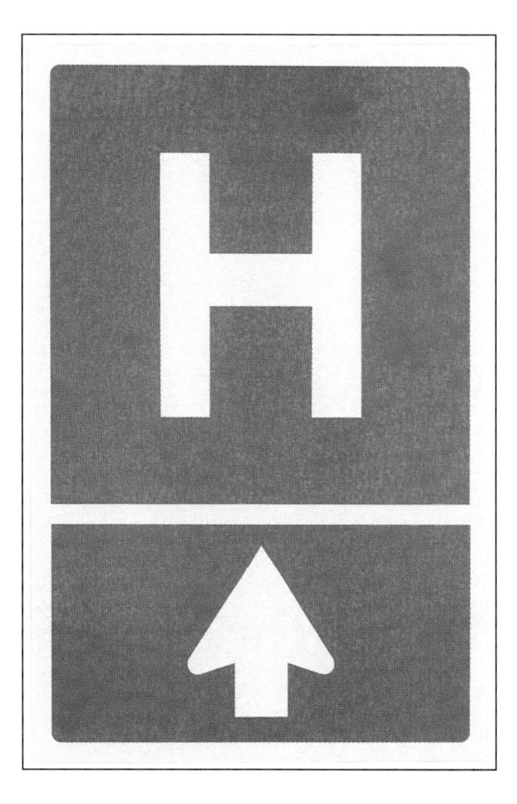

WedMD
http://www.webmd.com
Provides information for consumers, health professionals, and health care administrators. Also features lesson guides and other aids for those teaching health-related topics.

World Health Organization
http://www.who
Provides information on global health conditions, including communicable diseases. Also contains information on noncommunicable diseases, health policies, technology, lifestyles, and reproduction.

7 · Physics

The Year in Review

➤ *Alvin K. Benson*
Department of Geology
Brigham Young University

During 2000, a number of significant physics concepts and experiments were brought to various levels of fruition. Some of these had been pursued since the 1980's or even earlier. Ten of the top physics stories that developed in 2000 include discovery of a new state of matter, an update of the value for the lifetime of a neutron, production of the most proton-rich nucleus, generation of nuclear energy on a tabletop, creation of atom lasers, discovery of superconducting balls, slowdown of light to a crawl, conversion of sound energy into light, the creation of sense out of chaos, and more evidence for a fifth fundamental force in nature. Other important accomplishments reported in physics during 2000 but with a lesser degree of certainty include detection of the elementary tau neutrino particle, breakthroughs in early cancer detection by employing light-scattering spectroscopy, possible clues that even electrons may be made of other particles, and reports of speeds greater than the speed of light in vacuum.

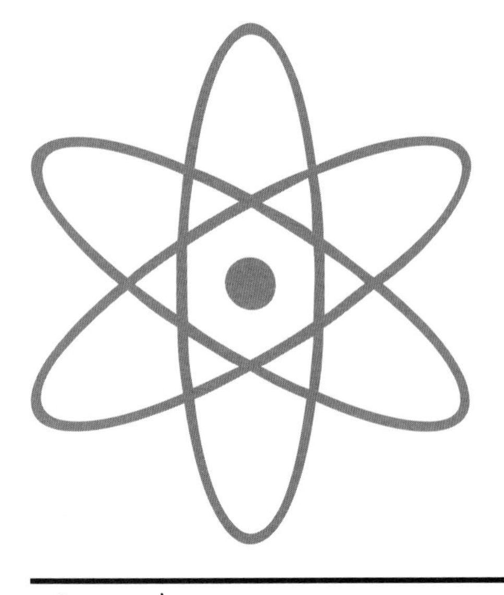

Elementary Particle Physics

In February, 2000, elementary particle physicists at the European Organization for Nuclear Research (CERN) facility in Switzerland announced corroborative evidence for the production of a novel state of nuclear matter. The conclusion was reached based on the results of seven experiments conducted at CERN between 1994 and 2000. By bombarding a target of lead or gold atoms with a high-energy beam of lead ions, high enough temperatures and nuclear densities were established to melt protons and neutrons into their constituent particles, thus simulating the conditions that apparently prevailed during the first few microseconds after the birth of the universe. The newly created nuclear state may be a manifestation of the much-sought swarm of free quarks and gluons, known as the quark-gluon plasma. Gluons normally bind the elementary quarks together to form protons and neutrons. Discovery of the quark-gluon plasma would extend under-

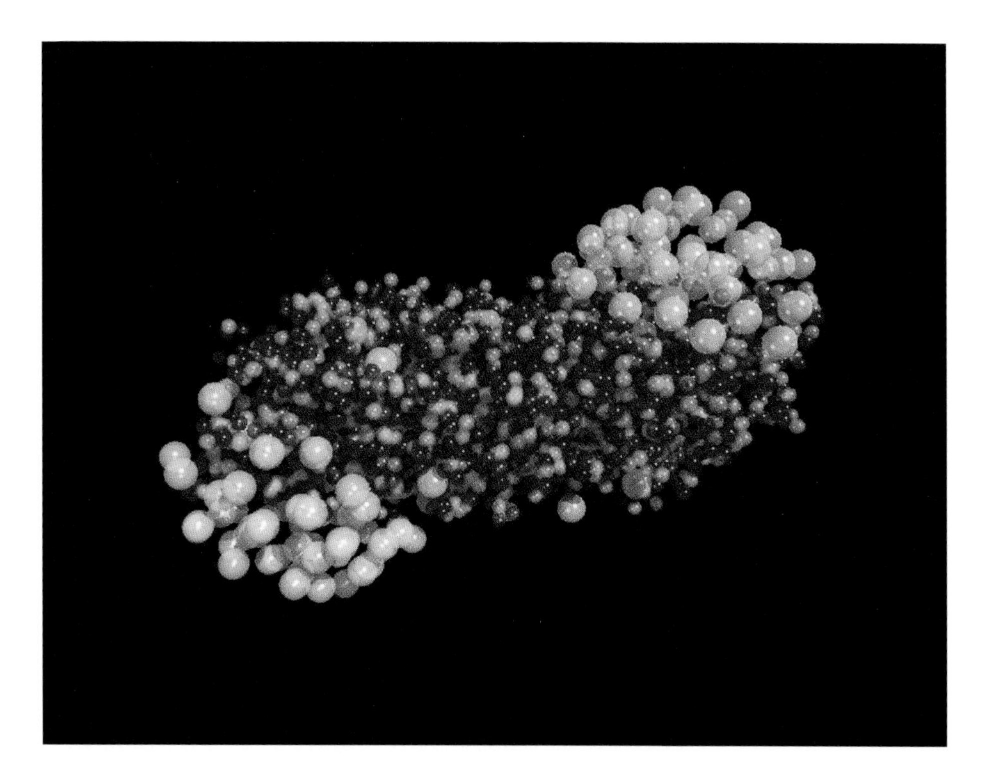

CERN scientists found evidence that quarks (green, red, and blue in the representation) free floated for a split second after the creation of the universe. The protons and neutrons (silver balls) of the two nuclei that collided in the big bang are shown moving away from each other. (AP/Wide World Photos)

standing of how the universe was created back to almost the moment of creation.

Precise measurements of the lifetime of a neutron are yielding clues to the way that subatomic particles, including the quark-gluon plasma, coalesced into the elements that formed the universe after the big bang. When neutrons exist outside an atom, as they did at the birth of the universe, they undergo radioactive decay. In January, 2000, a combined team of researchers from the National Institute of Standards and Technology in Maryland, the Institute Laue-Langevin in France, and the Petersburg Nuclear Physics Institute in Russia reported a more accurate determination of the neutron lifetime than ever before. The results should help refine models describing the early formation of the universe as well as clear up some discrepancies about the weak nuclear force. With an accurate value of the neutron lifetime, scientists can determine the initial ratio of hydrogen, helium, and other light elements in the universe. In addition, an accurate value of the neutron lifetime will allow physicists to determine the fundamental parameters describing the properties and asymmetry of the weak force.

In late July, 2000, the first detection of the elementary tau neutrino particle was announced. The evidence is slim

but impressive. The tau neutrino is the last theorized member of the family of elementary particles thought to constitute the universe. Smaller than electrons, neutrinos remain hidden inside neutrons until released by radioactive decay. Neutrinos come in three varieties, the electron neutrino, the muon neutrino, and the tau neutrino. To observe the tau variety, physicists at Fermi National Accelerator Laboratory in Batavia, Illinois, fired a highly energetic beam of protons at a tungsten target, whereupon some of the incoming energy produced other particles. A few events indicate that some of these particles decayed into tau neutrinos. Additional data necessary to confirm the discovery is being sought.

Nuclear Physics

In February, 2000, nuclear physicists at the Grande Accelerateur National d'Ions Lourdes (GANIL) in France announced the research details for the successful production of nickel 48, the most proton-rich nucleus ever observed. Although almost every nucleus in nature has more neutrons than protons, nickel 48 has eight more protons than neutrons. Because nickel 48 is doubly magic, meaning that it possesses a complete nuclear shell of both neutrons and protons, it can exist in a stable state for a short period of time. The mode of radioactive decay of this barely stable nucleus is evidently a form of radioactivity never before observed. This decay mode may answer the question of whether protons and neutrons act together in pairs within the nucleus as well as produce important insights into the fundamental nature and understanding of nuclei.

Also on the nuclear front in 2000, Todd Ditmire and associates at the Lawrence Livermore National Laboratory succeeded in using the most powerful lasers currently available to produce energy for the first time from nuclear fusion on a tabletop. Although this new approach may not be able to produce commercial power, it could provide a cheap source of neutrons for fusion reactors of the future. In addition, a group of researchers under the direction of Thomas E. Cowan at the Lawrence Livermore National Laboratory as well as another group at the Rutherford Appleton Laboratory in England also succeeded in using lasers to produce nuclear fission reactions on a tabletop. These tabletop nuclear experiments, which constitute a completely new approach to nuclear physics, are aimed at gaining insights into the fundamental interplay between light and solids and creating a new safe, clean source of energy with an abundant fuel supply that can replace the burning of fossil fuels.

Quantum Mechanics and Condensed Matter Physics

In 2000, a team of researchers led by Wolfgang Ketterle at the Massachusetts Institute of Technology (MIT) announced their success in producing a pulsed atom laser by using a Bose-Einstein condensate as the active medium. By shining a pair of optical laser pulses onto a Bose-Einstein condensate made from sodium, a weak matter wave was generated. This wave was then amplified in a way remarkably similar to how optical lasers augment an initial light wave. After the success at MIT, Japanese researchers at the University of Tokyo as well as German scientists at the Max Planck Institute for Quantum Optics in Garching and at the University of Munich announced similar results using a Bose-Einstein condensate made of rubidium. Because matter waves have much shorter wavelengths and move much more slowly than light waves, atom lasers

Because nickel 48 is doubly magic, meaning that it possesses a complete nuclear shell of both neutrons and protons, it can exist in a stable state for a short period of time.

should be orders of magnitude more sensitive than current devices based on optical lasers. Consequently, atom lasers may improve the performance of atom interferometers and also be used for high-resolution atom deposition on surfaces for the fabrication of novel materials and for the production of tiny nanostructures necessary for the production of future ultrafast computer circuits.

Another low-temperature experiment in 2000 led to the discovery of superconducting balls, a completely new phenomenon in physics. Rongjia Tao and his colleagues at Southern Illinois University discovered that a strong electric field can force the particles of a high-temperature superconductor into a bound ball. The ball forms quickly and is very sturdy, surviving repeated collisions with the electrodes that generate the electric field. The spherical shape of the ball indicates that the force holding it together must be a new type of surface tension related to superconductivity. Besides increasing the understanding of what causes superconductivity, the newly discovered surface energy of the particle ensemble may be used to lay down thin superconducting films on solid surfaces. These films would have applications in the fields of communication and energy transport, particularly the production of super speed computers.

Optical Physics

Sonoluminescence, the phenomenon of converting sound energy into light energy, created a stir in the physics community in 2000. Research groups led by Seth J. Putterman of the University of California at Los Angeles and Rainer Pecha of the University of Stuttgart in Germany aimed ultrasonic waves at air bubbles suspended in a tank of water. The sound waves caused the bubbles to stretch and compress, producing furious oscillations. During the compression phase, light emerged from the bubbles. The high temperatures generated during the process make sonoluminescence a possible candidate for triggering thermonuclear fusion in bubbles that contain heavy hydrogen atoms. If the sonoluminescence process could be used in this way, it might be another tabletop source for clean energy to drive turbines and generate electricity. In addition, the production of sonoluminescence by medical ultrasound devices may prove therapeutic to body tissues.

The high temperatures generated during the process make sonoluminescence a possible candidate for triggering thermonuclear fusion in bubbles that contain heavy hydrogen atoms.

Another optical phenomenon observed in 2000 was the slowing down of light to a mere crawl. Lene Hau of Harvard University and colleagues at the Rowland Institute for Science in Cambridge, Massachusetts, succeeded in slowing down light to a record-breaking speed of only one mile per hour (1.6 kilometers per hour). As in the development of atom lasers, the key to success was shining light onto a Bose-Einstein condensate. Hau and her associates cooled sodium atoms down to 50 billionths of a degree above absolute zero, which is one of the coldest temperatures ever achieved anywhere in a laboratory. Because the Bose-Einstein condensate state changed the index of refraction of the ultracold medium by approximately ten million times, exceptionally slow light speeds were recorded as light emerged from the condensate. The nonlinear optical effect observed in the experiment is directly applicable in producing a number of optical-electronic components, including slow light optical switches, memory chips, and delay lines. Computer systems could be made super fast by using optically switched logic gates instead of electronic logic gates.

The results of another interesting experiment with light were reported during the latter part of 2000. At the Nippon Electronics Corporation Research Institute in New Jersey, a pulse of light was passed through a chamber filled with cesium vapor and observed to travel greater than the speed of light in vacuum. It is true that "something" appeared to move faster than the speed of light, but the "something" carried no information. This result is predicted by Albert Einstein's special theory of relativity.

As the light pulse passed through the cesium vapor, it was transformed in a very particular way. The cesium gas preferentially absorbed some frequencies contained in the pulse of light and reemitted them at later times. Consequently, different components of the pulse were transmitted at different speeds so that the pulse that came out of the chamber was quite different from the pulse that went into the chamber. This process is referred to as anomalous dispersion. If the leading part of the outgoing pulse is transmitted faster than the rest of the pulse because of the frequencies that make it up, the pulse appears to exit the chamber before the peak of the entering pulse reaches the chamber. Therefore, the leading edge of the outgoing pulse travels at a speed greater than that of light in vacuum. However, to actually transmit information, enough of the outgoing pulse must be measured, which brings the total transmission speed back down below the speed of light in vacuum. Even though the upper limit of the speed of light in vacuum was not broken, the experiment was important because it represented the first time that anomalous dispersion had been demonstrated in a transparent medium.

It is true that "something" appeared to move faster than the speed of light, but the "something" carried no information.

Theoretical Physics

During 1999 and 2000, a group of researchers led by Seth Lloyd at the Massachusetts Institute of Technology analyzed how to minimize disorder in natural systems. They combined ideas from the fields of thermodynamics, information theory, and quantum mechanics to yield some quantitative answers about how to tame chaotic, disordered systems. The work of Lloyd and his group can be applied to a variety of physical systems, particularly the development of quantum logic computing systems and the reconstruction and enhancement of X-ray digital mammograms. Their ideas are also finding applications for enhancing radio and X-ray astronomy data, spectroscopic data, electron microscopic data, and geophysical data to obtain optimum images with minimum disorder.

Another important story in 2000 was reported by two physicists at the University of Pennsylvania, Paul G. Langacker and Jens Erler. They found possible evidence for a fifth fundamental force in nature. Their conclusions were based on the analysis of apparent anomalies found in data previously obtained from particle accelerator experiments conducted in Switzerland, California, and Colorado. In their investigation, Langacker and Erler concluded that if a fifth force is assumed to exist, the experimental data are best matched and understood. Their case was made stronger in September, 2000, when physicists at CERN released experimental evidence indicating the possible detection of the particle that would mediate the fifth force, the Z′ particle, or Higgs particle. If correct, Langacker, Erler, and the CERN researchers will have identified a feature of the elementary particle realm that is not predicted by the reigning theory of particle physics, known as the standard model of matter. Many theoretical physicists believe that the apparent experimental evidence for a fifth fundamental force may eventually lead to the long-sought grand unified theory of the fundamental forces of nature.

The year 2000 has again demonstrated the continued accelerating advancement of scientific knowledge. The year was filled with significant discoveries and explanations of important practical concepts in a number of areas in physics, particularly elementary particle, nuclear, quantum, condensed matter, optical, and theoretical physics. These new insights and developments are leading to many useful applications that enhance understanding of everything in the universe.

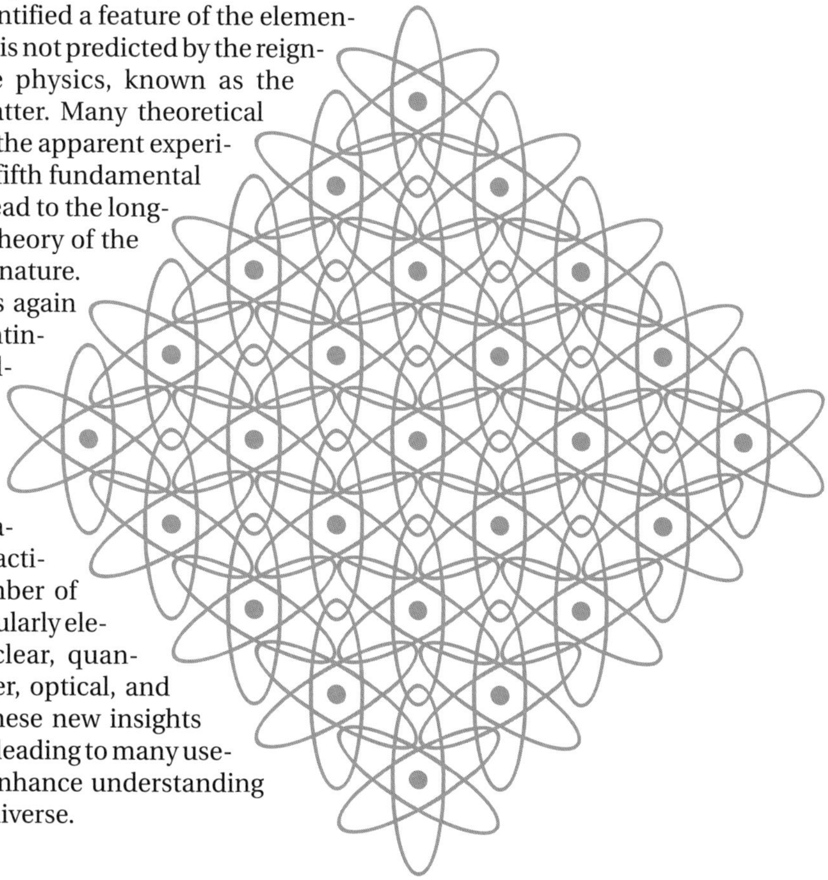

A Novel State of Matter

➤ *Elementary particle physicists announce corroborative evidence for the possible production of a quark-gluon plasma.*

Since the late 1930's, particle accelerators have made it possible to collide particles with great energy, allowing physicists to probe the inner parts of atoms and discover their elementary particle composition. Based on the most accepted theoretical model, quarks are the smallest building blocks of matter and are the foundation for modern theories of the universe. Quarks are bound together by the strong nuclear force, one of the four fundamental forces in nature. This force is thought to be produced when fundamental particles called gluons are exchanged between quarks on the submicroscopic level, conveying energy from one quark to another.

There are six postulated varieties of quarks: up and down quarks, which make up protons and neutrons; and top, bottom, strange, and charm quarks, which can combine in various ways to produce more exotic particles. The characteristic that distinguishes the six quarks from one another is termed color. For each quark, there is an antiquark, a particle like its quark counterpart but possessing opposite charge. The overall quark model of matter is quite complicated but very impressive in its internal consistency and ability to predict observed experimental results.

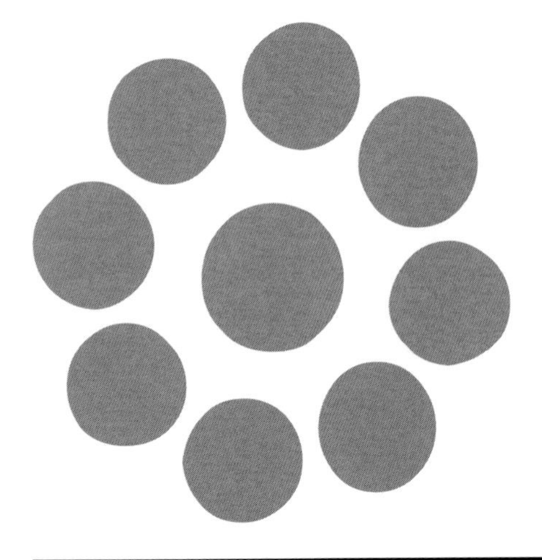

Although no one has succeeded in isolating a quark, the present understanding of the force that binds quarks together in the nucleus indicates that at sufficiently high temperatures and nuclear densities, there should be enough energy to melt protons and neutrons down into their constituent quarks and gluons, resulting in a new state of nuclear matter. Because this state is analogous to the condition of ionized atoms that form a plasma, the new state is termed a quark-gluon plasma. According to the big bang theory, the state would be similar to the condition that existed about ten microseconds after the birth of the universe, when the temperature and concentration of energy were so great that quarks and gluons could roam about freely in the form of a quark-gluon plasma. At that point in time, none of the particles that make up everyday matter had yet formed. The quarks and gluons, which in the modern cold universe are locked up inside protons and neutrons, would have been too hot to stick together.

Experimental Signals

On February 10, 2000, elementary particle physicists at the European Organization for Nuclear Research (CERN) facility, near Geneva, Switzerland, announced corroborative evidence for the possible production of a quark-gluon plasma. The conclusion for the possible production of a quark-gluon state of matter was reached based on the results of seven experiments that were conducted at CERN between 1994 and 2000 using the Super Proton Synchrotron (SPS) collider, a 6-kilometer (3.7-mile) circle of magnets. In the experiments, this ring of magnets accelerated lead ions up to very high energies. These ions were then smashed into fixed targets of lead or gold atoms. The temperature of the resulting nuclear collision fireball was estimated to be one hundred thousand times hotter than the sun's core and twenty times the density of an ordinary nucleus, conditions that are well beyond where a quark-gluon plasma should exist.

The temperature of the resulting nuclear collision fireball was estimated to be one hundred thousand times hotter than the sun's core and twenty times the density of an ordinary nucleus.

The main problem involved in observing the quark-gluon plasma is that physicists can see only the particles that escape from the nuclear fireball and reach a set of detectors. From these signals, what previously happened has to be reconstructed using the laws of physics. The experimental evidence necessary to confirm the production of the quark-gluon plasma is threefold. First, there should be an enhanced production of strange mesons. A meson is a particle composed of a quark bound with an antiquark, whereas a proton or a neutron is composed of three bound quarks. Most strange mesons, such as kaons and pions, have intermediate masses between a proton and an electron. Second, a decrease in the production of J-psi particles is expected. A J-psi particle is a heavy meson, more than three times as massive as a proton. Third, an increase in the creation of gamma rays and lepton-antilepton pairs should occur. Leptons are a group of elementary particles that include the electron, the muon (a heavy electron), and neutrinos. For each lepton, there is a corresponding antiparticle, or

One of six hodoscopes (a device for tracing the paths of ionized particles) in the NA50 detector, which is used in quark-gluon plasma research at CERN. (CERN)

antilepton, with the same mass but opposite electric charge. For example, a positron is the antilepton of an electron.

Current Status

Experimental results reported from CERN in 1999 and 2000 show an unexpectedly low production of J-psi particles. Many physicists believe that the reduction is caused by the destruction of J-psi particles in collisions with quarks and gluons. An alternative explanation suggests that other mechanisms, particularly collisions with particles that are flying away from the nuclear collision that are less exotic than quarks and gluons, could produce the low number of J-psi particles. In addition to the low number of J-psi particles, CERN scientists reported an equally striking increase in the number of strange mesons among the collision products. Alternatively, some proton-antiproton collision experiments have shown similar significant enhancement of strange mesons.

When CERN physicists searched their data for the expected increased gamma-ray production that should be emitted by the quark-gluon plasma, the data were not convincing. However, lepton-antilepton pairs, particularly electron-positron pairs, were observed emerging from the collisions. Putting all the pieces of the puzzle together that have accumulated from the seven CERN experiments, the conclusion is that a new state of matter has been generated in the SPS nuclear collisions. The question is whether this configuration is a manifestation of the much sought after quark-gluon plasma or some other state of dense nuclear matter.

Significance and Future Work

In the CERN experiments, later referred to as the "little bang," matter has been produced in a state never seen before, at energy densities twenty times higher than that inside the atomic nucleus, a state where quarks and gluons are no longer confined. This result verifies an important prediction of the present theory of the quark model of matter and the nature of the fundamental forces that bind quarks together. In addition, if the new state of nuclear matter is indeed the alleged quark-gluon plasma, then cosmic history has been reversed with protons and neutrons melting back into a soup of component quarks and gluons, similar to the state of the universe a tiny fraction of a second after the big bang. Consequently, for the first time, understanding of how the universe was created would be extended back in time to capture a glimpse very near the moment of creation.

The results from CERN present a strong incentive for future experiments. More direct confirmation of the production of a quark-gluon plasma is expected by observing particle jets and gamma ray radiation from higher-energy nuclear fireballs, where the plasma state will last a little longer than in the SPS experiments. In June, 2000, the new Relativistic Heavy Ion Collider (RHIC) at Brookhaven National Laboratory in Upton, New York, began preliminary operations with the necessary energy to directly seek the quark-gluon plasma state. In this collider, nuclei smash together with ten times greater energy than in the SPS collider. Satoshi Ozaki, the director of the RHIC, is encouraged by the research at CERN and looks forward to confirming the nature of this new state of nuclear matter in detail in the near future.

For the first time, understanding of how the universe was created would be extended back in time to capture a glimpse very near the moment of creation.

Matter Wave Lasers

➤A team of physicists announced that by using a Bose-Einstein condensate, they had produced a matter wave laser, which would be more sensitive than optical lasers

Conventional optical lasers are devices that convert light energy into a coherent beam of light, meaning that all the energy peaks and troughs are precisely in step. All the photons making up a beam of coherent light are identical, each one having the same frequency, phase, and direction. A beam of coherent light will not spread out or diffuse as it propagates. The process for producing a coherent light beam is termed light amplification by stimulated emission of radiation, from which the acronym "laser" is derived.

The optical laser process consists of light energy passing from a source into an active medium composed of atoms or molecules. Although the excited medium might emit light spontaneously, it can be stimulated into emitting duplicates of an incoming photon. When a photon with just the right frequency passes through the medium, the excited medium most likely will give up its stored energy by emitting an exact copy of the selected incoming photon. An optical resonator consisting of two mirrors causes the light to bounce back and forth through the active medium, amplifying the coherent beam. Thus, an initial photon becomes duplicated numerous times to produce amplified coherent light. Because one of the mirrors is semitransparent, some of the light emerges from the system as a laser beam.

Whereas an optical laser generates concentrated beams of coherent light, atom lasers produce amplified beams of coherent atoms that are moving along in lockstep.

Atom Lasers

In the early 1980's, physicists envisioned making a laser that used matter instead of light. In 2000, researchers at the Massachusetts Institute of Technology (MIT) announced their success in producing a pulsed atom laser, or matter wave laser. Led by physicist Wolfgang Ketterle, the group developed a device that increases the number of particles in a beam of atoms. Although Ketterle's research group initially developed a rudimentary form of an atom

Massachusetts Institute of Technology assistant professor Wolfgang Ketterle (right) and two graduate students examine the machine with which they demonstrated a rudimentary atom laser in 1997. Three years later, they successfully produced a pulsed atom laser. (AP/ Wide World Photos)

laser in 1997, the new device creates a seed wave that is amplified as it ripples along. Whereas an optical laser generates concentrated beams of coherent light, atom lasers produce amplified beams of coherent atoms that are moving along in lockstep. Instead of flying about and colliding randomly, the atoms display coordinated behavior, acting as if the entire group were one single entity.

Amplifying atoms is much more difficult than amplifying light waves because matter cannot be created, only changed in form. Consequently, the number of atoms must be conserved, whereas light waves can be generated from other forms of energy. The key to MIT's success was using an exotic form of matter known as a Bose-Einstein condensate as the active medium. To form the condensate, sodium atoms were cooled to temperatures slightly above absolute zero, making them slow to a crawl. Because all atoms obey Heisenberg's uncertainty principle, as the atoms slowed their momentum became more and more predictable, but as a result their location in space became less certain. Eventually, the location became so uncertain that the sodium atoms began to overlap, acting as though they were one coupled object. Such a quantum coupled system is termed a Bose-Einstein condensate. This medium provides a very narrow spread of velocities that can be transferred to an incoming beam of atoms in a stimulated way, similar to the operation of an optical laser.

Successful Experiments

In the MIT matter wave laser, the Bose-Einstein condensate was illuminated with two optical laser pulses tuned at slightly different frequencies. Initially, some of the atoms absorbed photons from the first laser beam. However, the second laser beam quickly stimulated the photons back out of the atoms and into its own beam. This caused the atoms to recoil with a momentum that sent a weak wave pulsing through the condensate. Then, by turning off the second laser and shining only the first laser on the condensate, the first laser's photons were diffracted by the initial wave. This process in turn directed more atoms into the wave. The result was an amplified beam of atoms that all marched in unison with the initial weak wave, but the beam was now thirty times stronger. Physicists worldwide were extremely surprised by the results. Most of them had speculated that it would be at least another twenty years before a matter wave amplifier would be made.

Each atom in the outgoing beam had exactly the same quantum mechanical wave formation as the ones that entered the condensate.

When the MIT researchers measured the interference pattern between the incoming matter wave and the amplified wave, the resulting pattern of dark and light bands proved that the input atoms had preserved their phase during the amplification process. Each atom in the outgoing beam had exactly the same quantum mechanical wave formation as the ones that entered the condensate, providing a rare glimpse of quantum mechanics working at a macroscopic scale. Because the phase-coherent atom amplifier outputs atoms acting together as one giant matter wave, it is the atomic counterpart of an optical laser.

After the success at MIT, Japanese researchers at the University of Tokyo announced similar results using a Bose-Einstein condensate of rubidium atoms. Shortly thereafter, German scientists at the Max Planck Institute for Quantum Optics in Garching and at the University of Munich developed an atom laser that emits a continuous rather than pulsed beam of matter waves. Using a rubidium condensate, they produced a continuous matter wave beam that was maintained for a tenth of a second.

Significance and Applications

Because matter waves have much shorter wavelengths and move much more slowly than light waves, atom lasers could be orders of magnitude more sensitive than current optical lasers. Similar to a beam of light, an atom laser beam can be focused and reflected by using lenses or mirrors constructed from a system of magnets or optical laser beams. It appears feasible to focus an atom laser beam to a spot the size of one nanometer, which is a thousand times smaller that the smallest focus of an optical laser beam. Consequently, atom lasers may improve the performance of atom interferometers, the analogs of optical interferometers, by making up for losses inside the device and by amplifying the output signal. Applications of atom interferometers include major improvements in precise rotation and gravity sensors that are used in navigation gyroscopes and geological exploration. An atom laser could be used to boost the signals in such devices, and the resulting patterns of two or more matter waves could then be compared.

Another very useful application of matter wave lasers would be as an energy source capable of high-resolution deposition of atoms on surfaces. This process could be used to fabricate novel materials and produce tiny nanostructures that are necessary for future computer circuits. Because of the high brightness and coherence of matter wave beams, many new developments are possible in atom optics. In particular, atom lasers may eventually replace conventional atomic beams where ultimate precision is necessary, such as in atomic clocks, in manipulating atoms, and in testing the fundamental laws of physics. One of the main problems to solve in making practical applications is that the power produced by amplified atom pulses is limited by the size of the Bose-Einstein condensate system. Such systems are presently very small. Much research is being done to address this problem.

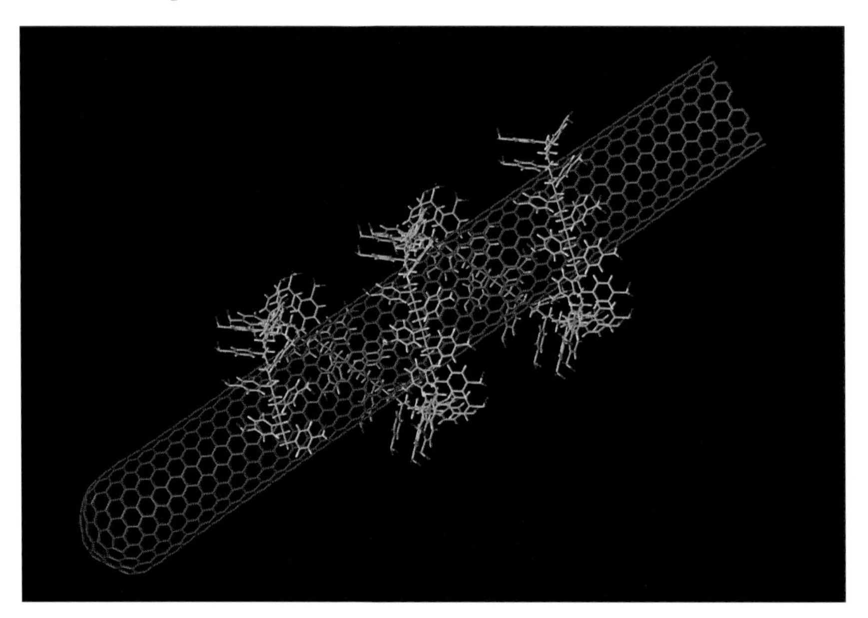

The Most Proton-Rich Nucleus

➤ *The successful production of nickel 48, the most proton-rich nucleus ever made, was confirmed and announced.*

Just as electrons fill different energy levels, or shells, outside the nucleus of an atom, neutrons and protons are arranged in energy shells inside the nucleus. For this shell model, the nucleus can be thought of as an onion consisting of layers of protons and neutrons, or nucleons. Each time a nuclear shell is filled with the maximum possible number of nucleons, the nucleus is particularly stable. As more shells are completed, the nucleus acquires even greater cohesion. These maximum numbers are termed "magic numbers." Based on the shell model of the nucleus and generally in agreement with experimental data, magic numbers occur when the number of neutrons or protons in the nucleus equals 2, 8, 20, 28, 50, 82, 114, 126, or 184.

When the number of protons and the number of neutrons are both magic numbers, the nuclei are doubly magic, possessing full shells of protons and full shells of neutrons. For example, the helium-4 nucleus with two protons and two neutrons is doubly magic and is quite stable. However, being doubly magic alone does not guarantee stability of a nucleus. For example, nickel 56 has twenty-eight protons and twenty-eight neutrons, making it doubly magic, but it is not stable. Until late 1999, of the approximately twenty-five hundred known nuclear isotopes (atoms of the same element that contain different numbers of neutrons), only nine doubly magic nuclei had been discovered by nuclear physicists. Scientists still sought the most interesting double-magic holdout, nickel 48, containing twenty neutrons and twenty-eight protons.

Until late 1999, of the approximately twenty-five hundred known nuclear isotopes, only nine doubly magic nuclei had been discovered.

For lighter elements, the relative numbers of protons and neutrons in a nucleus are typically equal. However, because of the repulsive electrical force between protons, as the size of a nucleus increases, more neutrons than protons are generally required in order to produce stability. For many years, nu-

clear physicists wondered what would happen when a larger nucleus was doubly magic, making it more stable than its closest isotope neighbors; but at the same time, the number of protons was greater than the number of neutrons, which would decrease the stability. The primary nuclear candidate meeting both conditions is nickel 48, but many previous nuclear models predicted that nickel 48 was highly unstable and would never be observed.

Magic Nickel 48 Produced

During the late 1990's, researchers at the Grande Accelerateur National d'Ions Lourdes (GANIL) in Caen, France, were successful in producing two of nickel's lightest doubly magic isotopes, but it was not until mid-September of 1999 that they were able to produce nickel 48. Results were confirmed and announced in February, 2000. The experiment at GANIL consisted of an ordinary nickel target that was bombarded with a beam of highly charged nickel-58 ions for a period of ten days. During this time, at least two, and possibly four, nickel 48 nuclei were observed as products of the nuclear collisions. The new nuclei were identified by their time of flight after emerging from the ordinary nickel target and traveling through a series of selectors and detectors. Further identification was provided by measuring their total energy and energy loss as they traveled through a stack of silicon detectors. Besides nickel 48, three other proton-rich isotopes, iron 45, chromium 42, and nickel 49, were also generated in the nuclear collisions at GANIL. About 50 iron-45, 290 chromium-42, and 100 nickel-49 nuclei were observed over the ten-day period.

In a plot of proton number versus neutron number for all known nuclei, nickel 48 falls on the extreme outer limit of nuclear stability, at the edge of where nuclei can no longer be held together by the strong nuclear force. The newly produced nickel-48 nucleus undergoes radioactive decay in a fraction of a second (a lower limit estimate of 0.5 microsecond), and if it had not been for the stability generated by its doubly magic numbers, it probably would have never been observed. Because only a few nickel-48 nuclei were observed, not enough events occurred to make a detailed comparison with existing nuclear models. This will require experiments with higher statistics so that the exact half-life of nickel 48 can be determined. The fact that nickel 48 was observed contradicts previous nuclear models. They predicted that nickel 48 was unstable and should last for less than one microsecond, the typical time it takes for collision products to travel from the stationary nickel target to the detectors in the GANIL experiment.

The fact that nickel 48 was observed contradicts previous nuclear models, which predicted that nickel 48 was unstable and should last for less than one microsecond.

Significance and Future Work

Although the role of magic numbers was established for stable nuclei in 1949, the production of nickel 48 extends the applicability of the magic number scheme to the extreme case of the most proton-rich nucleus ever observed. In addition, the production of nickel 48 allows nuclear physicists the opportunity to study a form of radioactive decay that has long been sought.

Because the nickel 48 experiment at GANIL was optimized to observe the doubly magic nucleus, the mode of decay was not detected. Single proton decay is energetically impossible for nickel 48, so it is expected to decay by emitting two protons, a form of radioactivity never before observed. Observing this decay mode may answer the question of whether protons and neutrons are correlated in pairs within the nucleus. Future experiments are planned to study the energy distributions and angular correlations of the two emitted protons, allowing researchers to determine whether they are correlated as they leave the nucleus. Furthermore, as advances lead to higher beam intensities and faster accelerations of the bombarding ions, corresponding higher production rates of nickel 48 will follow and allow accurate determination of the half-life and other important properties of the most proton-rich nucleus ever observed.

At present, existing particle accelerator facilities around the world are not capable of producing any other doubly magic nuclear configurations. Among the lighter nuclei, the remaining candidates would have too many protons compared with neutrons to produce bound stability. In addition, some of the lighter nuclei do not conform to the nuclear shell model with respect to magic numbers. For example, magnesium 32 has twenty neutrons, which is a magic number, but this isotope does not show closed-shell behavior. For the superheavy elements, magic numbers have not been accurately predicted. If and when they are, future experiments, similar to those at GANIL, will be designed to observe these nuclei and produce further insights into the fundamental nature and understanding of nuclei.

A Fifth Fundamental Force?

➤ *Two physicists at the University of Pennsylvania reported indications of a fifth fundamental force and the carrier particle associated with it.*

For hundreds of years, Isaac Newton's formula for the gravitational attraction between two bodies has helped scientists calculate the effects of the force of gravity, ranging from the orbits of planets to the trajectories of rockets. However, as physicists have tried to combine the four known fundamental forces of nature—gravitational, electromagnetic, weak nuclear, and strong nuclear—into one grand unified theory, they have been unable to incorporate gravity without postulating the existence of a fifth, undiscovered force. Because the force of gravity is so small in comparison to the other three known forces and has a much different mathematical structure, no one has yet succeeded in developing a complete quantum theory of gravity. A quantum theory would describe the gravitational force as originating from fundamental particles called gravitons. A fundamental particle is a particle with no internal structure. To formulate a unified theory of the forces of nature, a quantum theory of gravity is necessary.

The standard model of particle physics is the current theory of fundamental particles and how they interact. In the standard model, the fundamental particles are quarks, electrons, muons, neutrinos, photons, gluons, and W and Z particles. The model describes the carrier or mediator of the strong force as gluons, which are exchanged between quarks, the fundamental building blocks of the nucleus. In addition, the model includes a combined theory of the weak nuclear and electromagnetic forces, known as the electroweak theory. This theory describes photons as the mediators of the electromagnetic force and W and Z particles as the carriers of the weak force.

In the formulation of the standard model, at least one other fundamental force appears in addition to the four known forces. This fifth force is necessary to explain how the masses of the fundamental particles are generated. The difficulty of experimentally testing the standard model has produced a number of competing ideas of how this additional force operates. The simplest theoretical version introduces a fifth force and a new fundamental particle that is responsible for this force. Other models introduce additional extremely weak forces and more complicated explanations for the production of particle masses.

Indications of a Fifth Force

One proposal for the fifth force suggests that the force of attraction between two bodies is given by Newton's formula plus a much smaller fifth force that comes into play when objects are separated by distances on the order of 100 to 1,000 meters (109 to 1,094 yards). In the late 1980's, a number of experiments investigated this proposal by measuring the effect of gravity as a function of distance from the Earth's surface, and on this basis, some scientists claimed experimental evidence for a fifth fundamental force. For example, geophysicists in Australia measured a gravitational constant that was 0.7 percent greater than that measured in the laboratory, suggesting the presence of an additional force. However, because the behavior of gravity depends in detail on the density profiles of soils and rocks in the study area, all the observed effects could be explained by unexpected fissures, extra dense rocks, or unknown, deeply buried rocks without the necessity of a fifth fundamental force.

One glaring deficiency of string theory to date has been its total lack of experimental verification.

In 2000, two physicists at the University of Pennsylvania, Paul G. Langacker and Jens Erler, investigated apparent anomalies in existing data obtained from large particle accelerator experiments in Switzerland and California as well as tabletop atomic physics experiments in Colorado. They concluded that these data are best matched and understood if an additional force and its associated carrier particle are assumed to exist. Langacker and Erler suggest that the fifth force would be about one-hundredth as strong as the weak nuclear force. A Z′ particle, also known as the Higgs particle, would carry or mediate the force. This proposed particle would be about ten times as heavy as the Z particle. On the other hand, some physicists have suggested that the anomalies noted by Langacker and Erler may just be random errors that are generated by the measurement of many different parameters.

In September, 2000, elementary particle physicists at the European Organization for Nuclear Research (CERN)

Some scientists theorize that the force of attraction between two bodies is given by the formula derived by Sir Isaac Newton (pictured here) plus a much smaller fifth force that comes into play when objects are separated by distances on the order of 100 to 1,000 meters (109 to 1,094 yards). (Library of Congress)

facility in Switzerland announced possible experimental evidence for the Z′, or Higgs, particle. More conclusive confirmation that the new clues do indeed suggest the presence of a fifth force will be sought in particle accelerator experiments over the next few months to years, when higher energies will be available to generate the proposed Z′, or Higgs, particle.

Significance

Many scientists have proposed that in the big bang model of the universe, a single "superforce" ruled and then rapidly broke down into the forces observed today. If this superforce and the very earliest moments in the history of the universe are to be correctly understood, a description of gravity based on fundamental particles must be developed. In the beginning, the universe was a very dense fluid of very high-energy particles. At that time, gravitational interactions were comparable in strength to the other fundamental forces acting on the particles in that environment. Therefore, a consistent particle theory that can correctly treat both gravity and other particle interactions is necessary to understand that era of time. The combination of gravity and particle physics remains one of the major outstanding problems in physics. The connecting bridge could be the theory that includes a fifth fundamental force.

If the conclusions of Langacker and Erler and the discovery of the Higgs particle are proven to be correct, an additional fundamental force that originated from the superforce would join the four already known forces: gravity, electromagnetic, weak nuclear, and strong nuclear.

The proposed Z′, or Higgs, particle that generates the fifth force has also turned up in versions of a mathematical model of fundamental particles known as string theory. String theory incorporates relativity and quantum mechanics into a formulation in which the fundamental particles are viewed as extended objects that make up everything in the universe. One glaring deficiency of string theory to date has been its total lack of experimental verification. Confirmation of the discovery of the Higgs particle would lend strong support to the validity of the string theory model.

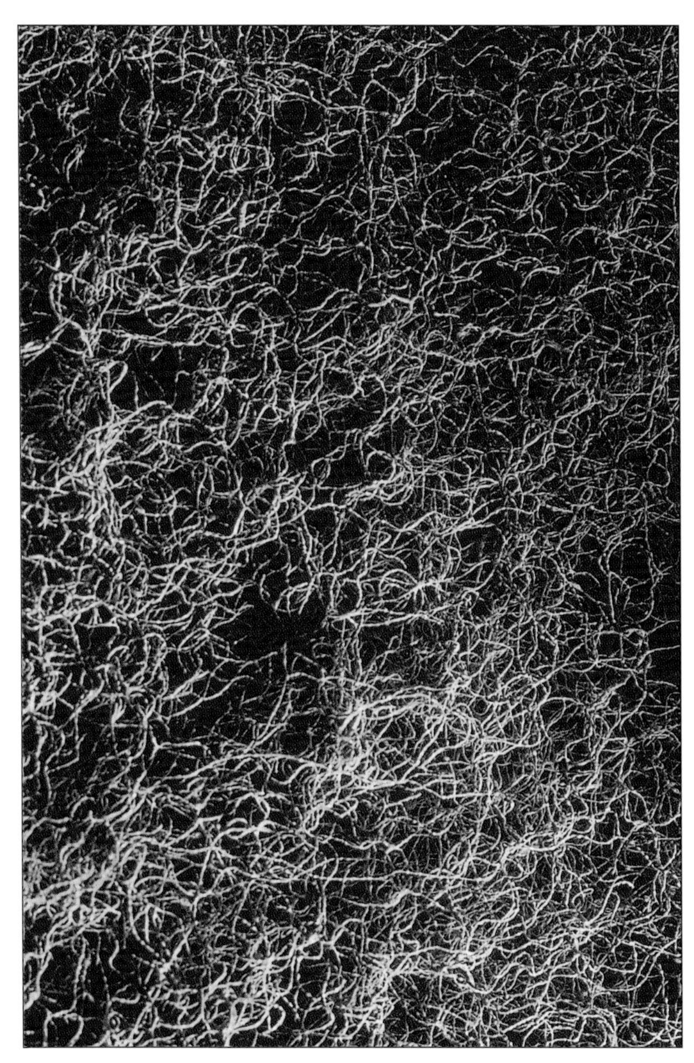

String theory predicts the correct force laws and the associated force-carrying particles, including gluons for the strong force, photons for the electromagnetic force, Z particles for the weak force, Z′ particles for a fifth force, and gravitons as the proposed carrier of the gravitational force. Using the idea that fundamental particles are not pointlike masses but rather small lines or loops of energy called "strings," string theory constitutes the most promising approach for developing an elementary particle theory of gravity. Consequently, many theoretical physicists are searching for experimental evidence of the fifth force, which will support the string approach and may finally lead to a grand unified theory of five fundamental forces of nature.

The Lifetime of a Neutron

➤*A combined team of researchers reported a more accurate determination of the neutron lifetime than ever before.*

A neutron is a common neutrally charged particle that is slightly more massive than a proton and exists in all nuclei except hydrogen. According to the standard model of matter, the current theory of fundamental particles and their interactions, a neutron is composed of quarks, the basic building blocks of matter. Of the six postulated quarks, known as up, down, top, bottom, strange, and charm, a neutron consists of two down quarks and one up quark. Quarks are bound together by other elementary particles called gluons to produce neutrons and protons, which along with electrons, make up every atom in the universe.

When neutrons exist inside an atomic nucleus, they are very stable. However, when neutrons are isolated outside the nucleus, as they were at the birth of the universe, they are unstable. They then radioactively decay into a proton, an electron, and an antineutrino, a particle that has no charge and almost zero mass. The slightly greater mass of a neutron compared with a proton represents the available energy for the neutron to undergo decay. The period of time that it takes a neutron to decay is termed the neutron lifetime.

In the late 1970's, the lifetime of a neutron was estimated to be 932 seconds. By the mid-1990's, that number had been reduced to 886.7 seconds. Research work in 2000 indicates that it may be on the order of 750 seconds. Because models of the universe and a better understanding of one of the four fundamental forces, the weak nuclear force, depend on the precise value of the neutron lifetime, many experiments have been devised to measure this important parameter.

Research work in 2000 indicates that the lifetime of a neutron may be on the order of 750 seconds.

Trapping Neutrons for a Lifetime

In January, 2000, a combined team of researchers from the National Institute of Standards and Technology in Maryland, the Institute Laue-Langevin in France, and the

Petersburg Nuclear Physics Institute in Russia reported that they had successfully confined neutrons in a magnetic trap in three dimensions. This entrapment allowed a much more accurate determination of the neutron lifetime than ever before. Astrophysicists and cosmologists have wanted a more exact and reliable value in order to refine models describing the early formation of the universe. In addition, elementary particle physicists have wanted a more exact value to clear up some discrepancies about the weak nuclear force.

Although physicists could trap atoms using magnetic fields in three dimensions since the early 1980's, neutrons had been successfully confined by magnetic fields in only two dimensions, leaving neutron lifetime measurements imprecise. The main problem with trapping neutrons to examine their lifetime is that they are electrically neutral, so they interact only weakly with other matter and light. Consequently, laser cooling and other atom-cooling techniques devised in the mid-1980's have not worked in attempts to trap neutrons in three dimensions.

In the late 1990's, a combined team of researchers from the United States National Institute of Standards and Technology (NIST), Los Alamos National Laboratory, Harvard University, and the Hahn-Meitner-Institut (HMI) in Berlin, Germany, overcame the problems associated with trapping neutrons by constructing a three-dimensional magnetic trap designed to take advantage of the weak magnetic interaction of a neutron with a confining magnetic field. At the NIST Center for Neutron Research in Gaithersburg, Maryland, a beam of cold neutrons liberated from uranium atoms inside a research nuclear reactor at NIST was directed into a bottle-shaped vessel surrounded by magnetic coils and filled with liquid helium at a temperature of less than 0.25 degree kelvin (–272.9 degrees Celsius, or –459.2 degrees Fahrenheit). The trapping magnetic field produced by the assemblage of superconducting magnetic coils held a fraction of the incoming neutrons at the center of the bottle until they radioactively decayed.

Improved measurements of the neutron lifetime will provide a clearer view of the weak nuclear force, which governs the process of radioactive decay.

The liquid helium slowed the neutrons so that a detector could register the subsequent decay by recording interactions that occurred between the helium atoms and the electrons emitted from the decaying neutrons. From these data, the neutron lifetime was calculated to be 750 seconds, with some rather large error bars. Researchers believe that the new technique of trapping neutrons in three dimensions will eventually yield a factor of a hundred or more improvement in the precision of the present value of the neutron lifetime.

Significance

Precise measurements of the neutron lifetime provide clues as to how subatomic particles combined into elements that formed the universe after the big bang. In the big bang model of the universe, quarks, electrons, photons, gluons, and other fundamental particles are believed to have coalesced into neutrons and protons a few milliseconds after the beginning of the universe. According to the theory, as the expansion of the universe continued, the temperature fell, and hydrogen and helium nuclei began to form. In the meantime, free neutrons began to decay.

Knowing the neutron lifetime allows scientists to determine the initial ratio and concentration of hydrogen, helium, and other light elements in the universe. Between three and thirty-five minutes after the initiation of the universe, the big bang theory postulates that the helium-to-hydrogen ratio became fixed at somewhere between 22 percent to 28 percent helium and 72 percent to 78 percent hydrogen. Measurements of these ratios billions of years later are used as evidence for the big bang scenario. The theoretical uncertainty in the predicted abundance of helium 4 depends on the uncertainty in the neutron lifetime.

Improved measurements of the neutron lifetime will also provide a clearer view of the weak nuclear force, which governs the process of radioactive decay. According to the standard model of matter, the weak force acts on particles in a single preferred orientation, referred to as the "left-handed" direction. A simple analogy is to consider a dinner party at a round table with water glasses symmetrically placed on the right-hand and left-hand side of each guest. The host chooses the water glass to his left, thereby influencing all the guests to take the glass on their left. The precise value of the neutron lifetime will shed light on whether the weak force has always been "left-handed" or if there was equal probability of it acting from the right or the left during the initial moments of the big bang. If the latter premise is true, then some symmetry-breaking event turned the weak force to the left. From measurements of the neutron lifetime and the asymmetry of the radioactive decay of a neutron, important fundamental parameters describing the weak force can be determined.

Making Sense of Chaos

➤ *During 1999 and 2000, a group of scientists at the Massachusetts Institute of Technology analyzed how to quantitatively minimize chaos in natural systems.*

Disorder is a fundamental part of nature. The disorder of molecular systems is expressed as a measure of the degree of randomness of motion of the molecules, the degree of mixing of different molecular species in the system, or the degree to which temperature differences have been eliminated by randomizing the energy among the molecules. The law of increasing disorder, also known as the second law of thermodynamics, requires that physical changes occurring in natural systems always proceed in such a manner that the total amount of disorder in the universe is increased. For ideal processes, the total disorder remains unchanged. If total disorder increases, the process is termed irreversible. As the amount of disorder in a physical system increases, the amount of information about the system decreases.

Scientists use a quantity called entropy to provide a mathematical measure of disorder. Simply stated, high entropy implies a state of high disorder, whereas low entropy implies a state of low disorder. One of the primary objectives of scientists and engineers is to acquire, process, and store data that capture the maximum amount of information about a system. Because many data sets are represented by digital data, the quantitative measure of the amount of disorder is often referred to as "digital entropy."

Order may be produced in a subsystem of the universe if a more than compensating amount of disorder is created elsewhere. Good examples are the operation of refrigerators and engines. If natural systems are left alone, they move toward a state of maximum disorder as time proceeds. Because the law of increasing disorder limits and controls people's actions and the processes that occur around them just as much as the laws of motion, force, or conservation do, physicists have found the study of disordered, chaotic systems to be a rich, timely area of research.

Taming Chaos

Exerting control over chaotic systems is particularly difficult because such systems constantly exhibit new de-

grees of uncertainty in their properties as a function of space and time. As reported in 2000, a group of researchers led by Seth Lloyd at the Massachusetts Institute of Technology analyzed how to minimize disorder in natural systems. They combined ideas from the fields of thermodynamics, information theory, and quantum mechanics to determine the amount of information needed to bring a disordered system under control. Their analysis provides some quantitative answers about how to tame chaotic, disordered systems. According to the principles of thermodynamics, the disorder, or entropy, must be decreased to control a system. For example, if the disorder of a hot gaseous system is reduced, the number of possible microscopic arrangements of the gas molecules is decreased. As a result, some of the uncertainty about the detailed properties and description of the system is removed.

Information theory employs probability and ergodic theory to study the statistical characteristics of data and communication systems. Applying these principles to a

This busy city street contains several examples of a familiar chaotic system: a person steering a car. (U.S. Landmarks and Travel)

hot gaseous system, Lloyd and his group deduced that when the entropy of the gas is reduced, the amount of information about the gas is increased. Furthermore, they found that when enough information is acquired about a system to keep the uncertainties in its properties at manageable levels, the entropy is lowered sufficiently so that the system can be controlled. For every single bit of information, there is a corresponding reduction in uncertainty.

Examples of chaotic systems vary from weather phenomena to turbulence encountered when flying to the safe manipulation of an automobile to Internet traffic to mammogram images. An excellent everyday example is represented by a person steering an automobile. A small movement of the steering wheel can result in a large change in the direction that the vehicle is traveling. If a blindfolded driver is within two feet of a curb, very small steering corrections can change the distance to 4 feet in one second, 8 feet in two seconds, and 16 feet in three seconds. If the driver receives information for making steering adjustments every second, the uncertainty in location relative to the curb can be maintained at 2 feet or less. The automobile is kept in control with minimum deviation from the curb. However, if instructions are received only every two seconds, the car will go out of control and either crash into or drift away from the curb. For either result, it takes twice as long to occur if the driver receives information for making steering adjustments.

Applications

The work of Lloyd and his group can be applied to a variety of physical systems. A chaotic system can be controlled by obtaining information about the system, processing it, and feeding it back into the system. Each single bit of information improves the level of control by one bit.

In quantum computing, researchers must determine how to control the basic interacting entities of a computer. These entities are called quantum bits. Whereas classical computers perform operations stored as classical bits, which can be in one of two discrete states, quantum computers perform operations on quantum bits, which can be put into any superposition of two quantum states. Because quantum bits can interact with one another on many different levels, they are powerful computing devices, but they are also very difficult to control. Lloyd is designing an experiment that uses one quantum bit as a sensor to collect information about another quantum bit in the computer system. The sensor then feeds the information back into the system to reduce the amount of disorder in the system. This approach could make extremely fast quantum computers practical by reducing the many redundant checks that are currently necessary to ensure that a quantum-based computer is providing reliable results and not deviating off in some erroneous direction.

Another important application of optimizing information by taming chaos is in the reconstruction and enhancement of X-ray digital mammograms. The goal is to reconstruct images without generating artifacts that could be misinterpreted as tumors or other biological irregularities. Applying the ideas of Lloyd and his group to process mammograms, the output images show better overall quality in terms of contrast, signal-to-noise ratio, and visibility of details than the input mammograms. Processed images reveal existing details that could not have been obtained by conventional mammogram analysis. In addition, these images display no artifacts. The process can also be applied to radio and X-ray astronomy data, spectroscopic data, electron microscopic data, and geophysical data to obtain optimum images with minimum disorder.

Because quantum bits can interact with one another on many different levels, they are powerful computing devices, but they are also very difficult to control.

Nuclear Energy on a Tabletop

➤ *Physicists succeeded in producing energy for the first time from nuclear fusion on a tabletop, and two other groups of researchers succeeded in using lasers to produce nuclear fission reactions on a tabletop.*

Nuclear energy research is focused on developing a reliable energy alternative to the burning of fossil fuels. Nuclear energy can be released by the processes of fission and fusion. Nuclear fusion combines, or fuses, small nuclei into larger nuclei and releases energy that could be used to generate electricity. The Sun, heated from within by continuous fusion reactions, is a huge fusion reactor. Controlled nuclear fusion offers a number of advantages over other energy sources. First, the fuel is widely available and virtually inexhaustible. Second, the process is environmentally favorable, producing no greenhouse gases and no long-lived radioactive byproducts. Third, it is inherently safe. A major pursuit of nuclear physicists is to develop nuclear fusion into an abundant source of energy for the world during the twenty-first century and beyond.

The main technical difficulty in producing controlled fusion is that the reacting nuclei initially repel each other because of the electrical interaction between like charges. The charged nuclei must approach each other fast enough so that the strong nuclear force is greater than the electrical repulsion. The interacting nuclei must then be kept together long enough at high enough temperatures for fusion to occur.

The results of tabletop fusion and fission show that lasers can be used to perform desktop physics experiments without scheduling a major multimillion-dollar particle accelerator facility.

When hot enough, the fusion fuel, typically consisting of hydrogen isotopes, becomes separated into its positively charged nuclei and negatively charged electrons, forming a plasma. One approach for containing the burning fusion fuel is inertial confinement. In this fusion process, a small pellet of fusion fuel is compressed and confined by its own inertia for a long enough time at a high enough temperature and density that fusion reactions occur.

Nuclear fission releases energy when heavy nuclei split into lighter nuclei. Controlled fission is initiated when a slow neutron strikes the heavy nucleus, typically ura-

The circular target chamber room appears in the lower right of this aerial view of the National Ignition Facility building. (Jacqueline McBride and Bryan Quintard, Lawrence Livermore National Laboratory)

nium. Commercial fission reactors depend on a chain reaction in which each fission releases neutrons that induce additional fissions. Fission has two major weaknesses as an alternative source of energy: some byproducts are intensely radioactive and long-lived, and the fuel is not plentiful.

Tabletop Fusion

In 2000, Todd Ditmire and some colleagues at the Lawrence Livermore National Laboratory used the most powerful lasers presently available to produce energy from nuclear fusion on a tabletop for the first time. The top of the table measured 4 feet by 11 feet (1.3 by 3.6 meters). Clusters of deuterium, or heavy hydrogen, molecules were bombarded with high-powered laser pulses. When the laser beam was focused on a very small volume of deuterium, the molecular clusters were heated to tens of millions of degrees Celsius, stripping electrons from the deuterium atoms to form a plasma. As the superheated clusters exploded, some of the fast-moving deuterium ions fused together with high enough velocity to form helium-3 nuclei plus energetic neutrons. For each helium ion formed, a neutron was also emitted.

Although this new approach to promoting fusion reactions probably cannot be scaled up to produce commercial power, it could provide a cheap source of neutrons. Currently, the laser inputs many times more energy into the process than the neutrons carry back out. The lost energy goes into heating ions and electrons and producing photons that make the plasma glow. For nuclear fusion to produce an energy gain, the fuel must be confined long enough to ignite a self-sustaining thermonuclear reaction. Currently, the plasma generated in tabletop fusion disperses too quickly.

To generate sustained fusion, the National Ignition Facility is being built at Livermore. The facility will be as large as a football stadium and contain 192 extremely powerful lasers. Complex computer programs have been developed to model the interaction of laser light and plasmas. An inertial confinement fusion reaction will be generated by focusing intense beams from

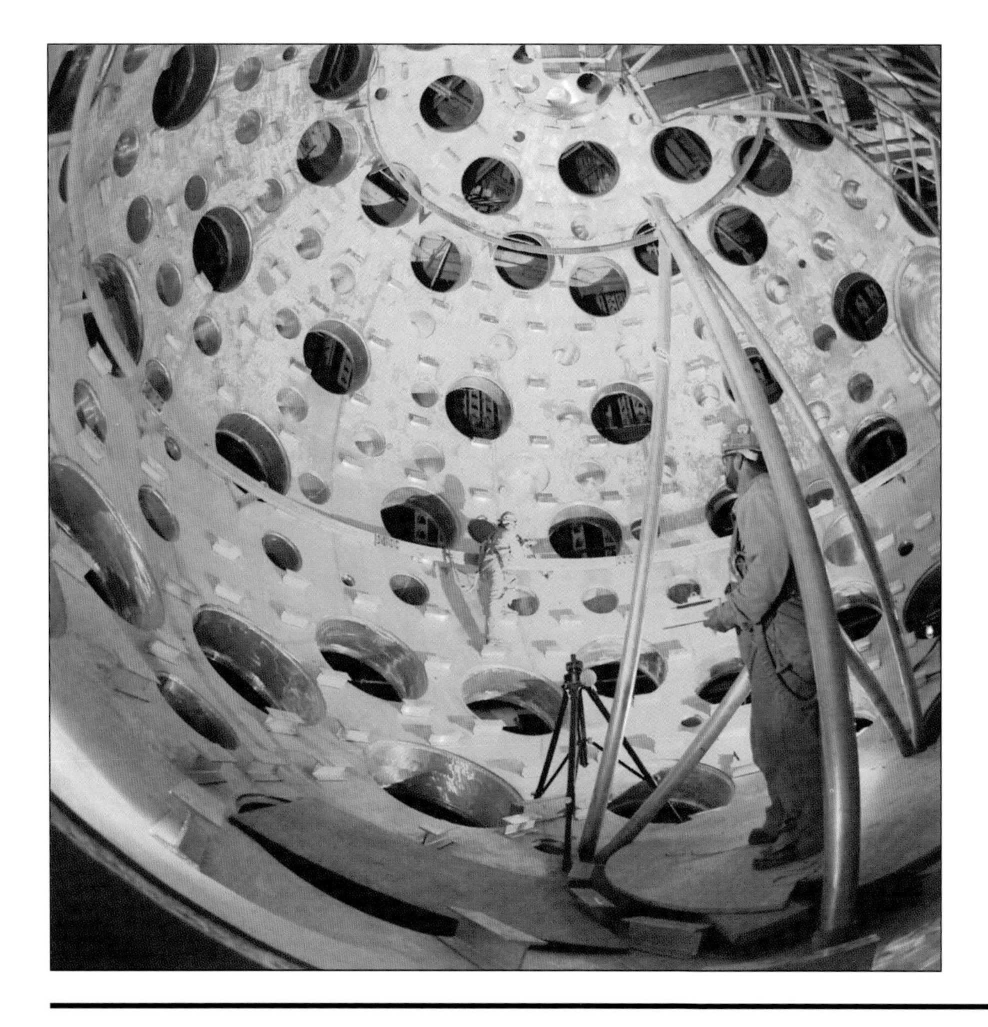

The NIF target chamber (inside view) is 30 feet in diameter and weighs one million pounds. (Jacqueline McBride and Bryan Quintard, Lawrence Livermore National Laboratory)

the lasers onto a pea-sized pellet consisting of deuterium and tritium, which are both isotopes of hydrogen. Plans are to detonate five pellets every second. The heat generated from these interactions will be used to drive turbines that generate about ten billion watts of electrical energy. For this prototype model to become a practical power plant, the costly lasers that initiate the fusion process must be reduced in size and complexity.

Tabletop Fission

In 2000, on another tabletop at Lawrence Livermore National Laboratory, Thomas E. Cowan and a team of researchers from the University of California produced nuclear fission reactions. A solid-gold target mounted on a sample holder containing uranium was first hit by a lower energy laser pulse to produce a swarm of electrons at the target's surface. These electrons were then accelerated to high energies by blasting the gold target with a high-power beam from the world's first thousand-trillion-watt laser.

Strategically placed detectors monitored the energies of the electrons escaping from the plasma. The accelerating electrons produced gamma rays that created electron-positron pairs and also liberated high-energy neutrons from the gold and from copper nuclei that were placed behind the gold. The neutrons in turn split some uranium-238 nuclei and produced other nuclear reactions as well. Later analysis of the target revealed the presence of radioactive isotopes of gold, copper, and other elements, which are all products of the nuclear reactions initiated by the high-energy neutrons. The quantities of nuclear isotopes that were produced agreed with the calculations based on the electron energy distributions that were measured by the detectors.

Researchers at the Rutherford Appleton Laboratory in Oxfordshire, England, obtained similar results, but their primary target was tantalum instead of gold. By placing different material behind the tantalum target, nuclear reactions were generated in potassium, zinc, and silver. Neutrons liberated from the tantalum caused uranium-238 samples to undergo fission.

Significance

Tabletop nuclear reactions initiate a new class of nuclear physics experiments. Tabletop fusion research is directed toward creating a new, safe, clean source of energy with an abundant fuel supply. Although not promising as a commercial energy source, tabletop fusion promises a sci-

entific payoff. This research is providing new information about the behavior of fusion plasmas and insights into the future design and development of commercial fusion power plants that can generate economical, reliable electricity.

Tabletop fission research has generated new insights into the fundamental interplay between light and solids, the interactions of gamma-ray photons with one another, and the physics of electron-positron plasmas. The results of tabletop fusion and fission show that lasers can be used to perform desktop physics experiments to study nuclear physics and astrophysical processes without scheduling a major multimillion-dollar particle accelerator facility for such work.

Converting Sound into Light

➤ *Research groups at two universities converted sound energy into light energy, a process called sonoluminescence.*

The important but little-understood phenomenon of converting sound energy into light energy is called sonoluminescence. It was first observed in an ultrasonic water bath in 1934 by researchers at the University of Cologne in France as an indirect result of wartime research involving applications of marine acoustic radar. This early work involved very strong ultrasonic fields of energy and yielded clouds of unpredictable light-flashing bubbles, now termed "multiple-bubble" sonoluminescence. Because of the instability, short lifetime, and chaos associated with this phenomenon, it received little scientific attention until 1988, when D. Felipe Gaitan succeeded in trapping a single light-flashing bubble at the center of a flask that was energized by sound energy.

During the late 1990's, sonoluminescence created a stir in the physics community. The experimental process of converting sound into light involves aiming ultrasonic waves at an air bubble in a small water cylinder. The sound waves cause the bubble to stretch and compress, producing furious oscillations. Initially, the bubble is about 5 millionths of a meter in diameter. It then expands to a maximum size of approximately 50 millionths of a meter in diameter. At maximum size, the inside of the bubble is a near vacuum because there are so few air molecules present. The low pressure, near-vacuum region of the bubble is surrounded on the outside by a much higher pressure region, which causes the bubble to dramatically collapse to between 0.1 to 1.0 millionth of a meter in diameter. During the compression phase, a flash of light emerges from the bubble.

How a low-energy density sound wave can concentrate enough energy into a small enough volume to cause the emission of light has been an intriguing mystery.

By early 2000, the study of sonoluminescence had yielded many interesting observations as well as many puzzles. The light flashes from the bubbles last an extremely short time, on the order of 50 trillionths of a second. The bubbles are very small when the light is actually emitted, on the order of 0.001 millimeter in diameter. Single-bubble sonoluminescence pulses are often very stable in time and in space. In fact, the frequency of the observed light flashes is sometimes more stable than the rated frequency stability of the oscillator that generates the ultrasonic waves that are converted into light. A recent, unexplained observation has shown that the addition of a small amount of an inert gas, such as helium, argon, or xenon, to the gas in the bubble dramatically increases the intensity of the emitted light.

The Explanation

How a low-energy density sound wave can concentrate enough energy into a small enough volume to cause the emission of light has been an intriguing mystery. More than a dozen reputable scientific theories have attempted to explain sonoluminescence. One proposed explanation of the phenomenon invokes the principles of quantum mechanics, the branch of physics that governs matter the size of an atom and smaller. One of the basic ideas of quantum mechanics is that empty space is never really empty. Instead, empty space, or the so-called quantum vacuum, is made up of fluctuating fields of energy that come in and out of existence. The fleeting fields can be viewed as virtual particles that are ordinarily not detected because of their short lifetimes. By adding a large amount of energy to the system, these virtual particles can become a visible part of the real world.

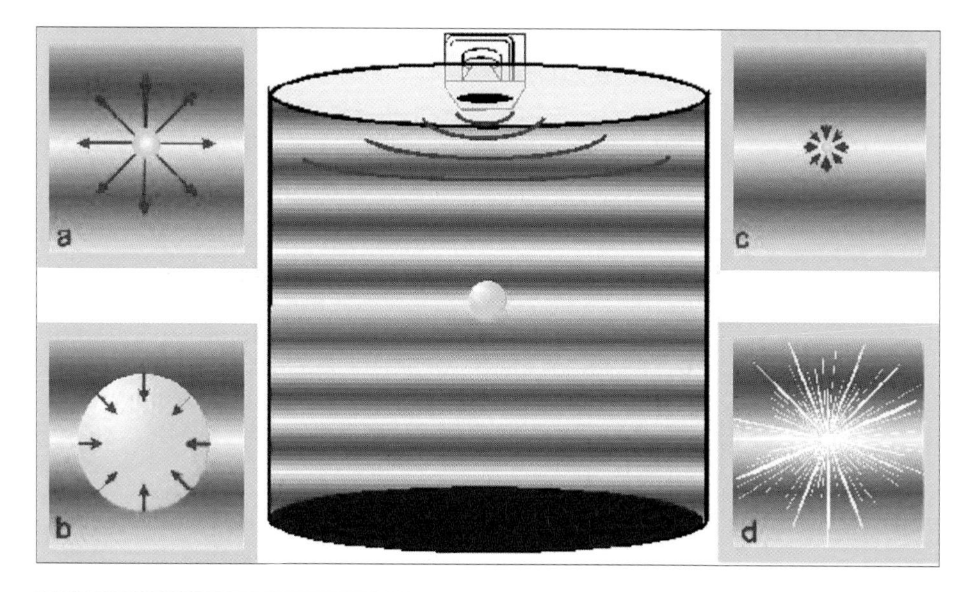

First, ultrasonic waves directed at an air bubble in a small water cylinder cause it to oscillate furiously. The 5-micron bubble (a) expands to about 50 microns (b), creating an internal near-vacuum. The difference between the internal low-pressure region and higher pressure outside causes the bubble to collapse to between 0.1 and 1 microns (c), during which time, a flash of light emerges from the bubble (d). (American Institute of Physics)

Applying this idea to the phenomenon of sonoluminescence, the bubble that is moving in atomic dimensions and lasting for trillionths of a second provides the right conditions for converting virtual light into real flashes of light. However, if this is the correct explanation for the source of light in sonoluminescence, some physicists have pointed out that it would probably take a bubble traveling faster than the speed of light to be able to pluck light from the quantum vacuum. As far as physicists know, this would be impossible because the speed of light is the upper speed limit for the transmission of information in the real world.

Another popular theory proposes that the bubble compresses at supersonic speeds, creating shock waves that concentrate to a radius as small as 20 billionths of a meter. In this extremely small area, energy could be concentrated to produce temperatures possibly reaching 15 million degrees Celsius (27 million degrees Fahrenheit). At these temperatures, hydrogen nuclei in the gas could fuse together by the same process that powers the sun.

In 2000, research groups led by Seth J. Putterman of the University of California at Los Angeles and Rainer Pecha of the University of Stuttgart in Germany turned the eye of a streak camera on individual bubbles levitated in water by a standing sound wave oscillating at 20,000 cycles per second. Both teams confirmed that the supersonic collapse of the bubble walls launches a shock wave at four times the speed of sound, but final temperature estimates have not been made. New experiments will directly measure the internal temperature of the bubble by tracking the thermal motion of electrons.

Significance

The process of converting sound energy into light energy requires that the sound energy be concentrated by a factor of approximately one trillion. In experimental observations, the wavelength of the resulting light is very short, with the spectrum extending well into the ultraviolet region. The bubble may even be radiating X rays. Shorter wavelengths of light mean higher frequency and energy. The observed spectrum of emitted light from the sonoluminescence process indicates a temperature in the bubble of at least 10,000 degrees Celsius (18,000 degrees Fahrenheit) and indicates that temperatures may rise as high as 100,000 degrees Celsius (180,000 degrees Fahrenheit) or greater. These high temperatures make sonoluminescence a possible candidate for thermonuclear fusion, a clean source of energy that could be used to generate electricity.

The proposed fusion of nuclei inside the sonoluminescence bubble is being termed "sonofusion." Because the temperatures and pressures may be at the threshold for fusing nuclei together, sonoluminescence may not generate very much power. Supercomputer simulations of sonoluminescence at the Lawrence Livermore National Laboratory in California have indicated that temperatures may be well over 100,000 degrees Celsius (180,000 degrees Fahrenheit), and pressures may be on the order of millions of atmospheres when the ultrasound is converted into light. If the temperature and pressure combinations could be raised by ten to one hundred times greater than the computer simulations indicate, then sonoluminescence could trigger fusion reactions in bubbles that contain deuterium or tritium atoms. Sonoluminescence in such bubbles has been recently demonstrated but not yet at the temperatures and pressures necessary for thermonuclear fusion to occur.

The sonoluminescence process also has some medical applications. Scientific evidence indicates that medical ultrasound devices can produce sonoluminescence. The resulting energy may produce therapeutic lesions in body tissues.

Electrical Superballs

➤ *The discovery of superconducting balls, a completely new phenomenon in physics, was made by a team of physicists at Southern Illinois University.*

In ordinary conductors of electricity, the electrons move in natural random, chaotic motion and collide with one another and with the stationary nuclei. When the collisions occur, electrons lose energy that they gained from the electric field produced by the power source. Kinetic energy is thereby transferred to the conductor as heat, subsequently reducing the amount of electric current. In contrast, for certain materials cooled to low enough temperatures, the electrons travel pathways so that the current does not decrease and no heat is generated in the material. These materials are called superconductors. Once electrical current is established in a superconductor, the current will persist indefinitely, even without an electric field.

Until 1986, it was thought that superconductivity worked only in certain materials at temperatures near absolute zero, or zero degrees Kelvin. Absolute zero is approximately –273 degrees Celsius (–459 degrees Fahrenheit). In 1986, superconductivity was achieved at 30 degrees Kelvin, and in 1987, it was accomplished in a compound of yttrium at 90 degrees Kelvin. Various ceramic oxides have since been found that behave as superconductors above 100 degrees Kelvin. These materials are referred to as high-temperature superconductors. Of the fifty or so high-temperature superconductors discovered so far, all are copper oxides. These materials exhibit many interesting properties that appear to be incompatible with conventional metal conductor and superconductor physics. The study of high-temperature superconductors expands ideas of what may be possible, compelling scientists to develop new theoretical concepts and experimental techniques.

Superconducting Balls

In nature, there are relatively few examples of granular particles aggregating together by themselves to form a round ball. Application of an electric field to the particles further reduces any such possibility. However, during

2000, Rongjia Tao and his colleagues at Southern Illinois University confirmed that a strong electric field can force the particles of a high-temperature superconductor into a remarkably sturdy bound ball about 0.25 millimeter (0.009 inch) across.

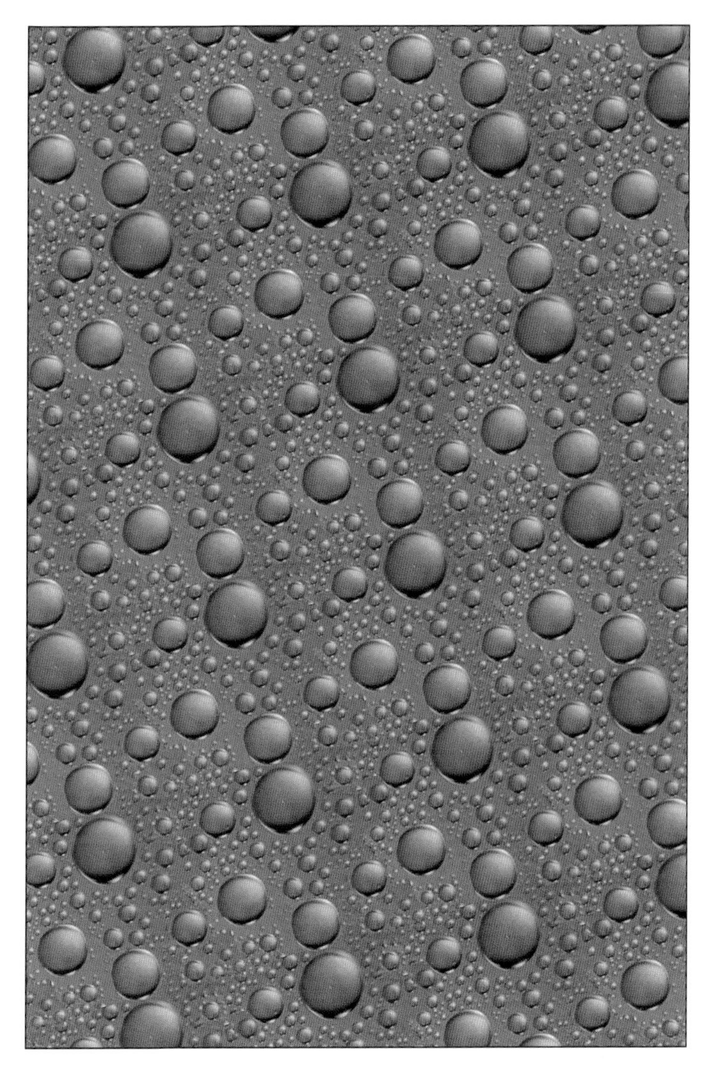

When starting the project in 1996, Tao and his associates were initially interested in investigating the motion of tiny superconducting particles of a copper oxide compound (bismuth-strontium-calcium-copper oxide) suspended in liquid nitrogen in an electric field. The boiling point of liquid nitrogen is 77 degrees Kelvin, which is below the transition temperature at which the copper oxide material becomes superconducting. Based on the principles of conventional physics, physicists expected the suspended particles to behave in one of two ways. The particles might bounce back and forth between the electrodes that established the guiding electric field, as bits of ordinary metal such as copper and aluminum do. Alternatively, the suspended particles might align in strings along the preferred direction established by the electric field, as some ceramic materials do. To the researchers' great surprise, several million particles packed themselves into a roughly spherical ball in a few milliseconds. The superconducting ball was stable enough to move as a single, electrically charged entity and survive many high-impact collisions with the electrodes. In some cases, two or more balls could be seen with a scanning electron microscope. It was obvious to Tao and his coworkers that some new, undiscovered physics must be involved.

In later experiments, Tao and his group suspended the superconducting particles in liquid argon. The boiling point of argon is 87.3 degrees Kelvin, which is just above the transition temperature at which the copper oxide becomes superconducting. As long as the temperature was

The National Science Foundation is researching the use of superconducting films to create petaflop computers, capable of performing one thousand-trillion floating point operations per second. Here, Paul Horn of IBM displays a mockup of a computer board the company plans to use in developing a petaflop computer. (AP/Wide World Photos)

below the transition temperature for the copper oxide compound, a macroscopic superconducting ball was formed in the liquid argon. The stable ball bounced back and forth between the electrodes.

When the temperature was raised above the superconductor's transition temperature, a high-speed camera caught the ball breaking into pieces just before colliding with an electrode. Tao and his group concluded that the formation of the ball must be a result of superconductivity. The spherical shape of the superconducting ball indicated that the force holding it together must be a new type of surface tension related to superconductivity. By self-assembling into a ball, the surface energy of the particle ensemble is minimized. This newly discovered surface energy is most likely related to the surface charges acquired on the superconducting particles, as well as the reactions between the layers forming the balls and the granular properties of the particles.

Significance

The discovery of superconducting balls is a completely new phenomenon in physics. Many physicists believe that it will lead to further insights into the causes of superconductivity. Tao and his collaborators are investigating possible practical uses of the properties of these balls. Applications include the fields of communication and energy transport.

It may be possible to take advantage of the newly discovered surface tension to lay down thin superconducting films on solid surfaces. The structure of these films is being studied in detail by Tao and his collaborators. Such superconducting films could be used to transmit electric fields that would make computers faster and communication more reliable. In particular, the National Science Foundation is researching petaflop computers capable of performing one thousand-trillion floating point operations per second. Today's fastest parallel computing operations have only reached teraflop speeds, or trillions of operations per second. It has been suggested that devices on the order of 50 billionths of a meter in size that will use unconventional switching mechanisms constructed with superconductors and superconducting films will be necessary to achieve these super speeds.

In the electronics industry, ultrahigh performance filters are now being built. By using superconductors and superconducting films as components of the filters, it may be possible to pass desired frequencies and block undesirable frequencies in cellular telephone systems. Superconductors and superconducting films may also play a role in Internet communications, with the development of superconducting digital routers for high-speed data communications. Because Internet data traffic is doubling every one hundred days or less, superconductor technology is being investigated to meet the ever-increasing data transmission rates. Another proposed application of superconducting films involves manufacturing ultrasensitive, ultrafast superconducting light detectors that can be adapted for use in telescopes because of their ability to detect a single photon of light. Other possible applications include use in superconducting electromagnetic systems that levitate high-speed trains for transportation systems, frictionless electric motors that rotate a million revolutions per minute while floating above a superconducting magnet, and superconducting storage facilities that reduce energy usage.

Superconducting films could be used to transmit electric fields that would make computers faster and communication more reliable.

Slowing Down Light

➤Research physicists observed the slowing of light to a record-breaking speed of only 1 mile (1.6 kilometers) per hour.

According to Albert Einstein's special theory of relativity, the speed of light in a vacuum has a constant value of 186,000 miles per second (299,330 kilometers per second), which is independent of the motion of the source and the observer. That is fast enough to go around the world seven times in the blink of an eye, or for light to travel 93 million miles (149 million kilometers) from the Sun to Earth in about 8.333 minutes. When light travels through a transparent medium, such as glass or water, it moves slightly more slowly than it does in a vacuum. As it moves from one medium into another, light slows as it bends, or refracts. Because light of different frequencies bends at different angles, white light is separated by a prism, or a raindrop, into its constituent colors, forming a rainbow.

Without refraction, lenses and eyeglasses would not work. The amount of refraction varies with the medium and is controlled by the medium's index of refraction. The speed of light in a medium is reduced by a factor equal to the index of refraction. When the index of refraction is 1.0, no bending occurs. The larger the value of the index of refraction, the more the light bends. In water, the index of refraction is 1.33, so light travels 25 percent slower in water than in a vacuum, which has an index of refraction of 1.0. The index of refraction of almost all known substances is only a little larger than 1.0, so in almost all cases, light is not significantly slowed down when traveling from one medium into another.

Bose-Einstein Condensate

Between 1998 and 2000, Lene Vestergaard Hau of Harvard University and some research colleagues at the Rowland Institute for Science in Cambridge, Massachusetts, succeeded in dramatically slowing down light by a factor of more than 20 million times. The key to their success was shining light on a new exotic form of matter known as a Bose-Einstein condensate. This is the same form of matter that was successfully employed in producing atom (or matter wave) lasers in 2000. The first Bose-Einstein condensate was formed in a laboratory in 1995. Hau and her associates cooled sodium atoms down to 50 billionths of a degree above absolute zero, one of the coldest temperatures ever achieved anywhere in a laboratory. Absolute zero is the lowest possible conceivable temperature, a temperature at which atoms and molecules stop their motions. It is about –273 degrees Celsius (–459 degrees Fahrenheit).

As the sodium atoms were cooled, they slowed down. Because all atoms obey Heisenberg's uncertainty principle, as the atoms slowed down, their momentum became more and more predictable, but as a result, their location in space became less certain. Eventually, the location became so uncertain that the sodium atoms began to overlap, acting as though they were one coupled object. Such a quantum coupled system is termed a Bose-Einstein condensate. When the quantum state of an atom is changed by absorbing energy, all the atoms change their state together because the atoms in a Bose-Einstein condensate act in unison. The index of refraction of a Bose-Einstein condensate is drastically higher than that of air.

In 1999, Hau reported that the speed of light had been slowed down to 38 miles per hour, the speed of a vehicle moving through a city or of a very energetic bicyclist.

Super Slow Light

In 1999, Hau reported that the speed of light had been slowed down to 38 miles per hour, the speed of a vehicle moving through a city or of a very energetic bicyclist. At that rate, it would take light almost three hundred years to travel from the Sun to Earth. Success was achieved by shining laser light of one particular frequency, or color, into a Bose-Einstein condensate produced from an ultracold gas of sodium atoms. First, a beam of yellow light was shined continuously at the sodium atoms. Because of the particular chosen frequency, the sodium atoms totally absorbed the yellow light.

Next, the researchers shined a second beam of laser light of a different frequency onto the extremely cold sodium atoms. This second beam altered the atomic properties of the sodium atoms by a process known as laser-induced transparency so that the original beam of yellow light was no longer totally absorbed. It was very important that the cloud be super cold for the second beam of laser light to interact effi-

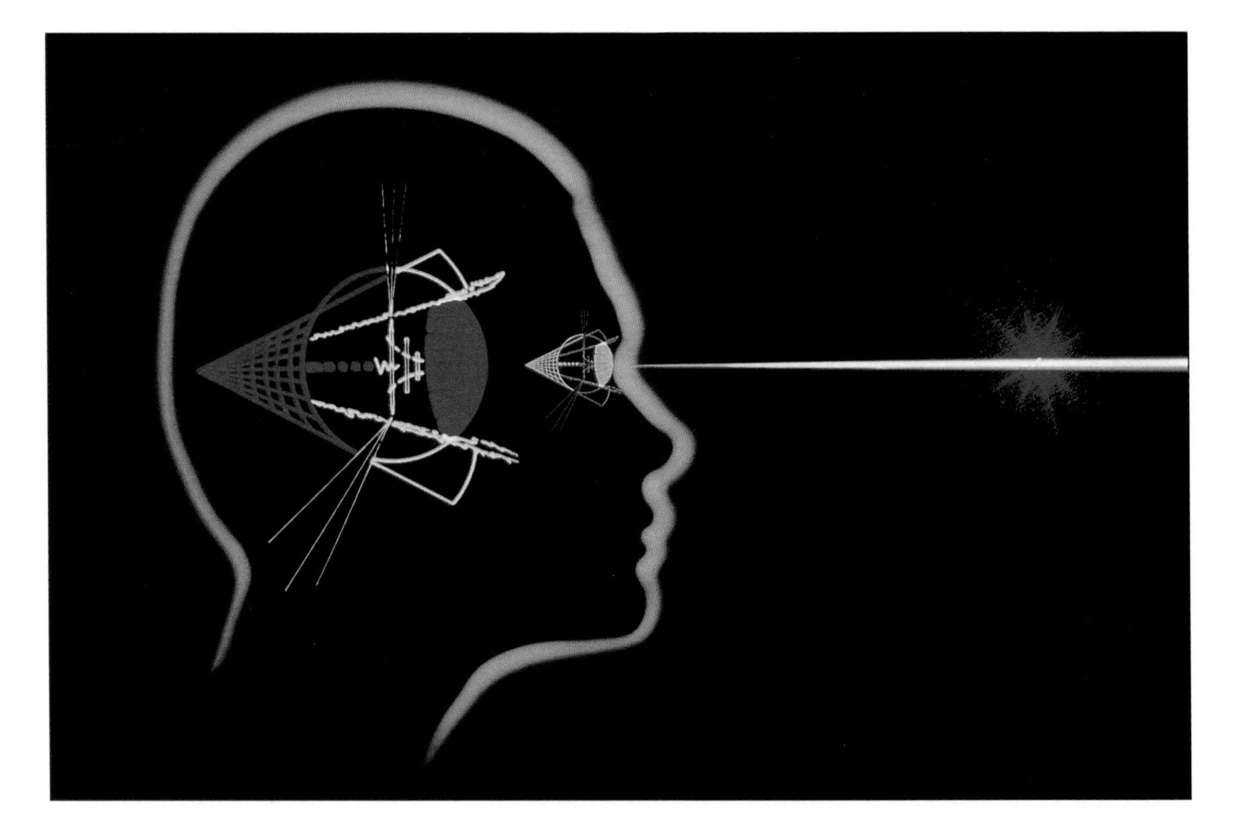

ciently with the sodium atoms. Under these conditions, the second laser beam cleared a pathway for some of the yellow light to make its way through the Bose-Einstein condensate. Even so, the sodium atoms still interacted strongly with the yellow light, slowing it down drastically without absorbing it. Approximately one-third of the yellow photons of light made it through the sodium atoms. Because the Bose-Einstein condensate state changed the index of refraction of the ultracold medium by 10 million times, exceptionally slow light speeds were observed.

After reducing the speed of light to 38 miles per hour, Hau reported in 2000 that she and her collaborators had succeeded in reducing the speed to 1 mile (1.6 kilometers) per hour by using a Bose-Einstein condensate. This established a new world record for the slowest light. (The record, however, was broken in January, 2001, by a group of researchers who were able to briefly stop a light beam.) The researchers also observed a large intensity-dependent transmission of the light, an unprecedented nonlinear optical effect. Hau's ultimate goal was to slow light to 1 centimeter (0.39 inch) per second or less. Her present experimental approach works for only one precise frequency of light. Other frequencies of light travel through the sodium atoms at nearly the speed of light in a vacuum.

Significance and Applications

Slowing down light to a realistic human dimension has produced an optical medium with very bizarre properties. If the cost of this process can be brought down sufficiently, many practical applications are possible. The nonlinear optical effect observed in the experiment is directly applicable in a number of optical-electronic components, including slow light optical switches, memory chips, and delay lines. It could also be used in converting light from one frequency to another.

Another application would be in the development of extremely fast computers based on slow light optical switches that would open and close under the control of very weak laser beams. Computer systems could be made super fast by using optically switched logic gates instead of the current technology of electronic logic gates. Slow light could also be used to filter noise from advanced optical communications systems. Other applications include low-power lasers, ultrasensitive night-vision glasses, laser light projectors that project very bright images, and improved television screens.

Resources for Students and Teachers

Books

Barbour, Julian. *The End of Time: The Next Revolution in Physics*. New York: Oxford University Press, 2000. Barbour presents evidence that time does not exist and explains what a timeless universe is like. His theories cast doubt on Albert Einstein's space-time continuum but work toward solving the chasm between classical and quantum physics.

Bernstein, Jeremy, Paul M. Fishbane, and Stephen Gasiorowicz. *Modern Physics*. Upper Saddle River, N.J.: Prentice Hall, 2000. A comprehensive treatment of general relativity and cosmology and their historical origins. Index.

Cutnell, John D., and Kenneth W. Johnson. *Phsycis*. 5th ed. New York: John Wiley & Sons, 2000. The authors of this college textbook offer clear explanations of physics concepts and real-world applications while encouraging sound reasoning skills.

Kirkpatrick, Larry D., and Gerald F. Wheeler. *Physics: A World View*. Fort Worth, Texas: Harcourt College Publishers, 2001. A physics textbook for nonscience majors at the college level that takes a conceptual approach to the subject.

Newton, David E. *Recent Advances and Issues in Physics*. Oryx Frontiers of Science Series. Phoenix: Oryx Press, 2000. Presents recent developments in basic and applied physics as well as areas of future interest. Describes social and legal factors affecting physics and provides biographical sketches of prominent physicists. Bibliography and index.

CD-ROMs and Videos

Physics: The Standard Deviants Core Curriculum. Video (10 vols.) and CD-ROM. Films for the Humanities and Sciences, 2000. Basic physics is the focus of this set of videotapes and accompanying CD-ROM. Topics covered include numbers, scalars, and vectors; circular, rotational, and projectile motion; Newton's laws of motion; friction, work, and energy; atomic structure; gravitation; harmonic motion and waves; heat; and thermodynamics.

PhysicsTutor. CD-ROM. Victory Multimedia, 2000. This interactive CD-ROM allows high school and college students to study physics, reinforcing material learned in class. It covers both quantitative and qualitative aspects of physics.

Reflections, Advice, and Diversions, or Falling Honey and Floating Logs. Video. University of California, Berkeley, Department of Physics, 2000. In a colloquium, University of California, Berkeley, physics professor J. D. Jackson describes flaws in physics education. He urges young physicists to be generalists who can explain all physical phenomenon.

Science Advantage 2000. CD-ROM. Encore Software, 2000. This set of five CD-ROMs is a self-paced study guide that focuses on physics, chemistry, and biology, designed with the Scholastic Achievement Tests in mind. Students can check their understanding of material learned through interactive study through practice problems.

Web Sites

About Physics
http://physics.about.com/science/physics/mbody.htm
Guide David Harris provides insights, discussions, and articles about physics concepts, as well as Internet links to current developments in physics. Topics addressed by Harris include atomic physics, chaos, condensed matter, electromagnetism, laser physics, nuclear physics, optics, particle physics, quantum physics, relativity, and thermodynamics. The latest news in physics can also be located at this site.

Glenbrook South Physics
http://www.glenbrook.k12.il.us/gbssci/phys/phys.html
Glenbrook South High School, Glenview, Illinois
This guide to high school physics contains a set of instructional pages written in clear, easy-to-understand language covering the topics in first-year high school physics courses. The "Multimedia Physics Studios" can be visited online to view animations and movies that illustrate the basic concepts of physics. The site provides an excellent prototype for physics teachers who want to develop their own physics sites on the Web.

Internet Pilot to Physics
http://physicsweb.org/TIPTOP
An international effort by scientists to disseminate information about physics research and education. It is aimed at promoting physics among scientists, industry, schools, students, and any interested individuals. It has an Internet search engine that focuses on physics sites and physics discussion forums.

Online Educational Resources for Physics Teachers
http://www.ba.infn.it/www/didattica.html
A plethora of useful links to physics information for teachers. Includes references to valuable instructional materials for teaching basic concepts in mechanics, heat and thermodynamics, electricity and magnetism, atomic and nuclear physics, quantum mechanics, engineering physics, chaos and fractals, and optics.

Physical Sciences Resource Center
http://www.psrc-online.org/newmain.html
American Association of Physics Teachers
Provides teaching resources for the entire spectrum of students from elementary school to college. Includes links to and examples of the latest teaching techniques in physics.

Physics Education
http://bubl.ac.uk/link/p/
Bulletin Board of Libraries
Provides a catalog of Internet resources in physics education. The site is indexed from A to Z and includes hyperlinks to Albert Einstein's contributions, how things in the world work, current developments in physics, interactive lessons in physics aimed at secondary school students, modern physics lecture notes, physics departments in the United States and worldwide, the physics of sports, and an encyclopedia of physics terms.

Physics News Update
http://www.aip.org/physnews/update/
American Institute of Physics
Provides a weekly digest of physics news items. Also describes current basic and applied research and development and the impact it has on society.

Physics Resources
http://www.ala.org/acrl/resmar00.html
Association of College and Research Libraries
Maintains an annotated list of key physics resources on the Internet. The sites identified here are some of the most useful physics sites on the Internet. Contains addresses for key sites exploring basic concepts and current developments in general physics, particle physics, relativity, quantum mechanics, physics research laboratories, electronic journals, and educational resources.

Physics 2000
http://www.colorado.edu/physics/2000/cover.html
University of Colorado at Boulder
Offers an introductory guide to physics and an interactive journey through current ideas of modern physics. Under the direction of physics professor Martin V. Goldman, this site furnishes visual and conceptual learning about twentieth century physics and high-tech devices. Explores the ideas and legacy left by Albert Einstein, including their application to the development of X-ray technology, microwave ovens, lasers, televisions, and computers. The basic principles underlying electromagnetic waves, quantum mechanics, the periodic table, Bose-Einstein condensates, the nucleus, and radioactivity are clearly illustrated.

PhysicsEd: Physics Education Resources
http://www-hpcc.astro.washington.edu/scied/physics.html
University of Washington
This site, maintained by Alan Cairns, provides information about education in physics, including research in physics education, physics education pages, physics textbooks, journals, and newsletters. Cotes many additional links to educational materials that illustrate basic concepts in physics.

Usenet Physics
http://math.ucr.edu/home/baez/physics/faq.html
Edited by Scott I. Chase and other physicists. Provides clear answers to frequently asked questions (FAQ) related to general physics, particle and nuclear physics, quantum physics, and relativity. Also includes discussions on accessing and using online physics resources, along with a links to other useful physics resources on the Web.

The Wonders of Physics
http://sprott.physics.wisc.edu/wop.html
University of Wisconsin at Madison
The center of the Wonders of Physics program, developed under the direction of physics professor Clint Sprott, is a fast-paced presentation of physics demonstrations carefully chosen for education and entertainment. A variety of educational materials, including printed handouts, videotapes, and computer software, have been developed to support the program. There are also useful lists of organizations and companies that supply physics demonstration equipment, audiovisual and computer materials, videotapes, and computer animations that teach the basic concepts of and current developments in physics.

8 · Applied Technology

The Year in Review

➤ *Bezaleel S. Benjamin*
Professor of Architecture and Architectural Engineering
University of Kansas

In the year 2000, applied technology made significant advances in a variety of fields, ranging from improvements in the design and function of automobiles and airplanes to a more in-depth understanding of the windchill and the weather. However, what has characterized the progress in most of these disciplines is the effect that the computer has had in enlarging, and sometimes exploding, scientific horizons for the benefit of humanity. What follows is less a summing up of the year's events than an examination of the technological advances in various disciplines in terms of the effects such progress has had on human knowledge, security, health and longevity, transportation, recreation, and the welfare of the disabled.

The year 2000 saw swirling controversy over the use of genetically altered (bioengineered) foods, but the advantages of genetically altered foods can hardly be minimized

The Human Aspect

Most of the scientific advances in human knowledge occur in the basic sciences rather than in applied technology. For these advances to have practical, useful applications to human society, they have to be researched to greater lengths with specific goals and objectives in mind, channeling knowledge for knowledge's sake into meaningful expression. One example can be found in nanotechnology. The basic premise that atoms could be manipulated to form any desired material—such as diamonds from coal—was first postulated in 1959. Forty-one years later, scientific breakthroughs in computer technology by IBM revealed that information can be transferred through solid molecules without the help of wires. These scientific advances will lead to the development of incredibly small and incredibly powerful nanocomputers and nanorobots that can clean up toxic waste dumps, solve the problems of hunger for Third World countries by turning grass clippings into bread, eliminate diseases by fighting bacteria and viruses within the bloodstream, enhance the delivery of drugs, and in the very distant future, with self-replication, build automobiles, roads, and skyscrapers. The enhancements made in the fields of nanocomputer technology will have a significant and lasting effect on the future of the planet.

Genetically altered foods were the subject of much debate in 2000. The associate director of the Center for Engineering Plants and Resistance Against Pathogens at the University of California at Davis displays genetically altered grape plants (left) and tomatoes. (AP/Wide World Photos)

None of the advances dreamed of with nanotechnology, however, will even have a chance to profit the human race without the security that it needs against rogue nations and governments, and the year 2000 saw some small progress as well as minor setbacks toward achieving that goal. U.S. president Bill Clinton delayed a decision on the National Missile Defense system following an unsuccessful test of the system's ability to shoot down a target missile. However, considerable scientific progress was made during the year in the defense application of computers to ensure the safety of both large-scale operations and the individual soldier.

Most of the advances in health, such as the development of new pharmaceutical drugs, occur in fields other than in applied technology. However, drugs are a bandage approach to the problem of disease in that they are applied after the illness has occurred. What most affects human health and longevity and quite possibly is a major cause of some diseases is the food people eat. The year 2000 saw swirling controversy over the use of genetically altered (bioengineered) foods. The advantages of genetically al-

tered foods can hardly be minimized: Genetically altered corn or wheat that resists spoilage or can withstand pesticides can provide poor countries with the sustenance they need for survival. What is as yet undetermined are the long-term effects, particularly regarding allergies, of eating such foods, and governments are taking a closer look at genetically altered foods. In the United States, for example, attention was drawn to bioengineered foods when genetically altered corn that had not been certified for human consumption was used to make taco shells. When the mistake was discovered, the taco shells were recalled by the manufacturer.

Transportation

Considerable developments took place in 2000 in three forms of transportation—automobiles, trains, and airplanes—and much of that development was fueled by the sharp increase in gas prices as a result of decisions made by the Organization of Petroleum Exporting Countries (OPEC). The most significant advance in the automobile industry was the resurgence of interest in the electric vehicle (EV). Two major car manufacturers have already put electric hybrids—powered by both batteries and small three-cylinder gas engines—on the market for use as general purpose cars (not just for city driving), and the other three major car companies have committed themselves to those same goals.

In 2000, Amtrak inaugurated Acela, its high-speed train with a speed of 150 miles per hour (240 kilometers per hour) in regional service linking Boston, New York, and Washington, D.C. The provision of such high-speed trains in corridors of heavy traffic, such as from San Diego to Los Angeles and San Francisco or from Chicago to St. Louis, has become economically viable.

The unfortunate loss of Alaska Airlines Flight 261 on January 31, 2000, off the coast of California as well as other plane crashes has prompted a closer look at the black boxes, or voice and flight data recorders. The United States employs an "autopsy" approach to flight safety in that these recorders are used to analyze what went wrong, and the knowledge is used to ensure the safety of future flights. In Europe, the "telemetry" approach, in which inboard recorders transmit the data directly to satellites as the flight is in progress, is finding increasing acceptance. The telemetry approach is capable of monitoring the performance of the vital components of the plane while in flight and providing this information to the pilot if action is necessary.

Although most significant government or university research has focused on transportation, not recreational travel, the potential for profit has motivated private interests to concentrate on satisfying people's recreational needs. Therefore, all the progress on providing a hotel in space has arisen out of private and commercial interests in the project. In 2000, the life of the Russian Mir space station was prolonged, with a view to using the station as a hotel in space, although by the end of the year, Russians planned for Mir's demise in February, 2001. Other projects for recreational travel into space have also appeared on the drawing board.

Technology for Humanity

In the final analysis, the humanity of a society can be judged by the care and attention it gives to its disabled, whether visually, hearing, or mobility impaired. The advances in computers and electronic chip technology made considerable progress in helping the disabled overcome their physical limitations. In 2000, most of this research was on the development of specific electronic guidance devices, which when wired into the brain, could promote sight, hearing, or muscle movement. One interesting development was the suggestion that a volunteer organization could be set up, via the Internet, to provide instant assistance to visually or hearing-impaired people through a social network of volunteer workers, given permission to offer such volunteer services even while at their place of work.

Nanotechnology

Molecular Computing Becomes a Reality

➤ *IBM physicists announced a breakthrough that would bring atomic-level circuitry into molecular computing. This nanotechnology will ultimately provide the power of a supercomputer in a size so small that it could be woven into garments and powered by body heat.*

Nanotechnology, by commonly accepted definition, is the creation of functional materials, devices, and systems through control of matter on the nanometer length scale and the exploitation of novel phenomena and properties (physical, chemical, biological) at that scale. A nanometer is one-billionth of a meter. By comparison, a nanometer is ten thousand times smaller than the diameter of a human hair. A typical virus is about one hundred nanometers in size, which is about the size at which nanotechnology functions.

All matter is made up of atoms. It is only in the arrangement of atoms that one material differs from another. For instance, coal and diamonds are both made from carbon atoms. If the atoms could be rearranged at the atomic level, it would be possible to make diamonds from coal. The possibilities are endless. It would be possible, for example, to rearrange the atoms in grass clippings to make bread, eliminating hunger in even the poorest of countries across the globe.

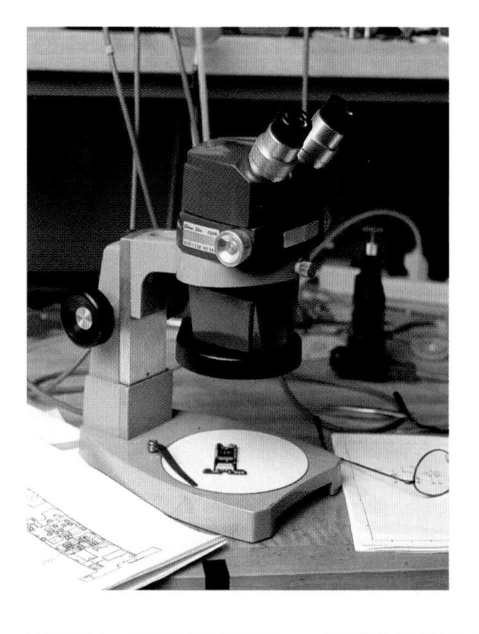

Nanorobots

The idea was first put forth by physicist Richard Feynman in 1959 in his lecture titled "There's Plenty of Room at the Bottom," in which he suggested that it would one day be possible to build machines so tiny that they would consist of only a few thousand atoms. These nanomachines or nanorobots could perform the primary task of rearranging the atoms to create the desired material. The idea gained popular appeal in the 1986 book *Engines of Creation* by K. Eric Drexler.

The idea has already been in use in biological structures for millions of years. Every cell is a superb, living example of nanotechnology. Each cell has a specific function. It is programmed to convert fuel into energy; can create proteins, enzymes, or hormones by rearranging the atomic structure of the foods that are eaten; and follows preprogrammed instructions encoded in the deoxyribonucleic acid (DNA). If the instructions require, the cell will even replicate itself. These would be precisely the same functions that nanorobots would perform. In biological construction, the living cells place layers upon layers of functional cells to eventually form huge structures such as sperm whales or giant Sequoia trees. This is bottom-up construction, as opposed to the prevailing form of construction in which large, fully formed clumps of steel, concrete, plastic, wood, or masonry are used as the basis for forming structures, roads, or automobiles.

However, to actually construct anything using nanotechnology, formidable problems must be overcome, the first of which is the creation of the nanorobot "finger." These fingers would be necessary to enable the nanorobots to manipulate tiny atoms and whole molecules. Research carried out in 1991 discovered hairlike carbon molecules called "nanotubes" that are one hundred times stronger than steel and ten thousand times thinner than a human hair. Nanotubes, which are rolled-up sheets of carbon hexagons, could serve as fingers to afford the nanorobots the dexterity necessary for construction activity.

The task is formidable. The nanorobots are incredibly small in size, yet trillions of atoms and molecules must be rearranged in a precise, meaningful way for activities such as cleaning up toxic waste dumps or building a roadway. This problem, nanotechnologists believe, can be solved by self-replication. Each nanorobot would be programmed to prepare two copies of itself. Through replication, trillions of nanorobots could be easily generated to perform the task. In addition, each nanorobot would need to be programmed in its duties. In the living cell, this programming occurs in the DNA of the cell. In the case of nanorobots, the instructions would be computerized. One danger is that these nanorobots might replicate until they are totally out of control, overwhelming the planet; however, nanotechnologists believe that nanorobots can be programmed to stop replication after a certain number of cycles or when too many of their fellow creatures are nearby. It is the strategy nature uses to keep bacteria in check. The prospect of nanorobot war between nations is another fearful scenario that clouds the horizon of nanotechnology.

Research carried out in 1991 discovered hairlike carbon molecules called "nanotubes" that are one hundred times stronger than steel and ten thousand times thinner than a human hair.

Promising Research

In 2000, the IBM research team showed that molecular atomic-level circuitry was a real possibility. Their research, dubbed a "quantum mirage," demonstrated that information can travel through solid substances without the benefit of wires. This would be a giant leap in the design of microprocessors, which at present require substantial electrical power to operate. Nanotechnology processing would not only reduce the power requirements of the microprocessor but also make it incredibly small. Such a molecular supercomputer could be injected into a patient's bloodstream as a diagnostic probe or provide the program information for each nanorobot. In medical applications, such nanorobots, injected into the body by the billions, could chip plaque from arteries, destroy bacteria and viruses, repair broken blood vessels, and even replace components in the human body to retard the aging process.

No one expects startling results from nanotechnology for many years, even decades, to come. However, some researchers are not convinced that research will produce even any meaningful, successful practical applications of nanotechnology any time soon. Steven Block, biophysicist at Stanford University, speaking to scientists and academics at a conference convened by the National Institutes of Health (NIH), said that one of the problems was that nanotechnology enthusiasts are typically less well informed about biology. Before biologists or nanotechnologists can manipulate or replace nature's machines, they will have to determine how they work. Despite some scientists' pessimism, U.S. president Bill Clinton earmarked about $500 million in funding for the National Nanotechnology Initiative (NNI) for research projects. Chad Mirkin, acting director of the Center for Nanofabrication and Molecular Assembly at Northwestern University, has said that nanotechnology has already had an impact in the field of diagnostic medicine in tests for colon cancer and in the field of drug delivery. The government-sponsored Interagency Working Group on Nanotechnology (IWGN) report said that the impact of nanotechnology on the health, wealth, and security of the world's people during the twenty-first century would rival that of antibiotics, the integrated circuit, and manmade polymers in the twentieth century.

In medical applications, nanorobots, injected into the body by the billions, could chip plaque from arteries, destroy bacteria and viruses, repair broken blood vessels, and even replace components in the human body to retard the aging process.

What Are We Eating?
Genetically Altered Foods

➤ *The United Nations produces an agreement on international trade in genetically altered foods and products; foods labeled "organic" are prohibited from containing genetically altered components.*

The food people eat has always been a product of the environment. In evolutionary terms, any changes that have taken place in the environment have resulted in genetic modifications in plants and animals. These genetic alterations have been accepted as natural, partly because they have been very slow and partly because they have not been the direct result of human intervention. With advances in genetics, however, for several years it has been possible to alter plants and animals at the genetic level, leading to genetically altered (GA) foods. Such foods are also sometimes called genetically modified foods, bioengineered foods, and even "Frankenstein" foods or "Franken" foods.

Nature has practiced genetic bioengineering for millions of years, and the process of natural selection has slowly made specific alterations more prominent in the following generations. Now, however, agricultural geneticists can speed up the process so that such alterations take place rapidly, with subsequent generations of plants and animals instantly showing the results of such changes. When nature took millions of years to effect the changes, the rest of the environment had those same millions of years to adjust to the genetic changes that were gradually taking place. However, the speed at which the new genetically altered foods have been developed has not provided the plants and animals in the environment with the time to make any necessary biological adaptations to the genetic changes.

Agricultural scientists can deliberately alter the deoxyribonucleic acid (DNA) structure of a plant or animal through the introduction of foreign genes in an effort to make the plant or animal better in some way. A plant, for example, can be made hardier against extreme cold or heat or more resistant to pests or disease; it can also be altered to produce a more nutritious grain or a longer-keeping fruit. Scientists isolate the gene affecting the trait and replace it with a gene with more desirable traits, thereby changing the characteristics of the plant. Future generations of the same plant then carry the more desirable gene and exhibit the more desirable trait. Many genetic alterations have been carried out in plants and affect the foods that people eat.

Altered Foods

Genetic alteration has affected numerous grains, including corn, which is sold in altered form in the U.S. market. About 35 percent of all corn, for both animal feed and human consumption, has been genetically altered by a bacterial gene that produces a toxic protein pesticide, *Bacillus thuringiensis*, designed to kill the corn borer and other organisms that damage the corn crop. In several countries, research on rice has produced genetically altered varieties with different specific traits. In Japan, genetically altered rice contains three times more dietary iron than conventional rice. The high-iron rice is made by inserting a soybean gene that produces a protein called ferritin into the rice DNA. Swiss and German researchers have inserted genes from daffodils and bacterium into rice to produce a variety with a higher vitamin A content. Soybeans are spliced with genes that make them resistant to herbicides used to kill weeds.

Genetic alteration has also been carried out with vegetables and fruits. Genetically altered potatoes contain a built-in bacterial toxin that prevents insect damage, eliminating the need for pesticides. Tomatoes have been genetically altered to ripen slowly so that they do not spoil by the time they reach the market. Squash and papaya have been made more resistant to viruses.

Beverages are also being affected by genetic alteration. Milk and dairy products can indirectly be classified as being genetically altered if they come from herds that have been given bovine growth hormone, a genetically engineered growth hormone that increases yields. In addition, a recent discovery by two research scientists in Japan and one in Scotland may lead to the production of caffeine-free tea and coffee through genetic alteration. The gene isolated by the researchers makes the enzyme that tea and coffee plants need to finish the last two steps in caf-

Danger in a Soybean

In the mid-1990's, agricultural seed company Pioneer Hi-Bred introduced a Brazil nut gene into soybeans in order to improve the protein content. Before putting the soybean on the market, however, the company turned to scientist Steve Taylor at the University of Nebraska to determine whether the new soybean would trigger allergic reactions in people allergic to nuts. Such allergies can cause reactions ranging from diarrhea and rhinitis to asthma and anaphylactic shock, which are potentially fatal. In Taylor's study, reported in *The New England Journal of Medicine* in 1996, he and his team of researchers found that the transferred protein in the new soybean caused allergic reactions in people. Pioneer immediately canceled the project. Although the soybean never reached the market, Taylor's findings indicated some of the risk associated with genetically altered foods.

feine production. If researchers find a way to remove the gene or prevent it from becoming active, the tea or coffee plants would not make the caffeine molecules, and the beverages created from them would naturally be free of caffeine.

These genetically altered foods or their derivatives are used in a vast variety of other edible products that the consumer would not regard as being genetically altered. For example, genetically altered corn is used in the manufacture of chips, cookies, candies, gum, bread, cereals, pickles, margarine, alcohol, enriched flours and pastas, salad dressings, and vanilla—all of which people are unlikely to link with genetic alteration.

Problems and Potential

Although most European countries have been quick to realize the dangers that genetically altered foods pose for human, animal, and insect populations, the United States has been slow to react. The dangers created by uncontrolled genetic alterations of plants and foods that find their way into the environment—or onto the dining table—cannot be ignored, particularly when tests on laboratory animals given such genetically altered foods show damaging effects.

In September, Kraft Foods recalled all the taco shells it sold in supermarkets under the Taco Bell brand after it was determined that they were made with genetically altered corn that had not been approved for human consumption. (AP/Wide World Photos)

A 1999 Cornell study linked genetically altered corn to the deaths of 20 percent of the larvae of the Monarch butterfly. The butterfly larvae were poisoned by feeding on milkweed leaves covered with pollen from the genetically altered corn. In October, 2000, the Environmental Protection Agency reviewed the scientific data from several studies on the butterflies and bioengineered corn and concluded that the corn posed no significant danger to the Monarchs. Another study showed that genetically altered soybeans have 12 percent to 14 percent less phytoestrogens—naturally occurring substances that help protect against cancer and heart disease—than normal soybeans. Laboratory studies in Scotland showed that feeding genetically altered potatoes to rats resulted in stunted growth and damage to major organs, including the kidneys and the stomach. There is growing concern that moving new genes into foods could transfer harmful allergens into those foods. In medical studies, the use of bovine growth hormone has been linked to cancer, so the European Union and, more recently, Canada have banned its use.

Although genetically altered foods can cause problems, they have great potential for doing good. Much of the uncertainty arises because of the lack of knowledge of the long-term effects of genetically altered foods. For example, rice that contains additional vitamin A could protect millions of children in poor countries from blindness. Speaking of genetically altered foods, Lester Crawford, director of the Center for Food and Nutrition Policy at Georgetown University, noted that many new technologies—including pasteurization, electricity, and fluoridated water—faced early, stiff opposition. "This is a war, not only of words but for the minds of the world population," he said. There is general agreement, however, that more research, along with correct labeling to inform the consumer of the precise nature of the food being eaten, is very necessary.

Labels as Protection

In 2000, United Nations talks produced an agreement on trade in genetically altered foods and products. It requires exporters to label shipments that contain genetically altered products such as corn or cotton. The United States was initially opposed to such restrictions but agreed to them. The rules are intended to protect the environment from damage caused by genetically modified organisms.

New rules written for organic foods will bar the use of genetically altered ingredients in foods labeled "organic." Steve Gilman, an organic farmer in Stillwater, New York, who supports the new labels, criticized the prevalence of genetically altered foodstuffs, saying,

> There's fish genes in fruit, poultry genes in fish, animal genes in plants, growth hormones in milk, insect genes in vegetables, tree genes in grain, and in the case of pork, human genes in meat.

Enjoying the View
A Hotel in Space?

➤*A hotel in space becomes a very real possibility, with several commercial plans for extending tourism to the final frontier being formulated.*

As the National Aeronautics and Space Administration (NASA) assembles the International Space Station high above Earth, many companies and hotel chains are seriously considering the future of space vacations. Hilton Hotels confirmed it was looking into the feasibility of a space hotel, and the company was far from being alone. Robert Bigelow, owner of the Las Vegas-based Budget Suites of America, has already committed $500 million in a new company, Bigelow Aerospace, with the goal of shuttling passengers to a luxury spacecraft permanently orbiting Earth.

The technical problems of operating a hotel in space are formidable. Jeanne Datz of Hilton Hotels said that although a space hotel could be built today, there were many unanswered questions regarding what people would do and eat once they were at the hotel. However, many of the answers are becoming available, and many technological experts are convinced that building and operating a space hotel would be just another challenge.

Building a Space Hotel

Two alternative plans for creating a hotel in space are being investigated. One would be to use an existing, abandoned space station and refurbish it as a hotel. The only vessel on the market is the Russian space station, Mir. The Russian government had all but given up on Mir, and Energia, the state-run enterprise that built Mir, had planned to let the space station destruct on reentering Earth's atmosphere. However, private investors led by U.S. telecommunications magnate Walt Anderson provided $20 million to form a new company, MirCorp, a partnership between the investors and Energia. Two cosmonauts have since reentered Mir with a view to making it habitable again. Mir will be used for a variety of commercial ventures, including space advertising, Internet camera hookups, and, perhaps, a hotel. The cost of operating Mir is estimated at $100 million a year. MirCorp and Energia signed an agreement to keep Mir active through the end of 2000.

In November, 2000, however, the Russian Space Agency announced its plan to take Mir out of service at the end of February, 2001, by sending it back through the atmosphere in a controlled fall into the Pacific Ocean. As a result, at the end of the year, MirCorp reportedly was considering focusing its commerical efforts on a module that would attach to the International Space Station.

Wimberly Allison Tong and Goo, a leading resort design firm, designed this orbiting one-hundred-room hotel, which incorporates empty space shuttle fuel tanks. (AP/Wide World Photos)

The other, more distant alternative is to specially design and build a space hotel. Bigelow plans a $2 billion hotel that will house one hundred guests and have a staff of fifty. Wimberly Allison Tong and Goo, one of the nation's leading resort design firms, has drawn up plans for a one-hundred-room hotel in space, complete with restaurant, observation deck, and recreation facilities. The design and construction of NASA's International Space Station has solved many of the problems associated with the building of hotels in space. Massive solar panels supply the electricity to power the space station, docking ports for the shuttle provide a link to Earth, and satellite dishes communicate constantly with the command center. Nevertheless, the problems experienced by space shuttles are magnified when viewed in the context of a hotel. One example is waste disposal. Astronauts may have no alternative but to recycle their urine to extract water for drinking purposes; hotel guests are likely to feel extremely squeamish about drinking the same recycled product.

Operating a Hotel in Space

The problems of operating a hotel in space are formidable, and nothing like it has ever been attempted, not even by NASA. Solving problems faced by trained military astronauts in space is different from catering to the needs and comforts of high-paying guests, each of whom may well have paid up to $1 million for their space vacation. Although astronauts endure space weightlessness for long periods of time, hotel guests, used to gravity, would need at least some measure of anchorage to the surface they are treading. Therefore, the space station would need to create a partial artificial gravity by a slow rotation of the hotel module. Centrifugal force would create an artificial gravity that would provide the comfort of weight to hotel guests. The use of magnetic strips on flooring and special magnetic shoes has been suggested as another alternative to provide some walking anchorage.

Many innovative ideas have been floated for recreational facilities in space, including golf ranges, at which putting would be carried out under zero-gravity conditions.

Then there is the question of food. Astronauts will eat whatever NASA provides and in any form provided—even as tablets or out of plastic bags. Guests may well expect the same gastronomical pleasures in space as they are accustomed to receiving in expensive hotels on Earth. Although this is not impossible, the transportation of supplies to the hotel would become a major operating cost of the venture.

NASA provides exercise facilities for astronauts in order to maintain their bone density, which deteriorates under long periods of weightlessness. Guests, however, would expect much more than just what is needed for good health, however important that might be. An observation deck for enjoying the views from space and a space-walking platform from which guests can experience the outside environment are a must. In addition, many innovative ideas have been floated for recreational facilities in space, including golf ranges, at which putting would be carried out under zero-gravity conditions. Perhaps the most exciting recreational activity for space hotel guests could be provided through the installation of both visual and radio telescopes. The visual telescopes would be used to probe the heavens, free of

Earth's atmosphere; the radio telescopes could be constantly monitored for signs of alien life.

Providing for the care and comfort of guests isolated in a space hotel is far more demanding than maintaining the rigorous military conditions under which astronauts function. Space hotel guests are likely to be of every adult age group and in neither the physical health nor mental condition of military astronauts. In addition to cooks, waiters, and housekeeping and managerial staff, the rotating service staff would need a doctor, a hair stylist, a physical trainer for the exercise room, and even possibly a librarian.

A vacation at a space hotel would be different from one at a resort hotel on Earth or even on a luxury cruise ship in that a guest cannot check out and leave the hotel at will. The guest is virtually a prisoner until the next shuttle flight leaves for Earth. Keeping the guests satisfied and comfortable is then of even greater importance to the management of the hotel.

Space Adventures at Present

Although a space hotel has become a real possibility, a grand opening is still some years away. However, space adventures are already available for the hardy—and the wealthy. Since 1997, Space Adventures of Arlington, Virginia, has been selling seats on the Russian space agency's IL-76, a jumbo jet stationed near Moscow. The jet flies in a parabola to create about thirty-second bursts of weightlessness, at a cost of about $5,000 per person, not including airfare to Russia.

Space Adventures also offers flights to the edge of space. Russian MiG-25 military jets, operating out of Russia's Zhukovsky Air Field, take passengers to a height of 85,000 feet (26 kilometers). From that altitude, it is possible to see the curvature of Earth.

The Windchill Index
A Cold Appraisal

➤ *The National Weather Service announced that it would undertake a review of the windchill index, which it adopted in 1973.*

In the winter of 1941, two Antarctic explorers, Paul Siple and Charles Passel, measured the time it took for a pan of water to freeze and found that the rate at which heat was lost from the pan depended not only on the temperature of the air but also on the wind speed. The wind helped "carry away" the heat faster. Army researchers, trying to design clothing that would keep soldiers warm in the extreme cold, then devised the basis for the present windchill index by using water-filled plastic cylinders with the wind speed measured at about 10 meters (30 feet) above ground level. This led to an empirical formula that determined the windchill index (WCI) by relating the temperature T (in degrees Fahrenheit) and the wind velocity V (in miles per hour):

$$WCI = 91.4 - (0.475 - 0.020V + 0.303\sqrt{V})(91.4 - T)$$

This rather simplistic formula led to the windchill index table developed by the National Oceanic and Atmospheric Administration, used to determine the windchill index for any given temperature and wind speed. A second table defines the range of coldness from "cold" to "extreme cold," based on the windchill index.

Note that although the windchill index is expressed in the same format as temperature, it is not a temperature. Water, for instance, will not freeze if the temperature is above freezing, regardless of how hard the wind is blowing. The windchill index may be well below freezing, but water will not freeze. The windchill index is, at best, a measure of the rate at which heat is lost from an object because of the presence of wind. In estimating heat loss from the human body, the index fails to take into account a number of variables. The level of comfort or discomfort felt by people depends on whether it is sunny or cloudy, whether the air is damp or dry, whether the person is exercising or resting, and how warmly the person is dressed.

Maurice Bluestein, a mechanical engineer who is very knowledgeable about heat transfer, is one of the harshest

COLDNESS RANGE

Windchill Index (°F)	Terminology
>15 to ≤ 32	Cold
>0 to ≤ 15	Very cold
>-20 to ≤ 0	Bitter cold
<-20	Extreme cold

critics of the current windchill index table. Bluestein, together with his colleague Jack Zecher, has devised a table that more accurately reflects the heat lost from the human body taking into account the thermal properties of skin and modern heat transfer theory. For any given wind speed and air temperature, Bluestein and Zecher's table gives a higher (warmer) windchill index than that produced by the windchill index table used by the weather service.

WINDCHILL INDEX

Wind Speed (mph)	Air Temperature (degrees Fahrenheit)																
	35	30	25	20	15	10	5	0	−5	−10	−15	−20	−25	−30	−35	−40	−45
4	35	30	25	20	15	10	5	0	−5	−10	−15	−20	−25	−30	−35	−40	−45
5	32	27	22	16	11	6	0	−5	−10	−15	−21	−26	−31	−36	−42	−47	−52
10	22	16	10	3	−3	−9	−15	−22	−27	−34	−40	−46	−52	−58	−64	−71	−77
15	16	9	2	−5	−11	−18	−25	−31	−38	−45	−51	−58	−65	−72	−78	−85	−92
20	12	4	−3	−10	−17	−24	−31	−39	−46	−53	−60	−67	−74	−81	−88	−95	*
25	8	1	−7	−15	−22	−29	−36	−44	−51	−59	−66	−74	−81	−88	−96	*	*
30	6	−2	−10	−18	−25	−33	−41	−49	−56	−64	−71	−79	−86	−93	*	*	*
35	4	−4	−12	−20	−27	−35	−43	−52	−58	−67	−74	−82	−89	−97	*	*	*
40	3	−5	−13	−21	−29	−37	−45	−53	−60	−69	−76	−84	−92	*	*	*	*

An Appraisal of the Windchill Index

The windchill index was adopted by the National Weather Service in 1973. For the past decade, it has been criticized by scientists for overstating how cold people really feel. However, the National Weather Service has no plans to abandon it, though a review is planned. Mike Matthews, a spokesperson for the Office of Meteorology in Silver Spring, Maryland, said that the weather service would continue using the windchill index until the scientific community could develop a formula that could be shown to be better.

At the heart of this resistance to change is human safety. Phil Johnson, a meteorologist for KDLH-TV in Duluth, Minnesota, says that the windchill index may not be perfect, but it is valuable for people in his city, where the mercury can fall to 40 degrees below zero Fahrenheit (40 degrees below zero Celsius) and the windchill commonly plunges to –60 or –70. Although the windchill equivalents might not be exact, Johnson believes they help people understand that they have to take extra precautions when they are outdoors.

Maurice Bluestein, engineering professor at Purdue University, has asked the National Weather Service to reassess the windchill index, which he believes is inaccurate. (AP/Wide World Photos)

Others disagree. Edwin Kessler, retired director of the National Severe Storms Laboratory, in Norman, Oklahoma, has long been skeptical of the accuracy and the value of the windchill index. He believes that although travelers need to take precautions in extreme cold, most people are actually outdoors for short durations. Kessler said there is a tremendous difference between a person dashing out to retrieve the morning newspaper and a soldier on sentry duty for eight hours in 20-degree-below-zero weather. Kessler agrees with Bluestein, who believes that overstating the cold—as the present windchill index table does—does the public a disservice by convincing people that they can safely endure extreme cold. In addition, crying wolf with exaggerated windchill factors can cause people to not take the threat of cold seriously.

In the face of such criticism, Canada switched to a dif-

A COMPARISON OF CURRENT AND PROPOSED WINDCHILL INDICES
(air temperature = 10 degrees Fahrenheit)

Wind speed (mph)	Current windchill index	Proposed windchill index
5	6	10
10	–9	0
15	–18	–8
20	–24	–14
25	–29	–19
30	–33	–23
35	–35	–26
40	–37	–29

ferent system with an index in watts per square meter, but the public refuses to accept it. Bruce Paruk, a Canadian government meteorologist, said that the public wants windchill equivalents instead, and Canada is considering going back to the old windchill index, but using the newer models.

Cars of the Future

New Power Sources Fight Gas Hikes, Pollution

➤ *Car manufacturers, both domestic and foreign, announced their commitment to developing the electric vehicle, in some form, to reduce reliance on fossil fuels and eliminate pollution, leading to a cleaner environment.*

At the Woodburn, Oregon, Electric Drags in August, 1997, a Dodge V10 Viper was soundly beaten by a homemade electric dragster. The dragster went from 0 to 145 kilometers per hour (0 to 90 miles per hour) in 12 seconds. Every year since then, that record has been improved, with electric vehicle (EV) dragsters now achieving 0 to 200 kilometers per hour (0 to 125 miles per hour) in 10.5 seconds, thus exploding the myth that EVs cannot go fast. Speed is not the issue in the debate about the practical use of the EV, but range is sacrificed if the EV is designed only with speed in mind.

The EV has an electric motor driven by power from fuel cells, commonly known as electric batteries, and what makes the EV so promising is the breakthrough progress in recent years in fuel cell technology. A battery essentially works through a process of reverse electrolysis. In electrolysis, the application of an electric current can break water into its original elements of hydrogen and oxygen. In reverse electrolysis, the battery has an anode and a cathode with an electrolyte between them. During the electricity-producing reaction, the hydrogen atoms release electrons at the anode and enter the electrolyte as hydrogen ions. These electrons travel through the external circuit (powering the EV in the process) to reach the cathode—if the circuit is complete. At the cathode, the hydrogen ions and the electrons combine with the oxygen to form water again, giving off heat in the process.

The lead-acid battery, commonly used in cars, does not produce much energy for its weight, in comparison with gasoline, and has to be continuously recharged by the gas-operated engine to maintain sufficient charge. However, considerable progress has been made in the development of new nickel cadmium, nickel metal-hydride (NiMH), and lithium-ion batteries. Using NiMH batteries, a jellybean-shaped Sunrise sedan was driven 340 kilometers (211 miles) from Boston to New York City on a single charge. The Sunrise was built mainly from composite materials to make it ultralight and cost

$100,000. In most commercially manufactured cars, the range would be considerably less than that but still very practical. John Wallace of the Ford Motor Company's Alternative-Fuel Vehicle programs believes that a real-world range of 160 kilometers (100 miles) is the upper limit "for a very long time." The development of hydrogen fuel cells and flywheel energy storage are exciting prospects for the future.

The Honda Insight, a hybrid car, gets 61 miles per gallon in city driving and 70 miles per gallon on the highway. (AP/ Wide World Photos)

Range and Maintenance

Using conventional, cost-effective, lead-acid batteries, it is realistically possible to get between 40 kilometers (25 miles) and 80 kilometers (50 miles) in the city on a single charge. The range depends on the voltage of the batteries used. With relatively light 12-volt batteries, a range of 40 kilometers (25 miles) is easily possible. With 6-volt batteries, a range of 80 kilometers (50 miles) is possible. However, the EV needs multiple batteries, which results in a heavier car. This affects hill-climbing ability and acceleration, though severe hills and freeway speeds are still within the performance capabilities of the EV. These between-charge ranges may appear to be small, but consider the driving habits of most Americans. It has been estimated that 70 percent of all Americans drive less than 40 kilometers (25 miles) in a day, and 90 percent drive less than 80 kilometers (50 miles) in a single day. Half of all urban trips are less than

8 kilometers (5 miles), with the car on the road less than ten minutes on each trip. Converted into real time, therefore, the range of an EV in city driving is about 60 minutes between charges. While this is certainly not to be recommended for long-distance commuting, most Americans could safely complete all their city driving with an EV. As the first car, for city driving, the EV may well become the preferred car of the American family.

The engine of the Toyota Prius, a hybrid car, switches from electric power to gasoline after the car picks up speed.

The batteries of the EV—whether lead-acid or NiMH—need to be charged between trips. Steep hills or high speeds drain the batteries faster than city driving. A fully depleted battery pack can be charged overnight, in about eight to ten hours, though it is better to charge whenever the opportunity arises. The EV usually comes equipped with a standard 110-volt battery charger, so the battery can be charged at any standard outlet. Charging is much faster with 220-volt battery chargers, but these are bulky and are best left in the EV owner's garage.

The maintenance of an EV is simpler than that of a gas-powered car. The EV's tires, lights, brakes, suspension, transmission, and body need the same maintenance as the same components in regular cars. Other than regular bat-

tery care, very little maintenance is needed to keep the EV in top condition because the electric motor has very few moving parts—a drive shaft and bearings—compared with the hundreds of such parts for the internal combustion engine. An EV motor is expected to last for more than a million miles of travel.

Hybrids

The EV is making great progress toward becoming the car of the future. However, the slow development of better batteries leading to increased range has led manufacturers to develop hybrids that give efficient gas mileage without sacrificing range of travel. The Honda Insight and Toyota Prius are hybrid electric cars that use both gas engines and electric batteries. General Motors, Ford, and Chrysler have each announced plans to produce hybrid vehicles. In normal travel, the Honda Insight, for example, uses a three-cylinder internal combustion engine that supplies power to the wheels and charges the batteries at the same time. The car uses both modes when climbing uphill or for acceleration.

It has been argued that the emissions are merely shifted from the tailpipe to the power generating plant that burns fossil fuels rather than a hydroelectric source.

Such cars do have their drawbacks. The Insight, for example, seats only two, has minimal trunk space, and being made of aluminum, is light and susceptible to crosswinds. It does, however, ride smoothly at 128 kilometers per hour (80 miles per hour), gives a gas mileage of about 97 kilometers per gallon (60 miles per gallon) in the city, and emits 75 percent less ozone-damaging carbon dioxide.

Advantages of the EV

There are two major advantages of the EV over regular gas-operated vehicles with internal combustion engines. The first is reduced pollution. Gas vehicles emit more air pollution than any other manmade machine. These emissions, which have long been recognized as major health hazards and a danger to the environment, contain chlorofluorocarbons (CFCs), carbon dioxide (CO_2), carbon monoxide (CO), and methane (CH_4). These emissions contribute to the greenhouse effect, leading to global warming and depletion of the ozone layer. The EV, however, has no tailpipe emissions. It has been argued that the emissions are merely shifted from the tailpipe to the power generating plant that burns fossil fuels rather than a hydroelectric

source. However, in the Pacific Northwest, for example, more than 90 percent of the power is hydroelectric, eliminating emissions at the power plant. Even in the power plants burning fossil fuels, it is easier to control emissions at a central source than from millions of tailpipes. Batteries do contain lead, but this lead is in a stable form that is easily recycled. On the other hand, even trace amounts of lead in gasoline do find their way into the environment.

The second major advantage of EVs is that they reduce people's reliance on fossil fuels. The American consumer is in many ways captive to the actions of OPEC and the oil industry, with fluctuating crude oil prices reflected at the gasoline pump. The internal combustion engine, at optimum performance when running hot, is only 20 percent efficient at converting energy into motion. The EV, on the other hand, is 70 percent efficient at converting the energy in its batteries into motion at the wheels. For most city travel, the internal combustion engine is rarely operating at optimum performance, further adding to its inefficiency.

In addition, the EV is noticeably quiet in its operation and very dependable. The EV first became popular as the preferred type of motorized transport in the 1920's. Even after its popularity waned, Clara Ford, wife of Henry Ford, who could have had any car she desired, still treasured her Detroit EV.

High-Speed Trains
Mass Transport in the New Millennium

➤ *Amtrak takes the first steps toward high-speed train travel in the United States.*

In December, 2000, the inauguration of Acela regional service from Boston through New York to Washington, D.C., marked the beginning of all-electric, high-speed train travel in the United States. The trains travel at 177 kilometers per hour (110 miles per hour) but with new signaling equipment will hit 200 kilometers per hour (125 miles per hour) in regional service and 240 kilometers per hour (150 miles per hour) in express service along the Northeast Corridor. Although this may well be a milestone in train travel in the United States, high-speed trains have been used for many years in other parts of the world, particularly in France and Japan. The technology used in the French system, Train à Grande Vitesse (TGV), has already been exported to Spain, South Korea, the United States, Taiwan, and China.

TGV

The abbreviation TGV can refer to the trains themselves, the high-speed lines on which the trains run, or the entire French high-speed rail system. TGV is a technological system with specialized tracks, all-electric locomotives (also called power cars) receiving power through pantographs from the overhead electric catenary, cars (also called trailers) with special articulation between them, braking systems, lightweight construction, and innovative signaling. These technological innovations enable the trains to reach speeds of 300 kilometers per hour (186 miles per hour) under commercial operating conditions and of 515 kilometers per hour (320 miles per hour) under test conditions.

High-speed trains travel at 177 kilometers per hour (110 miles per hour) but with new signaling equipment will hit 200 kilometers per hour (125 miles per hour) in regional service and 240 kilometers per hour (150 miles per hour) in express service along the Northeast Corridor.

The specialized tracks are welded rails laid on hybrid steel and concrete ties over a thicker-than-usual bed of ballast. However, the greatest difference between these tracks and normal tracks is the combination of curve radii and superelevation that makes such high speeds possible. The track centers are also spaced farther apart from each other to reduce the blast of two trains passing in opposite directions. With safety in mind, high-speed tracks are completely fenced off.

The Locomotive and Trailers

The power is supplied to the locomotive through a fixed overhead line aligned with the track. Electrical contact is maintained through a pantograph, which for high speeds is designed with a top linkage member that operates like a hydraulic damper with a short stroke to maintain intimate contact at all times with the overhead conductor, without bouncing away from it. Although a locomotive is at each end of the train, only the rear pantograph collects the power, feeding it through a cable on the roof of the train to the forward locomotive. This single pantograph arrangement prevents one pantograph from disturbing the wire for the following one. The contact wire pressure is about 70 newtons (16 pounds force). The power in the conductor is supplied at 25,000 volts (25 kilovolts), but this is converted to 1,500 volts by the main transformer in the locomotive.

The transformer is one of the heaviest parts of the lo-

Acela, Amtrak's high-speed train, passes through the borough of Queens in New York City. (AP/Wide World Photos)

comotive. It weighs 8 metric tons (17,700 pounds) and sits in a bath of oil circulated by pumps and cooled by fans. The power from the output of the transformer is then passed through a thyristor (basically a very large switch), passing current in only the "forward" direction when a suitable voltage is applied at the "gate" of the thyristor. The direct current (DC) power is then converted into a three-phase, variable frequency alternating current (AC) that runs the synchronous AC traction motors that are slung from the vehicle body between the main frames of the locomotive and level with the axles. The final step in the transmission of power is a gear train that rides on the axle itself and transfers power to the wheels. Each motor can develop 1,100 kilowatts (1,475 horsepower). Each locomotive has four such traction motors, developing a total of 4,400 kilowatts (5,900 horsepower). The locomotive weighs just 68 metric tons (150,000 pounds).

At very high speeds, the safety and comfort of the passengers becomes critical. Many innovations in locomotive cab design, including aerodynamic styling of the nose, reduce buffeting of wind pressure on the train. However, the most important innovation is the articulation between cars. In normal trains, two adjacent cars sit on independent axles and are coupled together. In high-speed trains, the two adjacent cars are permanently coupled together because they sit on a common two-axle assembly. It is more appropriate, therefore, to speak of them as "trailers" rather than "cars." Such a design reduces interior noise levels, improves aerodynamics, and provides more space for the suspension system. It also permits clean, quiet passage from one trailer to the next. In addition, in collisions, this design prevents trailers from jackknifing, which often happens with conventional train cars with standard coupling.

For such high speeds to be achieved, the entire construction of the locomotives and the trailers must be as light as possible. The primary structural members of the locomotives are of high-strength steel in a rigid space frame structure designed by computer for optimum weight. The trailers are built of extruded aluminum in a shell construction, leading to a 20 percent reduction in overall weight. This monocoque design is used in the TGV duplex, which has passenger seating at two levels, with 45 percent more passenger capacity than a single-level TGV. In the near future, composite materials will be used in trailer construction, leading to yet further weight reductions.

Braking and Signaling

High-speed trains use more than one braking system to dissipate the kinetic energy of the train. In conventional braking systems, such as disc brakes, this energy is dissipated as heat. However, for the very high speeds of TGV, these braking systems alone cannot stop the train in a reasonable distance or perform emergency stops because the kinetic energy of the train increases as the square of the velocity.

In addition to conventional braking methods, high-speed trains use the magnetic induction brake. This entirely new method of braking is suitable for use only at speeds above 220 kilometers per hour (137 miles per hour). The magnetic induction brake uses the current generated by the traction motors to create a magnetic field that interacts with the eddy currents induced in the rail by the motion of the train to produce a retardation of the train. The energy is dissipated as heat generated in the rail. The brake shoe, mounted under the side frames of the locomotive, rides 10 centimeters (4 inches) above the rail. When the brake is applied, the shoe skims the rail without actually touching it. The enormous current of the traction motors is then used to create a strong magnetic field, which when cut by the high-speed motion of the train, causes eddy currents inside the rail. This creates a retarding force on the original field (the train), producing great heat in the rail. It also produces an upward force on the track, for which the track and bed have to be designed.

At such high speeds, stationary signals on the side of the track, like those used for conventional trains, are not effective. The signals, therefore, are sent through the rails to signaling antennae mounted underneath the lead locomotive. These antennae read the signals and relay them to the locomotive's central computer, which displays the information to the engineer. The on-board computer also manages all the subsystems of the train. It can diagnose faults and prepare a maintenance report that is transmitted to the maintenance shop before the train's arrival.

Mass Transport of the Future

As roads grow more clogged and airspace more crowded, train travel is likely to become the favored mode of mass transport, particularly over medium distances. With cruising speeds now approaching 320 kilometers per hour (200 miles per hour), it would be possible to travel from Los Angeles to San Francisco in less than three hours. When real times to reach the airport, check in, board the plane, fly to the destination, await baggage, and reach the city center are taken into account, high-speed trains become far more convenient a mode of transport than either car or plane. In France, when TGVs were first introduced on the Paris-Lyon route, the airline business between these two major French cities was gutted. The same may well prove true in the United States, if investments are made in high-speed trains.

Light from Black Boxes
Ensuring the Safety of Air Travel

➤ *Voice and flight data recorders recovered from the wreck of an aircraft reveal the cause of the crash and help determine future protective measures.*

On January 31, 2000, at approximately 4:21 P.M., Alaska Airlines Flight 261 plunged into the Pacific Ocean near Port Hueneme, just northwest of Los Angeles off the coast of California. Several days later, the black boxes were recovered and told a grim tale of the last thirty minutes of the ill-fated flight. The flight originated in Puerto Vallarta, Mexico, and was on its way to San Francisco and Seattle. The MD-83 aircraft had eighty-eight people on board. There were no survivors. Transcripts of recorded conversations between the pilot and air traffic controllers show that about ten minutes before the fatal crash, about 4:10 P.M., the pilot had informed controllers that the plane was "in a dive."

"Say again?" the controller asked.

"Yeah, we're out of 26,000 feet," one of the pilots replied. "We're in a vertical dive—not a dive yet—but, uh, we've lost vertical control of our airplane."

At that point, the pilots were able to reestablish control of the plane. Reporting about four minutes later, the crew described the problem with the tail stabilizer and announced their intention of making an emergency landing at Los Angeles International Airport. The pilot said they were going to "reconfigure the jet," presumably trimming the jet for landing, by deploying flaps and preparing to drop the plane for an approach to the airport. Two other jets were in the area at the time and provided visual confirmation of the behavior of the MD-83. One was Sky West Flight 5154; the other was a corporate jet.

In the "telemetry" approach the flight data information is automatically transmitted, on an instant-by-instant basis, by satellite to computers, where it is analyzed and sent back to the pilots.

Four minutes later, the big jet nosed over a second time. This time, there would be no recovery. At 4:21 P.M., both the corporate jet and Sky West aircraft reported that they had seen the Alaska Airlines jet roll over onto its back and plunge nose-first into the sea.

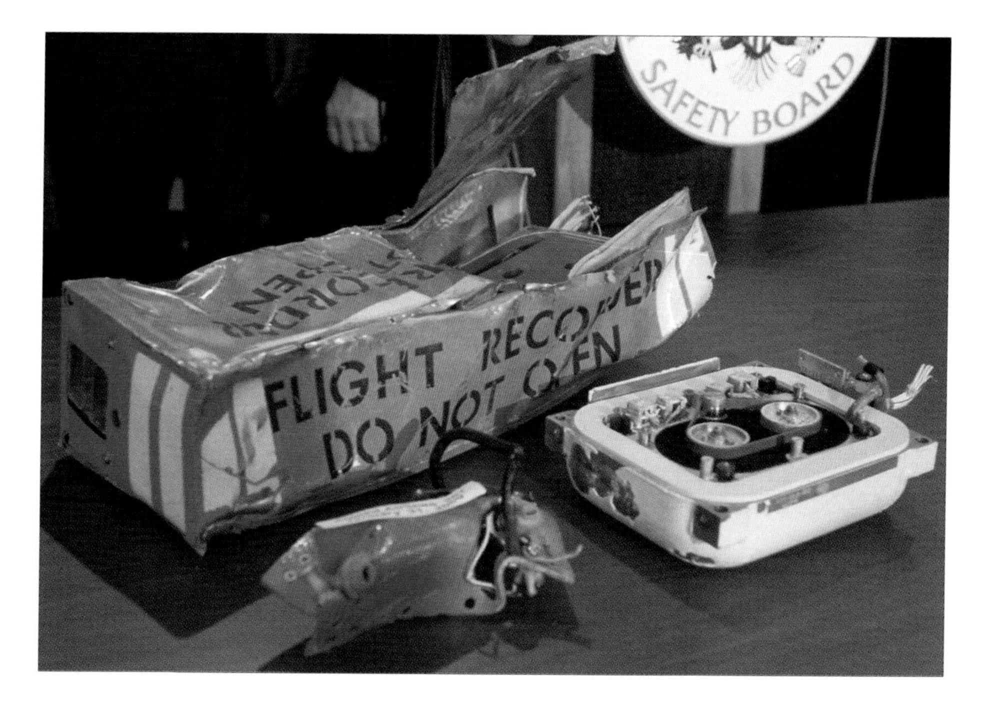

The U.S. Navy helped recover the first black box, the voice data recorder (center), from the wreck of Alaska Airlines Flight 261. (AP/Wide World Photos)

Black Boxes

There are two black boxes on each commercial aircraft. The cockpit voice recorder tapes all the sounds and voices in the cockpit, including engine sounds and conversations between crew members and between crew and air traffic control. The sensitive cockpit microphone picks up the noises and relays them to the cockpit voice recorder. In the case of Flight 261, for example, a flight attendant is heard telling the pilots about a loud noise from the rear of the jet. "The crew acknowledged they had heard it too," said John Hammerschmidt, a member of the National Transportation Safety Board (NTSB). Minutes later, toward the end of the tape, a second loud noise is heard.

The flight data recorder collects and records continuous data from sensors located in various parts of the plane. The MD-83 has forty-eight such sensors. The data recorded include the positions of the controls in the cockpit; the positions of the stabilizers, wing flaps, and ailerons; complete engine performance; and speed and altitude at every instant of the flight.

Both the cockpit and flight recorders are placed for safety in the upper fuselage at the rear of the plane. They are crash-protected by being sealed in insulated titanium boxes that are virtually indestructible and can protect the data recorded from impact and fire or water damage. Both recorders have pingers that are activated automatically.

The beacons transmit pulses for up to thirty days and have a range of about 2 miles (3.2 kilometers). On recovery, the black boxes are sent to NTSB headquarters in Washington, D.C., for analysis. Representatives of the NTSB, the Federal Aviation Administration (FAA), and aircraft and engine manufacturers create written transcripts of the voice recorder tapes, which are then synchronized with time-coded air traffic control tapes. The data from the flight recorder are used to create a computerized, animated simulation of the final minutes of the flight.

The "Autopsy" Approach

Both voice and flight recorders from Flight 261 were recovered from the Pacific Ocean and analyzed to re-create the events leading up to the crash. They told a slightly different story from what was already known. What they showed was that the pilots became aware of the stabilizer problem and were grappling with it for at least thirty minutes before the crash, not just the ten minutes after they reported the first dive to air traffic control. If the crew of Flight 261 had had the information contained in the flight data recorder thirty minutes before the crash, they would have been better informed in making their decision on whether to set the plane down immediately at the first available airfield or attempt to solve the problem and head for Los Angeles International Airport.

This tragic crash highlights the radically different approaches taken by the United States and Britain to the data contained in the black boxes. In the United States, experts take what is essentially an "autopsy" approach. After the crash has taken place, the black boxes are analyzed to determine what actually occurred and identify the cause in an effort to ensure that such a crash does not occur again. This approach can lead to redesign of plane or engine components or changes in the airline's overhaul and maintenance procedures. What it cannot do is to prevent the crash that has already taken place. No analysis is made of flight recorder data if the plane reaches its destination safely.

The autopsy approach has the further disadvantage that some of the older planes have flight data recorders that monitor as few as eleven sensors. These planes met the government requirements at the time that they were put into service, but the lack of sensors makes it difficult to accurately detail the performance of the components of the plane just before the crash. In some of these crashes, the aircraft manufacturer and the NTSB arrive at differing causes of the crash, leaving the matter of crash responsibility to be handled in the courts. In the case of Flight 261, however, a lack of lubricating grease on the jackscrew operating the rear stabilizer was identified as the cause of the crash.

The "Telemetry" Approach

The approach in Britain is radically different. In the "telemetry" approach, the information being recorded by the sensors of the flight data recorder is automatically transmitted, on an instant-by-instant basis, by satellite to computers, where it is analyzed and sent back to the pilots, so that the crew has full information of the ongoing performance of all vital components of the plane. This type of flight data recorder, called the in-flight quick access recorder, is used by British Airways. It potentially could prevent a crash, thus enabling safer travel in the skies for the flight that is being recorded as well as for future flights. If Alaska Airlines Flight 261 had had the in-flight quick access recorder, not only would the pilots have been aware of the magnitude of the stabilizer problem, but also it is very likely that ground control computers, monitoring the data, would have sounded the alarm in the cockpit, mandating immediate descent to the nearest civilian or military airfield.

Missile Defense Systems
The Race Against Time

➤ *In a test, the National Missile Defense system fails to shoot down a target missile, and the United States delays deployment of the system.*

With the proliferation of missile technology, particularly in the area of long range, intercontinental ballistic missiles (ICBMs), the need for a defense system that guards against nuclear, chemical, or biological attack from rogue states such as Iraq, Iran, Libya, Syria, and North Korea becomes very real. In addition, there is always the possibility of an accidental launch of an ICBM carrying a nuclear warhead from friendly countries such as Russia, China, India, and Pakistan. To guard against such possibilities, the Pentagon has been working on a National Missile Defense (NMD) system that would intercept an incoming missile in space and destroy it. Critics argue that such a system will not work because of the technological problems involved, that solving the problems would be prohibitively expensive, and that working on such a defense system is counterproductive to global security because it encourages a weapons race.

The NMD has several component systems that act together to protect the United States against missile attack. In all these systems, which can be land-based, sea-based, or space-based, the key to successful interception and destruction of the incoming missile is early detection and warning. This is to be accomplished by the construction of an advanced radar installation in Alaska.

Land-Based Systems

This approach creates about a twenty-eight-minute window in which to detect the launch of an enemy missile, track and project the course of the missile, and launch a missile that will intercept and destroy the incoming missile or its warhead. An unsuccessful test of the system was carried out on January 18, 2000. At 8:19 P.M. central standard time, a Minuteman II intercontinental ballistic target missile lifted off from Vandenburg Air Force Base in California and exited the atmosphere over the Pacific Ocean. Some twenty-one minutes later, an Exoatmospheric Kill Vehicle (EKV) interceptor on a rocket booster was

The Pentagon displayed a miniature Exoatmospheric Kill Vehicle during a news conference before a test of the National Missile Defense system. (AP/Wide World Photos)

launched from the U.S. Army missile range on Kwajalein Atoll in the Marshall Islands. A radar system on a nearby island sent a signal to the EKV, giving it more precise data on the flight path of the target ICBM. At about 1,400 miles (2,250 kilometers) from the target, the EKV, weighing 121 pounds (55 kilograms) and nicknamed "smart rock," separated from the booster rocket and guided itself, using infrared sensors, toward the target. In the meantime, the target separated from the ICBM, releasing both a dummy warhead and a decoy balloon. Everything worked perfectly for about eight minutes. About six seconds before impact, however, both of the heat-seeking infrared sensors on the EKV failed, essentially "blinding" the EKV in the final maneuvers necessary to destroy the warhead, about 140 miles (225 kilometers) above the Pacific Ocean.

The test indicates the complexity of the tasks that need to be successfully completed. An enemy ICBM can be fired from anywhere in the world; however, it must be within tracking range of U.S. radar. This is possible only if the Alaska-based ground radar works in association with Space-Based Infrared System (SBIRS) sensor satellites or with other ground-based radars scattered across the globe.

When tracking occurs, and information is supplied to the EKV, a cat-and-mouse game develops between interceptor and target. There are three phases of target flight during which interception can occur: boost, mid-course, and terminal. In the boost phase of the target rocket, the target warhead is most vulnerable because the rocket is slow, emits great heat, and has all its warheads still contained within the nose of the ICBM. In the mid-course phase, the warheads and the decoys have all been deployed above Earth's atmosphere, and the interceptor then has to distinguish between the real reentry vehicle (RV) and the decoys. To complicate matters further, a target rocket could contain multiple warheads and therefore more than one RV. In the terminal phase of the attack, the atmosphere does slow the decoys, making it easier to distinguish them from the RVs, but the warheads themselves may be capable of maneuvering. Further, destroying the warheads in the terminal phase may be too late to prevent their lethal effects.

Sea and Space Systems

Clearly, destroying an enemy missile in its mid-course or terminal phase is far more complex than destroying it in the boost phase. However, attacking an enemy rocket while all its warheads are still contained within the nose of the ICBM requires that the interceptor missiles be located within easy range of the enemy bases that could fire such ICBMs. For this reason, the sea-based missile defense system becomes very important, at least in the short term. In this system, ground radar and SBIRS sensor satellites would provide immediate data to U.S. Navy ships stationed around the globe. Smaller, modified, ship-based interceptor missiles would then attack and destroy the enemy missile. If the enemy base from which the missile originated was located near the sea, interception would occur in the boost phase of the enemy missile; if the enemy base was in the interior of a land mass, interception would take place in the immediate post-boost phase, when the enemy warhead vehicle has detached from the ICBM but is still on its upward trajectory and therefore still contains all the multiple warhead RVs and the decoys. The Heritage Foundation has estimated that upgrading twenty-two existing U.S. Navy Aegis-equipped cruisers to function in this sea-based system would cost roughly $3 billion.

Recent success with the tactical high energy laser in blowing up two battlefield rockets in simultaneous flight encourages research into space-based lasers.

In the space system, space-based lasers would attack and destroy enemy missiles in the boost phase. Recent success with the tactical high energy laser (THEL) in blowing up two battlefield rockets in simultaneous flight encourages this research. The laser destroyed the rockets by heating them until they exploded. However, THEL's range is only a few miles. In the future, actual space-based interceptors might perform in much the same way as ground-based interceptors but would attack the enemy warheads much earlier, in the post-boost phase.

Status of the NMD System

Speaking at Georgetown University on September 1, 2000, President Bill Clinton announced that he would delay deployment of the NMD system and leave the decision regarding its future to his successor. He called the technology promising but unproven. He said more tests and more simulations were necessary before the system could be deployed. Reaction to the decision was mixed. At the Russian Defense Ministry in Moscow, Colonel General Leonid Ivashov called the president's decision "constructive." Vice President Al Gore, Democratic presidential candidate, endorsed the postponement, saying that if elected, he would pursue the same policies as President Clinton. Governor George W. Bush, Republican presidential candidate, criticized the decision, saying he would develop and deploy a larger missile defense system that would protect not only the fifty states but also U.S. soldiers abroad and the nation's allies.

Training Computers for War

Chips in the Battlefield

➤ *The use of computers in the battlefield as thinking machines for generals or as parts of a sophisticated uniform for the infantry soldier became a very real possibility.*

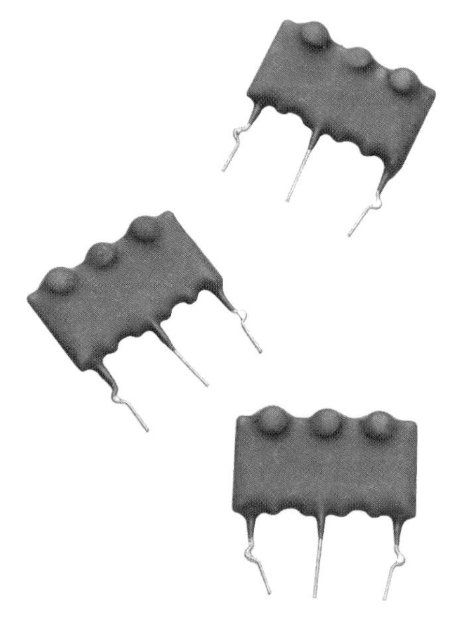

For many years, computers have been an important part of the military machine, whether used as guidance systems for cruise missiles in the air or in radar installations on the ground. However, such computers have been used for very specific tasks. For example, the computer in cruise missiles was programmed to photograph ground detail, match the images with inboard photographic detail already programmed into the computer, and locate the target. Computers were never used as thinking machines; all thought was left to the human mind. However, sometimes calculations must be performed at lightning speed for the correct decision to be made on the battlefield. In modern warfare, the scenario is very different from earlier times. Jets and radar-guided missiles are in the air and on their way to targets, satellites are feeding streams of electronic information into the command control center, and ground motion detectors and military commanders are relaying minute-by-minute logistic information of the movements of tanks and heavy artillery in the battlefield. Life-or-death decisions have to be made in seconds in the midst of the action, and the human mind is simply not capable of making the myriad calculations that may be necessary.

Costas Tsatsoulis and Douglas Niehaus, professors in the Department of Electrical Engineering and Computer Science at Kansas University, have obtained research grants from the Defense Advanced Research Projects Agency (DARPA) to turn networked military computers into thinking machines that determine the best course of action with little or no human assistance. The task assigned to the computers is simple, finding a solution that is "good enough soon enough," according to Tsatsoulis, who said the team is using techniques from artificial intelligence to build intelligent systems.

A Network of ANTs

The heart of this networking idea is the Autonomous Negotiating Team (ANT). Every entity in the military has

These computerized dog tags (personal information carriers), which contain an individual soldier's X rays, immunization record, and complete medical history, are part of the computerization of the military. (AP/ Wide World Photos)

an ANT, which could be something as large as a brigade or an AEGIS cruiser to something as small as a single soldier. The various ANTs negotiate resources, authorizations, capabilities, actions, and plans with each other, seeking the best solutions, with the billions of calculations necessary being performed in milliseconds. The way this case-based reflective negotiation works can be seen in the following two examples.

For example, satellite-based radar informs ground control of two targets. One is a fast-moving target at long range; the other is a slow-moving target at long range. The fast target requires frequent measurements that will require increased computer capability. The fact that it is at long range means ground control will need information from closer sensors to track it. The slow target, on the other hand, will require less frequent measurements, though because it is at long range, it will still need the closer sensors for tracking. If the computers are networked, the ANTs begin negotiating with the available resources for action. The fast-moving target can afford little negotiating time, whereas the slow-moving target gives more negotiating time available for decision making.

In the case of an AEGIS cruiser, an ANT is created at the first sensing of a potential threat. The ANT negotiates with other networked computer systems for resources for the targeting and elimination of the threat. The ANT pro-

vides all information needed to target and destroy the threat. Finally, the ANT assesses battle damage and repeats its actions until all threats are eliminated, after which the ANT is dissolved.

The research on networked computers and case-based negotiation could also lead to applications in civilian emergency management situations or other situations in which the country has fewer instant resources than required for all possible problems. For example, Tsatsoulis said that a system could be developed that gave messages from the Federal Emergency Management Agency priority over the telephones lines during emergency situations.

Computer Technology for Soldiers and Equipment

A soldier's uniform being tested at Fort Bragg, North Carolina, has a 2.2 pound (1 kilogram) computer that fits into the back of the uniform. The primary focus of the Land

Warrior Uniform is to increase the soldier's effectiveness on the battlefield. The computer, which is powered by two small batteries, provides every soldier with a virtual briefing room by enabling the soldier to receive briefings from the command post. It runs Windows 2000 and has a Pentium processor, with 128 megabytes of RAM (random-access memory). The wires are sewn into the uniform. The miniature monitor is mounted on the helmet and can be lowered over either eye to study maps or read e-mail. The display can also show the position of each soldier in the company. When not in use, the monitor is slid up on the helmet. The soldier can roll the mouse on either shoulder or on the weapon and can click letters on the keyboard mounted on the wrist. Lieutenant Colonel Scott Crizer, product manager with Soldier Electronics, in Fort Belvoir, Virginia, said that this uniform is designed to modernize the infantry soldier.

A computer powered by two small batteries will provide every soldier with a virtual briefing room by enabling the soldier to receive briefings from the command post.

A vexing problem with military equipment has been the ability to keep track of it. Not only is the unexplained loss of such equipment, big or small, wasteful to the military, it can also be dangerous if such equipment falls into the wrong hands. To combat this problem chips have been incorporated into a small tag, the size of a child's bandage, equipped with an antenna. A radio frequency tag reflects waves being sent to it from a nearby computer. Because it does not produce the waves, the battery can last longer, up to ten years. The tags let the military inventory, monitor, and even communicate with devices using the tags—and cost only ten dollars each. The development of the tags was funded by the Defense Threat Reduction Agency for use with night-vision goggles, but Ron Gilbert, senior scientist with Pacific Northwest National Laboratory in Richland, Washington, said the technology could be used to tag "anything from hockey sticks to aircraft carriers."

And the Blind Shall See
Computer Chips Ease Disabilities

➤ *Progress in computer chip technology helped ease the lives of people with disabilities, enabling a blind man to see and a paraplegic to walk, although normal functioning was not restored.*

Ever since the advent of computer technology, researchers have been looking for meaningful ways in which exciting developments in that field could be translated into electronic devices that would benefit the disabled, particularly the visually impaired. In 1978, a blind man known as Jerry volunteered for a study in which researchers implanted electrodes into his brain in an effort to restore his sight. Since then, scientists have been working to improve the software and enable him to "see." In 2000, before a group of reporters, the sixty-two-year-old man, who has been blind since the age of thirty-six, demonstrated his ability to see and find a mannequin in a room, walk to a black stocking cap hanging on a white wall, return to the mannequin, and plop the cap on its head. He can also recognize 2-inch-tall letters from a distance of 5 feet.

In order to see, Jerry wears glasses with a tiny pinhole camera mounted on one lens and an ultrasonic range finder on the other. The camera is wired to the electrodes that were implanted into his brain, which makes it the first artificial eye to provide useful vision. Jerry has confirmed that he does not actually see an image. What he sees are one hundred specks of light that appear and disappear like stars that come and go behind passing clouds as his field of vision shifts. However, these specks are sufficient for him to identify objects and read large letters.

Computerized Travel Aids

In those cases in which such artificial eyes cannot help the visually impaired person because the problem lies not in the eye but in the brain, researchers strive to improve the individual's ability to get around and to travel. Computer chip technology has made significant progress in helping such mobility. Sharga Shoval of the Technion in Israel, together with Johann Borenstein and Yoram Koren, both of the University of Michigan, developed the NavBelt, a computerized travel aid for the visually impaired, based on mobile robotics technology. The device consists of a

Researchers are developing global navigation aids for the visually impaired that rely on the Global Positioning System (GPS). Here, a Boeing Delta II rocket lifts off carrying a U.S. Air Force GPS navigation satellite. (AP/Wide World Photos)

portable computer worn as a backpack, the NavBelt which is worn around the waist, and stereophonic headphones. The NavBelt has eight ultrasonic sensors, each covering a sector of 15 degrees, for a total scan surveillance of 120 degrees. The sensors, several of which fire simultaneously, form the heart of the obstacle avoidance system (OAS). The computer uses a stereophonic imaging technique to

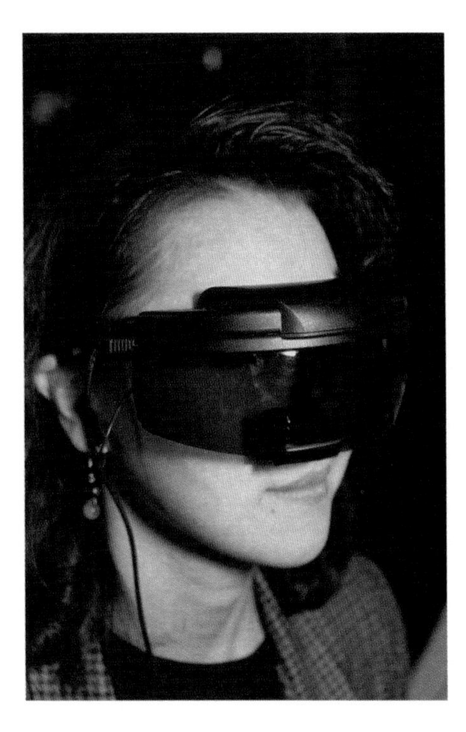

process the signals from the ultrasonic sensors and relays the information to the user via the stereophonic headphones. The acoustic signals are transmitted as discrete beeps or continuous sounds. The direction of the obstacle is indicated by the spatial direction of the signal, and the distance to the obstacle is indicated by the signal's pitch and volume. The closer the obstacle, the higher are the pitch and the volume. The visually impaired person can then interpret the information as an acoustic "picture" of the surroundings.

However, what happens if a visually impaired person gets lost and has no one around to ask for help? Research has been ongoing in global navigation aids. Such aids aim at providing the absolute position of the person in relation to specific detail such as the intersection of two streets, the entrance to a public building, or a bus stop. For such pinpoint accuracy, these aids rely on the Global Positioning System (GPS), which uses radio signals from satellites, to provide the computer of the user with updated information about the surroundings. The computer findings are then converted into voice information relayed to the visually impaired user.

The social conscience of the human race has always motivated people to provide assistance—often on a volunteer basis—to those not quite as fortunate as themselves. Researcher David Dewhurst has proposed a Vuphonics Distance Assisted System (DAS) that would provide visually impaired (or hearing-impaired people) with assistance from volunteers on a planned or ad-hoc basis by matching, through the Internet, disabled clients to volunteer assistants located around the world. Such a social network of volunteer workers could help visually impaired clients by performing identification, mobility, or other vision-related tasks and transmitting the information, almost instantly, to the visually impaired client by a friendly and human voice. In the case of a hearing-impaired client, the information would be displayed on a computer screen.

Chips, Muscles, and Cells

Marc Merger, a thirty-nine-year-old French financial consultant, was involved in a car crash that paralyzed him from the waist down. In December, 1999, surgeons implanted fifteen electrodes on nerves and muscles in Merger's legs, connecting them to a computer chip embedded in his abdomen. Problems arose, and the process had to be repeated in February, 2000. In early March,

Merger was able to stand by himself and since then has taken his first steps. Professor Pierre Rabischong of Montpellier University in France, a coordinator of the European Union project, said that the implanted computer chip allows Merger to create artificial muscle movement. Researchers, he said, are trying to reproduce what happens in the brain with electrodes, thus enabling patients to stand using their own muscles. The scientists transmit instructions to the chip via computer. These signals are then transmitted to the electrodes in Merger's legs and converted into muscle movement. The ultimate goal is for patients to be able to control their own movements, by pressing buttons on a walking cane, now under development, that will act as a remote control.

The ultimate goal of all such research, whether for the disabled or for other health purposes, is to mate the computer chip with the human cell so that they can actually communicate with each other. The first steps to making this possible occurred when Boris Rubinsky, professor of Mechanical Engineering at the University of California, Berkeley, mated a healthy human cell with an electronic chip in a tiny device that is smaller and thinner than a human hair. By controlling the activity of the chip via computer, scientists can control the activity of the cell. The research focuses on a long-known phenomenon that cell membranes become permeable when exposed to certain voltages. Researchers hope to develop cell-chips tuned to the precise voltage to activate different body tissues, whether in the bone, muscle, or brain.

The most exciting research, however, is being carried out in biochemistry, in which researchers are attempting to get cells and chips to talk to each other. Nerve cells called neurons carry messages to and from the brain in the form of electrical impulses. These same electrical impulses can be transmitted from the chip to the neuron. Researchers Peter Fromherz and Alfred Stett showed that by building up a voltage on the surface of the neuron (thereby not damaging the cell), it was possible to fire the neuron—and receive an impulse from the neuron as well. The rewards, in the fields of medicine, of establishing communication between cells and chips would be breathtaking. With such development, the brain of the disabled person would receive messages from the computer as if they came from the eyes or the limbs, and messages coming from the brain would be transmitted to the computer in an interactive process that would make them act as one.

Expo 2000

➤ *Hanover, Germany, hosts a fair at which many nations showcase how humankind can use high technology in environmentally friendly ways.*

It has become customary to chronicle the achievements of humankind, on a continuing basis, by holding expositions in nations across the world. In 2000, Hanover, Germany, hosted the biggest world's fair, with a common theme for the millennium of Humankind-Nature-Technology: A New World Arising. Nearly two hundred countries were represented, including most of the major industrial nations of the world and many of the poorer ones. Fifty national pavilions served to showcase the art, culture, technology, and lifestyle of the nations involved for the millions of visitors that visited Expo 2000. The United States, however, did not have a pavilion, reportedly because organizers failed to raise enough private funds to cover the related costs.

The Architecture

Every pavilion was designed with the environmental theme in mind. The striking Japanese pavilion was a huge, arched structure made almost entirely of recycled paper. Other pavilions were made of glass and wood; shapes included a wing, a vase, and an African hut. The pavilion of the Netherlands had trees growing out of its middle and water tumbling down its sides. It used wind generators to produce electricity. The Nepal pavilion was constructed without the use of modern machinery.

Fifty national pavilions served to showcase the art, culture, technology, and lifestyle of the nations involved for the millions of visitors that visited Expo 2000.

The exposition used Hanover's existing trade fair halls, but a number of buildings were constructed for the event. Five interconnected theme park buildings were built around such concepts as humankind, nutrition, energy, and other environmentally conscious themes, all geared toward the new millennium. In addition, two churches, restaurants, hotels, office buildings, a new train station, a gondola, and even a heliport were part of the fair site. The most striking building was the six-story bright yellow mailbox towering over the entire grounds.

Expo 2000 visitors examine a large globe at the Planet of Visions exhibit. (AP/Wide World Photos)

However, this mailbox was not intended to be in competition with other towering structures of past expositions, such as the Eiffel Tower (Paris, 1900) or the Space Needle (Seattle, 1962). All those expositions, starting with the first World's Fair in London in 1851, attempted to showcase the latest feats of technology and science. Expo 2000 director

Birgit Breuel said, however, that this fair would show how humans and technology can exist and help improve the natural world.

The Experience

The highly interactive exhibits in the theme parks, designed by artists, architects, and even choreographers, included virtual tours through human chromosomes, an opportunity to sample the experiences of a space traveler, and a tour that ended in a giant egg. One interactive experience, billed as a "wonder of the world," required visitors to wear bathing suits and swim underwater to reach it.

The many cultural events included performances and pop concerts from around the world in a disco hall with a 5,000-person capacity. In addition, many of the music performances were held at the Preussag Arena, with a capacity of 13,500. This arena is a permanent addition to the city of Hanover and, like every other permanent structure built for Expo 2000, was built with a post-fair purpose in mind. The Preussag Arena will host half the matches of the World Ice Hockey Championships.

The Economics

Expo 2000 opened to the public on June 1 and closed its doors on October 31. Although the initial hope was that it would draw 40 million visitors at the rate of about 250,000 a day, the actual initial attendance was far less at only 60,000 a day, which caused the organizers to slash ticket prices to encourage visitors. The fair needed an attendance of at least 150,000 a day to break even. The result has been that the private company, largely funded by the German government, that subsidized Expo 2000 incurred a loss of about $200 million on an initial outlay of $1.5 billion. However, the benefits of such a fair cannot be judged merely by whether it makes a profit or loss. As Maren Brandt, spokesperson for Expo 2000, pointed out, the fair has brought benefits to the community that far outweigh the loss incurred. In a country with about 10 percent unemployment, Expo 2000 created 100,000 temporary jobs.

One interactive experience, billed as a "wonder of the world," required visitors to wear bathing suits and swim underwater to reach it.

The city of Hanover has benefited greatly from the facelift it received as the host city. More than $5 billion in public and private funds were spent in improvements to the city, including an expansion of the subway lines, a new subway station built just for Expo 2000, renovation of the city hall—one of the few historic buildings still standing after the Allied bombing of World War II—and a new airport terminal. In addition, freeways were repaired, and the German railway system built a new train station at the site of Expo 2000 to bring in visitors by high-speed trains from big cities such as Hamburg, about one hour away, and Berlin, about one hour and forty minutes away.

Many of the city's museums offered special exhibitions for visitors and tourists. These included an exhibition by the Kestner Museum on a century of design. The Hanover/German Museum for Caricature and Critical Graphics hosted "Big City Fever: Seventy-five Years of the New Yorker," featuring the magazine's most famous covers. The city also got permission to expand its shopping hours to 10 P.M. on weekdays during the months of the fair. This is a rarity in Germany, where shops are legally required to close by 8 P.M. on weekdays.

Expo 2000 lived up to expectations. Although it lost money for its organizers, it did expand the breadth of human vision by bringing the nations of the world together in the common goal of preserving the natural environment for the benefit of humankind.

The final "Grand Parade of Cultures" held on closing day in Hanover. (AP/Wide World Photo)

Resources for Students and Teachers

Books

Cummins, Ronnie, Ben Lilliston, and Andrew Kimbrell. *Genetically Engineered Foods: A Self-Defense Guide for Consumers*. New York: Marlowe, 2000. The authors argue that bioengineered foods constitute a hazard to people's health, their environment, and even the agricultural industry. They present specific information on which products contain genetically altered components and urge consumers to make their opinions heard.

Dole, Charles E., and James E. Lewis. *Flight Theory and Aerodynamics: A Practical Guide for Operational Safety*. 2d ed. New York: John Wiley & Sons, 2000. This volume, used by the United States Air Force, describes the basic principles of aerodynamics and physics. Designed for students with little engineering background, it explains what lifts and propels an aircraft in a simple, concise manner.

Job, MacArthur. *Air Disasters*. Vol. 4. Fyshwick, Australian Capital Territory: Australian Aviation, 2001. The fourth volume in the series by an Austalian aviation expert takes an in-depth look at several major air disasters, describing what happened and why.

Lyshevski, Sergey Edward. *Nano- and Microelectromechanical Systems: Fundamental of Nano- and Microengineering*. Boca Raton, Fla.: CRC Press, 2000. Lyshevski discusses the techniques and technologies involved in nanoengineering. He also examines applications of the technology and its impact in fields from the environment to national security.

McHughen, Alan. *Pandora's Picnic Basket: The Potential and Hazards of Genetically Modified Foods*. New York: Oxford University Press, 2000. A Canadian molecular biologist argues that genetic modification of foods is a safe process and that much of the consumer concern is caused by misunderstanding of the genetic processes involved and a distrust of agribusiness.

Motavalli, Jim. *Forward Drive: The Race to Build the Car of the Future*. San Francisco: Sierra Club Books, 2000. Syndicated

auto columnist and environmental reporter Motavalli writes of the development of electric cars, hybrid cars, and other fuel-efficient, low-polluting cars.

Wright, Michael, and M. N. Patel, eds. *Scientific American: How Things Work Today.* New York: Crown, 2000. *Scientific American,* a popular science magazine, explains how modern technology—from escalators to credit card readers to the Internet—works in easy-to-understand language. Contains more than six hundred illustrations and photographs.

Ziman, J. M., ed. *Technological Innovation as an Evolutionary Process.* Cambridge, Mass.: Cambridge University Press, 2000. Scientists from many fields discuss how inventions such as axes, medicines, and aircraft go through a cyclic process of variation and selection similar to the evolutionary process undergone by living organisms.

CD-ROMs and Videos

Building Big: Explore the Greatest Engineering Feats of Modern Times. Video. 5 vols. WGBH Boston Video, 2000. This series of videos deals with some of the largest engineering structures built, including segments on bridges, domes, skyscrapers, and dams.

Science and Technology: One Hundred Years of Progress. Video. Films for the Humanities and Sciences, 2000. In this video produced by ABC News, narrator Peter Jennings examines the technological innovations and discoveries in science that took place over the last one hundred years.

Transistorized. Video. Public Broadcasting Service, 2000. This documentary examines the development of the transistor and its effects on society. It also looks at the three men who invented the transistor at Bell Labs in the 1940's: William Shockley, John Bardeen, and Walter Brattain.

Web Sites

Agricultural and Food Engineering Around the World
http://aginfo.snu.ac.kr/ipforum/fdengin2.htm
Contains an extensive listing of Web pages on agriculture and food engineering, mostly academic.

Ask an Engineer
http://expage.com/page/askanengineer
Pittsburgh, Pennsylvania, Section of the Society of Women Engineers
The society provides a site designed to help teachers, students, and their parents understand what the field of engineering entails and what an engineer does.

CORA
http://cora.whizbang.com
This research paper search engine contains links to papers on computer science, including artificial intelligence and human-computer interaction.

CrossRail
www.crossrail.com
Information exchange offers services and solutions for the railway industry, including the latest news.

Edinburgh Engineering Virtual Library (EEVL)
http://www.eevl.ac.uk/
This United Kingdom site provides access to engineering information on the Internet. It provides links to and descriptions of engineering sites by subject area as well as links to news sources, organizations for engineers, and information for teachers.

Engineering Electronic Library Sweden (EELS)
http://eels.lub.lu.se/
Swedish Universities of Technology Libraries
This sites provides links to Web sites in various field of engineering, including chemical, civil, electrical, mechanical, and mining.

Engineering Resources Online
http://www.motionnet.com
Provides a quick guide to engineering search engines, trade shows, books, and journals.

Market Station
www.marketstation.com
Covers railroad and train related information for fans of railroading.

Mechanical Engineering, World Wide Web Virtual Library
http://www.vlme.com
Provides searchable links to a wide variety of Web sites on mechanical engineering.

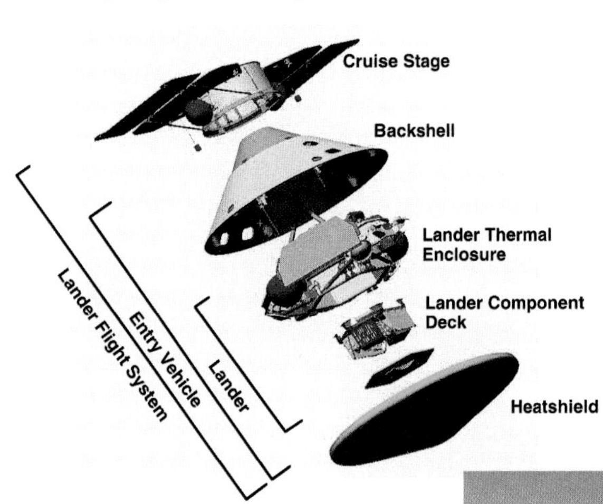

In March, the National Aeronautics and Space Administration officially abandoned any attempts to talk to the crashed Mars Polar Lander. (NASA)

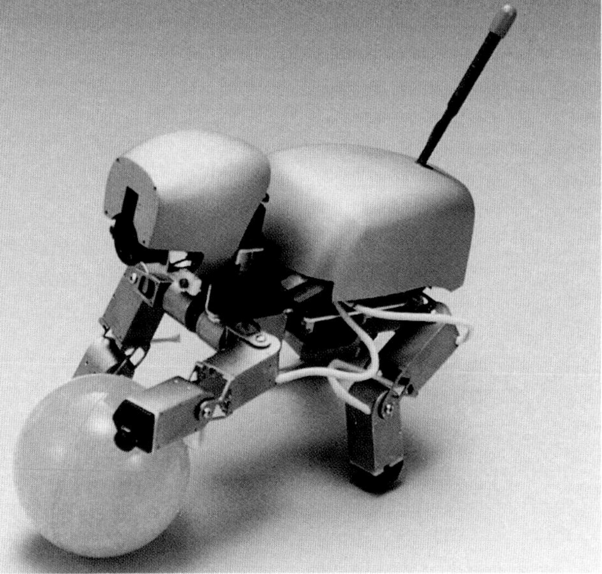

In September, Sony reports that it has taught a robot dog to speak–to recognize objects and say their names. (AP/Wide World Photos)

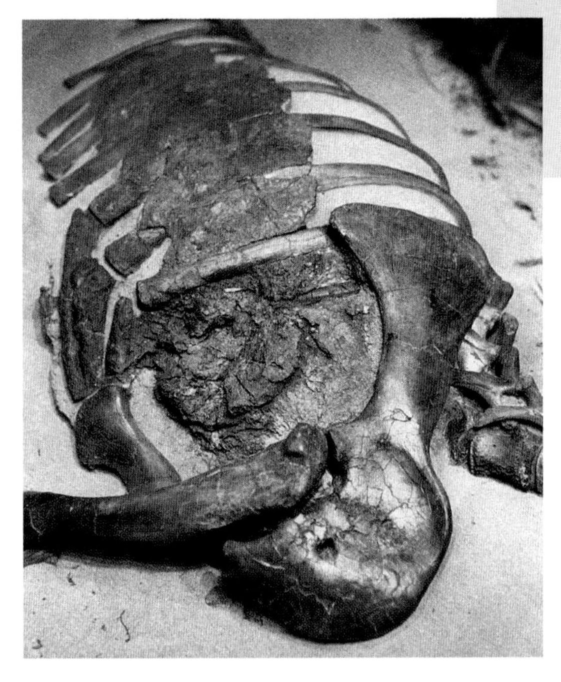

In April, scientists find the first fossilized heart of a dinosaur (lower center, between an apex of shoulder bones) in these 66-million-year-old remains on exhibit at the North Carolina Museum of National Sciences in Raleigh, North Carolina. (AP/Wide World Photos)

The Year's Events in Science

➤ *Note: Results or analyses published in monthly magazines are dated as if they appeared on the first of the month of publication.*

Date	Science	Event
Jan. 1	Information technology	Despite forecasts of large-scale malfunctions in computers worldwide, the transition from 1999 to 2000 passes relatively smoothly. The "millennium bug," which causes computers to apply the wrong year to time-sensitive functions, has only small, localized effects.
Jan. 1	Mathematics	In *Annals of Mathematics*, number theorist Ken Ono proves that infinite examples of Ramanujan congruencies exist, long thought to be a limited set of numerical oddities. The discovery opens new prospects for applying numerical relations called partitions.
Jan. 1	Mathematics	*Notices of the American Mathematics Society* reports that three mathematicians proved the local Langlands correspondence, a conjecture concerning a relation between prime numbers and squares, cubes, and other powers of numbers.
Jan. 1	Medicine	Two studies published in *Nature Medicine* find that accumulation of glutamate on nerve cells is involved in multiple sclerosis and suggest development of a medication that blocks the glutamate.
Jan. 3	Space	Passing Jupiter's moon Europa at 350 kilometers (218 miles), the space probe Galileo detects signs that the north pole's magnetic field reverses about every five hours, evidence that a conductive liquid, such as salt water, lies below the ice-bound surface.
Jan. 4	Astronomy	In the *Proceedings of the National Academy of Sciences*, two researchers explain a well-known illusion about the size of the moon when on the horizon: The brain misinterprets reference points on earth to mean that the moon is farther away than it actually is.

Date	Science	Event
Jan. 5	Medicine	An article in the *Journal of the National Cancer Institute* shows that the gastrin-releasing peptide receptor (GRPR) gene, which nicotine activates, is twice as common in women smokers than in men smokers, possibly explaining the higher rate of lung cancer among women.
Jan. 6	Health	A study in *Nature* establishes that blooms of the diatom *Pseudo-nitzschia* in the ocean can put toxins deadly to mammals into the food chain. The toxins, when in shellfish eaten by people, can cause neurological damage and death.
Jan. 6	Life sciences	*Nature* reports that marine iguanas of the Galápagos Islands can regrow bones that shrank during famine. They are the only vertebrate known to have that capability. It is hoped that the finding can help medical researchers develop a way to reverse the effects of osteoporosis in humans.
Jan. 10	Astronomy	Based on observations made with the Chandra X-Ray Observatory, two teams of astronomers report finding supermassive black holes at the cores of galaxies, sources contributing to the pervasive X-ray background in outer space.
Jan. 10	Genetics	Celera Genomics, a biotechnology company, announces that its scientists had sequenced 97 percent of the human genome.
Jan. 10	Physics	In *Physical Review Letters*, two researchers claim to have found evidence for a fifth force of nature in addition to electromagnetism, gravity, and the weak and strong nuclear forces, based on anomalies uncovered in previous particle physics experiments.
Jan. 12	Environmental science	A National Academy of Sciences panel issues a report estimating that beyond any doubt the earth's average temperature increased between 0.7 percent and 1.4 percent during the twentieth century, a 30 percent increase over previous estimates.
Jan. 14	Genetics	A study of stickleback fish and fruit flies published in *Science* provides evidence that evolution can occur quickly and predictably.
Jan. 14	Genetics	The journal *Science* reports that researchers at the Oregon Health Science University cloned rhesus monkeys using a technique called embryo splitting, its first successful application to a primate species.

Date	Science	Event
Jan. 15	Astronomy	Images from the Chandra X-Ray Observatory suggest that supermassive black holes are far more common than previous evidence indicated, according to speakers at a meeting of the American Astronomical Society.
Jan. 15	Medicine	A report in *Neurology* supplies evidence that an echovirus, a variety of enterovirus, may be partly responsible for amyotrophic lateral sclerosis (ALS, also known as Lou Gehrig's disease) by damaging neurons.
Jan. 16	Space	MirCorp, a Bermuda-based company, claims to have paid Russia to keep the Mir space station operational as government funding ends, a rare instance of private investors rescuing a space program.
Jan. 17	Chemistry	In *Angewandte Chemie*, chemists Philip E. Eaton and Mao-Xi Zhang announce synthesis of octanitrocubane, a molecular cube with carbon atoms at each corner. The substance promises to be 20 percent to 25 percent more powerful than existing military high explosives.
Jan. 17	Environmental science	According to a study published in *Nature*, the twentieth century was the warmest of the last five centuries, having an average temperature 1.8 degrees Fahrenheit higher than that of the 1500's.
Jan. 18	Astronomy	Geologist Charles F. Roots spots a meteor descending to Earth. Found days later, it is a rare carbonaceous chondrite, and its lack of contamination by terrestrial chemistry makes it a precious find for scientists studying the development of the solar system.
Jan. 19	Space	Lockheed Martin's new Atlas III successfully launches, becoming the first American rocket to be powered by a Russian engine, developed by NPO Energomash.
Jan. 20	Physics	Researchers at the National Institute of Standards and Technology report in *Nature* that they quantified the degree of decoherence for an electron with superimposed spin states in relation to the states' physical distance from each other, an important measurement for quantum mechanics.
Jan. 27	Health	*The New England Journal of Medicine* reports the first proven case in which a mortician was infected with tuberculosis from a corpse, explaining the notoriously high rate of TB inflection among funeral home workers.

Date	Science	Event
Jan. 28	Chemistry	French researchers report in *Science* that they can identify the origin of gemstones from the mixture of oxygen isotopes specific to each. They used the technique to help answer historical questions about several famous precious gems.
Feb. 1	Health	The U.S. Department of Agriculture permits the use of ionizing radiation to kill such pathogens in raw meat as *Escherichia coli 0157:H7* and *Salmonella*.
Feb. 1	Physiology	In *Nature Neuroscience*, University of Miami scientist Nirupa Chaudhari finds that specific receptors in taste buds exist for L-glutamate, vindicating earlier claims that there is a fifth taste (called the "umami taste") in addition to those for sweet, sour, bitter, and salty.
Feb. 1	Applied technology	An Illinois Institute of Technology team invented a flame-retardant substance that can prevent lithium-ion batteries from overheating and igniting, which may make larger, more powerful batteries possible for a wide range of applications, according to *Electrochemical and Solid-State Letters*.
Feb. 3	Medicine	Japanese scientists announce discovery that some cases of type I (juvenile-onset) diabetes are not caused by an autoimmune reaction. Instead, they suspect environmental chemical toxins or a virus, they report in *The New England Journal of Medicine*.
Feb. 4	Biotechnology	*Science* reports that medical researchers have successfully controlled diabetes in mice with a genetic implant that stores insulin in cells so that it can be released as needed. The technique is thought to hold promise for human diabetics.
Feb. 5	Environment	Fifty nations sign the Cartagena protocol on biosafety, a treaty to protect global biodiversity and set standards for international trade in bioengineered organisms.
Feb. 10	Physics	Scientists announce that the particle accelerator at the European Center for Nuclear Research (CERN, near Geneva, Switzerland) has created a state of matter that has not existed since immediately after the big bang. It consists of free-moving quarks and gluons.
Feb. 14	Space	After a four-year voyage, the Near Earth Asteroid Rendezvous (NEAR) spacecraft orbits the asteroid Eros on Valentine's Day. It is the first probe to orbit an asteroid and begins a year of mapping the surface and measuring its magnetic field.

Date	Science	Event
Feb. 15	Chemistry	In *Proceedings of the National Academy of Sciences*, scientists describe their isolation of a compound in barberry plants, 5'-methoxyhydnocarpin, which used in combination with antibiotics kills drug-resistant bacteria.
Feb. 15	Environment	Scientists from France, Germany, and Japan release a study in *Environmental Science and Technology* showing that volcanoes are minor sources of chlorofluorocarbons (CFCs), a class of greenhouse gases, contradicting an influential earlier study and supporting the view that most CFCs in the atmosphere are human-made.
Feb. 16	Information technology	In a controversial press release, a Stanford University study finds that people who use the Internet more than ten hours weekly, a growing percentage of users, increase their isolation, a trend that could decrease overall community involvement.
Feb. 16	Health	Echoing a study published in the January issue of *The New England Journal of Medicine*, researchers report in the *Journal of the National Cancer Institute* that by taking the hormones estrogen and progestin, women increase the risk of contracting breast cancer. Women commonly take the hormones after menopause to fend off heart disease and bone degeneration.
Feb. 18	Life sciences	At the annual meeting of the American Association for the Advancement of Science, Celera Genomics scientist J. Craig Venter announces that his team sequenced the entire genome of the fruit fly *Drosophila melanogaster* using a controversial technique known as shotgun sequencing.
Feb. 18	Medicine	The Food and Drug Administration approves Prevnar, the first vaccine designed to protect children younger than five from the pneumococcus bacterium, a leading cause of ear infections, meningitis, and blood poisoning.
Feb. 21	Health	During the annual meeting of the American Association for the Advancement of Science, scientists unveil findings that the antioxidant compounds called flavonoids present in chocolate and cocoa powder can help protect chocolate lovers from cardiovascular disease.
Feb. 22	Environment	At the annual meeting of the American Association for the Advancement of Science, a National Aeronautics and Space Administration researcher reveals evidence that growing cities are covering some of Earth's best croplands, reducing photosynthesis productivity, and affecting biodiversity.

Date	Science	Event
Feb. 22	Earth sciences	The space shuttle *Endeavour* lands after spending a week collecting radar images of Earth that will be converted to make the most finely detailed three-dimensional maps of the surface ever produced.
Feb. 25	Physics	Experimenters describe in *Science* magazine successful use of a single photon to control a single atom by quantum mechanical means. The "atom-cavity microscope" creating the interaction may lead to quantum computers and telecommunication devices.
Feb. 29	Information technology	The President's Council on Year 2000 Conversion is demobilized after helping government agencies and businesses avoid computer malfunctions as they converted from the year 1999 to the year 2000.
Mar. 1	Medicine	*Nature Medicine* publishes a study in which the effects of type I (juvenile-onset) diabetes are reversed in mice by transplantation of insulin-producing beta cells from healthy, genetically identical mice. No tissue rejection occurred in the tests, raising hopes a similar procedure can be developed for people.
Mar. 1	Medicine	*Nature Genetics* reports success in efforts to control hemophilia by using a virus to repair genetic errors that cause the disease.
Mar. 3	Medicine	In *Science*, U.S. Army researchers reveal discovery of five types of antibodies that may protect against the deadly Ebola virus. Infected mice injected with the antibodies had an 88 percent survival rate after exposure to the disease; all untreated mice died.
Mar. 5	Biotechnology	The Scottish company PPL Therapeutics, which first successfully cloned a sheep, announces the birth of the first cloned piglets, five females. PPL claims the births to be the first step toward pigs genetically altered so that their organs and cells can be transplanted to cure human diseases without tissue rejection.
Mar. 9	Space	The National Aeronautics and Space Administration officially abandons attempts to communicate with the Mars Polar Lander, which apparent crashed during its descent to Mars on December 3, 1999.

Date	Science	Event
Mar. 10	Astronomy	In *Science*, astronomers describe types of ripples that show up on the surface of the sun before eruptions that broadcast energy and particles of a solar storm. The storms can damage satellites and power grids on Earth. The discovery will help astronomers forecast the eruptions and issue warnings of danger as much as two weeks in advance.
Mar. 10	Life sciences	An Argentine paleontologist announces discovery of the largest carnivorous dinosaur yet found. Thriving during the Cretaceous period, the newly discovered species was 10 percent larger than the previous record holder, *Gigantosaurus carolinii*, which could grow to forty-five feet in length. Confounding a long-held theory about large predator dinosaurs, the species appears to have hunted in packs.
Mar. 14	Medicine	A report in *Neurology* describes a surgical technique to transfer brain cells from pigs into the brains of people suffering from advanced Parkinson's disease. The pig cells manufacture dopamine, the neurotransmitter lacking in victims' brains. The experimental surgery helped relieve Parkinson's symptoms in some test patients.
Mar. 16	Life sciences	Paleontologists announce in *Nature* discovery of the smallest known primate. The lemur-like species, long extinct, lived 42 million years ago in central China and weighed no more than one-half ounce.
Mar. 16	Applied technology	*Nature* describes a recently developed fuel cell that produces electricity through electrochemical reactions of hydrocarbons, such as natural gas or diesel. Operating at safer temperatures than previous fuel cells, it produces water and carbon dioxide as byproducts and could be used to generate commercial electrical power or to run automobiles.
Mar. 16	Chemistry	Researchers report in *Nature* that they used a special technique, flash chemical vapor deposition, to grow microscopic cones of carbon. The technique may allow production of nanotubes, which are expected to be key components of nanotechnology.
Mar. 16	Physics	A study published in *Nature* reveals that scientists succeeded in coaxing four beryllium ions into quantum entanglement, considered an important step in creation of a quantum computer.

Date	Science	Event
Mar. 17	Earth sciences	In *Scientists*, atmospheric scientists argue that a short-term climate cycle that starts in the Indian Ocean, called the Madden-Julian Oscillation, is associated with the formation of hurricanes in the Gulf of Mexico and western Caribbean and can be used to predict hurricanes.
Mar. 20	Astronomy	In a study posted on the Internet, astronomers supply evidence that the Milky Way acquired its vast halo of stars by cannibalizing nearby galaxies.
Mar. 22	Earth sciences	The National Science Foundation announces that a record-size section of Antarctica's Ross Ice Shelf has broken off. It is 177 miles long and 22 miles wide.
Mar. 23	Genetics	A study published in *Nature* finds that lesbians, like men, have shorter index fingers than ring fingers, whereas in heterosexual women the two fingers are about the same length. The researchers believe that the male hormone androgen may be responsible for the difference and that the findings support the view that homosexuality is at least partly a biological phenomenon.
Mar. 27	Environment	At the annual meeting of the American Chemical Society, scientists produce evidence that pollution by pharmaceuticals, such as aspirin and antibiotics, is an increasing danger to water purity.
Mar. 28	Chemistry	In the *Proceedings of the National Academy of Sciences*, scientists announce they found fullerenes, a spherical carbon molecule, of extraterrestrial origin. Discovered near ancient meteor impact sites, the molecules contained noble gases of unearthly isotope mixture and demonstrate that organic molecules survive catastrophic meteor and asteroid impacts.
Mar. 28	Biotechnology	A research team reports in the *Proceedings of the National Academy of Sciences* that its genetically engineered version of a rabies virus produced massive amounts for antibodies against human immunodeficiency virus (HIV) in mice. Moreover, this potential vaccine does not destroy HIV-infected cells, as is the case with other types of experimental vaccine.
Mar. 28	Health	A study by the U.S. Department of Agriculture finds that 28 percent of cattle in large slaughterhouses are contaminated by the *Escherichia coli* strain O157:H7, which can cause severe food poisoning. Previously, only 2 percent of cattle were thought to be tainted by the bacteria. Reported in the *Proceedings of the National Academy of Sciences*.

Date	Science	Event
Mar. 28	Space	National Aeronautics and Space Administration investigators announce that a software error probably caused the landing engines on the Mars Polar Lander to shut down prematurely so that the craft crashed and was destroyed. They blame the error on cost-cutting that left too little money for thorough testing.
Mar. 29	Astronomy	At a National Aeronautics and Space Administration news conference, astronomers report detection of two planets orbiting sunlike stars. The existence of the planets, both smaller than Saturn and the smallest thus far discovered, boosts hopes that Earth-size planets exist outside the solar system.
Apr. 1	Mathematics	In *Physical Review E*, Japanese mathematician Takashi Nagatani describes computer simulations showing that by speeding up and slowing down randomly, a single driver can cause traffic jams in moderate traffic as large as those normally associated with much heavier traffic.
Apr. 1	Astronomy	The *Monthly Notices of the Royal Astronomical Society* publishes an article describing the first direct evidence that supermassive black holes at the centers of galaxies grow as their galaxies grow, which is by acquiring mass from galactic mergers.
Apr. 3	Medicine	At a meeting of the American Association for Cancer Research, Genta company scientists report the first success with "anti-sense technology" in inhibiting cellular production of a protein that fosters cancer growth. The technique helped shrink melanomas in some test patients.
Apr. 4	Medicine	In *Nature*, researchers report creation of a synthetic peptide, called beta-peptide, that in laboratory experiments proved effective in killing drug-resistant strains of bacteria that cause infections in the skin, bones, and intestinal and urinary tracts.
Apr. 5	Health	The National Academy of Sciences releases a report concluding that genetically engineered crops are safe but calls on the Environmental Protection Agency, Food and Drug Administration, and Department of Agriculture to work together more closely in testing new bioengineered plants.
Apr. 7	Applied technology	A team of researchers announces in *Science* invention of a low voltage electro-optical modulator that may improve data transfer in computers, fiber-optic communications, and radar systems.

Date	Science	Event
Apr. 11	Earth sciences	In the *Proceedings of the National Academy of Sciences*, geologists propose that cycles of moon-caused tides, and not sunspots, bring about fluctuations from warm periods to cold periods on Earth and may start ice ages.
Apr. 11	Life sciences	A theory that the original genetic molecule was peptide nucleic acid (PNA) receives support when an article in the *Proceedings of the National Academy of Sciences* describes an experiment in which PNA was created under conditions that mimicked those of Earth's primordial atmosphere.
Apr. 13	Environment	*Nature* publishes a large-scale study supporting the contention that on a global scale, amphibian populations have declined in recent decades by about 2 percent per year.
Apr. 13	Physics	A report in *Nature* describes a technique whereby scientists can position individual atoms with a scanning tunneling electron microscope at room temperatures. The capability, previously possible only at near absolute zero, is important to the development of nanotechnology.
Apr. 14	Health	Papers delivered at a convention of the American Chemical Society tout white tea and unprocessed olive oil for their ability to combat cancer and heart disease with antioxidants.
Apr. 14	Life sciences	Swiss scientists reveal in *Science* that they have found a molecular reaction causing micrometer-size silicon strips to bend when they bind to deoxyribonucleic acid (DNA) strands. A reaction-driven device powered by the effect could serve as a microscopic sensor or as machines operating inside the body.
Apr. 14	Astronomy	A Princeton University astrophysicist, participating in the Sloan Digital Sky Survey, finds a quasar 12 billion light-years from Earth, the most distant object ever detected; in November the distance is recalculated to be only 9.8 billion light-years.
Apr. 16	Health	Papers delivered at the Experimental Biology 2000 meeting reveal evidence that consumption of calcium strongly affects how quickly weight can be lost during dieting. The reports refer to studies of mice and women.
Apr. 17	Economics	In a study financed by the United Nations and World Bank, the World Resources Institute calls on governments and businesses to rethink basic economic assumptions in order to reverse the decline in ecosystems. Otherwise, the report claims, humanity could face "devastating implications."

Date	Science	Event
Apr. 18	Environment	The National Oceanic and Atmospheric Administration announces that the first quarter of 2000 is the warmest on record. The average daily temperature of 41.7 degrees Fahrenheit (5.4 degrees Celsius) is one degree higher than the previous record, set in 1990.
Apr. 21	Life sciences	*Science* reports discovery by a father-son team of the first fossilized dinosaur heart. Studies of it suggest that the species *Thescelosaurus* was warm-blooded and had a high rate of metabolism.
Apr. 21	Life sciences	In *Science*, linguists advance the theory that baby talk resembles the first human language. The theory builds on studies of baby talk in seven countries with different languages.
Apr. 27	Astronomy	*Nature* displays images of the universe when it was only 300,000 years old. Collected by a thirty-six-member international team, the data supplies direct evidence for the theory of inflation and suggests that the universe is flat.
Apr. 27	Medicine	Doctors claim in *Science* that they achieved the first true success for gene therapy when they cured two infants of severe combined immunodeficiency by replacing defective genes with working genes.
Apr. 27	Life sciences	Researchers at Advanced Cell Technology report in *Science* that they succeeded in reversing the aging process in cloned cells. One of the biggest problems in cloning is that the cloned cell is the same biological age as the parent cell.
Apr. 28	Chemistry	*Science* relates that after a twenty-year effort, Stanford University scientists deciphered the core structure of polymerase II, which unbinds deoxyribonucleic acid (DNA) in a cell's nucleus and makes possible a ribonucleic acid (RNA) transcript of it for protein synthesis.
Apr. 29	Astronomy	After a decade-long search, scientists report at an American Physical Society meeting that they discovered unexpectedly strong magnetic fields in intergalactic space, possibly relics of the early universe.
Apr. 29	Health	A study reported in *Lancet* reveals that beer increases concentrations of vitamin B_6 in the blood. The vitamin processes amino acids necessary for brain function and also breaks down homocystine, an amino acid associated with heart disease.

Date	Science	Event
May 1	Biotechnology	Researchers announce they have grown crops of herbicide-resistant corn created by altering a gene in the corn with a ribonucleic acid (RNA)-deoxyribonucleic acid (DNA) chimera. The announcement of the new genetic engineering technique for plants appears in *Nature Biotechnology*.
May 1	Biotechnology	In tests with hemophiliac mice, according to *Nature Genetics*, Stanford University researchers succeeded in using transposons, mobile sections of deoxyribonucleic acid (DNA), to deliver a functional bioengineered gene. After receiving the new gene, originally from a fish, the mice had greatly improved blood coagulation.
May 1	Earth sciences	Scientists warn in *Geology* that sections of seafloor on the edge of the continental shelf off the U.S. East Coast are unstable and could slough off. The resulting underwater landslides would generate tsunami waves as much as fifteen feet (more than three meters) high.
May 1	Astronomy	In *Astronomical Journal Letters*, a team of observers reports evidence of vast clouds of hydrogen in intergalactic space. Detected indirectly with the Hubble Space Telescope, the hydrogen may help solve the "missing matter" problem of cosmology. According to cosmologists, 90 percent of the matter in the universe is unaccounted for.
May 1	Physics	At a meeting of the American Physical Society, two University of Washington scientists report that Earth is billions of metric tons lighter than previously thought. They base their calculations, which they call preliminary, on a revaluation of the gravitational constant *G* at 6.6742×10^{-11} cubic meters per kilogram-second.
May 1	Applied technology	By presidential order, the United States no longer blocks civilian access to the full capability of the Global Positioning System (GPS), allowing GPS navigation devices to be as much as ten times more accurate. The system can locate receivers to within thirty to sixty feet, a degree of precision hitherto available only to the U.S. military.
May 3	Information technology	The ILOVEYOU computer virus, originating in the Philippines, cripples hundreds of thousands of e-mail and computer systems worldwide by targeting Microsoft Outlook software.

Date	Science	Event
May 8	Genetics	The Human Genome Project announces that a German-Japanese team completely sequenced the genes on human chromosome 21, the smallest chromosome. Defective genes on it are associated with Down syndrome, epilepsy, amyotrophic lateral sclerosis (ALS, or Lou Gehrig's disease), and Alzheimer's disease.
May 11	Applied technology	The Powell Structural Research Laboratory in San Diego tests a two-story house specially designed to survive earthquakes. Subjected to a simulated quake measuring 6.7 on the Richter scale, the house suffers only minor cracks, although its furniture and appliances are largely destroyed.
May 12	Life sciences	*Science* reports discovery of two 1.7 million-year-old fossilized human skulls in the Republic of Georgia. Identified as the species *Homo ergaster*, the skulls are the earliest evidence of humans outside of Africa and may be from the first species that migrated to Eurasia.
May 15	Medicine	Researchers report a possible treatment for multiple sclerosis (MS) in the *Journal of Clinical Investigation*. The treatment, a combination of the hormone estrogen and a T-cell receptor vaccine, prevents MS-like symptoms in laboratory mice.
May 15	Computer science	*Physical Review Letters* publishes a report by University of Innsbruck, Austria, scientists in which they describe construction of a microchip based upon the flow of whole atoms instead of electrons. The achievement is part of the quest to develop ultrafast quantum computing.
May 16	Physics	Legendary mathematical physicist Freeman J. Dyson receives the Templeton Prize for his books arguing that religion and science both should be respected as ways of understanding the universe and for insisting that ethics guide technological development.
May 17	Information technology	At the Ninth International World Wide Web Conference, International Business Machines scientists announce development of the first comprehensive map of the web, which suggests that it has a more intricate structure than previously thought.
May 17	Astronomy	Mercury, Venus, Mars, Jupiter, and Saturn all line up on the far side of the sun from Earth, a rare alignment that inspires Internet doomsayers to forecast catastrophe.

Date	Science	Event
May 20	Biotechnology	In the journal *Human Gene Therapy*, researchers report that they successfully used a virus to insert a gene into mice and rat cells to stimulate bone-growing proteins. The technique could provide treatment for fractures in human bones that now require surgery.
May 20	Astronomy	Beating a record established just a month earlier, astronomers announce detection of a galaxy 13.6 billion light-years from Earth. The observation team used the Keck II telescope in Hawaii to spot it.
May 22	Physics	Italian physicists report in *Physical Review Letters* an experiment in which microwaves appear to exceed the speed of light. Results of a similar experiment are described in the May 30 *New York Times*.
May 23	Applied technology	During the annual meeting of the American Society for Microbiology, researchers report successful testing of a pen-size device that uses electricity to quickly kill viruses, bacteria, and parasites in water. Miox Corporation of Albuquerque, New Mexico, manufactures the water-purification device.
May 25	Health	Two studies reported in *Nature* find that existing data show cell phones are safe to use but warn that radiation from the phones can spur physiological changes with unknown health effects.
May 25	Space	The space shuttle *Atlantis* nudges the International Space Station (ISS) out of a dangerously low orbit. Increased solar activity caused Earth's atmosphere to expand and ISS to drop to a 203-mile low, where increasing atmospheric drag threatened the station.
May 25	Applied technology	In *Nature*, doctors at the Royal London Hospital report successful use of a pill-sized camera that, when swallowed, can image the entire gastrointestinal tract. Developed by Israel's Given Imaging, the tiny probe can detect lesions and tumors in the small intestine, where current diagnostic tools, endoscopes, cannot reach.
May 26	Space	National Aeronautics and Space Administration controllers send the malfunctioning Compton Gamma Ray Observatory into the atmosphere so that it will burn up safely over an ocean. The move ends nine years of gamma-ray source mapping that greatly expanded astronomers' understanding of gamma-ray bursts and high-energy emissions from the sun.

Date	Science	Event
May 29	Applied technology	A novel device that transmits ultrashort laser pulses through a special fiber precisely splits the light into evenly spaced frequencies, according to an article in *Physical Review Letters*. The device dramatically lowers the cost of precision electromagnetic measurements, an important process in developing technology and basic research.
June 1	Applied technology	In the *Journal of the American Chemical Society*, University of California, San Diego, chemists describe a sensor capable of detecting such chemical warfare agents as the nerve gases sarin, soman, and cyclohexyl methylphosphonofluoridate (GF). The sensor employs a porous silicon interferometer.
June 5	Astronomy	In a controversial report to the American Astronomical Society, a group of astronomers claims that at least fifteen of the fifty extrasolar planets detected thus far are really brown dwarfs or small stars instead of planets.
June 6	Astronomy	A report delivered to the American Astronomical Society supplies evidence that all galaxies with a central bulge, such as the Milky Way, contain a black hole, usually accounting for 0.2 percent of the bulge's mass.
June 6	Environment	A study published online in the *Proceedings of the National Academy of Sciences* finds that a variety of corn genetically engineered to produce its own pesticide protection against corn borers does not harm butterflies. Suspicions of a corn-butterfly connection inspired widescale protests over genetically engineered crops.
June 7	Environment	At a meeting of the National Ground Water Association, state, federal, and university scientists report that pharmaceuticals excreted with human waste are often still biologically active and pose a threat to wildlife, especially fish.
June 8	Astronomy	At a meeting of the American Astronomical Society, scientists discuss evidence that vacuum energy creates most of the gravitation in the universe. The evidence comes from observation of 100,000 galaxies by the Two Degree Field Galaxy Redshift Survey.
June 12	Applied technology	*Physical Review of Letters* describes the successful test of a system called quantum-key distribution to encrypt signals. The system, applying quantum mechanics principles, is designed for use in satellite communications and is billed as unbreakably secure.

Date	Science	Event
June 15	Astronomy	The National Radio Astronomy Observatory announces detection of a sugar, glycolaldehyde, in interstellar space. The molecule, detected by a radio telescope atop Kitt Peak in Arizona, is potentially a biological building block.
June 22	Space	At a National Aeronautics and Space Administration news conference, researchers release photographs from the Mars Global Surveyor that suggest liquid water exists near the surface of Mars. The photographs show remnants of mudflows that occurred in the recent past at 120 locations.
June 22	Chemistry	French scientists demonstrate in *Nature* that Earth's magnetic field influences which enantiomers of molecules are preferentially incorporated into compounds. The discovery sheds light on a 150-year-old mystery concerning the left-handedness of amino acids and the right-handedness of sugars.
June 26	Genetics	Leaders of the national Human Genome Research Institute and Celera Genomics announce that together they have deciphered the sequence of codons in all 23 chromosomes making up the human genome. This major achievement helps scientist locate genes and determine their function.
June 27	Computer science	In a press release, scientists from the University of Iowa and Argonne National Laboratory announce they have a solution for the thirty-two-year-old NUG30 quadratic assignment problem. It entails assigning thirty facilities to thirty fixed locations in such a way as to minimize the cost of transferring material among them. The solution required more than one thousand computers to calculate for a week.
July 1	Computer science	International Business Machines prepares to market a 75-gigabyte hard drive for personal computers (PCs), while Maxtor readies a 60-gigabyte drive. Both have far more memory capacity than existing PC drives.
July 3	Information technology	President Bill Clinton signs legislation authorizing use of electronic signatures (e-signatures) in commercial transactions, making possible such new technology as smart cards, server-based personal identification, and biometric systems.
July 6	Biotechnology	The *Journal of Infectious Diseases* reports successful testing of potatoes genetically modified to immunize people against a viral disease. All but one of the volunteers who ate the vaccine-potatoes for three weeks had elevated levels of antibodies combating Norwalk virus, which causes diarrhea, nausea, and stomach cramps.

Date	Science	Event
July 11	Applied technology	The U.S. Food and Drug Administration approves the first robotic device for surgery, the Da Vinci Surgical System. Seated behind a video monitor, the surgeon controls the robot through a computer. The device allows small incisions, even for major operations such as cardiac surgery, and is expected to reduce recovery times.
July 15	Astronomy	As reported in *Geophysical Research Letters*, a reexamination of mineral evidence from Mars leads to the conclusion that two to three times more water may exist underground on Mars than previously estimated.
July 15	Life sciences	In *Genes and Development*, Japanese scientists reveal the mechanism by which the deadly *Escherichia coli* strain O157:H7 causes symptoms. One of its two toxic chemicals, called verotoxins, disables a protein that prevents cells from self-destructing.
July 18	Environment	The National Research Council predicts that synthetic chemical pesticides will continue to play an important role in U.S. agriculture until at least 2010.
July 18	Applied technology	In *Proceedings of the National Academy of Sciences*, a team of German scientists reports development of a microscope capable of resolving details as small as two hundred nanometers in size. The stimulated emission depletion microscope can peer inside fluorescent-labeled living cells without harming them.
July 21	Physics	An international research team announces detection of the elusive tau neutrino, the last of the theorized twelve basic building blocks of matter to be confirmed experimentally. The detection was made at the Fermi National Accelerator Laboratory (Fermilab) in Illinois.
July 21	Earth sciences	A study in *Science* says that although most of Greenland's vast ice cover is unchanged, considerable thinning of it exists near the coasts, contributing to rising sea levels.
July 21	Information technology	The administration of President Bill Clinton eases restrictions on 128-bit encryption technology, allowing U.S. companies to sell it in some European and Pacific Rim countries.
July 25	Information technology	At the annual SIGGRAPH conference, Bell Laboratory scientists describe a novel technique that makes it practical to compress and transmit detailed three-dimensional images via the Internet and view them on a personal computer.

Date	Science	Event
July 26	Space	The Zvezda service module, built by Russia, docks with the International Space Station. Long delayed, the addition of Zvezda, a major component, provides living quarters and steering control for the station's first crew.
July 28	Computer science	In *Science*, Japanese researchers describe a process to make nearly perfect crystal-like structures known as photonic crystals. Such crystals are needed to construct microchips that operate with light instead of electricity.
Aug. 1	Medicine	A long-term study published in the *Journal of the American Academy of Child and Adolescent Psychiatry* finds that most children nine to sixteen years old who take Ritalin or similar stimulants do not suffer from attention-deficit hyperactivity disorder (ADHD). ADHD is the only condition for which the drugs are approved.
Aug. 1	Environment	A study conducted in California and published in *Environmental Science and Technology*, suggests that automobiles have overtaken livestock as the principal source of atmospheric ammonia pollution.
Aug. 1	Computer science	A study in the *American Journal of Roentgenology* finds that computers running neural network scans equal human performance in reading X-ray images of the body and interpreting the information for signs of disease.
Aug. 1	Medicine	In the *Journal of Neuroscience Research*, scientists report that they induced bone marrow cells to turn into nerve cells, raising the possibility that pristine brain cells can be cultured for treatment of neurological disorders such as Alzheimer's and Parkinson's diseases.
Aug. 1	Environment	In *Environmental Science and Technology*, Chinese scientists reveal that Chinese industry and energy production put 305 tons of mercury into the environment in 1995, one-twentieth of the world's mercury pollution, and the nation's emissions grow about 5 percent yearly.
Aug. 3	Life sciences	Scientists announce in *Nature* that they have sequenced the 3,885 genes in the two circular chromosomes of *Vibrio cholera*, a step toward developing a vaccine against the bacterium, which kills thousands of people worldwide every year.
Aug. 4	Chemistry	A study published in *Science* demonstrates that a group of ceramics called fluorites are nearly impervious to radiation damage. They may prove to be safe materials for nuclear waste containers.

Date	Science	Event
Aug. 15	Medicine	A clinical study published in *Annals of Internal Medicine* confirms earlier reports that people can halve the length of a head cold by taking zinc pills every few hours when they first feel symptoms.
Aug. 18	Astronomy	The Spacewatch telescope atop Kitt Peak in Arizona finds yet another moon of Jupiter, the seventeenth. The moon, provisionally designed UX_{18}, is about three miles in diameter.
Aug. 18	Earth sciences	Tourists aboard a Russian icebreaker visiting the Arctic report that the North Pole has thin ice and even open water instead of the normal ice cover several feet thick. The report is taken as evidence of the effects of global warming on the oceans.
Aug. 20	Computer science	A paper delivered at the Hot Chips 2000 conference describes a computer, composed of a few atoms, which demonstrated that quantum computation techniques can solve some mathematical problems more efficiently than can conventional computers.
Aug. 21	Applied technology	American University scientist Paul F. Waters reports at an American Chemical Society meeting that adding the polymer polyisobutylene to gasoline increases the performance of automobile engines while reducing emissions dramatically. The polymer is usually used to make synthetic rubber.
Aug. 23	Medicine	The National Institutes of Health rules that it can provide federal funds to scientists for research involving stem cells from human embryos, provided that they are embryos already scheduled for destruction at fertility clinics.
Aug. 24	Chemistry	In *Nature*, Finnish scientists recount how they constructed a molecule with argon in it, argon fluorohydride (HArF), at a temperature near absolute zero. It is the first time argon, a noble gas, is coaxed into bonding with other atoms.
Aug. 25	Biotechnology	In *Science*, researchers reveal that they have bioengineered the common *Streptococcus gordonii* bacterium to fight other disease-causing organisms. In tests *S. gordonii* cleared female mice of *Candida albicans*, which causes vaginal yeast infections.

Date	Science	Event
Aug. 25	Life sciences	An article in *Science*, along with one published later in *Behavioral and Brain Sciences*, advances evidence that bottlenose dolphins, killer whales, sperm whales, and humpback whales (the best-studied cetacean species) all convey cultural traditions of communication, feeding, mating, and training of the young from one generation to the next.
Aug. 26	Medicine	Investigators report in *Lancet* that reimplanted stem cells from the bone marrow of seven patients with a severe form of the autoimmune disease lupus apparently cured six of them, some of whom were near death.
Aug. 28	Environment	An English scientific team reports in *Science* on a newly discovered greenhouse gas, trifluoromethyl sulfur pentafluoride (SF_5CF_3), which is increasing in concentration at about 6 percent yearly. The gas absorbs infrared light from the sun and contributes to atmospheric warming.
Aug. 28	Physics	In *Science*, University of California, Berkeley, scientists describe their fabrication of a multiwall nanotube only about one hundred atoms wide. They suggest the nanotube could serve as springs or bearings in ultrasmall machines.
Aug. 31	Applied technology	In *Nature*, Brandeis University scientists announce that they have built the first robot that can design other robots on its own and instruct humans how to build them.
Aug. 31	Computer science	Taking into account only the limitations of physical constants, theorist Seth Lloyd of the Massachusetts Institute of Technology claims that the operational maxima for a laptop-size computer are 10^{51} operations per second and a 10^{31} bit memory capacity.
Sept. 1	Medicine	A study published in *Stroke* asserts that carotid endarterectomy, a surgical procedure to clear plaque from the carotid artery in the neck, wards off strokes better than medications.
Sept. 1	Medicine	Researchers announce in the *American Journal of Psychiatry* that they have devised a test to reveal who is most likely to develop Alzheimer's disease. The test requires identification of forty odors. Those with the lowest scores are most at risk for Alzheimer's.

Date	Science	Event
Sept. 1	Medicine	A genetic susceptibility triggered by a streptococcal infection such as strep throat can create pediatric autoimmune neuropsychiatric disorders (PANDAS) in children, says an article in the *Journal of the American Academy of Child and Adolescent Psychiatry*. Among the PANDAS are obsessive-compulsive disorder and Tourette's syndrome.
Sept. 1	Medicine	*Nature Medicine* publishes a study that suggests most occurrences of narcolepsy are caused by the loss of brain cells, raising the possibility of a treatment by implantation of new cells. Another study in *Neuron* confirms the finding.
Sept. 1	Astronomy	In *Astrophysical Journal Letters*, an observational team offers evidence of Lense-Thirring precession, a bizarre helical twisting of space-time hinted at in Albert Einstein's general theory of relativity. The effect was detected near rapidly spinning neutron stars.
Sept. 5	Medicine	In the *Journal of Experimental Medicine*, researchers warn of a new tactic of the human immunodeficiency virus (HIV). Test tube studies show that the HIV can attach itself to B cells, antibody factories for the body's immune system, and thereby remain safely hidden until an opportunity comes to attack T cells, whose destruction may lead to development of acquired immunodeficiency syndrome (AIDS).
Sept. 7	Chemistry	German scientists announce in *Nature* that they synthesized $C_{20,}$ the smallest of a class of carbon molecules known as fullerenes.
Sept. 8	Information technology	RSA Security releases to the public domain its widely used RSA public-key encryption algorithm, which dominated the field of encryption for seventeen years. The move comes a week before the patent on the algorithm is due to expire.
Sept. 11	Medicine	At a meeting of the Royal Society in London, England, Claudio Basilico of New York University Medical Center reports that a polio vaccine used more than forty years ago in Africa was not tainted with any viruses or chimpanzee deoxyribonucleic acid (DNA). The finding appears to refute the theory of journalist Edward Hooper that the human immunodeficiency virus (HIV) was transmitted to humans via polio vaccinations.

Date	Science	Event
Sept. 12	Life sciences	An insect study at Sweden's University of Umeå, presented in *Proceedings of the National Academy of Sciences*, finds that species with high degrees of mating conflict split into new species more often than do species with low degrees of mating conflict.
Sept. 14	Chemistry	In the *Journal of Physical Chemistry B*, an international research team describes a solar-powered hydrogen fuel cell that is 50 percent more efficient than previous systems. The development is part of the attempt to replace fossil fuels in automobiles and other machinery.
Sept. 14	Physics	The European Center for Nuclear Research (CERN) near Geneva, Switzerland, announces it will extend operation of the Large Electron-Positron collider (LEP) past its scheduled shutdown because experiments with it have uncovered hints of the Higgs boson. Researchers hope further runs of LEP will supply more evidence of the existence of the long-sought particle.
Sept. 14	Physics	Researchers announce in *Nature* that by replacing certain yttrium ions with calcium ions in an yttrium-barium-copper oxide system, they can achieve superconduction temperatures high enough to make superconductors commercially practical.
Sept. 21	Chemistry	A novel method for purifying titanium from titanium dioxide promises to make the metal as common as steel, according to an article in *Nature*. Corrosion-resistant and lightweight, cheap titanium would have many practical applications, such as in automobile parts.
Sept. 22	Health	Kraft announces recall of tacos that contain a variety of bioengineered corn not approved for human consumption, fanning a controversy about food purity and genetically altered plants.
Sept. 23	Applied technology	*New Scientist* reports that Sony's Computer Science Laboratory taught the company's robotic dog to recognize objects and say their names, a step in the development of an autonomous robot that can interact with people.
Sept. 25	Computer science	Eight weeks after entering the race to produce "superchips," high-tech giant Texas Instruments announces it is abandoning the effort. Superchips combine the functions of processors, read channels, support logic, and memory for use in hard drives.

Date	Science	Event
Sept. 28	Medicine	An article in the *New England Journal of Medicine* finds that interferon-beta-1a can delay and possibly prevent multiple sclerosis in people with early-warning indications of the disease.
Sept. 29	Astronomy	Observers using the Canada-France-Hawaii telescope on the island of Hawaii spot an unidentified object in space, as large as 230 feet across, that stands a one-in-five-hundred chance of colliding with Earth in September, 2030. The object could be an asteroid or the hulk of a Saturn V moon rocket. The assessment is later modified to a one-in-one-thousand chance during 2071.
Oct. 1	Applied technology	The *Journal of the Acoustical Society of America* reports development of software that should make it possible for portable sonic scanners to detect flaws, such as corrosion, in aircraft fuselages with greatly increased sensitivity. Such a scanner could help prevent devastating aircraft skin and structural failures.
Oct. 1	Applied technology	A new method for microscopic imaging, reported in *Biophysical Journal*, improves resolution as much as ten thousand times over alternative techniques. This Fourier imaging correlation spectroscopy permitted scientists to glimpse a new process within mitochondria.
Oct. 1	Life sciences	Two experimental drugs designed to deactivate free radicals increase the life span of nematodes 50 percent, providing evidence for a popular theory linking free radicals and aging, an article in *Science* reports.
Oct. 1	Genetics	A report in *Nature Genetics* links a newly discovered gene, *CAPN10* on chromosome 2, to diabetes. Normally coding for proteases, the gene can produce polymorphisms that increase the risk for type II (adult-onset) diabetes threefold.
Oct. 2	Computer science	The National Institute of Standards and Technology announces the winner of a worldwide contest for the U.S. government's Advanced Encryption Standard. The winner, Rijndael, developed in Belgium, permits encryption keys of 128, 192, or 256 bits.
Oct. 5	Environment	A scientist at New Zealand's National Institute of Water and Atmospheric Research announces that the hole in the ozone layer over Antarctica has reached record size—11.4 million square miles (29.5 million square kilometers)—and on September 9 and 10 for the first time extended over a large city, Punta Arenas, in southernmost Chile.

Date	Science	Event
Oct. 5	Space	A hardware problem turns up involving the link between the Cassini spacecraft, due to explore Saturn in 2002, and the small Huygens probe aboard it that is to descend into Titan's atmosphere, reports the Jet Propulsion Laboratory. If not corrected, the problem could prevent Cassini from relaying much of the data sent to it by Huygens.
Oct. 6	Astronomy	In *Science*, astronomers announce detection of eighteen free-roaming bodies that could be nascent planets in the young star cluster sigma Orionis, twelve hundred light-years away. The discovery prompts theorists to rethink the definition of a planet.
Oct. 9	Medicine	The Nobel Prize in Physiology or Medicine is won by Paul Greengard and Eric R. Kandel of the United States and Arvid Carlsson of Sweden for explaining how dopamine transmits messages among brain cells, work that led to antidepressant drugs and treatments for Parkinson's disease.
Oct. 9	Space	The National Aeronautics and Space Administration launches the High-Energy Transient Explorer, an orbital observatory that is expected to record gamma-ray bursts at higher rates than before and help astronomers solve the mystery of their sources.
Oct. 10	Physics	American Jack Kilby wins half the Nobel Prize in Physics for inventing the integrated circuit in 1958. The second half of the prize is shared by American Herbert Kroemer and Russian Zhores Alferov, both of whom helped develop the tiny semiconductor heterostructure laser, crucial for cell phone technology, compact disc players, and satellite communications.
Oct. 10	Chemistry	Alan Heeger and Alan MacDiarmid of the United States and Hideki Shirakawa of Japan share the Nobel Prize in Chemistry for creating polyacetylene, a polymer that conducts electric current. These "brilliant plastics" found applications in information display screens, photography, and surgical materials.
Oct. 11	Economics	Americans James J. Heckman and Daniel L. McFadden receive the Nobel Memorial Prize in Economic Sciences for their research in microeconomics. Heckman is cited for developing a theory and methods for analyzing selective samples; McFadden for a theory and methods for analyzing discrete choice.

Date	Science	Event
Oct. 12	Applied technology	An article in *Nature* describes a new rechargeable magnesium battery produced by Israel's Doron Aurbach company. It is environmentally benign and could replace the lead batteries used in automobiles and other large machines.
Oct. 14	Applied technology	Micro-Craft, a San Diego aerospace company, announces that it has successfully tested a nine-inch spy plane. Front-line troops can fly the unmanned aerial vehicle (UAV) by remote control to observe enemy positions and movements.
Oct. 15	Environment	A report in *Environmental Science and Technology* finds that computer monitors and color television sets can leach lead into ground water and calls for the government to classify discarded units as toxic waste.
Oct. 18	Life sciences	According to a report in *Nature*, scientists succeed in reviving a bacterium embedded in a salt crystal for 250 million years. The crystal was excavated from a depth of 1,850 feet near Carlsbad, New Mexico.
Oct. 18	Environment	British Petroleum, Shell International, Du Pont, Suncor, Ontario Power Generation, Alcan of Canada, and Pechiney join in the Partnership for Climate Action, vowing to reduce their yearly emissions of total greenhouse gases by 80 million metric tons before 2010.
Oct. 20	Health	North Carolina health officials find a crow infected with West Nile virus, the furthest south in the United States that the deadly virus has been discovered.
Oct. 20	Medicine	In *Science*, Harvard Medical School researchers report tests on rhesus monkeys with a vaccine for acquired immunodeficiency syndrome (AIDS). The vaccine does not prevent infection by human immunodeficiency virus (HIV), but it does stimulate the immune system enough to hold the virus in check.
Oct. 21	Medicine	A large study reported in *Lancet* finds that it is safer to deliver babies positioned feet-down in the womb (breech position) by caesarian section rather than vaginally.
Oct. 21	Space	Space shuttle *Discovery* undocks from the International Space Station after an eleven-day mission during which crew members completed the final construction and transfer of supplies before the arrival of the first permanent crew for the station.

Date	Science	Event
Oct. 24	Medicine	An article in the *Proceedings of the National Academy of Sciences* announces that University of North Carolina, Chapel Hill, investigators are the first to identify and purify liver stem cells. The discovery, they hope, will lead to liver regeneration therapy as a replacement for liver transplantation.
Oct. 25	Applied technology	The Pacific Northwest National Laboratory unveils the Timed Neutron Detector, which can quickly find landmines, even those made without metal, in a one-hundred-square-foot area.
Oct. 25	Environment	The Intergovernmental Panel on Climate Change, sponsored by the United Nations, issues a report citing new evidence that human-generated pollution has contributed substantially to global warming. Moreover, the report concludes that Earth will grow warmer than previously predicted—almost eleven degrees Fahrenheit by the end of the twenty-first century.
Oct. 25	Environment	Scientists attending the International Coral Reef Symposium in Bali say that as much as one quarter of the world's reefs were killed in recent years by rising ocean temperatures. They call on governments to take measures to check and reverse global warming.
Oct. 25	Astronomy	At a meeting of the International Astronomical Union, French astronomers announce discovery of another moon orbiting Saturn. The planet's twenty-two moons are the most in the solar system, followed by Uranus with twenty-one.
Oct. 26	Space	Following a six-month review prompted by the failure of the Mars Polar Lander, the National Aeronautics and Space Administration releases its revamped schedule for Mars exploration up to 2020. It calls for orbiters, landers, rovers, and sample-return missions, including a new class of small "scout" probes for science missions.
Oct. 27	Life sciences	A study reported in *Science* indicates that the mutation rate in round worms is one hundred times higher than previously thought and suggests that the same may be true for other animals.
Oct. 27	Biotechnology	*Science* announces that researchers successfully tested in monkeys a gene therapy to relieve the symptoms of Parkinson's disease. The scientists introduced a virus genetically altered to carry a human gene coding for glial-derived neurotrophic factor, which stimulates production of dopamine, a key neurotransmitter.

Date	Science	Event
Oct. 28	Computer science	Microsoft reveals that hackers broke into the company's computers and for six weeks had access to highly valuable source codes for future software products.
Oct. 31	Health	At a meeting of the American College of Rheumatology, Swedish researcher Harald P. Roos reports that sports injuries in teenagers increases the risk of developing arthritis later in life. His team studied young women and found them more vulnerable to the sports injury-arthritis connection than are men.
Nov. 1	Medicine	Researchers report in *Nature Medicine* the ability to reverse prostate cancer in mice. Their technique uses prostate-specific antigen, which is naturally abundant in the prostate gland, to fight tumors.
Nov. 1	Health	An article in *Nature Neuroscience* advances evidence that long-term exposure to rotenone, a widely used pesticide, causes symptoms associated with Parkinson's disease in rats. The result supports the theory that Parkinson's disease in humans has environmental causes.
Nov. 1	Health	In *Cancer Research*, doctors describe a sputum test that detects genetic markers signaling a risk for lung cancer. The test could help physicians screen people for the disease.
Nov. 1	Biotechnology	Researchers at the University of Pennsylvania announce success in delivering genes to arterial cells with a deoxyribonucleic acid (DNA) impregnated polymer-coated stent. The genes prevent excessive arterial cell growth. Appearing in *Nature Biotechnology*, the report discussed the device as an aid to keeping coronary arteries unclogged.
Nov. 1	Applied technology	A British study published in *Human Nature* finds that many men treat their cell phones as "lekking devices"—that is, a means to increase their sexual appeal to women.
Nov. 2	Space	American astronaut William M. Shepherd and Russian cosmonauts Yuri Gidzenko and Sergei Krikalev take up residence in the International Space Station as its first crew.
Nov. 2	Physics	The European Center for Nuclear Research (CERN) announces it is shutting down the Large Electron-Proton collider, taking itself out of the race to find the elusive Higgs boson until it can bring online the more powerful Large Hadron Collider.

Date	Science	Event
Nov. 2	Applied technology	The National Aeronautics and Space Administration successfully tests the X-38 prototype crew return vehicle for the International Space Station. It is to replace the Russian Soyuz capsule as the station's lifeboat and is one of a new generation of reusable manned spacecraft.
Nov. 3	Life sciences	Paleontologists report in *Science* discovery of a 280-million-year-old lizard capable of running on its high legs. The ten-inch-long reptile is the earliest known bipedal animal.
Nov. 5	Space	Vice President Hu Hongfu of China Aerospace Science Technology Corporation announces at a press conference that the People's Republic of China will soon launch its own astronauts into space aboard a Chinese-built spacecraft.
Nov. 8	Space	A severe solar storm threatens satellites and the crew of the International Space Station with high radiation levels. Although the space station crew takes protective measures, ground controllers insist they are not in danger.
Nov. 9	Medicine	At a symposium in the Netherlands, researchers release the results of a much anticipated clinical trial of the anticancer drug endostatin. It inhibits angiogenesis, the blood-vessel building process whereby tumors receive blood. The study proves the drug safe and beneficial, but it does not turn out to be the cancer killer that many hoped it would be.
Nov. 9	Medicine	In the *New England Journal of Medicine*, researchers from Johns Hopkins Oncology Center describe a genetic test that predicts which patients with gliomas, a type of brain tumor, will respond to chemotherapy.
Nov. 9	Applied technology	At the 2000 International Mechanical Engineering Congress and Exposition, Purdue University scientists unveil a prototype chair that uses sensors and computer processing to analyze and adjust to a sitter's posture.
Nov. 9	Environment	In *Nature*, scientists offer genetic confirmation that a recently discovered foreign algae invading California waters is *Caulerpa taxifolia*, which is also devastating Mediterranean seabeds.
Nov. 10	Information technology	For the first time, World Wide Web domain names are registered in languages other than English, ending its dominance of the Internet. New software makes it possible to list web sites in Chinese characters, Arabic script, and other written forms.

Date	Science	Event
Nov. 10	Astronomy	The online magazine *Space.com* reports that the Chandra X-Ray Observatory recorded temperatures for young stars in the Orion nebula that are twice the expected level. Combined with earlier observations of anomalous stars, the measurements of the Orion group provide evidence for greater stellar variety than astronomers expected.
Nov. 10	Genetics	A study in *Science* reveals that about 80 percent of European men are descendants of humans who moved into the region 40,000 years ago and that four out of five of them share a common ancestor. The study focused on patterns in the Y chromosome.
Nov. 13	Astronomy	In *Earth, Moon and Planets*, National Aeronautics and Space Administration scientists conclude that meteors do not become as hot as previously thought when speeding through the atmosphere and that organic matter could survive the fall to Earth with them. The finding lends credence to the theory that meteors and comets seeded Earth with biological molecules.
Nov. 13	Medicine	A course of statin, which lowers harmful cholesterol, combined with niacin, which boosts beneficial cholesterol, reduces risk of hospitalization for cardiac disease by 70 percent, according to a study that University of Washington researchers present to a meeting of the American Heart Association.
Nov. 13	Environment	Federal officials place the wild Atlantic salmon in eight Maine rivers on the endangered species list despite objections by state leaders who say the move will cripple the fishing economy.
Nov. 14	Health	In *Neurology,* Mayo Clinic investigators trace a link between coffee consumption and the risk for Parkinson's disease. Although heavier coffee drinkers appear less likely to develop the disease, the study's authors are not convinced that coffee affords direct protection.
Nov. 14	Biology	David A. Grimaldi and Donat Agosti report discovery of a 92-million-year-old ant embedded in amber. Their finding, appearing in *Proceedings of the National Academy of Sciences*, demonstrates that the specimen's subfamily is 40 million years older then previously thought.

Date	Science	Event
Nov. 16	Applied technology	An article in *Nature* describes a working connection between a monkey's brain and a robotic arm. At the Massachusettes Institute of Technology, scientists used a computer to interpret the monkey's brain waves and convert the signal to a command moving the robotic arm. A similar connection between a lamprey brain and a robot is reported a week earlier at a meeting of the Society for Neuroscience.
Nov. 28	Biology	A study reported at the Radiological Society of North America reveals that women use both sides of their brains when listening to conversations, while men usually use only their left side; moreover, women can listen to two conversations at once.
Nov. 28	Earth sciences	In *Eos*, scientists predict that Alaska's Columbia Glacier, the fastest moving in the world, will disintegrate in fifty years, leaving behind a deep fjord.
Nov. 30	Life sciences	Pennsylvania State University scientists publish evidence in *Nature* that organisms lived on land as much as 2.6 billion years ago, much earlier than paleontologists previously estimated.
Dec. 1	Applied technology	*Measurement Science and Technology* describes a device able to detect the vapors from potatoes with soft rot long before the disease is widespread; the sensor could help reduce spoilage.
Dec. 1	Astronomy	A new National Science Foundation-sponsored study of ALH89001, the celebrated meteorite from Mars, concludes that Martian microorganisms could have made some of the crystal materials found in it. The study appears in *Geochimica et Costnochimica Acta*.
Dec. 3	Space	Astronauts install the first solar panel on the International Space Station. The second is added the next day. They are the largest ever deployed in space.
Dec. 4	Applied technology	Los Alamos National Laboratory announces that its scientists have developed a process for rapidly making high-performance superconducting tape capable of carrying two hundred times more electrical current than copper wire.
Dec. 5	Earth sciences	Oceanographers aboard the research vessel *Atlantis* announce discovery of a spectacular hydrothermal vent field deep underwater on the Mid-Atlantic Ridge. They call the field "The Lost City" because of towering spires formed by the vents.

Date	Science	Event
Dec. 5	Health	The Centers for Disease Control announces that 18 percent of American women and 8 percent of American men carry the sexually transmitted human papilloma virus, which is associated with most cervical cancers.
Dec. 6	Earth sciences	An article in *Nature* argues that Earth's granite surface formed quickly—in 1,000 to 100,000 years—instead of over millions of years. Professor Alexander Cruden, University of Toronto, proposes that the granite was relatively runny and able to channel up to the surface through the Earth's crust.
Dec. 8	Astronomy	In *Science*, planetologists report discovery of rock stratification on the surface of Mars, evidence that standing water was once widespread there. Spotted in photographs taken by the Mars Global Surveyor, the layers probably came from slowly deposited sediments and occur in most Martian regions.
Dec. 11	Computer science	Intel announces fabrication of a transistor six times smaller than the industry standard; some of its structures are only 30 nanometers wide. By about 2005, the company plans to use 400 million of the transistors in a computer chip capable of operating at speeds up to 10 gigahertz.
Dec. 11	Medicine	The online issue of *Proceedings of the National Academy of Sciences* reveals that biologists have uncovered a new step in the process whereby malaria parasites infect humans; moreover, the study's authors were able to block the spread of the parasite in the blood stream with a protease inhibitor, raising hopes of a vaccine for the disease.
Dec. 11	Health	In a finding likely to surprise no one, an *Archives* of *Internal Medicine* article presents evidence that quarrels between husbands and wives significantly raise blood pressure in those who are already hypertensive.
Dec. 13	Genetics	At a news conference, American, European, and Japanese scientists reveal that they have completely sequenced the genome of thale cress (*Arabidopsis thaliana*), a common weed related to mustard. The achievement, the first sequencing of a plant's genome, provides a precisely defined laboratory specimen for agricultural and medical research.

Date	Science	Event
Dec. 15	Genetics	University of Connecticut geneticists report in *Science* that they have found a mutated gene in some fruit flies that can double an individual's life span. Dubbed *Indy* (for "I'm not dead yet"), the mutation appears to reduce production of a protein involved in digesting nutrients; this action is consonant with the theory that reduced caloric intake extends life spans.
Dec. 15	Astronomy	Data collected by the Galileo space probe suggests that Jupiter's moon Ganymede, the largest in the Solar System, has a subterranean saltwater ocean, according to a report presented at a meeting of the American Geophysical Union.
Dec. 20	Health	Two small studies published in the *Journal of the American Medical Association* find no correlation between use of cell phones and risk of brain cancer. However, the studies call for further research.
Dec. 21	Medicine	University of Toronto scientists describe in *Nature* successful tests on mice of a vaccine against Alzheimer's disease. The vaccine prevents the onset of memory loss and behavioral changes associated with Alzheimer's.
Dec. 25	Astronomy	A 72 percent eclipse of the sun is visible in the United States, Canada, Mexico, and the Caribbean.
Dec. 25	Life sciences	The *Journal of Cell Biology* reports that an Iowa State University group discovered a second microtubule spindle that participates in cell division.
Dec. 28	Astronomy	Australian astronomers say they found evidence of a new kind of galaxy that contains stars much older than any seen before. The galaxies, five to eleven billion light-years from Earth, were detected by Australian radio and optical telescopes and the Hubble Space Telescope.

Roger Smith

Obituaries

Ahrens, Edward Hamblin, medical researcher, died 12/9/00, age eighty-five. Ahrens established that the eating of saturated animal fats such as butter or lard will raise a person's blood cholesterol, while the consumption of unsaturated vegetable fats such as olive oil or margarine can have the opposite effect. High cholesterol is a problem because it causes a deposit on the walls of arteries that can increase the likelihood of a stroke and heart attack. Ahrens also showed that high-cholesterol foods such as egg yolks do not raise the blood cholesterol because the body compensates by producing less of it in the liver.

Aldrich, Michael Sherman, neurologist, died 7/18/00, age fifty-one. Aldrich was an internationally recognized authority on sleep disorders, including apnea, which temporarily stops a person's breathing during sleep, and narcolepsy, an overpowering desire to sleep. He approached sleep disorders as a neurological problem rather than a psychological condition. He founded the Sleep Disorders Laboratory at the University of Michigan, which became a center for the study and diagnosis of insomnia.

Bascom, Willard, mining engineer, died 9/20/00, age eighty-three. Bascom was a leader in exploring the ocean bottom. In the 1960's, he directed the Mohole project, designed to drill through the ocean floor to bring up rock cores from two miles below the water surface. He founded a company that prospected for tin, gold, and other minerals underseas but later became skeptical about the economic feasibility of underseas mining.

Bell, George Irving, biophysicist, died 5/28/00, age seventy-three. Bell's professional career of more than forty years was spent at the Los Alamos National Laboratory in New Mexico. His work included nuclear reactor design, thermonuclear weapons, and most recently the Human Genome Project. The goal of this project is to decode human deoxyribonucleic acid (DNA) so that the genetic basis of diseases can be understood and treated. Bell published more than a hundred papers in physics, immunology, and biophysics.

Bellak, Leopold, psychiatrist, died 3/24/00, age eighty-three. Bellak was widely known for developing the Thematic Apperception Test (TAT), including a version for children, which uses pictures to explore a patient's unconscious fantasies. He investigated attention deficit disorder, which is found in adults as well as in children. He wrote more than two hundred articles and some forty books, appeared frequently on television talk shows, and narrated a Public Broadcasting Service series entitled *The Human Mind*.

Benesch, Ruth Erica, biochemist, died 3/25/00, age seventy-five. Benesch, together with her husband who was also a biochemist, did fundamental research on the mechanism by which hemoglobin in the blood transfers oxygen molecules to living tissue. They published a paper in 1966 showing how a particular phosphorous compound loosened the intermolecular bond to allow the oxygen to be released from hemoglobin. Their discovery contributed greatly to an increased knowledge of the respiratory system and stimulated an outpouring of new research by scientists from around the world.

Bloch, Konrad Emil, biochemist, died 10/15/00, age eighty-eight. Bloch received the Nobel Prize in 1964 for his work on cholesterol. He identified a complex sequence of more than thirty biochemical reactions by which cholesterol molecules are produced. Cholesterol is formed in the liver and is essential for life, but an excess in the bloodstream can cause narrowing of arteries, possibly leading to a heart attack or stroke. Bloch was a professor of biochemistry at Harvard University for twenty-eight years.

Fales, De Coursey, Jr., archaeologist, died 4/12/00, age eighty-two. Fales was an expert on a single, unique Greek vase from the sixth century B.C. This two-feet high vase can be seen at the Archaeological Museum in Florence, Italy. It is decorated with more than two hundred figures, including Achilles and other warriors from the Trojan War. Fales wrote interpretive articles about the vase and the artist who painted its mythological scenes.

Favaloro, Rene Geronimo, heart-bypass surgeon, died 7/29/00, age seventy-seven. Favaloro was a pioneer in heart-bypass surgery. In 1967 at the Cleveland Medical Clinic, he operated on a patient with a blocked artery, using a vein from the leg to bypass the blockage. Within the next year, the clinic performed more than one hundred bypasses. Favaloro returned as a national hero to his native Argentina, where he founded a world-class heart clinic to treat patients and to train surgeons.

Friedman, Herbert, X-ray astronomer, died 9/9/00, age eighty-four. Friedman was a pioneer in the detection of X rays coming from space, using rockets and satellites. In the 1950's, he showed that the disruption of radio communication on earth during solar flares was caused by X rays. He studied more distant X-ray sources, such as the Crab Nebula, the remnant of an ancient supernova explosion. He published some three hundred technical articles and a popular National Geographic book entitled *The Amazing Universe* (1975).

Goetz, Robert Hans, heart surgeon, died 12/15/00, age ninety. Goetz was a pioneer in open heart surgery. In 1960, he initiated a procedure called coronary bypass surgery, in which a partially blocked artery around the heart muscle is replaced by a new blood vessel usually taken from the patient's own leg. Later the technique was extended to double, triple, and quadruple bypass surgery, replacing several blood vessels at one time. One of his famous students was the physician Christiaan Barnard, who performed the first human heart transplant in 1967.

Gottlieb, Melvin Burt, physicist, died 12/1/00, age eighty-three. Gottlieb was a leader in the effort to harness nuclear fusion energy for electric power production. For more than twenty years, he headed the plasma physics laboratory at Princeton University. A plasma is a very hot gas of hydrogen ions and electrons that releases energy when the ions fuse together into helium. Gottlieb strongly promoted cooperation with the Soviet Union because the apparatus for this type of research is very expensive and he believed that the practical results should be made available worldwide.

Hamilton, William Donald, biologist, died 3/7/00, age sixty-three. Hamilton is known for his concept of "inclusive fitness," which provides a genetic basis to explain why some animals, such as bees, sacrifice their lives to ensure the survival of their kinship group. A popularization of this concept is found in Richard Dawkins's best-selling book, *The Selfish Gene* (1989). Hamilton's work helped to unify naturalist Charles Darwin's principle of natural selection with Mendelian genetics.

Hockett, Charles Francis, anthropologist, died 11/4/00, age eighty-four. Hockett was an expert on linguistics, known for his careful analysis of languages as viewed in their social context. He was a strong critic of Noam Chomsky, who sought to describe human language as a biological process subject to mathematical rules, without reference to the anthropological environment. Hockett authored several widely used textbooks including *A Course in Modern Linguistics* (1955) and *Man's Place in Nature* (1973).

Kabat, Elvin Abraham, microbiologist, died 6/16/00, age eighty-five. Kabat was a pioneer in the field of immunology. He analyzed the structure of antibodies such as gamma globulin, which are produced by the body to counteract infectious diseases such as measles and hepatitis. He studied multiple sclerosis, in which the immune system attacks other parts of the body. In 1991, he received the National Medal of Science, the highest award for scientific achievement, for his role in bringing immunology to its present prominence.

Keenan, Philip C., astronomer, died 4/20/00, age ninety-two. Keenan was the coauthor of *An Atlas of Stellar Spectra* (1943), which classifies stars by the wavelengths of light that they emit. Each star, including the Sun, emits light that looks like a continuous spectrum from red to violet but actually has a unique pattern of superimposed dark lines. These lines provide information about the chemical makeup and temperature of stars. Keenan's system of classifying stars is widely used by astronomers.

Knipling, Edward Fred, entomologist, died 3/17/00, age ninety. Knipling was an expert on agricultural insect pests, especially the screw worm fly, which was a scourge of American cattle farmers. In the 1950's, Knipling developed the sterile male technique, in which millions of male flies were grown in cages, irradiated with X rays, then released to breed with females without producing any offspring. The procedure was successful in eradicating screw worms without using insecticides.

Knobil, Ernst, physiologist, died 4/13/00, age seventy-three. Knobil discovered a hormone produced by the brain that regulates the menstrual cycle in women. By synthesizing this hormone, scientists developed an effective treatment for infertility. Knobil also studied the function of the pituitary gland, which eventually led to a successful treatment for dwarfs, who have a deficiency in the hormone that controls body growth. He was coeditor of a four-volume *Encyclopedia of Reproduction* (1998).

Miller, Merton H., economist, died 6/3/00, age seventy-seven. Miller received the Nobel Prize in Economic Sciences in 1990 for his pioneering work on financial guidelines to help corporations determine an appropriate mix of stocks and bonds that would appeal to investors. He grew up during the Depression, when Keynesian economics advocated a major role for government programs to stimulate economic development. In contrast with economist John Maynard Keynes, Miller believed strongly in the power of free markets to optimize economic progress, with minimum governmental regulations.

Morse, Roger Alfred, entomologist, died 5/12/00, age seventy-two. Morse was an international expert on bees. He authored a book that is popular with both amateur and commercial producers of honey, *The Complete Guide to Beekeeping* (1972). He emphasized that bees have a vital role in agriculture as pollinators for various crops. Morse also studied the so-called killer bee, an aggressive species from Brazil that arrived in the southern United States in the early 1990's.

Murphy, Gerald Patrick, urologist, died 1/22/00, age sixty-five. Murphy's research established the presence of prostate specific antigens (PSAs) that develop in response to a cancerous tumor of the prostate. Since the 1980's, the PSA test has been used to detect possible prostate cancer in millions of men so it can be treated at an early stage. Murphy helped lobby the U.S. Congress in 1971 to pass the National Cancer Act, which provides federal funding for cancer research centers.

Neel, James V., geneticist, died 2/1/00, age eighty-four. In 1947, Neel was a member of the Atomic Bomb Casualty Commission, which examined survivors from Hiroshima and Nagasaki to look for genetic mutations caused by radiation. No immediate effects were found, but years later, the incidence of leukemia and tumors was discovered to be higher in these survivors. Neel did early research on the genetic basis for sickle cell anemia, a blood disease. He was a leading proponent of the theory that viruses may cause genetic damage that leads to cancer.

Neer, Eva Julia, biochemist, died 2/20/00, age sixty-two. Neer's research focused on neurotransmitters, which carry nerve impulses that are stimulated by sensory organs in the body or by hormones and drugs. She was known for her work on the interaction between cells in the brain and the production of subsequent responses in the body. She received the National Institutes of Health Merit Award in 1989.

Nierenberg, William Aaron, physicist, died 9/10/00, age eighty-one. During World War II, Nierenberg worked on the development of the atomic bomb. Later he did extensive research with atomic beams and then became director of the Scripps Oceanographic Institute in San Diego for more than twenty years, where drilling surveys of the ocean floor were conducted. He served on numerous governmental committees, providing advice on diverse issues such as antisubmarine warfare, acid rain, ballistic missile strategy, global warming, and pollution.

Oliphant, Marcus Laurence, physicist, died 7/14/00, age ninety-eight. Oliphant was Australia's leading nuclear physicist. In the 1930's, he was codiscoverer of a rare form of hydrogen gas called tritium. During World War II, he led the British scientists who helped develop the atomic bomb. After the war, he supported peaceful uses for nuclear energy, such as electric power plants. However, he became an outspoken critic of the nuclear arms race, stating that the use of nuclear weapons could not be justified under any circumstances.

Orne, Martin Theodore, psychiatrist, died 2/11/00, age seventy-two. Orne was an expert on hypnosis, multiple personality disorders, and "brainwashing." He was critical of using hypnosis in criminal trials to get witnesses to "remember" incriminating evidence. In 1974, he testified for the defense in the trial of newspaper heiress Patricia Hearst, who had participated in a bank robbery with the Symbionese Liberation Army. In another high-profile trial, he assisted the prosecution in convicting Kenneth Bianchi, the so-called Hillside Strangler of Los Angeles.

Pais, Abraham, physicist and historian, died 7/29/00, age eighty-two. Pais was born in Holland of Jewish parents, narrowly escaping deportation by the Gestapo during World War II. He received a Ph.D. from the University of Amsterdam, later specializing in the study of subatomic particles. He wrote a biography of Albert Einstein entitled *Subtle Is the Lord* (1982), which was widely recognized as a masterpiece of science history. He also published a definitive biography of Niels Bohr, who originated the atom model with electrons orbiting around a nucleus.

Piore, Emanuel Ruben, physicist, died 5/9/00, age ninety-one. Piore was director of research and later chief scientist at International Business Machines (IBM), where he contributed to the development of digital computers. His other management responsibilities included directing the television laboratory at the Columbia Broadcasting System, heading the Office of Naval Research, helping to found the National Science Foundation, and serving on the Science Advisory Committees for Presidents Dwight D. Eisenhower and John F. Kennedy, Jr.

Reemtsma, Keith, organ transplant surgeon, died 6/23/00, age seventy-four. Reemtsma was chief of surgery at the University of Utah in 1982 when the first artificial heart was implanted in a patient named Barney Clark. In the 1960's, Reemtsma transplanted a kidney from a chimpanzee to a woman who then lived with it for nine months. He felt that transplants from animals to humans were necessary because the supply from human donors was insufficient. This brought him into conflict with animal rights advocates.

Rossi, Harald Hermann, biophysicist, died 1/1/00, age eighty-two. Rossi was known for establishing the field of microdosimetry, which measures radiation doses used in cancer therapy. He designed several instruments, including the Rossi counter, which is still used to determine the energy absorbed by cells during irradiation. He helped to establish radiation safety regulations for workers in hospitals and nuclear laboratories. Rossi was a professor at Columbia University until his retirement in 1987.

Schilling, Martin, missile engineer, died 4/30/00, age eighty-eight. During World War II, Schilling worked on the German V-2 missile under the leadership of Wernher von Braun. He was responsible for the design of control vanes that could survive in the hot exhaust gases from the rocket engines. After the war, he was brought to the United States to work on the U.S. satellite program. In 1958, Schilling was awarded the Exceptional Civilian Service Award for his contribution to the U.S. Army's Sidewinder and Patriot missiles.

Simpson, John Alexander, physicist, died 8/31/00, age eighty-three. During World War II, Simpson worked on the atomic bomb project. After the war, he helped Senator Brian McMahon write legislation for civilian control of peaceful applications for atomic energy. In later research, he became a leader in the study of cosmic radiation in outer space, being selected as principal investigator for dozens of space missions to measure radiation and dust particles from Mercury, Venus, Mars, Jupiter, Saturn, and Halley's comet.

Smith, Michael, chemist and biotechnologist, died 10/4/00, age sixty-eight. In the early 1980's, Smith developed one of the basic techniques of genetic engineering. He was able to change the genetic code sequence in human deoxyribonucleic acid (DNA) so that the rearranged DNA then could produce altered proteins in living organisms. This procedure is the basis for gene therapy, which has the potential to correct harmful mutations that cause genetic diseases. He was cowinner of the Nobel Prize in Chemistry in 1993.

Snow, George A., physicist, died 6/24/00, age seventy-three. Snow made both theoretical and experimental contributions to an understanding of the two families of subnuclear particles called quarks and leptons. With his colleagues, he showed how high-energy collision experiments at large accelerators can yield information about the new particles that were being produced. He helped to establish the nature of the two nuclear types of force, called strong and weak. Snow was an early advocate for increasing the number of women in the sciences.

Soffen, Gerald Alan, biologist, died 11/22/00, age seventy-four. Soffen was the chief scientist for the National Aeronautics and Space Administration (NASA) in the 1970's when two Viking missions were sent to explore the surface of Mars for evidence of life. The Viking spacecraft landed and operated successfully but found no evidence of biologically significant molecules. Soffen also developed instruments for space flight and for medical care of astronauts in the space shuttle program. He helped initiate the program of observations of Earth from space.

Stebbins, George Ledyard, botanist, died 1/19/00, age ninety-four. Stebbins was an expert on plant life. The field of genetics was dominated by animal studies until the 1950's, but Stebbins insisted that botany and zoology have equal importance in biological evolution. He had an encyclopedic knowledge of plant species and was an early leader in the conservation movement. He published more than 250 articles and books, including a popular work entitled *From Darwin to DNA* (1982).

Strauch, Karl, high energy physicist, died 1/3/00, age seventy-seven. Strauch was known for his research on the fundamental structure of matter, particularly on the existence of quarks. The new particles are detected through an analysis of the debris created when a beam of nuclear particles from an accelerator collides with a target. Strauch was a member of the Harvard University physics faculty for forty years. He frequently participated in colliding-beam experiments at Stanford University and at the Nuclear Research Center in Geneva, Switzerland.

Tanaka, Toyoichi, biophysicist, died 5/20/00, age fifty-four. Tanaka was a specialist in the creation of polymers, long chains of molecules that can be linked together to form a gel. These gels can absorb or release large quantities of fluid, leading to useful applications such as the cleanup of toxic spills, the slow release of insulin in the body, and fluid absorption in disposable diapers. Tanaka's research explored other potential uses for gels in medicine and industry.

Thompson, Homer Armstrong, archaeologist, died 5/7/00, age ninety-three. For thirty-nine years, Thompson was a leading investigator in the excavation of the Agora, the civic center of ancient Athens. Thousands of relics gradually were uncovered in the digs. To build a museum in Athens for housing these artifacts, Thompson reconstructed the floor plan of an authentic two-story stoa, a form of ancient Greek public building. His work helped to clarify daily life in the Athenian civilization.

Tukey, John Wilder, statistician, died 7/26/00, age eighty-five. Tukey specialized in statistical analysis of numerical data using computers. He coined the word "software" to describe computer programs. He took an active part in social controversies, criticizing the 1948 and 1953 Kinsey Reports on sexual behavior for flawed statistics, warning about the danger of aerosol cans to the ozone layer, and recommending that the U.S. Census should be statistically adjusted to account for urban poor. In 1973, he received the National Medal of Science from President Richard M. Nixon.

Washburn, Sherwood Larned, anthropologist, died 4/16/00, age eighty-eight. **Washburn** was an anthropologist who linked the evolution of human behavior to the activities observed in apes and monkeys. Among his published books was *Ape into Human: A Study of Human Evolution* (1980). He showed how the specific anatomy of bones and muscles relates to social behavior in humans and other primates. He pioneered the study of monkeys in their natural habitat in Africa, Thailand, and elsewhere.

Weber, Joseph, physicist, died 9/30/00, age eighty-one. In the 1950's, Weber originated the basic concept of coordinated light emission from many separate molecules, which later led to the invention of the laser with its many applications. He also originated the experimental search for gravitational waves, which were predicted by Albert Einstein's theory of general relativity. His two major research accomplishments are coming together in the construction of a multimillion dollar laser gravitational wave observatory.

Weinstein, Louis, infectious disease specialist, died 3/16/00, age ninety-two. Measles, tuberculosis, diphtheria, and polio are infectious diseases that spread by contact. Weinstein was a leader in establishing guidelines and training doctors in the appropriate use of new antibiotics such as penicillin when they were first introduced. He was an outspoken critic of the overuse of antibiotics, warning that they could kill off the body's normal microbes and cause serious side-effects.

Whyte, William Foote, sociologist, died 7/16/00, age eighty-six. Whyte combined academic scholarship with a passion for social change. In 1943, he published a study of Italian-American gangs in Boston entitled *Street Corner Society*, based on his role as participant-observer at pool halls and other hangouts. His book later was translated into six languages and sold more than 250,000 copies. Other research included restaurant workers in Chicago and peasant villages in Peru, showing his concern for the gap between rich and poor.

Widdowson, Elsie May, nutritionist, died 6/14/00, age ninety-three. Widdowson was coauthor of a book entitled *The Chemical Composition of Foods* (1940), which became known as the dietitian's bible and had reached its sixth edition by the time of her death. Her research on the body's need for salt led to an improved treatment for patients in a diabetic coma. During World War II, she convinced the British government that lack of calcium can be overcome by adding ground chalk to bread flour. After the war, she recommended a diet for concentration camp survivors to restore their health.

Wilson, Robert R., physicist, died 1/16/00, age eighty-five. Wilson earned international fame as the founding director of Fermilab, near Chicago, where the world's highest energy nuclear particle accelerator was completed in 1972. He was instrumental in designing the two-mile circular ring of magnets around which the proton beam must travel millions of times as it gains energy with each successive orbit. At Fermilab, he developed a strong outreach program to inform students and the general public about research.

Zinn, Walter H., nuclear engineer, died 2/14/00, age ninety-three. Zinn was the first director of Argonne National Laboratory near Chicago and a pioneer in the development of commercial nuclear power plants. In 1942, he was part of the team that constructed the first nuclear reactor using uranium fission, a necessary step in the atomic bomb project. After the war, he promoted research on new reactor types suitable for electric power plants. He received the Atoms-for-Peace award in 1960.

Hans G. Graetzer

Books of the Year

➤ *This annotated list of the best science books of 2000 groups the works by subject area. They range from histories of science to controversial new theories.*

General Studies

Bragg, Melvyn. *On Giants' Shoulders: Great Scientists and Their Discoveries from Archimedes to DNA*. New York: John Wiley & Sons, 2000. The host of the British Broadcasting Corporation's radio show *Start of the Week* publishes his interviews with modern scientists about their most famous predecessors. Among the interviewees are Stephen Jay Gould, Oliver Sacks, and Richard Dawkins; among the "greats" are Michael Faraday, Sigmund Freud, and Charles Darwin.

Casti, John L. *Paradigms Regained: A Further Exploration of the Mysteries of Modern Science*. New York: William Morrow, 2000. A sequel to *Paradigms Lost* (1990), this book also uses a mock trial as the framing device for examination of such scientific debates as those concerning the origin of evolution, language acquisition, human perception, and extraterrestrial life. In some cases, Casti changes his judgment about the future direction of the topics.

Downey, Roger. *Riddle of the Bones: Politics, Science, Race, and the Story of Kennewick Man*. New York: Copernicus, 2000. Recounts the discovery of an ancient skull near the Columbia River and the subsequent battle for its ownership between the anthropologists who wanted to study it and the Native Americans who claimed it as an ancestor. Downey considers the views of both sides, the influence of the media, and the potential for scientific insight into the past afforded by the skull.

Eldredge, Niles. *The Triumph of Evolution: And the Failure of Creationism*. New York: Freeman, 2000. Eldredge argues that the aim of creationist theories of life is political and endangers science education in the United States by presenting a distorted message about the nature of science.

Fuller, Stephen. *Thomas Kuhn: A Philosophical History of Our Times*. Chicago: University of Chicago Press, 2000. A critical biography of Thomas Kuhn (1922-1996), the preeminent philosopher of science during the twentieth century, which places his thought in its historical and philosophical context.

Gardner, Martin. *Did Adam and Eve Have Navels?: Discourses on Reflexology, Numerology, Urine Therapy, and Other Dubious Subjects*. New York: W. W. Norton, 2000. Collected essays from the author's *Skeptical Inquiry* column debunking pseudoscience, although the last philosophizes on science and the unknowable.

Kuhn, Thomas S. *The Road Since Structure: Philosophical Essays, 1970-1993, with an Autobiographical Interview*. Chicago: University of Chicago Press, 2000. Eleven essays that defend and modify Kuhn's seminal *The Structure of Scientific Revolutions* (1962).

Newbold, Heather. *Life Stories: World-Renowned Scientists Reflect on Their Lives and the Future of Life on Earth.* Berkeley: University of California Press, 2000. A collection of autobiographical essays, including those by Thomas Lovejoy, James Lovelock, Ruth Patrick, David Suzuki, and Paul Ehrlich, that also discusses environment and biological diversity.

Pais, Abraham. *The Genius of Science: A Portrait Gallery.* New York: Oxford University Press, 2000. A celebrated physicist and biographer profiles sixteen of his contemporaries, many of whom he knew well, explaining the achievements of such thinkers as Albert Einstein, Niels Bohr, and John von Neumann. He includes personal anecdotes as well.

Rudacille, Deborah. *The Scalpel and the Butterfly: The War Between Animal Research and Animal Protection.* New York: Farrar, Straus and Giroux, 2000. Examining 150 years of animal experimentation, Rudacille points out the ethical dilemmas and the increasingly vociferous objections to them, including the efforts of People for the Ethical Treatment of Animals and the Animal Liberation Front.

Schweber, S. S. *In the Shadow of the Bomb: Oppenheimer, Bethe, and the Moral Responsibility of the Scientist.* Princeton, N.J.: Princeton University Press, 2000. Examines the careers of three leaders in the Manhattan Project in order to discuss the role of scientists in a democracy.

Thomas, David Hurst. *Skull Wars: Kennewick Man, Archaeology, and the Battle for Native American Identity.* New York: Basic Books/Peter N. Nevraumont, 2000. Thomas recounts the legal and political battles that erupted between Native American tribes and scientists after a 9,500-year-old skeleton was found near the Columbia River. Thomas harshly criticizes archaeologists and calls for greater cooperation with Native Americans.

Wallace, Joseph. *A Gathering of Wonders: Behind the Scenes at the American Museum of Natural History.* New York: St. Martin's Press, 2000. Wallace recounts the history behind some of the famous exhibits at New York's American Museum of Natural History and comments on the importance of its preservation programs.

White, Michael. *Leonardo: The First Scientist.* New York: St. Martin's Press, 2000. White brings balance to Leonardo da Vinci's legacy by emphasizing his studies in optics, anatomy, flight, astronomy, and weapons technology, but he also recounts the details of Leonardo's life and associations with contemporary intellectuals.

Wise, Steven M. *Rattling the Cage: Toward Legal Rights for Animals.* New York: Perseus, 2000. The author, a law professor, promotes a case for considering primates as free persons, thus banning ownership of them by laboratories and preventing their use in experiments.

Wright, Robert. *Nonzero: The Logic of Human Destiny.* New York: Pantheon, 2000. Applies game theory to biological and cultural evolution. The author believes the increasing complexity of evolution produces cooperation and reflects an underlying purpose in the universe.

Astronomy, Cosmology, and Space

Bartusiak, Marcia. *Einstein's Unfinished Symphony: Listening to the Sounds of Space-*

Time. Washington, D.C.: Joseph P. Henry Press, 2000. A distinguished science writer uses the analogy of music to elucidate the exotic properties of relativity, particularly gravity waves and their relation to such phenomena as black holes.

Burnham, Robert. *Great Comets.* Cambridge, England: Cambridge University Press, 2000. Burnham addresses both the science and social effects of comets as he recounts recent appearances of them and past and planned space missions to study them up close.

Christianson, John Robert. *On Tycho's Island: Tycho Brahe and His Assistants, 1570-1601.* Cambridge, England: Cambridge University Press, 2000. Recounts how the aristocratic Tycho Brahe prevailed on the king of Denmark to build an observatory, Uraniborg, and drew to him some of the leading minds of the day, including Johannes Kepler.

Consolmagno, Guy. *Brother Astronomer: Adventures of a Vatican Scientist.* New York: McGraw-Hill, 2000. The autobiography of an American astronomer who took up duties in the pope's country house-observatory and participated in the search for meteorites of Martian origin in Antarctica. The author also discusses the Catholic Church's treatment of Galileo.

Godwin, Robert. *Mars: The NASA Mission Reports.* Croydon, England: Apogee, 2000. A collection of press kits and mission reports for every mission to Mars from Mariner 4 in 1964 to Mars Global Surveyor in 1998.

Goldsmith, Donald. *The Runaway Universe: The Race to Find the Future of the Cosmos.* New York: Perseus, 2000. Goldsmith traces the history of cosmological expansion from Albert Einstein's general theory of relativity to the evidence that the expansion is accelerating. He discusses rate of expansion, a controversial topic, and such matters as the universe's age and ultimate fate.

Gribbin, John. *The Birth of Time: How Astronomers Measured the Age of the Universe.* New Haven, Conn.: Yale University Press, 2000. A noted scientist and science writer begins with Edwin Hubble and Allan Sandage and reviews the development of cosmological measurements, including those that produced the current value of 13 billion years for the universe's age.

Hoyle, Fred, Geoffrey Burbidge, and Jayant V. Narlikar. *A Different Approach to Cosmology: From a Static Universe Through the Big Bang Towards Reality.* Cambridge, England: Cambridge University Press, 2000. Describes the quasi-steady-state theory, the most developed alternative theory to the prevailing Big Bang model of the universe.

Krantz, Gene. *Failure Is Not an Option: Mission Control from Mercury to Apollo 13 and Beyond.* New York: Simon & Schuster, 2000. A leader in the National Aeronautics and Space Administration's Mission Control Center during the 1960's, Krantz participated in the often heroic efforts of controllers to keep astronauts safe. He offers an insider's account of the glory years of the space program and the moon shots in particular.

Krauss, Lawrence M. *Quintessence: The Mystery of Missing Mass in the Universe.* New York: Basic Books, 2000. A guide to the modern cosmological problem of missing mass, the book discusses dark matter and nonzero vacuum energy as governing the universe's density and development.

Levy, David H. *Shoemaker by Levy: The Man Who Made an Impact.* Princeton, N.J.: Princeton University Press, 2000. The author, half of the team that discovered the comet Shoemaker-Levy, describes the achievements and personality of his partner, Eugene M. Shoemaker.

Lidsey, James E. *The Bigger Bang*. Cambridge, England: Cambridge University Press, 2000. Following the history of the universe to its first instant, the author explains in simple terms such topics as superstrings, multiple universes, quantum physics, and astrophysics.

Linenger, Jerry M. *Off the Planet: Surviving Five Perilous Months Aboard the Space Station Mir*. New York: HarperCollins, 2000. The author recounts his harrowing stay aboard Russia's Mir station, during which fire and other emergencies threatened the crew's lives.

Livio, Mario. *The Accelerating Universe: Infinite Expansion, the Cosmological Constant, and the Beauty of the Cosmos*. New York: John Wiley & Sons, 2000. Livio addresses the latest research in cosmology and its implications, including such humanistic concerns as symmetry, simplicity, and self-understanding.

Maor, Eli. *June 8, 2004: Venus in Transit*. Princeton, N.J.: Princeton University Press, 2000. Maor recounts the last five transits of Venus across the sun and the events on Earth that were concurrent with them. He also discusses the transits in 2004 and 2012.

Rees, Martin. *Just Six Numbers: The Deep Forces That Shape the Universe*. New York: Basic Books, 2000. England's Astronomer Royal discusses the six physical constants that determined how the universe evolved to its present form.

Standage, Tom. *The Neptune File: A Study of Astronomical Rivalry and the Pioneers of Planet Hunting*. New York: Walker, 2000. After an eighth planet was predicted, the race was on to find it, and Standage tells the dramatic story of Neptune's detection in 1841.

Tribble, Alan C. *A Tribble's Guide to Space: How to Get to Space and What to Do When You're There*. Princeton, N.J.: Princeton University Press, 2000. A rocket scientist, Tribble uses the fictional setting of *Star Trek* to explain navigation, life support, and the physics of space travel.

Wheeler, J. Craig. *Cosmic Catastrophes: Supernovae, Gamma-Ray Bursts, and Adventures in Hyperspace*. Cambridge, England: Cambridge University Press, 2000. Explains the nature of astrophysical phenomena such as black holes and gamma-ray sources and discusses such cutting-edge theories and speculations as string theory, quantum gravity, wormholes, and time machines.

Chemistry

Emsley, John. *The Thirteenth Element: The Sordid Tale of Murder, Fire, and Phosphorus*. New York: John Wiley & Sons, 2000. Tells of the discovery of phosphorus by Hennig Brandt in the seventeenth century and its subsequent importance in weapons, biochemistry, and consumer goods.

Perkowitz, Sidney. *Universal Foam: From Cappuccino to the Cosmos*. New York: Walker, 2000. Among the commonest forms of matter, foam (solids or liquids perfused with gas bubbles) is crucial to technology and medicine, as the author relates in addition to elucidating the chemical and physical principles of foam.

Computer Science and Information Technology

Brown, Julian. *Minds, Machines, and the Multiverse: The Quest for the Quantum Com-*

puter. New York: Simon & Schuster, 2000. Brown recounts the efforts to build computer chips exploiting quantum mechanical effects, such as interference and entanglement, in order to create small, extremely powerful computers.

Davis, Martin. *The Universal Computer: The Road from Leibniz to Turing.* New York: W. W. Norton, 2000. Traces the development of the mathematical logic that modern computers use through the work of Gottfried Leibniz, George Boole, Gottlob Frege, Georg Cantor, David Hilbert, Kurt Gödel, and Alan Turing.

Earth Sciences and the Environment

Alt, David, and Donald W. Hyndman. *Roadside Geology of Northern and Central California.* Missoula, Mont.: Mountain Press, 2000. The latest in a series of books dedicated to the curious amateur geologist. The authors cover, for example, the tufa mounds at Mono Lake, coastal formations, the basalt Modoc Plateau, and major faults.

Bailey, Ronald, ed. *Earth Report 2000: Revisiting the True State of the Planet.* New York: McGraw-Hill, 2000. The essays in this book, sponsored by the Competitive Enterprises Institute, argue that free enterprise will bring forth solutions to such environmental problems as pollution and declining biodiversity, problems that, they assert further, are exaggerated.

Cattermole, Peter. *Building Planet Earth: Five Billion Years of Earth History.* Cambridge, England: Cambridge University Press, 2000. Supported by color photographs and illustrations, Cattermole explains the astrophysical and geological processes that shaped the Earth, starting with the big bang.

Childs, Craig. *The Secret Knowledge of Water: Discovering the Essence of the American Desert.* Seattle, Wash.: Sasquatch, 2000. Interviews with scientists and the author's personal experiences in deserts create a portrait of desert terrain, water sources, and geology.

Davies, Pete. *Inside the Hurricane: Face to Face with Nature's Deadliest Storms.* New York: Henry Holt, 2000. The author presents the science of hurricanes as he describes his encounter with two deadly ones aboard a research aircraft.

McNeill, J. R. *Something New Under the Sun: An Environmental History of the Twentieth Century.* New York: W. W. Norton, 2000. A history professor identifies key changes in the past one hundred years that transformed the world environment. The result of social, political, and cultural patterns, these changes, he argues, will be the twentieth century's greatest influence on future generations.

Prager, Ellen J. *Furious Earth: The Science and Nature of Earthquakes, Volcanoes, and Tsunamis.* New York: McGraw-Hill, 2000. Drawing from interviews with leading scientists, Prager describes the natural forces creating these catastrophic phenomena and discusses the technology evolved to protect people from them.

Prager, Ellen J., and Sylvia Earle. *The Oceans.* New York: McGraw-Hill, 2000. The authors discuss the origins of oceans, the life forms there, important oceanographic expeditions, and the oceans' impact on weather and people.

Sussman, Art. *Dr. Art's Guide to Planet Earth: For Earthlings 12 to 120.* White River Junction, Vt.: Chelsea Green, 2000. Sussman tells readers about Earth systems science, an integrated approach to understanding the interconnectedness of materials, energy, and life.

Thompson, Dick. *Volcano Cowboys: The Rocky Evolution of a Dangerous Science.*

New York: Thomas Dunne, 2000. The story of United States Geological Service scientists who risked death to collect data during eruptions and the discoveries made because of their efforts.

Thornton, Joe. *Pandora's Poison: Chlorine, Health, and a New Environmental Strategy.* Cambridge, Mass.: MIT Press, 2000. Thornton, once a leading researcher for Greenpeace, argues that toxic chlorine residues from water treatment, paper manufacture, solvents, and refrigerants cause cancer and endocrine disorders and calls for use of alternative chemicals.

Upgren, Arthur, and Jurgen Stock. *Weather: How It Works and Why It Matters.* Cambridge, Mass.: Perseus, 2000. Basic information about weather processes, global warming, and climate alteration since the Ice Age, including current trends.

Life Sciences and Genetics

Alexander, Shana. *The Astonishing Elephant.* New York: Random House, 2000. A natural and social history of elephants, including tales about circus elephants.

Allport, Susan. *Primal Feast: Food, Sex, Foraging, and Love.* New York: Crown, 2000. Allport explores the relation among food, reproduction, and social power in animals, particularly chimpanzees, and humans.

Benvie, Sam. *An Encyclopedia of North American Trees.* Willowdale, Canada: Firefly, 2000. Listings citing the names, growing habits, environmental importance, and key identifying characteristics of 287 species of trees, accompanied by color photographs.

Burnham, Terry, and Jay Phelan. *Mean Genes: From Sex to Money to Food—Taming Our Primal Instincts.* Cambridge, Mass.: Perseus, 2000. Billed as an owner's manual for the human brain, the book advises readers how to control behavior that evolved for primitive humans but is unsuited to a healthy life in modern society.

Cokinos, Christopher. *Hope Is the Thing with Feathers: A Personal Chronicle of Vanished Birds.* New York: Penguin Putnam, 2000. The story of six bird species made extinct by humans, written in the hope that people will take better care of living species.

Edelman, Gerald M., and Giulio Tononi. *A Universe of Consciousness: How Matter Becomes Imagination.* New York: Basic Books, 2000. Describes the structure of the human brain and advances the dynamic core hypothesis to account for consciousness as an emergent property of neuronal groups.

Fowler, Brenda. *Iceman: Uncovering the Life and Times of a Prehistoric Man Found in an Alpine Glacier.* New York: Random House, 2000. Fowler, a journalist, recounts the discovery of a well-preserved 5,300-year-old corpse in the Austrian Alps, the bickering between Austria and Italy over ownership, and the conflicting explanations about how the man died.

Goff, M. Lee. *A Fly for the Prosecution: How Insect Evidence Helps Solve Crimes.* Cambridge, Mass.: Harvard University Press, 2000. A history and explanation of forensic entomology, the use of evidence left by insects to determine the time, place, and cause of murders.

Gould, Stephen Jay. *The Lying Stones of Marrakech: Penultimate Reflections on Natural History.* New York: Crown, 2000. The ninth anthology of Gould's essays published in *Natural History.* Many spring from biographical portraits of scientists to the cultural impact of their ideas, while others explore the history of science and the conflict between science and religion.

Hauser, Marc D. *Wild Minds: What Animals Really Think.* New York: Henry Holt, 2000. A psychology professor finds that all animals possess mental tools enabling them to solve social and environmental problems, capabilities he illustrates by describing the activities of primates, lions, birds, and insects.

Jaffe, Mark. *The Gilded Dinosaur: The Fossil War Between E. D. Cope and O. C. Marsh.* New York: Crown, 2000. Recounts the beginnings of American paleontology by focusing on the nineteenth century rivalry of Edward Drinker Cope and Othniel Charles Marsh about the nature of dinosaur fossils.

Jones, Steve. *Darwin's Ghost.* New York: Random House, 2000. An homage to Charles Darwin's *Origin of the Species* that explains the central ideas of heredity, variation, and natural selection in light of modern genetics and molecular biology.

Kaufman, Wallace. *Coming Out of the Woods: The Solitary Life of a Maverick Naturalist.* Cambridge, Mass.: Perseus, 2000. The author's account of his life in wild North Carolina woods. He argues that technology and society can contribute to the preservation of wilderness.

Klawans, Harold. *Defending the Cavewoman: And Other Tales of Evolutionary Neurology.* New York: W. W. Norton, 2000. A neurologist proposes that women were most likely the ones to foster the development of complex language, which resulted in the evolutionary development of the brain. Klawans also describes dramatic instances of brain impairment.

Koerner, David, and Simon LeVay. *Here Be Dragons: The Scientific Quest for Exobiology.* New York: Oxford University Press, 2000. The book ponders a discipline with no hard evidence of its subject: life beyond Earth. The authors cover the search for extraterrestrial intelligence (SETI), the basic requirements of life and the conditions for evolution on other planets, and the stellar processes that could produce those conditions.

Lewontin, Richard. *The Triple Helix: Gene, Organism, and Environment.* Cambridge, Mass.: Harvard University Press, 2000. Lewontin takes fellow biologists to task, especially geneticists, for growing so specialized that they neglect environmental influences on the development of individual organisms.

Low, Bobbi S. *Why Sex Matters: A Darwinian Look at Human Behavior.* Princeton, N.J.: Princeton University Press, 2000. The author draws upon behavioral ecology to argue that organisms evolve by competing for environmental resources in order to reproduce as much as possible and that inherited traits govern sexual behavior.

McGavin, George C. *Insects: Spiders and other Terrestrial Arthropods.* New York: Dorling Kindersley, 2000. An encyclopedia, flush with color photographs, describing the taxonomy, life cycle, body dimensions, feeding habits, and identification traits of 550 insects.

McKee, Jeffrey K. *The Riddled Chain: Chance, Coincidence, and Chaos in Human Evolution.* Piscataway, N.J.: Rutgers University Press, 2000. Proposes a "bottom-up" model of evolution to explain how it produced humans and predicts that the race may evolve at an unprecedented pace.

Mayor, Adrienne. *The First Fossil Hunters: Paleontology in Greek and Roman Times.* Princeton, N.J.: Princeton University Press, 2000. A classicist points out that the ancients studied and wrote about the fossils of extinct animals, including mammoths and mastodons, and so were the first paleontologists; she reviews their contributions.

Miller, Geoffrey F. *The Mating Mind: How Sexual Choice Shaped the Evolution of Human Nature.* New York: Doubleday, 2000. Drawing from recent theories in evolutionary psychology and biology, the author associates human behaviors, including the urge to create, with the competition to attract sexual partners and reproduce.

Ridley, Matt. *Genome: The Autobiography of a Species in Twenty-three Chapters.* New York: HarperCollins, 2000. Ridley regards the human genome as a historical document. He picks one gene from each chromosome and uses it to examine the genetic bases for physiological and behavioral traits.

Russon, Anne E. *Orangutans: Wizards of the Rain Forest.* Willowdale, Canada: Firefly, 2000. The author spent ten years studying orangutans in Borneo, and the book relates her experiences with the animals and their remarkable communication and learning skills as well the dangers to them from the expanding human presence.

Souder, William. *A Plague of Frogs: The Horrifying True Story.* New York: Hyperion, 2000. Souder recounts the discovery of deformed frogs in Minnesota and elsewhere in 1995 and surveys the theories of scientists about what environmental factors caused the mutations and the implications for people.

Tattersall, Ian, and Jeffrey H. Schwartz. *Extinct Humans.* Boulder, Colo.: Westview Press, 2000. Based on exhaustive examinations of the hominid fossil record, the authors conclude that human evolution, like that of all Earth's fauna, involved experimentation, diversification, and extinction.

Turner, J. Scott. *The Extended Organism: The Physiology of Animal-Built Structures.* Cambridge, Mass.: Harvard University Press, 2000. Turner contends that such edifices as tunnels, mounds, webs, and reefs should be considered extensions of the organisms that build them and describes how they control the flow of matter and energy in the builders' environments.

Tyson, Peter. *The Eighth Continent: Life, Death, and Discovery in the Lost World of Madagascar.* New York: William Morrow, 2000. A historical and biological survey of the world's fourth largest island.

Ward, Peter D., and Donald Brownlee. *Rare Earth: Why Complex Life Is Uncommon in the Universe.* New York: Copernicus, 2000. Professors of geological sciences and astronomy advance their Rare Earth Hypothesis, holding that Earth-like conditions seldom develop in the universe and so life is rare, in contrast with the popular view among scientists that life must be abundant.

Watson, Lyall. *Jacobson's Organ: And the Remarkable Nature of Smell.* New York: W. W. Norton, 2000. Watson explains how scents, smells, and odors are information for humans and animals, affecting emotions and basic behavior.

Weinberg, Samantha. *A Fish Caught in Time: The Search for the Coelacanth.* New York: HarperCollins, 2000. Accounts of the unexpected 1938 discovery of a coelacanth, a primitive fish thought to be long extinct, and another specimen found in 1998, to the joy of zoologists.

Wills, Christopher, and Jeffrey Bada. *The Spark of Life: Darwin and the Primeval Soup.* Cambridge, Mass.: Perseus, 2000. The authors argue that life began on

Earth's surface, rather than underwater or underground, and critique other theories on the origin of species.

Wilmut, Ian, Keith Campbell, and Colin Tudge. *The Second Creation: Dolly and the Age of Biological Control.* New York: Farrar, Straus and Giroux, 2000. With the help of Tudge, a science writer, the scientists who first cloned a sheep tell the story of Dolly and another achievement, Polly, a transgenic sheep. The authors consider how cloned animals can aid in the battle against human disease.

Mathematics, Statistics, and Economics

Berlinski, David. *The Advent of the Algorithm: The Idea That Rules the World.* New York: Harcourt, 2000. Recounts the mathematical, logical, and philosophical roots of the algorithm and its use as a step-by-step guide for performing complex operations in many disciplines.

Bunch, Bryan. *The Kingdom of Infinite Number: A Field Guide.* New York: W. H. Freeman, 2000. A series of essays about individual numbers and their membership among the categories of natural, integral, rational, real, and complex numbers.

Doxiadis, Apostolos K. *Uncle Petros and Goldbach's Conjecture.* London: Bloomsbury, 2000. A fictional but historically and mathematically realistic tale of a man who solves the famous conjecture that every even number is the sum of two primes.

Fink, Thomas, and Yong Mao. *The Eighty-five Ways to Tie a Tie: The Science and Aesthetics of Knots.* New York: Broadway, 2000. Two theoretical physicists offer a primer to practical topology, following a sketch history of ties, cravats, and ascots, well illustrated.

Ifrah, Georges. *Universal History of Numbers: From Prehistory to the Invention of the Number.* New York: John Wiley & Sons, 2000. A historical encyclopedia, beginning with prehistoric body-centered counting systems and proceeding through Jewish, Muslim, and Mayan time calculations to the birth of numbers in India.

Orkin, Mike. *What Are the Odds: Chance in Everyday Life.* New York: Freeman, 2000. A math-based approach to assessing the hazards of life, accompanied by explanations of relevant statistical and mathematical analytical tools.

Seife, Charles. *Zero: The Biography of a Dangerous Idea.* New York: Viking, 2000. Readers discover in this book that the zero was intensely controversial among philosophers, clergy, and scientists ever since its invention in ancient times, and it remains problematic today, for example in the year 2000 computer dating problem. Seife also addresses the history of the idea of infinity.

Medicine and Health

Ashcroft, Frances. *Life at the Extremes: The Science of Survival.* Berkeley: University of California Press, 2000. The author defines the physiological responses to harsh environments, such as the bends in deep diving and frostbite, and summarizes theories about life under extreme conditions.

Cooper, David K. C., and Robert P. Lanza. *Xeno: The Promise of Transplanting Animal Organs into Humans.* New York: Oxford University Press, 2000. To solve the shortage of human organs for transplantation, say the authors, organs from transgenic animals could be the answer. They consider the medical and ethical problems attendant on such an approach to curing human disease.

Ewald, Paul W. *Plague Time: How Stealth Infections Cause Cancer, Heart Disease, and Other Deadly Ailments.* New York: Free Press, 2000. The author, a biology professor, argues that physicians understand the evolution of microbes poorly and that infections by them cause many chronic ailments.

Garrett, Laurie. *Betrayal of Trust: The Collapse of Global Public Health.* New York: Hyperion, 2000. Garrett suggests that the declining investment in basic public health infrastructure and disease control, coupled with bioengineered biological warfare agents, may result in devastating superplagues.

Greenfield, Susan. *The Private Life of the Brain: Emotions, Consciousness, and the Secret of the Self.* New York: John Wiley & Sons, 2000. The director of Great Britain's Royal Institution, a neuroscientist, discourses on her theory of the interrelations of the brain, personality, mental well-being, pleasure, and emotion, also considering the effects of recreational drug use.

Klaidman, Stephen. *Saving the Heart: The Battle to Conquer Coronary Disease.* New York: Oxford University Press, 2000. A history of efforts to treat coronary artery disease and a critique, often incisive, of modern cardiology, surgery, and heart research.

Lutz, Tom. *Crying: The Natural and Cultural History of Tears.* New York: W. W. Norton, 2000. Relates the physiological importance of tears, the differing hormonal content in the three types of tears—reflex, basal, and psychic—and the variation between men and woman and from culture to culture.

Nuland, Sherwin B. *The Mysteries Within: A Surgeon Reflects on Medical Myths.* New York: Simon & Schuster, 2000. A celebrated medical writer considers both the ancient understanding of the uterus, heart, stomach, liver, and spleen and modern medical knowledge. He addresses modern myths concerning the organs and the peril of trusting New Age remedies.

Provine, Robert R. *Laughter: A Scientific Investigation.* New York: Viking, 2000. Explores the complex physiological and social reasons for laughter, including its renowned healing power.

Regush, Nicholas. *The Virus Within: A Coming Epidemic.* New York: Dutton, 2000. An American Broadcasting Corporation correspondent acquaints readers with human herpes virus-6 (HHV6) and the controversy about its possible role in various human diseases, including acquired immunodeficiency syndrome (AIDS) and multiple sclerosis.

Wall, Patrick. *Pain: The Science of Suffering.* New York: Columbia University Press, 2000. A physiologist explains the neurological gate-control theory of pain, offers advice on pain management, and relates some of the odd ways people and animals react to pain.

Wolpert, Lewis. *Malignant Sadness: The Anatomy of Depression.* New York: Free Press, 2000. A biology professor examines the psychological and biological theories concerning depression and its psychoanalytic and pharmaceutical treatments.

Zimmer, Carl. *Parasite Rex: Inside the Bizarre World of Nature's Most Dangerous Creatures.* New York: Free Press, 2000. The human and natural history of parasites that prey on people, which concludes by exploring the notion that humans are uncontrolled parasites to the world's ecosystem.

Physics

Bodanis, David. *E=MC²: A Biography of the World's Most Famous Equation*. New York: Walker, 2000. Bodanis anatomizes Albert Einstein's famous mass-energy equation, supplies its historical context and place in relativity theory, and recounts it practical applications, including the development of the atomic bomb.

Close, Frank. *Lucifer's Legacy: The Meaning of Asymmetry*. New York: Oxford University Press, 2000. An exploration of the effect of symmetry, asymmetry, and supersymmetry in the structure of matter, and the reason for the greater abundance of matter than antimatter.

Fraser, Gordon. *Antimatter: The Ultimate Mirror*. Cambridge, England: Cambridge University Press, 2000. Fraser unfolds the scientific understanding of antimatter from Paul Dirac's first theoretical conception of it to synthesis of the first antihydrogen atom in 1996.

Kane, Gordon. *Supersymmetry: Unveiling the Ultimate Laws of Nature*. Cambridge, Mass.: Perseus, 2000. Kane explains the next advance in particle physics, a theory holding that every known particle has an exotic twin. The author believes that supersymmetry, if verified, would solve many mysteries, such as the roles of dark matter and antimatter.

Newton, Roger G. *Thinking About Physics*. Princeton, N.J.: Princeton University Press, 2000. A physicist urges colleagues to ponder the philosophical implications of their theories as he surveys such concepts as symmetry, time, causality, and the fundamental entity of quantum physics.

Parker, Barry. *Einstein's Brainchild: Relativity Made Relatively Easy!* Buffalo, N.Y.: Prometheus, 2000. Without recourse to advanced math, the author addresses the special and general theories of relativity, aided by diagrams, and discusses such implications as black holes, curved space-time, and time travel, as well as the elaborations of later scientists.

Applied Technology

Ballard, Robert D., with Will Hively. *The Eternal Darkness: A Personal History of Deep-Sea Exploration*. Princeton, N.J.: Princeton University Press, 2000. Tales drawn from Ballard's hundred-plus dives in special submarines to the deep ocean floor and discussions about diving history and technology.

Brockman, John, ed. *The Greatest Inventions of the Past 2,000 Years: Today's Leading Thinkers Choose the Creations That Shaped Our World*. New York: Simon & Schuster, 2000. Texts of 109 answers to a question posted on the Internet: "What is the most important invention of the last two thousand years and why?" Responses are from scientists of all disciplines, artists, and others.

Buderi, Robert. *Engines of Tomorrow: How the World's Best Companies Are Using Their Research Labs to Win the Future*. New York: Simon & Schuster, 2000. Buderi finds that a resurgence in research and development in nine top high-tech companies holds the promise of revolutionary new technology, such as holographic computer memory and ultrasmall sensors, all with vast economic impact.

Lindsay, David. *House of Invention: The Secret Life of Everyday Products*. New York: Lyons Press, 2000. Proceeding from room to room in the typical American house, Lindsay points out common items, such as razors and frozen foods, and describes their creation and history.

Lunt, Karl. *Build Your Own Robot!* Natick, Mass.: A. K. Peters, 2000. Drawn from the author's column in *Nuts and Bolts,* the book provides how-to guidance to hobbyists in robotics, a field predicted soon to become as revolutionary in daily life as computers were in the 1990's.

Menzel, Peter, and Faith D'Aluisio. *Robo sapiens: Evolution of a New Species.* Cambridge, Mass.: MIT Press, 2000. Considers the possibility of a hybrid species of humans and robots as part of a survey of robotic technology presented in short chapters and sidebars, accompanied by many photographs.

Morton, David. *Off the Record: The Technology and Culture of Sound Recording in America.* Piscataway, N.J.: Rutgers University Press, 2000. Morton considers the economic and cultural influence of the recording industry as he lays out its technical history, beginning with Thomas Edison. He includes anecdotes concerning dictation and answering machines.

Rybczynski, Witold. *One Good Turn: A Natural History of the Screwdriver and the Screw.* New York: Scribner, 2000. Traces the history of the screwdriver, which the author considers the best tool of the second millennium, back to the fifteenth century.

Taylor, Nick. *Laser: The Invention, The Nobel Laureate, and the Thirty-Year Patent War.* New York: Simon & Schuster, 2000. A history of the laser centered on the legal wrangle between Charles Townes (the Nobel laureate) and Gordon Gould over who thought of it first and is entitled to the basic patent for it.

Wolfe, Maynard Frank. *Rube Goldberg: Inventions.* New York: Simon & Schuster, 2000. Biography of Goldberg and collection of his famous cartoons illustrating wacky, impossibly complicated contraptions.

Roger Smith

Journals and Magazines

➤ *The following is an annotated list, by subject, of some of the best science journals and magazines that are appropriate resources for students and teachers. Each entry includes publication information, the publication's ISSN, and a code that indicates appropriate audience levels. The audience level code is as follows: MS (middle school), HS (high school), U (undergraduate), and GP (general public).*

General Science and Technology

American Scientist. Six times per year, first published 1942. ISSN: 0003-0996. U, GP. Official publication of the Society of Sigma Xi, a scientific research society. Covers a broad scope of scientific developments and news written by scholars in the field.

Current Science. Eighteen times per year. ISSN: 0011-3905. MS, HS. This Weekly Reader publication covers a wide range of scientific and health topics with a focus on how they affect young people. Liberally illustrated; typically includes puzzles, activities, and games. A teacher's edition is available.

Discover. Monthly, first published 1980. ISSN: 0274-7529. MS, HS, U, GP. An excellent all-around science magazine appropriate for all levels of readership. Covers such diverse topics as conservation, genetics, and robotics.

Exploring. Quarterly, first published 1992. ISSN: 0889-8197. MS, HS. Designed to interest young people in science. Each issue examines a different scientific topic from nutrition to engineering. A recent issue, for example, examined the structure of suspension bridges. Formerly called *Exploratorium.*

Natural History. Eleven times per year, first published 1919. ISSN: 0028-0712. HS, U, GP. While focusing on the anthropological, biological, natural, and earth sciences, no scientific topic is off limits in this beautifully illustrated, widely read publication of the American Museum of Natural History.

Nature. Weekly, first published 1869. ISSN: 0028-0836. U, GP. Includes review articles and reports on research in all the sciences. Authoritative and informative.

New Scientist. Weekly, first published 1971. ISSN: 0262-4079. HS, U, GP. British weekly that covers international scientific news and developments in the world of technology. An excellent source for up-to-the-minute updates on all aspects of the scientific world.

Science. Weekly, first published 1880. ISSN: 0036-8075. U, GP. This publication of the American Association for the Advancement of Science reports on current discoveries and developments in all areas of science. Scholarly, yet accessible.

Science News. Weekly, first published 1966. ISSN: 0036-8423. MS, HS. This publication of Science Service, the sponsors of the Discovery Young Scientist Challenge and the Intel Science Talent Search, emphasizes science news for young people.

Science Teacher. Nine times per year, first published 1936. ISSN: 0036-8555. HS, U. Strives to "promote excellence and innovation in science teaching and learning." Written largely by teachers who are sharing their classroom experiences.

Scientific American. Monthly, first published 1845. ISSN: 0036-8733. HS, U, GP. Covers the gamut of scientific topics from astronomy to anthropology in an authoritative manner. Although primarily written by scientists, articles in this long-established journal are accessible to the layman. Liberally illustrated.

Astronomy and Space

Astronomy. Monthly, first published 1973. ISSN: 0091-6358. HS, U, GP. Well-written feature articles covering all aspects of astronomy for the layperson. Beautifully illustrated; includes diagrams, sky maps, and a monthly sky almanac. Departments include AstroNews and Star Stuff.

Odyssey. Nine times per year, first published 1979. ISSN: 0163-0946. MS. Designed to help upper elementary and middle school students understand astronomy, outer space, and space exploration. Regular features include projects, puzzles, and contests. Colorfully illustrated.

Sky and Telescope. Monthly, first published 1941. ISSN: 0037-6604. HS, U, GP. Written with amateur astronomers in mind, this long-established monthly has also earned the respect of professionals. Liberally and colorfully illustrated with regular features such as News Notes and Deep-Sky Notebook.

Chemistry

Chemical & Engineering News. Weekly, first published 1942. ISSN: 0009-2347. U, GP. The self-proclaimed "newsmagazine of the chemical world," this publication of the American Chemical Society is devoted to covering current news and information on the chemical industry. Regular sections include Business, Government and Policy, and Science/Technology.

Journal of Chemical Education. Monthly, first published 1924. ISSN: 0021-9584. HS, U, GP. This American Chemical Society publication is aimed primarily at high school students and teachers but is also of interest to practitioners and the educated layperson. Includes classroom activities, laboratory experiments, and information on current findings in the field of chemistry.

Computer Science and Information Technology

(Note that many magazines on computers have gone to online formats.)

PC World. Monthly, first published 1983. ISSN: 0737-8939. HS, U, GP. Coverage includes all aspects of IBM-type PC computing. Often provides comparisons of hardware, peripherals, and software. Useful for all PC users.

Technology Review. Six times per year, first published under its current name in 1899. ISSN: 1099-274X. U, GP. This Massachusetts Institute of Technology publication provides a good overview of advances in the scientific and technological world.

Earth Sciences and the Environment

Amicus Journal. Four times per year, first published 1980. ISSN: 0276-7201. U, GP. A well-researched journal published by the Natural Resources Defense Council that focuses on a wide variety of environmental issues, usually from a legal or political point of view.

Audubon. Six times per year, first published 1899. ISSN: 0097-7136. MS, HS, U, GP. This National Audubon Society publication covers a wide range of environmental topics from wildlife to nature. Also reports extensively on environmental problems. A classic that has stood the test of time. Printed on recycled paper.

Calypso Log. Six times per year, first published 1974. ISSN: 8756-6354. MS, HS, GP. This Cousteau Society magazine focuses on marine and freshwater environmental issues. Beautifully illustrated, printed on recycled paper.

Dolphin Log. Six times per year, first published 1981. ISSN: 8756-6362. MS. This Cousteau Society publication teaches late elementary to middle school students about science and the environment as they relate to the sea. Regularly includes news about the latest Cousteau expeditions.

National Wildlife. Six times per year, first published 1963. ISSN: 0028-0402. MS, HS, U, GP. This widely read National Wildlife Federation publication focuses on wildlife topics and other issues of broad environmental concern, primarily in the United States.

Rocks and Minerals. Six times per year, first published 1926. ISSN: 0035-7529. HS, U, GP. Includes articles on mineralogy, geology, and paleontology on a popular level. Written primarily for amateurs and hobbyists. Includes museum notes.

Sierra. Six times per year, first published 1977. ISSN: 0161-7362. HS, U, GP. A publication for the outdoor enthusiast that focuses on conservation, wilderness and energy issues, and outdoor recreation. Also includes book reviews, fiction, poetry, political updates, and photographs and essays contributed by readers.

Life Sciences and Genetics

American Biology Teacher. Eight times per year, first published 1938. ISSN: 0002-7685. MS, HS, U. Authored primarily by biology teachers, this publication of the National Association of Biology Teachers provides ideas for the classroom, including laboratory projects, research updates, and teaching methods.

American Naturalist. Monthly, first published 1867. ISSN: 0003-0147. U, GP. The official journal of the American Society of Naturalists, this long-established publication aims to "advance and diffuse the knowledge of organic evolution and other broad biological principles." Reports on current research in the study of ecology and evolution as they relate to plant and animal communities.

Bioscience. Monthly, first published 1964. ISSN: 0006-3568. U, GP. This publication of the American Institute of Biological Sciences focuses on current topics in biology and their impact on society. Besides feature articles on timely topics, the magazine includes opinion pieces and essays, software and equipment reviews, and relevant political news.

Mathematics, Statistics, and Economics

Chance. Four times per year, first published 1988. ISSN: 0933-2480. HS, U, GP. This publication of the American Statistical Association appeals to anyone who needs to analyze data or enjoys statistical problems. Successfully illustrates statistical applications in daily life by including articles on such topics as sports and public policy.

The Economist. Weekly, first published 1843. ISSN: 0013-0613. HS, U, GP. Analysis and commentary on the economic state of the world. Also reports on international news.

Mathematics Magazine. Five times per year, first published 1947. ISSN: 0025-570X. HS, U, GP. This publication of the Mathematical Association of America includes current topics in math, articles on the history of math and its application, and sample problems and solutions.

Mathematics Teacher. Nine times per year, first published 1908. ISSN: 0025-5769. MS, HS, U. This publication of the National Council of Teachers of Mathematics is devoted to the improvement of mathematics instruction in junior and senior high schools. Provides excellent teaching aids.

Medicine and Health

American Health. Ten times per year, first published 1982. ISSN: 0730-7004. HS, GP. Emphasizes current information on personal health and wellness. Covers such topics as nutrition, fitness, psychology, and health and beauty.

FDA Consumer. Ten times per year, first published 1972. ISSN: 0362-1332. HS, U, GP. This well-written, informative magazine published by the U.S. Department of Health and Human Services provides consumers with access to the research of the Food and Drug Administration. Articles discuss such wide-ranging topics as food safety, new surgical techniques, and prescription drugs.

Harvard Health Letter. Monthly, first published 1990. ISSN: 1052-1577. HS, U, GP. This Harvard Medical School publication provides health information for a general audience. Advises on wellness issues and how to cope with a wide variety of medical problems.

Mayo Clinic Health Letter. Monthly, first published 1983. ISSN: 0741-6245. HS, U, GP. A valuable resource for anyone needing health information, this Mayo Clinic newsletter offers reliable information on a wide variety of medical topics.

University of California, Berkeley, Wellness Letter. Monthly, first published 1984. ISSN: 0748-9234. HS, U, GP. This popular monthly newsletter strives to provide current information on a variety of topics that will help its readers be wise medical consumers.

Physics

Physics Teacher. Nine times per year, first published 1963. ISSN: 0031-921X. HS, U. Aimed at high school physics teachers, this journal published by the American Association of Physics Teachers includes articles on teaching methods, laboratory demonstrations and experiments, and equipment and apparatus.

Physics Today. Monthly, first published 1948. ISSN: 0031-9228. HS, U, GP. This American Institute of Physics publication reports on physics research and information in a manner accessible to both the layperson and the professional. Colorfully illustrated and engagingly written.

Jane Marie Smith

Audiovisual and CD-ROM Resources

➤ *The following annotated list presents a selection of the year's science audiovisuals, including videotapes, CD-ROMs, and DVDs, grouped by subject. The audiovisual materials range from learning aids to programs designed for the general public.*

General Science

The Amateur Scientist: The Complete Twentieth Century Collection. CD-ROM. Tinker's Guild, 2000. This electronic archive contains all "The Amateur Scientist" columns published in *Scientific American* from 1928 through 1999. These columns feature practical how-to projects for creating scientific instruments and apparatus.

I Love Science! 2000. CD-ROM. DK Multimedia, 2000. This interactive tool lets students practice and review basic scientific concepts in an entertaining manner. Students can run experiments and study topics at their own rate. For ages seven to eleven.

Integrated Science. Multimedia. Carolina Academic Press, 2000. The set features a student text, two CD-ROMs, a teacher guide, and a teacher resource for each of these three topics: constancy and change, interactions and limits, and patterns and cycles. Brings together various fields of scientific study. For juveniles.

Learning That Works: Science in the Real World. Video. 3 vols. WGBH Boston Video, 2000. These tapes, accompanied by a guidebook, are for science teachers to use in developing educational plans. The three videos cover lessons from the field, community projects, and partners in education.

Science 2000: The Multimedia Inquiry Approach to Science. Laser optical disks. Decision Development Corporation, 2000. This interactive resource covers the fields of physical, life, earth, and environmental sciences.

Scientific Problem Solving. Video. Understanding Science series. Educational Video Network, 2000. This educational video discusses scientific methodology and problem solving as well as scientific experimentation.

Teaching High School Science. Video. 6 vols. WGBH Boston and Annenberg/CPB Project, 2000. These six tapes and accompanying guidebook contain ideas for high school science teachers. Topics include thinking like scientists, chemical reactions, investigating crickets, exploring Mars, and the physics of optics.

Applied Technology

Overview of Biotechnology. Video. Films for the Humanities and Sciences, 2000. Examines biotechnology and how it relates to such varied disciplines as medicine, health care, ergonomics, and communications. Industry employees offer comments on their work and the industry.

Science and Technology: One Hundred Years of Progress. Video. Films for the Humanities and Sciences, 2000. Produced by ABC News and narrated by Peter Jennings, this video looks at the technological innovations and discoveries in science during the last one hundred years.

Technological Change. Video. Films for the Humanities and Sciences, 2000. Describes the effect of information technology on the airline business in Hong Kong and that of the Internet on markets and business organization. Also looks at how a traditional industry—tomato growing in Iceland—is using technological advances.

Astronomy

Astronomy. Video. Prentice Hall, 2000. This video, designed for elementary and secondary school students, covers aspects of astronomy such as the space program, the northern lights, the electromagnetic spectrum, the Hubble Space Telescope, and the stars.

Astronomy: The Earth and Beyond. CD-ROM. Films for the Humanities and Sciences, 2000. Covers core elements of astronomy and earth science curriculums. Provides information on the origin of the universe and the life and death of a star as well as a guided tour of the solar system. An affiliated Web site contains a downloadable eclipse simulator.

The Beginning. Video. Lights in the Sky series. Phoenix Multimedia, 2000. This video, the first in a series on the history of astronomy, examines the beginning of the universe.

Inside the Space Station. Video or DVD. Discovery, 2000. In addition to presenting the structure and function of the International Space Station, this documentary covers the engineers involved in its creation and the astronauts who plan to live in the finished station.

Meteors, Asteroids, and Comets. Video. Lights in the Sky series. Phoenix Multimedia, 2000. Examines the history of astronomy, focusing on meteors, asteroids, and comets.

The Milky Way's Invisible Light. Video. NASA Goddard Space Flight Center, AT&T, Digital Media Center, 2000. This video deals with three-dimensional imaging in astronomy and the Milky Way.

Planets and Moons: Our Solar System. Video. Lights in the Sky series. Phoenix Multimedia, 2000. This installment in a series on the history of astronomy deals with the planets of the solar system.

Runaway Universe. Video. Nova, 2000. This video presents the Nova program on the universe.

The Sky Above: A First Look. Video. Rainbow Educational Video, 2000. Written and produced by Peter Cochran, this video examines the sky, Moon, and solar system.

Space Travel: The History. Video. Lights in the Sky series. Phoenix Multimedia, 2000. This video, the fifth in a series on the history of astronomy, covers space flight.

Stargaze: Hubble's View of the Universe. DVD. Alpha, 2000. Provides narrated views of the universe in a widescreen format.

Sun and Stars. Video. Lights in the Sky series. Phoenix Multimedia, 2000. The second video in a series on the history of astronomy deals with the Sun and stars.

The Universe: A Guided Tour. DVD. Films for the Humanities and Sciences, 2000. This two-part series is a digital video disc presentation of *The Complete Cosmos* video series. The twenty-five ten-minute programs cover the solar system and astronomy and space explorations. Subjects include the Hubble Space Telescope, black holes, and the big bang theory.

Computers

Computer Vocabulary for the Classroom. Video. Educational Video Network, 2000. This educational video provides information on common terms used in information technology.

Digital Divide: Technology and Our Future. Video. 2 parts. Films for the Humanities and Sciences, 2000. These videos, produced by Studio Miramar in association with the Independent Television Service, cover computer-assisted education and virtual diversity.

Future of Digital Music: Senate Judiciary Committee. Video. 2 vols. C-SPAN Archives, 2000. These two videotapes cover the Senate hearing taped on July 11, 2000, dealing with the legal issues surrounding digital music. Committee members include Orrin Hatch and Patrick Leahy.

Technoculture: Finding Our Way in the Terra Incognita. Video. Living in the Brave New World series. Films for the Humanities and Sciences, 2000. This video, produced by Galafilm in association with Electronic Post Office and CBC Newsworld, deals with the interactions between technology and society.

Technoscience: Blurring the Line Between Man and Machine. Video. Living in the Brave New World series. Films for the Humanities and Sciences, 2000. Produced by Galafilm in association with Electronic Post Office and CBC Newsworld, this video looks at the scientific and social effects produced by interactions between people and computers.

Wired for What? The Dividends of Universal Access. Video. Films for the Humanities and Sciences, 2000. This film, originally part of the Public Broadcasting Service television series *Digital Divide,* deals with Internet access, computers, and the education of children.

Earth Sciences

Conserving Earth's Biodiversity with E. O. Wilson. CD-ROM. Island Press, 2000. Evolutionary biologist E. O. Wilson and photographer Dan L. Perlman worked together to create this interactive CD-ROM on conservation and environmental science. For high school students and undergraduates.

Earth and Fire. Video. One Hundred Percent Educational Videos, 2000. This educational earth sciences video covers land forms and geomorphology in its discussion of the earth and fire.

Earth Science in Action. Video. 14 vols. Schlessinger Media, 2000. This fourteen-volume set of educational videos designed for juveniles covers earthquakes, fossil fuels, fossils, geological history, land formations, minerals, natural resources, oceans, rocks, soil, topography, volcanoes, the water cycle, and weathering and erosion. Contains teacher's guide, which is also available online.

Savage Planet. Video. 4 vols. MPI Home Video, 2000. The four videos in this program created by Granada Television, Thirteen, WNET, deal with volcanic killers, storms of the century, deadly skies, and weather extremes.

Stormchasers. Video. MacGillivray Freeman Films, 2000. This documentary about storm chasers, people who pursue storms in order to study them, is narrated by Hal Holbrook.

Understanding the Weather. Video. Educational Video Network, 2000. This video is a teaching aid for those seeking to teach students about the weather.

Weather: A First Look. Video. Rainbow Education Video, 2000. This educational video, produced by Peter Cochran, focuses on the weather. Contains twelve activity pages and solutions.

Weather and Climate. Video. Earth Science Video series. Prentice Hall, 2000. This educational video, meant for elementary and secondary school students, looks at the weather and climate, covering topics such as hot air balloons, changes in the climate, the greenhouse effect, heat, and violent storms.

What in the World? The Earth Explained—Weather and Climate. Video. Science Is Elementary series. SVE & Churchill Media, 2000. This video, one in a series on science, focuses on weather and climate.

What's Up with the Weather? Video. WGBH Boston Video, 2000. This video, produced and directed by Jon Palfreman, was a NOVA/Frontline special report. It examined global warming and the greenhouse effect, climatology, long-range weather forecasting, and meteorology.

Health and Medicine

Age Happens: Psychological and Physiological Aspects of Aging. Video. Aquarius Health Care Videos, 2001. This video, originally broadcast as part of the television series *The Human Condition,* focuses on aging and the field of geriatrics.

Highlights from the Thirteenth International AIDS Conference. Video. The Bureau, 2000. This video presents the most important findings from the AIDS conference held in Durban, South Africa, in July, 2000.

Leptin and the Neural Circuit Regulating Body Weight. Video. National Institutes of Health, 2000. This video, part of the National Institutes of Health director's Wednesday afternoon lecture series, deals with the role of leptin in metabolism and in controlling body weight.

Living Longer . . . Living Better? Video. Aging: Growing Old in a Youth-Centered Culture series. Films for the Humanities and Sciences, 2000. Experts on aging and medical ethics examine quality of life issues presented by the technological advances that allow people to live longer.

The Quality Gap: Medicine's Secret Killer. Video. Films for the Humanities and Sciences, 2000. Pulitzer Prize-winning journalist Hedrick Smith investigates the quality of health care being received in several states in the United States. He looks at performance reports on physicians and facilities doing open-heart surgery in New York and a Kentucky veterans hospital's efforts to lessen medical errors.

Safety of Gene Therapy Senate Committee. Video. Nation Cable Satellite, 2000. This video, created by the Public Health Subcommittee of the Health, Education, Labor, and Pension Committee of the U.S. Senate, examines the safety of gene therapy, a procedure used experimentally in the United States.

Twenty-first Century Medicine. Video. Scientific American Frontiers series. PBS Home Video, 2000. Originally produced for Connecticut Public Television, this program, hosted by Alan Alda, looks at the future of medicine.

Ultimate Human Body 2.0. DVD. DK Interactive, 2000. This updated version of the 1994 best-seller includes the American Medical Association Family Medical Guide. It allows users age ten and up to explore the body in three dimensions and see how the body works. Contains descriptions of every part of the body and their functions.

Weighing In: Problems of Obesity. Video. Aquarius Health Care Videos, 2001. This video, originally part of the television series, *The Human Condition*, deals with the physiological aspects of obesity and weight loss.

Life Sciences

After Darwin: Genetics, Eugenics, and the Human Genome. Video. 2 vols. Films for the Humanities and Sciences, 2000. Examines human genetics, particularly the ethical issues involved in the study of the human genome.

The Burning Sands. Video. 3 vols. Ambrose Video Publishing, 2000. An examination of deserts and desert biology.

Dawn of Man. Video. 4 vols. TLC Video, 2000. These videos, produced by the British Broadcasting Service and the Learning Channel, examine human evolution from its earliest roots.

Dinosaur Giants Found! Video. National Geographic Video, 2000. This National Geographic video examines the dinosaurs of Africa, including Sue the Tyrannosaurus Rex.

Killing Coyote. Video. High Plains Films, 2000. This Ecology Center production by Doug Hawes-Davis examines coyotes and the government policies that affect them and their environment.

The Origin of Species: An Illustrated Guide. Video. Films for the Humanities and Sciences, 2000. This BBC Worldwide production narrated by David Attenborough looks at Charles Darwin and the theory of evolution.

Plant Life in Action. Video. 6 vols. Schlessinger Media, 2000. Produced and directed by Stone House Productions, LLC, and First Light Pictures, this video examines photosynthesis, plants and animal interdependency, plant biodiversity, plant reproduction, plant structure and growth, and plants and people.

Prehistoric Giants. Video. 2 vols. TLC Video, 2000. These videos examine the dinosaurs from their origin to demise, focusing on the brontosaurus and killer raptor.

Raising the Mammoth. DVD. Discovery Channel, 2000. A team led by a French archaeologist went to Siberia in search of a possible buried woolly mammoth in the permafrost. This DVD follows their efforts to find and extract the animal as well as information on the species.

Science, Religion, and Evolution: The Controversy Continues. Video. Eugenie Scott, 2000. Eugenie Scott gives a lecture on religion and science, touching on creationism and evolution.

Spares or Repairs: Applications and Implications of Cloning. Video. Films for the Humanities and Sciences, 2000. Prominent genetics researchers such as Ian Wilmut of the Roslin Institute are featured in this video on cloning animals and specialized cells. Discusses the ethical issues and the future of tissue engineering.

Walking with Dinosaurs. Video or DVD. Twentieth Century Fox, 2000. This BBC, Discovery, and TV Asahi production, narrated by Kenneth Branagh, is an expanded video edition of the original six-part television series on dinosaurs produced and directed by Jasper James. Features computer-generated effects and sophisticated animatronic models.

When Dinosaurs Ruled. DVD. 3 vols. Madacy Records, 2000. This three-part DVD series includes "At the Ends of the Earth," "Birth of the Giants," and "The Real Jurassic Park."

Mathematics and Economics

Amazon.com. Video. Films for the Humanities & Sciences, 2000. This program, originally produced for the *NewsHour with Jim Lehrer*, deals with Amazon.com and its leader Jeff Bezos, booksellers, and Internet marketing.

Internet Shopping in the Twenty-First Century. Video. PBS Home Video, 2000. This discussion features leaders of prominent Internet businesses including Amazon.com's Jeff Bezos.

Internet Shopping: Interactive or Invasive? Video. Films for the Humanities and Sciences, 2000. This program, originally broadcast on the *NewsHour with Jim Lehrer*, looks at Internet companies, electronic commerce, and marketing and advertising on the Web.

Math Advantage 2000. CD-ROM. 3 vols. Encore Software, 2000. This interactive learning aid covers pre-algebra, algebra I, geometry, measurement, statistics, and probability. For students age eleven and up.

The Web World of Business. Video. Television Education Network, 2000. This video deals with business on the Web, also known as e-commerce.

Physics

Bikes and Cars: Centripetal Acceleration. Video. Films for the Humanities and Sciences, 2000. Written and presented by Tim David, this video examines mathematical models and mathematical physics through a study of centripetal acceleration.

Einstein's Relativity and the Quantum Revolution: Modern Physics for Nonscientists. Video. 4 vols. 2d ed. Teaching Company, 2000. These four videotapes, accompanied by two booklets, focus on Albert Einstein's theory of relativity and quantum mechanics.

Physical Science in Action. Video. 15 vols. Schlessinger Media, 2000. This set of videos covers topics such as force and energy, matter, electricity, magnetism, and sound. Contains a teacher's guide.

Physics: The Standard Deviants Core Curriculum. Video (10 vols.) and CD-ROM. Films for the Humanities and Sciences, 2000. This set of videotapes and accompanying CD-ROM provide information on basic physics. Includes sections on numbers, scalars, and vectors; circular, rotational, and projectile motion; Newton's laws of motion; friction, work, and energy; atomic structure; gravitation; harmonic motion and waves; heat; and thermodynamics.

Reflections, Advice, and Diversions, or Falling Honey and Floating Logs. Video. University of California, Berkeley, Department of Physics, 2000. University of California physics professor J. D. Jackson describes flaws in physics education at a colloquium. He cautions against overspecialization.

Rowena Wildin

Organizations, Agencies, and Archives

➤ *The organizations listed were selected because of their usefulness to students, educators, and the general public. All home pages have additional links and search engines that allow quick access to hidden resources. Many organizations do not have an e-mail address for general questions; however, they often have a link that allows people to send them a message.*

General

Eisenhower National Clearinghouse (ENC)
Eisenhower National Clearinghouse for Mathematics and Science Education
U.S. Department of Education
555 New Jersey Avenue, NW
Washington, D.C. 20208-5645
Telephone: (202) 219-2210
E-mail: web@enc.org
Home page: http://www.enc.org/

ENC's mission is to identify effective curriculum resources and disseminate useful information and products to improve elementary and secondary mathematics and science teaching and learning. See the organization's Internet home page for lesson plans, teaching tips, and free materials. "ENC partners" leads to more than 150 access centers scattered across the nation.

Library of Congress
First and Independence Avenues, SE
Washington, D.C. 20001-0000
Telephone: (202) 863-0320
Home page: http://www.loc.gov/

The Library of Congress is the largest library in the world, with nearly 119 million items on approximately 530 miles of bookshelves. The collections include 18 million books, 2 million recordings, 12 million photographs, 4 million maps, and 53 million manuscripts. Scientific and technical information make up roughly one-fourth of the total book and journal collection. The agency's Internet home page provides links to an online catalog of 12 million records, patent information, several online exhibitions, and various features for students of all ages. The agency provides an extensive collection of links to electronic text collections on the Internet at http://www.loc.gov/global/etext/etext.html. Hard copies of other books are available on interlibrary loan.

The Congressional Research Service operates under the Library of Congress and provides confidential, factual, nonpartisan research and analysis on issues as requested by members of Congress. These reports should be very useful to students and educators. Many of them are available free online through the National Council for Science and the Environment at http://www.cnie.org/ or the Federation of American Scientists at http://www.fas.org/.

National Academies
2101 Constitution Avenue, NW
Washington, D.C. 20418-0006
Telephone: (202) 334-2000
Home page: http://www.nas.edu/

The National Academy of Science was founded by an act of Congress in 1863 to serve as official adviser to the federal government on scientific and technical matters. The original academy has expanded into four organizations: The National Academy of Sciences, the National Academy of Engineering, the Institute of Medicine, and the National Research Council. The academies are private, honorary organizations dedicated to the furtherance of science, engineering, and medicine and whose members are elected in recognition of their accomplishments. Research, policy studies, and the compilation of technical reports is done under the direction of the National Research Council. The Internet home page for the four academies includes news notes and a subject index. There are links to each of the following subjects: agriculture, behavioral and social sciences, biology, business and economics, chemistry, computers and technology, earth sciences, education, engineering, environmental issues, health and medicine, mathematics and physics, national and international policy issues, space, and transportation.

More than 1,500 reports are available free online. The "Publications" link leads to *Beyond Discovery*, which tells fascinating stories of how various discoveries or advances were made. *Proceedings of the National Academy of Sciences* reports work done by leading researchers and is the second-most-cited scientific serial in the world. Current issues, preprints, and selected back issues are available free online. *Proceedings* can also be accessed directly at http://www.pnas.org/.

University of Pennsylvania Library
Van Pelt-Dietrich Library Center
3420 Walnut Street
Philadelphia, Pennsylvania 19104-6206
Telephone: (215) 898-7555
Home page: http://www.library.upenn.edu/

An extensive university library at fifteen locations. Its Internet home page has links to online catalogs, an online reference shelf, and an excellent list of online journals.

Astronomy and Space

Ames Research Center
National Aeronautics and Space Administration (NASA)
Moffet Field, California 94035
Telephone: (650) 604-5000
Home page: http://www.arc.nasa.gov/

The Ames Research Center specializes in research geared toward creating new knowledge and new technologies that span the spectrum of NASA interests, including supercomputing, neural networking, artificial intelligence, software development, aviation operations systems, and astrobiology. Its Internet home page is linked to educational resources for "Kids and Teachers" and "Astrobiology." There are articles, projects, and interactive simulations designed to interest students in science, mathematics, and technology.

Jet Propulsion Laboratory (JPL)
4800 Oak Grove Drive
Pasadena, California 91109
Telephone: (818) 354-4321
Home page: http://www.jpl.nasa.gov/

JPL is the National Aeronautics and Space Administration's leading center for the robotic exploration of the solar system, including exploring Earth from space. Its Internet home page has news articles on interesting subjects such as frost on Mars and a small, hopping robot. The "Technology" link leads to information and pictures of various spacecraft and their instruments. The "Education" link leads to a wealth of images, articles, tutorials, fact sheets, and activities.

Malin Space Science Systems (MSSS)
San Diego, California
E-mail: info@msss.com
Home page: http://www.msss.com/

MSSS designs, develops, and operates instruments that fly on robotic spacecraft. The Mars orbiter camera on the Mars Global Surveyor is probably its most successful instrument. Its Internet home page leads to an extensive gallery of images made with the Mars orbiter camera. The "Education and Public Outreach" link leads to the "Planetary Society Marslink Materials," an excellent study unit on various aspects of Mars.

National Aeronautics and Space Administration (NASA)
NASA Center for Aerospace-Public Affairs
300 East Street, SW
Washington, D.C. 20024-0321
Telephone: (202) 358-0000
NASA/Johnson Space Center
2101 Nasa Road 1
Mail Code Bd 35
Houston, Texas 77058
Telephone: (281) 483-3734
Home page: http://www.nasa.gov/

Established by the federal government in 1958, NASA has been charged with the exploration of space and the exploration of Earth from space. NASA is committed to developing practical applications such as weather satellites and communication satellites and to public education. In addition to the locations listed above, NASA maintains a number of facilities throughout the country, most of which offer tours and displays. Speakers can be requested from the nearest NASA center. Its Internet home page is a gateway to numerous sources of information. "Education Resources" includes Internet and multimedia resources for both teachers and students. The "Astronomy Picture of the Day" at http://antwrp.gsfc.nasa.gov/apod/ is particularly interesting. This site's archive contains the largest collection of annotated astronomical images on the Internet. NASA's "Space Science Education Resource Directory" is at http://teachspacescience.stsci.edu.

Planetary Society
65 North Catalina Avenue
Pasadena, California 91106
Telephone: (626) 793-5100
E-mail: tps@planetary.org
Home page: http://www.planetary.org/

The society bills itself as "the largest space interest group on Earth," with more than 100,000 members from 140 countries. A nonprofit organization dedicated to the exploration of the planets, the solar system, and the search for extraterrestrial life, the society supports and funds projects, provides public information, and supports educational activities. The Web site contains more than 2,000 pages of news and information about space exploration along with hundreds of photos and artwork. There are links and updates to the SETI@home project (enlisting users' home computers in the search for extraterrestrial intelligence). Users can submit their entries for the hundred Great Questions about life or the universe to be included in the PlanetTrek solar system scale model. Both students and teachers will find plenty to interest them in the Mars Millennium Project to design a community on Mars for the year 2030.

Solar and Heliospheric Observatory (SOHO) Mission
NASA Goddard Space Flight Center
Solar Physics Branch/Code 682
Greenbelt, Maryland 20771
E-mail: Dr. SOHO at letters@sohops.gsfc.nasa.gov
Home page: http://sohowww.nascom.nasa.gov/

The SOHO mission observes the Sun continuously and is a joint project of the European Space Agency (ESA) and the U.S. National Aeronautics and Space Administration (NASA). Its Internet home page has an excellent picture gallery and teacher resources, but the real excitement is at the SOHO sungrazer site, http://sungrazer.nascom.nasa.gov/. Sungrazers are an important and interesting class of comets. The general public can aid in their discovery by monitoring the realtime SOHO images at the sungrazer site. A new comet is being discovered every few days in this way.

Students for the Exploration and Development of Space, Arizona chapter (UA SEDS)
Box 174 Space Sciences Building
University of Arizona
Tucson, Arizona 85721
Telephone: (520) 621-9790

E-mail: seds@seds.org
Home page: http://www.seds.org

SEDS is an independent, student-based organization that promotes the exploration and development of space by educating people about the benefits of space and by involving students in space-related projects. Chapters exist in various parts of the world. The Arizona chapter sponsors one of the best general astronomy sites on the Web. Its home page contains numerous links to images and articles on subjects from constellations to space colonies. One of the best links is the "Nine Planets Solar System Tour." The images are spectacular, and the explanations are well-written, complete, accurate, and frequently updated.

Chemistry

American Chemical Society (ACS)
1155 Sixteenth Street, NW
Washington, D.C. 20036
Telephone: (202) 872-4600; (800) 227-5558
Home page: http://www.acs.org/

The ACS is a self-governed individual membership organization consisting of 161,000 members at all degree levels. Thirty-four divisions represent a wide range of disciplines for chemists, chemical engineers, and technicians. The ACS promotes public education, advises the federal government, and seeks to enrich the experience of all who are involved with chemistry. Its Internet home page provides information and links to the many ACS activities and resources. National, regional, and local section meetings provide opportunities for peer interaction and for education. Speakers can be requested for local meetings. Articles and job postings aid career development. There are searchable data bases on chemicals, lab ware, and professional journals. "Recommended Chemistry Sites" includes links to help sites and tutorials for students in high school and college. There are educator resources for teachers at all levels, links to online Material Safety Data Sheets, and links to "Green Chemistry" (environmental issues).

WebElements
Chemistry Department
University of Sheffield
Western Bank, Sheffield S10 2TN, United Kingdom
Telephone: +44 114 222 2000
Home page: http://www.webelements.com/

A marvelous compilation of information on the chemical elements. Its Internet home page provides easy access to information on the elements including description, history, uses, geology, biology, physical properties, and chemical compounds. "Chemdex" is an international listing of universities, government agencies, and companies that deal with chemistry.

Computer Science and Information Technology

American Association for Artificial Intelligence (AAAI)
445 Burgess Drive
Menlo Park, California 94025-3442

Telephone: (650) 328-3212
E-mail: info@aaai.org
Home page: http://www.aaai.org/

AAAI is a nonprofit scientific society devoted to advancing the scientific understanding of the mechanisms of thought and intelligent behavior and their embodiment in machines. The quick link "AI Topics" on the Internet home page allows users to reach a series of easily understood explanations of the important aspects of artificial intelligence.

Association for Women in Computing (AWC)
41 Sutter Street, Suite 1006
San Francisco, California 94104
Telephone: (415) 905-4663
E-mail: awc@awc-hq.org
Home page: http://www.awc-hq.org/

AWC is a nonprofit professional organization for individuals with an interest in information technology. It seeks to promote communication among women in computing and further their professional development and advancement. It promotes the education of women of all ages in computing. Its Internet home page contains links of interest to both students and professionals.

Computing Research Association (CRA)
1100 Seventeenth Street, NW, Suite 507
Washington, D.C. 20036-4632
Telephone: (202) 234-2111
E-mail: info@cra.org
Home page: http://www.cra.org/

CRA is an association of more than 180 academic departments and research laboratories. Its mission is to strengthen research and education in the computing field. Its Internet home page leads to information on many topics, but it may be most useful for students looking for strong academic programs or looking for employment in computing.

Tech Corps
Two Clock Tower Place, Suite 230
Maynard, Massachusetts 01754
Telephone: (978) 897-8282
E-mail: info@techcorps.org
Home page: http://techcorps.org/

Inspired by the Peace Corps, the Tech Corps is a program that helps teachers acquire and master computer technology with the help of volunteers from the technology community. Its Internet home page leads to information on the volunteer help program. In addition, the "Web Teacher" link yields instructions from beginning to advanced web surfing, and the "Resources" link leads to a plethora of education-related web sites.

Earth Sciences and the Environment
Ecological Society of America (ESA)
1707 H Street, NW, Suite 400
Washington, D.C. 20006
Telephone: (202) 833-8773
E-mail: esahq@esa.org
Home page: http://www.sdsc.edu/esa/

The ESA is a nonprofit organization of scientists founded in 1915 to promote ecological science by raising public awareness, doing research, teaching, and working to provide the knowledge needed to solve environmental problems. The "Education" and the "Public Affairs" links on its Internet home page lead to a number of reports and activities that may be of interest to teachers and students.

Federation of American Scientists (FAS)
307 Massachusetts Avenue, NE
Washington, D.C. 20002
Telephone: (202) 675-1010
E-mail: fas@fas.org
Home page: http://www.fas.org/

FAS is a privately funded nonprofit policy organization whose board of sponsors includes fifty-one of the United States' Nobel laureates in the sciences. FAS conducts analysis and advocacy of science, technology, and public policy, including cyberstrategy, environmental issues, national security, nuclear weapons, arms sales, biological hazards, secrecy, and space policy. Its Internet home page has a search engine that accesses congressional research reports as well as FAS documents. It features current topics and contains links to other topics such as "Disease Surveillance" (infectious animal diseases, worldwide) and "Special Weapons," which links to nuclear weapons and missile defense systems. The information given is well- researched and easy to understand. "CyberStrategy" calls up an extensive guide to authoring Web pages and converting documents from word processing files into web documents. It provides tips on software and helpful web sites.

Greenpeace International
Greenpeace USA
702 H Street, NW, Suite 300
Washington, D.C. 20001
Telephone: 800-326-0959 or (202) 462-1177
Home page: http://www.greenpeace.org/

An international environmental organization, Greenpeace lobbies governments, publishes information, and stages protests on behalf of environmental issues. Its Internet home page lists news stories and allows access to various countries' Greenpeace home pages. There are additional links to information and news regarding areas of concern such as climate, nuclear power and waste, and genetic engineering.

National Council for Science and the Environment (NCSE)
1725 K Street, NW, Suite 212

Washington, D.C. 20006
Telephone: (202) 530-5810
E-mail: cnie@cnie.org
Home page: http://www.cnie.org/

NCSE is a nonprofit organization funded by various universities, companies, and individuals. Its purpose is to educate the public and decision-makers and thereby "improve the scientific basis of environmental decision making." Its Internet home page features the National Library for the Environment. Key links are "Population & Environment Linkages," online books, reports, data sets, and more; "Environmental Journals Online," 330 journals grouped according to the amount of free online content; "Educational Resources & Directories," directories of programs, syllabi, and more; "Congressional Research Service Reports," more than 750 reports classified by topic; and "Congressional Research Service Issue Briefing Books," including one on global climate change.

National Geophysical Data Center (NGDC)
NOAA, Mail Code E/GC
325 Broadway
Boulder, Colorado 80303-3328
Telephone: (303) 497-6826
E-mail: info@ngdc.noaa.gov
Home page: http://www.ngdc.noaa.gov/

NGDC is the national repository for geophysical data and is part of the National Oceanic and Atmospheric Administration (NOAA). Its Internet home page leads to both general level and technical data on satellite data and imagery, glaciology, marine geology and geophysics, paleoclimatology, solar-terrestrial physics, and solid-earth geophysics (including the environment and biosphere). Teachers may wish to go to "Paleoclimatology" and then to "Education and Outreach," where they may access an extensive list of resources for all educational levels.

National Oceanic and Atmospheric Administration (NOAA)
NOAA Office of Public Affairs
Fourteenth Street and Constitution Avenue, NW, Room 6013
Washington, D.C. 20230
Telephone: (202) 482-6090
Home page: http://www.noaa.gov/

NOAA conducts research and gathers data about the oceans, atmosphere, space, and the Sun. It warns of dangerous weather, charts the seas and skies, and guides the use and protection of coastal resources. NOAA is a Commerce Department agency and works through five organizations: the National Weather Service, the National Ocean Service, the National Marine Fisheries Service, the National Environmental Satellite Data and Information Service, and Office of Oceanic and Atmospheric Research. Its Internet home page displays news stories, pictures, and links leading to a vast variety of information and pictures: current weather conditions, watches and warnings; current tide levels; information on marine sanctuaries, diving, and oil spills; real-time satellite imagery; fishery market news; information on global warming, drought, climate predictions, and paleoclimatology; and nautical and aeronautical charts, including historical charts. “Educational Resources” has special sites for teachers, for students, and “cool sites for everyone.”

Sierra Club
85 Second Street, Second Floor
San Francisco, California 94105-3441
Telephone: (415) 977-5500
E-mail: information@sierraclub.org
Home page: http://www.sierraclub.org/

The Sierra Club has more than 600,000 members. Its goals include enjoying and protecting the wild places on Earth and educating and enlisting humanity to protect and restore the quality of the natural and human environment. Its Internet home page is the portal to a wealth of information on the club's history and activities. There are links to news stories and to background articles on topics such as global warming, clean air, genetic engineering, nuclear waste, and pollution.

Life Sciences and Genetics

American Institute of Biological Sciences (AIBS)
AIBS Headquarters
1441 I Street, NW, Suite 200
Washington, D.C. 20005
Telephone: (202) 628-1500
Home page: http://www.aibs.org/

AIBS is the national umbrella organization for biologists and biological scientists. Both individuals and scientific societies may join. Its public outreach activities include congressional briefings, scientific roundtables, lectures, workshops, and conferences. Its Internet home page links to articles, policy statements, and career information. Most important, it also links to about seventy member organizations whose interests span all of biological science.

Botanical Society of America (BSA)
1735 Neil Avenue
Columbus, Ohio 43210-1293
Telephone: (614) 292-3519
E-mail: hiser.3@osu.edu
Home page: http://www.botany.org/

BSA promotes botany, the study and inquiry into the form, function, diversity, reproduction, evolution, and uses of plants. It is an umbrella organization for fifteen special interest sections of the society. Its Internet home page carries news items and links to numerous botany Web sites. BSA sponsors publications, regional and national meetings, symposia, field trips, and workshops. The “BSA Online Teaching Images” should be of particular interest to educators.

Genetics Society of America
9650 Rockville Pike
Bethesda, Maryland 20814-3998
Telephone: (301) 571-1825

Home page: http://www.faseb.org/genetics/gsa/gsamenu.htm

A professional society to promote research, advise Congress, and educate the public about genetics. Its Internet home page provides access to news notes, career information, and position papers on genetically modified organisms and on education in evolution.

National Association of Biology Teachers (NABT)
12030 Sunrise Valley Drive, Suite 110
Reston, Virginia 20191
Telephone: (703) 264-9696 or (800) 406-0775
E-mail: nabter@aol.com
Home page: http://www.nabt.org/

NABT is primarily for teachers of life sciences in elementary and secondary schools. It provides resources and helps teachers keep up with trends and developments in the field as well as enabling them to grow professionally. Its Internet home page has news of publications and meetings. "NABT Resources" includes links to "Spellex," a spelling list of 200,000 biotechnical words; "Position Statements" on evolution and the use of animals in education; and "Online Resources." The latter has numerous links—including "Ken's Bio-Web Resources," 800 Web sites for biology teachers and students—and lesson plans under "Biology Lessons for Prospective and Practicing Teachers."

Mathematics and Economics

American Mathematical Society (AMS)
P.O. Box 6248
Providence, Rhode Island 0294-6248
Telephone: (800) 321-4267 or (401) 455-4000
E-mail: ams@ams.org
Home page: http://www.ams.org/

AMS is a professional society with about 30,000 members. Its purposes are to promote mathematical research, to increase the awareness of the value of mathematics to society, and to foster excellence in mathematics education. Teachers may wish to use its Internet home page link "Government Affairs and Education." Students will find the link "What's New in Mathematics" more interesting. It leads to "Math in the Media" and to "Math Resources." The latter includes research topic suggestions for math students and some fun online math museums.

Econophysics Forum
Home page: http://www.econophysics.org/

The Econophysics Forum is a Web-based organization designed to facilitate communication among its members. Its home page leads to articles, books, and conferences on econophysics. A beginner will not find much that is helpful.

MacTutor History of Mathematics Archive
School of Mathematics and Statistics
Mathematical Institute
North Haugh

St Andrews, Fife KY16 9SS
Scotland
Telephone: +44 1334 46 3744
Home page: http://www-groups.dcs.st-and.ac.uk/

The Internet home page is that of the Turnbull WWW Server maintained by the School of Mathematical and Computational Sciences of the University of St. Andrews. It links to a computer algebra system and to the MacTutor History of Mathematics archive. The archive has more than 1,000 biographies and historical articles pertaining to mathematics.

Mathematical Association of America (MAA)
1529 Eighteenth Street, NW
Washington, D.C. 20036-1385
Telephone: (800) 331-1622 and (202) 387-5200
Home page: http://www.maa.org/

The MAA stimulates interest in mathematics through articles on contemporary mathematics and recent development. Its more than 30,000 members include college faculty and high school teachers. The "Student and Student Chapters" link on its Internet home page leads to a page for students that includes links to sites of interest to high school students and others for college students. The home page link "Columns" leads to interesting articles generally involving math in common situations such as different systems that might be used to determine the winner of an election.

Medicine and Health

American Medical Association (AMA)
AMA Headquarters
515 North State Street
Chicago, Illinois 60610
Telephone: (312) 464-5000
Home page: http://www.ama-assn.org/

The AMA is a professional organization for physicians, and they characterize the AMA as the "nation's leader in promoting professionalism in medicine and setting standards for medical education, practice, and ethics." The "Patient" link on its Internet home page will probably be of most interest to the general public. It leads to a "Doctor Finder," "News and Events," "Public Health," and "Specific Conditions." The latter two links lead to helpful and authoritative information on various health issues and conditions.

Centers for Disease Control and Prevention (CDC)
1600 Clifton Road
Atlanta, Georgia 30333
Telephone: (404) 639-3311
Home page: http://www.cdc.gov/

The CDC is an agency of the U.S. Department of Health and Human Services. Its mission is to promote health and quality of life by preventing and controlling disease, injury, and disability. Opening the links on its Internet home page releases floods of useful information. For example, "Hoaxes and Rumors" provides authoritative information on current health-related hoaxes and rumors. "Data and Statistics" gives the facts. "Health Topics A-Z" lists about 400 topics, and each topic may lead to several fact sheets, pamphlets, or books—all online.

Food and Drug Administration (FDA)
5600 Fishers Lane
Rockville, MD 20857-001
Telephone: (888) 463-6332
Home page: http://www.fda.gov/

The FDA is the agency of the U.S. Department of Health and Human Services charged with the mission of promoting and protecting public health by helping safe and effective products reach the market in a timely fashion and by monitoring products for continued safety after they are in use. Its Internet home page contains news notes and has links to safety alerts and product approvals. It also links to information packets designed for consumers, patients, women, seniors, and children.

World Health Organization (WHO)
WHO Headquarters Office
Avenue Appia 20
1211 Geneva 27
Switzerland
Telephone: (+00 41 22) 791 21 11
E-mail: info@who.ch
Home page: http://www.who.int/

WHO is a specialized agency of the United Nations. It promotes technical cooperation for health among nations, carries out programs to control and eradicate disease, and strives to improve the quality of human life. Its Internet home page "Reports" link lists available reports and information on disease outbreaks. The "Health Topics" link leads to numerous files on various diseases and conditions. WHOSIS (WHO Statistical Information System), which can be found at http://www.who.int/whosis/, provides worldwide health statistics.

Physics

American Association of Physics Teachers (AAPT)
One Physics Ellipse
College Park, Maryland 20740-3845
Telephone: (301) 209-3300
Home page: http://www.aapt.org/

The AAPT provides a forum for the advancement of the educational aspects of physics and emphasizes its place in general culture. Its Internet home page links to the "Physical Science Resource Center," which in turn links to classroom demonstrations and laboratory experiments.

American Institute of Physics (AIP)
One Physics Ellipse
College Park, Maryland 20740-3843
Telephone: (301) 209-3000

Home page: http://www.aip.org/

AIP is the umbrella organization for several member societies and organizations. Its purpose is to promote the advancement and diffusion of the knowledge of physics and its application to human welfare. Its Internet home page includes links to information on all aspects of physics, including government policies, environmental issues, the history of physics, and employment. The current issue of *Physics Today* is available online as are selected articles from previous issues.

American Physical Society (APS)
One Physics Ellipse
College Park, Maryland 20740-3844
Telephone: (301)-209-3200
Home page: http://www.aps.org/

The APS is a professional society with more than 42,000 members worldwide. It is dedicated to the advancement and dissemination of the knowledge of physics. Its Internet home page contains links to recent issues of the research journals, physics in the news, and a wonderful list of physics Internet resources.

Colour Museum
Echo Productions
Zollikerstrasse 195
CH-8008 Zurich, Switzerland
Telephone: ++41 1 3830406
Home page: http://www.colorsystem.com/

The Colour Museum is a Web-based educational resource. Its Internet home page leads to fifty-nine illustrated color theories from antiquity to modern times. Clicking on small illustrations enlarges them. Other essays discuss the special meanings that colors have in some cultures.

Fermilab Education Office
Education Office
Fermilab MS 226
Box 500
Batavia, Illinois 60510
Telephone: (630) 840-3092
Home page: http://www-ed.fnal.gov/

The Fermi National Accelerator Laboratory is the home of the world's most powerful particle accelerator. The mission of the Education Office is to enhance mathematics and science education. There are numerous resources that can be accessed from the "Student" and "Educator" links on the laboratory's Internet home page. One module allows students to use Fermilab data to see the effect of special relativity on the lifetime of a subatomic particle.

Applied Technology
American Nuclear Society (ANS)
555 North Kensington Avenue
LaGrange Park, Illinois 60526

Telephone: (708) 352-6611
E-mail: outreach@ans.org
Home page: http://www.ans.org/

A nonprofit international, scientific, and educational organization composed of about 11,000 engineers, scientists, administrators, and educators, the ANS promotes the advancement of engineering and science relating to the atomic nucleus by encouraging research, disseminating information, holding meetings and workshops, and establishing scholarships. The ANS periodically holds workshops and provides Geiger counters and educational materials for teachers of grades seven through twelve. *ReActions* is a newsletter about applications of radiation technology such as using X-ray fluorescence to detect a fake antique astrolabe. It is available online or by mail. Its Internet home page links to meetings, announcements, news releases, educational resources, and background information on nuclear power, the effects of radiation, and associated topics.

Energy Information Administration (EIA)
Energy Information Administration, EI 30
1000 Independence Avenue, SW
Washington, D.C. 20585
Telephone: (202) 586-8800
E-mail: infoctr@eia.doe.gov
Home page: http://www.eia.doe.gov/

The EIA is a statistical agency of the U.S. Department of Energy. It is charged with providing policy-independent data, forecasts, and analyses regarding energy and its interaction with the economy and the environment. Its Internet home page is the gateway to authoritative numbers on all forms of energy, including petroleum, coal, nuclear, renewable energy, and alternative fuels. There is also a children's page featuring Energy Ant.

Milnet
E-mail: milnet@mcint.com
Home page: http://www.milnet.com/

Milnet is a labor of love by a small group of volunteers headed by Michael Crawford, the chief editor. Their work is constantly reviewed by several "military open source intelligence" organizations that send updates and point out errors. They have been on the Internet since 1985, and the information they provide is generally excellent. The group's home page links to information on intelligence agencies, the military, terrorism, and recent conflicts. The information on the United States is the most comprehensive, but several other countries such as the former Soviet Union and China are also covered. "Military Information" includes a comprehensive list of weapons, along with their descriptions and how they work. Milnet has one of the most detailed descriptions of nuclear weapons (including their physics and technology) available.

Society of Women Engineers (SWE)
120 Wall Street, Eleventh Floor
New York, New York 10005-3902

Telephone: (212) 509-9577
E-mail: hq@swe.org
Home page: http://www.swe.org/

The mission of SWE is to stimulate women to achieve their full potential in careers as engineers and leaders. Career guidance programs help women prepare for engineering careers and also assist women in readying themselves for a return to active work after temporary retirement. Links on its Internet home page lead to information modules and kits to introduce engineering careers to grade school, junior high, and high school students. High school students may request an engineer e-mail mentor. Other information and activities will be helpful to college students and graduates.

Technology and Applications Program (TAP): NASA
4800 Oak Grove Drive
Pasadena, California 91109
Telephone: (818) 354-4321
Home page: http://technology.jpl.nasa.gov/nasa/nasa.html

TAP is a special division of the Jet Propulsion Laboratory (JPL) that develops and applies advanced technologies for JPL. Its Internet home page links to descriptions of current projects such as solar electric spacecraft power, robotics, space inflatables, and microspacecraft. The education link leads to interesting materials for various age groups and to the rules for the FIRST Robotics Competition. The FIRST (For Inspiration and Recognition of Science and Technology) competition is a national engineering contest that "immerses high school students in the exciting world of engineering."

Union of Concerned Scientists
2 Brattle Square
Cambridge, Massachusetts 02238
Telephone: (617) 547-5552
E-mail: ucs@ucsusa.org
Home page: http://ucsusa.org/

UCS is an independent nonprofit alliance of 50,000 concerned citizens and scientists across the United States. They hope to combine rigorous scientific analysis with innovative thinking and committed citizen advocacy to build a better world. Its Internet home page leads to briefings, updates, FAQs (frequently asked questions), fact sheets, guides, and analyses on topics such as arms control, transportation, energy, health, and green living.

United States Department of Energy (DOE)
U.S. Department of Energy, Headquarters
Forrestal Building
1000 Independence Avenue, S.W.
Washington, D.C. 20585
Telephone: (202) 586-5000
Home page: http://www.doe.gov/

DOE's four missions involve national security, science and technology, energy security, and environmental quality. Its national security responsibilities are to maintain the safety, security and reliability of the nuclear stockpile; dismantle surplus nuclear weapons; and assist in curbing global proliferation of weapons of mass destruction. Its science and technology duties are to conduct breakthrough research in energy sciences and technology, high energy physics, superconducting materials, and material and environmental science. Its energy security responsibilities are to ensure that people have clean, affordable, and dependable supplies of energy now and in the future and to bring renewable energy sources into the market. Its environmental quality charges include to clean up the environmental legacy of the nuclear weapons program and provide for the safe treatment, storage, and final disposal of radioactive wastes. Its Internet home page has interesting science news features. The Science Education link leads to "Science Pages for Kids" and "Lab/Computer Equipment Donation Programs." The "Content Map" is the easiest way to browse deeper links. The "DOE Information Bridge" at http://www.osti.gov/bridge/ provides access to full-text DOE research and development reports. The "Energy Files" link on the "Information Bridge" page provides search and full-text retrieval of a vast number of articles on various aspects of energy.

Uranium Information Centre (UIC)
GPO Box 1649N
Melbourne 3001, Australia
Telephone: (03) 9629-7744
E-mail: uic@mpx.com.au
Home page: http://www.uic.com.au/

The UIC was set up to increase Australian public understanding of uranium mining and nuclear electricity generation. *Encyclopedia Britannica* selected UIC's home page as one of the most valuable and reliable on the Internet. It provides access to an excellent set of educational resource papers and other useful information.

Charles W. Rogers

Web Sites

➤ *This appendix includes a variety of Web sites that range from providing basic information to exploring specific topics in depth. Some of the sites were created to address science and technology issues that were significant in 2000, such as genome sequencing, space satellites and exploratory probes, and paleontology discoveries. Many of the sites are interactive. Web sites for scientific societies, foundations, and competitions can be found in the section on science awards.*

General Science Sites

Discover Magazine
http://discover.com/

Discovery Channel
http://discovery.com/

Explorezone.com
http://explorezone.com/

National Geographic Society
http://www.nationalgeographic.com

National Science Foundation
http://www.nsf.gov

Nature Magazine
http://www.nature.com

New Scientist
http://www.keysites.com

Nova (Public Broadcasting Service)
http://www.pbs.org/wgbh/nova/

Odyssey Magazine
http://www.odysseymagazine.com

SciCentral
http://www.SciQuest.com/

Scientific American
http://www.sciam.com

Yes Mag: Canada's Science Magazine for Kids
http://www.islandnet.com/~yesmag/

News and Links

Biology in the News
http://www.nbii.gov/bionews/

The Environment News Network (ENN)
http://www.enn.com

The New York Times Science and Technology News
http://www.nytimes.com/pages/science/index.html

Popular Science Magazine
http://www.popularscience.com

Science Daily
http://www.sciencedaily.com

Science News Magazine
http://www.sciencenews.org

Science Online
http://www.scienceonline.org

The Scout Report for Science and Engineering
http://scout.cs.wisc.edu/

Yahoo! Science and Technology News
http://dailynews.yahoo.com/h/sc/?u

The Why Files: The Science Behind the News
http://whyfiles.news.wisc.edu/

Reference Sites

Academic Press Dictionary of Science and Technology
http://www.harcourt.com/dictionary/

Biotech Life Science Dictionary
http://biotech.icmb.utexas.edu/pages/dictionary.html

Conversion Factors
http://www.wsdot.wa.gov/Metrics/factors.htm

A Dictionary of Scientific Quotations
http://naturalscience.com/dsqhome.html

A Dictionary of Units of Measurement
http://www.unc.edu/~rowlett/units/

Glossary of Microscopy and Microanalytical Terms
http://www.mwrn.com/feature/glossary.htm

How Stuff Works
http://www.howstuffworks.com/

Internet Public Library Science and Technology Resources
http://www.ipl.org/ref/rr/static/sci00.00.00.html

Internet Public Library Science Organization Links
http://www.ipl.org/ref/AON/

The Library of Congress Science Reading Room
http://lcweb.loc.gov/rr/scitech/

MagPortal.com: Magazine Articles on Science and Technology
http://www.MagPortal.com/c/sci/

Martindale's Health Science Guide
http://www-sci.lib.uci.edu/HSG/HSGuide.html

Martindale's The Reference Desk
http://www-sci.lib.uci.edu/HSG/Ref.html

On Being a Scientist: Responsible Conduct in Research
http://www.nap.edu/readingroom/books/obas/

The On-line Medical Dictionary
http://www.graylab.ac.uk/omd/

Scholarly Societies Project
http://www.lib.uwaterloo.ca/society/overview.html

Science Service Historical Image Collection
http://americanhistory.si.edu/scienceservice/

Statistical Reports of U.S. Science and Engineering
http://www.nsf.gov/sbe/srs/stats.htm

Temperature Conversion Calculator
http://www.cchem.berkeley.edu/ChemResources/temperature.html

Science Resources

Biographies

African Americans in Science
http://www.princeton.edu/~mcbrown/display/faces.html

American Indian Science and Engineering Society
http://www.aises.org

Biographies of Physicists
http://hermes.astro.washington.edu/scied/physics/physbio.html

4,000 Years of Women in Science
http://www.astr.ua.edu/4000WS/4000WS.html

Galileo and Einstein Home Page
http://galileoandeinstein.physics.virginia.edu/

The National Academy of Engineering (NAE) Celebration of Women in Engineering
http://www.nae.edu/cwe

The Nobel Channel
http://www.nobelchannel.com

The Nobel Prize Internet Archive
http://www.almaz.com/

Women and Minorities in Science and Engineering Resources
http://www.mills.edu/ACAD_INFO/MCS/SPERTUS/Gender/wom_and_min.html

Educational Resources

Bill Nye the Science Guy's Nye Labs Online
http://nyelabs.kcts.org/

Eisenhower National Clearinghouse Resource Finder
http://www.enc.org/rf/nf_index.htm#rf

Electronic Journal of Science Education
http://unr.edu/homepage/jcannon/ejse/ejse.html

Intel Scholarships and Grants
http://www.intel.com/education/GRANTS.HTM

Journal of Young Investigators
http://www.jyi.org

National Aeronautics and Space Administration (NASA) Education Resource Page
http://education.nasa.gov/

National Institute for Science Education
http://www.wcer.wisc.edu/nise/

National Science Teachers Association
http://www.nsta.org

Newton's Apple Index
http://ericir.syr.edu/Projects/Newton/

Resources in Science and Engineering Education
http://www2.ncsu.edu/unity/lockers/users/f/felder/public/RMF.html

Science Learning Network (SLN)
http://www.sln.org/

Science/Nature for Kids
http://kidscience.about.com

Society for College Science Teachers
http://science.clayton.edu/scst/

Governmental Agencies for Science and Space

Department of Energy Office of Science and Technical Information
http://www.osti.gov/

U.S. House of Representatives Committee on Science
http://www.house.gov/science/welcome.htm

The U.S. Senate Committee on Commerce, Science and Transportation
www.senate.gov/~commerce/

History, Society, and Culture of Science

The American Physical Society: A Century of Physics
http://timeline.aps.org/APS/home_HighRes.html

The Art of Renaissance Science
http://www.pd.astro.it/ars/arshtml/arstoc.html

Case Studies in Science
http://ublib.buffalo.edu/libraries/projects/cases/case.html

History of the Light Microscope
http://www.utmem.edu/personal/thjones/hist/hist_mic.htm

Important Historical Inventions and Inventors
http://www.lib.lsu.edu/sci/chem/patent/srs136_text.html

Internet History of Science Sourcebook
http://www.fordham.edu/halsall/science/sciencesbook.html

Links to Science, Technology, and Society-Related Information Sources
http://www2.ncsu.edu/ncsu/chass/mds/stslinks.html

Science in Our Daily Lives
http://www.lib.virginia.edu/science/events/Sci_Daily_Life.html

Inventions

Invent America!
http://www.inventamerica.com

Invention Convention/National Congress of Inventor Organizations
http://www.inventionconvention.com

The Inventors Museum
http://www.inventorsmuseum.com/

Inventors Web Site
http://inventors.about.com

Kids Inventor Resources
http://www.InventorEd.org/k-12/becameinv.html

The Lemelson-MIT Awards Program Invention Dimension website
http://web.mit.edu/invent/index.html

National Collegiate Inventors and Innovators Alliance
http://www.nciia.org/

National Inventors Hall of Fame
http://www.invent.org/book/index.html

Tips for Parents of Young Inventors (Young Inventors Fair website)
http://www.ecsu.k12.mn.us/yif/tips.html

United States Patent and Trademark Office
http://www.uspto.gov

"What It Takes to Be an Inventor," 3M Collaborative Invention Unit
http://mustang.coled.umn.edu/inventing/Inventing.html

Museums
American Museum of Natural History
http://www.amnh.org

American Museum of Science and Energy, Oak Ridge, Tennessee
http://www.amse.org

California Science Center
http://www.casciencectr.org

Carnegie Museum of Natural History
http://www.CarnegieMuseums.org/cmnh

DNA Learning Center, Cold Spring Harbor
http://vector.cshl.org

The Exploratorium
http://www.exploratorium.edu/

Field Museum of Natural History in Chicago
http://www.fmnh.org

Franklin Institute Science Museum
http://sln.fi.edu/tfi/welcome.html

Further Explorations
http://www.explorations.org/further_explorations.html

Museum of Science and Industry
http://www.msichicago.org

National Air and Space Museum
http://www.nasm.si.edu

Natural History Museum, London
http://www.nhm.ac.uk/

Peabody Museum of Natural History
http://www.peabody.yale.edu/

Science Museum of Minnesota, Thinking Fountain
http://www.sci.mus.mn.us/

SciTech, the Science and Technology Interactive Center
http://scitech.mus.il.us/

Smithsonian Institution
http://www.si.edu

The U.S. Space & Rocket Center
http://www.spacefun.com

WebExhibits
http://www.webexhibits.com/

Science Fair Information

Experimental Science Projects: An Introductory Level Guide
http://www.isd77.k12.mn.us/resources/cf/SciProjIntro.html

Intel Science Talent Search
http://www.sciserv.org/sts/

International Chemistry Olympiad
http://www.fcho.schule.de/ForYou/ForYou.html

Internet Public Library
http://www.ipl.org/youth/projectguide/

National Gallery for America's Young Inventors
http://www.pafinc.com/nat_gal.htm

The Science Club
http://www.halcyon.com/sciclub/

Science Fair Projects: A Resource for Students and Teachers
http://www4.umdnj.edu/camlbweb/scifair.html

The Ultimate Science Fair Resource
http://www.scifair.org

The World-Wide Web Virtual Library: Science Fairs
http://physics1.usc.edu/~gould/ScienceFairs/

Topics in Science

Archaeology

Ancient Technologies and Archaeological Materials
http://www.uiuc.edu/unit/ATAM/index.html

Archaeology Magazine Online
http://www.archaeology.org/main.html

Web Info Radiocarbon Dating
http://www.c14dating.com

The WWWorld of Archaeology
http://www.archaeology.org/wwwarky/wwwarky.html

Astronomy and Aerospace

About.com Aerospace and Astronomy Sites
http://kidsastronomy.about.com
http://space.about.com

The American Meteor Society
http://www.amsmeteors.org/

Arctic Asteroid!
http://science.msfc.nasa.gov/headlines/y2000/ast01jun_1m.htm

AstroWeb—Astronomy Resources on the World Wide Web
http://www.stsci.edu/science/net-resources.html

Encyclopedia Astronautica
http://www.friends-partners.org/~mwade/spaceflt.htm

Far Ultraviolet Spectroscopic Explorer (FUSE)
http://fuse.pha.jhu.edu/

GPS Applications Exchange
http://gpshome.ssc.nasa.gov/

High Redshift Supernova Search Home Page of the Supernova Cosmology Project (Lawrence Berkeley National Laboratories)
http://www-supernova.lbl.gov/

Intelligent Satellite Data Information System (ISIS)
http://isis.dlr.de/

J-Track Satellite Tracker
http://liftoff.msfc.nasa.gov/RealTime/JTrack/Spacecraft.html

The Minor Planet Center's Sky Coverage Plots
http://cfaps8.harvard.edu/~cgi/SkyCoverage.html

National Aeronautics and Space Administration (NASA)
http://www.nasa.gov

NASA Earth Observatory
http://earthobservatory.nasa.gov/

NASA Human Spaceflight
http://spaceflight.nasa.gov/station/

NASA: Space Environment Center
http://www.sec.noaa.gov/

Orbital Elements
http://spaceflight.nasa.gov/realdata/elements/index.html

Recent Images Suggesting Liquid Water on Mars
http://www.msss.com/mars_images/moc/june2000/

Robotic Antarctic Meteorite Search: Antarctica 2000
http://www.frc.ri.cmu.edu/projects/meteorobot2000/

Science News About the Sun-Earth Environments
http://www.spaceweather.com

Sky and Telescope magazine
http://www.skypub.com/skytel/skytel.shtml

Skyview
http://skyview.gsfc.nasa.gov/

Solar Max 2000 (Exploratorium)
http://www.solarmax2000.com/

Solar Web Guide
http://www.lmsal.com/SXT/html2/list.html

Space News, Games, Entertainment
http://www.space.com

Space Science News
http://www.spacescience.com

Space Telescope Science Institute (Home of the Hubble Space Telescope)
http://www.stsci.edu/

The Spacewatch Project, University of Arizona Lunar and Planetary Laboratory
http://www.lpl.arizona.edu/spacewatch/index.html

Sunspot Cycle Predictions
http://science.nasa.gov/ssl/PAD/SOLAR/predict.htm

Tagish Lake Meteorite
http://phobos.astro.uwo.ca/~pbrown/tagish/

Terra: The EOS Flagship
http://terra.nasa.gov/

2001 Mars Odyssey
http://mars.jpl.nasa.gov/2001/

U.S. Naval Observatory Astronomical Tables
http://aa.usno.navy.mil/AA/AAmap.html

Wide Web of Astronomy
http://georgenet.net/astronomy.html

Biology

Association for Biology Laboratory Education "Hot" Biology Web Sites
http://www.zoo.toronto.edu/able/hotsites/hotsites.htm

Biolinks.com
http://www.biolinks.com

Cells Alive
http://www.cellsalive.com

Computer Enhanced Science Education, Whole Frog Project
http://george.lbl.gov/ITG.hm.pg.docs/Whole.Frog/Whole.Frog.html

An Electronic Introduction to Molecular Virology
http://www.uct.ac.za/microbiology/tutorial/virtut1.html

The e-Skeletons Project
http://www.eSkeletons.org/

Evolution Website (BBC Education)
http://www.bbc.co.uk/education/darwin/index.shtml

Harvard University Department of Molecular and Cellular Biology—Biology Links
http://mcb.harvard.edu/BioLinks.html

Microbe Zoo–Digital Learning Center for Microbial Ecology
http://commtechlab.msu.edu/sites/dlc-mi/zoo/

Microbiology Education Library
http://www.microbelibrary.org/

MicroWorlds (Lawrence Berkeley Laboratories)
http://www.lbl.gov/MicroWorlds/

Ongoing Biology: A Research Communication Service for Biologists
http://www.tilgher.it/ongoing.html

UCMP Exhibit Hall: Evolution Wing
http://www.ucmp.berkeley.edu/history/evolution.html

Botany

Agricultural Research Service Science 4 Kids
http://www.ars.usda.gov/is/kids/

Ancient Bristlecone Pine
http://www.sonic.net/bristlecone/intro.html

Biotechnology: An Information Resource
http://www.nal.usda.gov/bic/

Carnivorous Plants
http://www.sarracenia.com/cp.html

ForestWorld
http://www.forestworld.com/

Fungus 2000 Database
http://194.131.255.3/bmspages/Fungus2000/Fungus2000.htm

The International Organization for Plant Information Database of Plant Databases
http://iopi.csu.edu.au/iopi/iopidpd1.html

Introduction to the Plant Kingdom
http://scitec.uwichill.edu.bb/bcs/bl14apl/bl14apl.htm

Realizing the Potential of Plant Genomics: From Model Systems to the Understanding of Diversity
http://www.nsf.gov/pubs/2001/bio011/start.htm

What Is Photosynthesis?
http://photoscience.la.asu.edu/photosyn/education/learn.html

Chemistry
Chem4Kids
http://www.chem4kids.com/

Chemical Education Resource Shelf
http://www.umsl.edu/~chemist/books/

Chemical Health and Safety Data
http://ntp-server.niehs.nih.gov/Main_Pages/Chem-HS.html

Chemical Heritage Foundation
http://www.chemheritage.org/

Chemistry Teaching Resources
http://www.anachem.umu.se/eks/pointers.htm#Curriculum

ChemPen 3D—Classic Organic Reactions
http://home.ici.net/~hfevans/reactions.htm

ChemWeb.com
http://chemweb.com/

General Chemistry Site
http://chemistry.about.com

An Introduction to Surface Chemistry
http://www.chem.qmw.ac.uk/surfaces/scc/

IUPAC Compendium of Chemical Terminology
http://www.chemsoc.org/chembytes/goldbook/index.htm

Molecule of the Month
http://www.bris.ac.uk/Depts/Chemistry/MOTM/motm.htm

On-line Introductory Chemistry
http://www.scidiv.bcc.ctc.edu/wv/101-online.html

Polymers and Liquid Crystals
http://plc.cwru.edu/

Science Is Fun
http://scifun.chem.wisc.edu/scifun.html

Earth Sciences
ARGO: Observing the Ocean in Real Time
http://www.argo.ucsd.edu/

Digital Tectonic Activity Map
http://denali.gsfc.nasa.gov/dtam/

Global Ice-Core Research
http://id.water.usgs.gov/projects/icecore/

Hydrologic Information Center: Current Hydrologic Conditions
http://www.nws.noaa.gov/oh/hic/current/

Igneous Rocks Tour
http://seis.natsci.csulb.edu/basicgeo/IGNEOUS_TOUR.html

138 Years of Agricultural Research—United States Department of Agriculture
http://www.ars.usda.gov/is/timeline/

Soil Biological Communities—Bureau of Land Management
http://www.id.blm.gov/soils/index.html

Topozone
http://www.topozone.com/

Tundra-Cam : Institute of Arctic and Alpine Research
http://tundracam.colorado.edu/

United States Geological Survey (USGS)
http://www.usgs.gov

Virtual Geosciences Professor
http://www.uh.edu/~jbutler/anon/anonfield.html

Energy, Ecology, and Environmental Sites

American Rivers Online
http://www.amrivers.org/

Audubon Magazine's Resolutions for a New Millennium
http://magazine.audubon.org/resolutions.html

Bioenergy Information Network
http://www.esd.ornl.gov/bfdp/

Clean Energy Basics (National Renewable Energy Laboratory)
http://www.nrel.gov/clean_energy/

DOE EnergyFiles: Virtual Library of Energy Science and Technology
http://www.doe.gov/EnergyFiles

eNature.com Field Guides
http://www.enature.com/

Friends of the Earth Home Page
http://www.foe.co.uk

Geothermal Energy Program
http://www.eren.doe.gov/geothermal/

Global Water Sampling Project
http://www.k12science.org/curriculum/waterproj/

Green Buildings
http://www.sustainable.doe.gov/buildings/gbintro.shtml

Gridded Population of the World (GPW)
http://sedac.ciesin.org/plue/gpw/

Interactive Health Ecology Access Links (IHEAL)
http://mole.utsa.edu/~matserv/iheal/

Live from the Estuary
http://www.estuarylive.org/

Millennium Ecosystem Assessment (WRI)
http://www.ma-secretariat.org/

Recycle City (United States Environmental Protection Agency)
http://www.epa.gov/recyclecity/

The REINAS Project: Real-time Environmental Information Network and Analysis System
http://csl.cse.ucsc.edu/projects/reinas/

A Resource Guide of Solid Waste Educational Materials
http://www.epa.gov/epaoswer/general/bibligr/educatn.htm

River Watch (National Weather Service)
http://www.riverwatch.noaa.gov/

The Sunwise School Program: UV Index (EPA)
http://www.epa.gov/sunwise/uvindex/index.html

World Resources 2000-2001—People and Ecosystems: The Fraying Web of Life
http://www.wri.org/wri/wrr2000/

Engineering, Computer Science, and the Internet

American Society for Engineering Education
http://www.asee.org/

Brain Spin: Technology for Students
http://www.att.com/technology/forstudents/brainspin/

The Canada Centre for Remote Sensing
http://www.ccrs.nrcan.gc.ca

Circuits Archive
http://www.ee.washington.edu/circuit_archive/

Composite Materials Handbook
http://mil-17.udel.edu/

The Computer Vision Handbook
http://www.cs.hmc.edu/~fleck/computer-vision-handbook/

Engineering Links
http://www.englinks.com/

Greatest Engineering Achievements of the Twentieth Century
http://www.greatachievements.org/

History of the Web
http://dbhs.wvusd.k12.ca.us/Chem-History/Hist-of-Web.html

iCivilEngineer
http://www.icivilengineer.com/

NEEDS, A National Digital Library for Engineering Education
http://www.needs.org/

Netdictionary
http://www.netdictionary.com

NetLib Repository
http://www.netlib.org/

Resource Center for Cyberculture Studies
http://otal.umd.edu/~rccs/

The Robotics Institute
http://www.ri.cmu.edu/

Thinkquest
http://library.thinkquest.org

Virtual Reality
http://www.cms.dmu.ac.uk/~cph/VR/whatisvr.html

Yahoo! Engineering Links
http://www.yahoo.com/r/eg

Genetics

The Comprehensive Microbial Resource (CMR) Home Page/
The Institute for Genomic Research (TIGR)
http://www.tigr.org/tigr-scripts/CMR2/CMRHomePage.spl

The DNA Learning Center
http://vector.cshl.org/resources/resources.html

Ergito—GENES 2000
http://www.genes.net/

GeneCards: human genes, maps, proteins and diseases
http://bioinformatics.weizmann.ac.il/cards/

Genetic Science Learning Center (GSLC)
http://gslc.genetics.utah.edu/

Genetics Education Center
http://www.kumc.edu/gec/

Human Genome Central
http://www.ensembl.org/genome/central/

Human Genome Project Information
http://www.ornl.gov/TechResources/Human_Genome/
home.html

Science Functional Genomic
http://www.sciencegenomics.org

Mathematics

Clay Mathematics Institute—Millennium Prize Problems
http://www.claymath.org/prize_problems/index.htm

General Mathematics Site
http://math.about.com

Geometry in Action
http://www.ics.uci.edu/~eppstein/geom.html

Interactive Parabola
http://www-groups.dcs.st-andrews.ac.uk/~history/Curves/Parabola.html

Math and Physics Help
http://www2.ncsu.edu/unity/lockers/users/f/felder/public/kenny/home.html

Math Archives
http://archives.math.utk.edu/

Math Goodies Newsletter
http://www.mathgoodies.com/newsletter/

MathCounts Coaching and Competition Program
http://www.MathCounts.org

Mathematical Sciences Research Institute (MSRI)
http://www.msri.org/index.html

Metamath Proof Explorer
http://www1.shore.net/~ndm/java/mmexplorer1/mmset.html

Principles and Standards for School Mathematics updated April 2000
http://standards.nctm.org/

Sloane's On-line Encyclopedia of Integer Sequences
http://www.research.att.com/~njas/sequences/

Medicine and Nutrition

Antibiotics
http://ericir.syr.edu/Projects/Newton/12/Lessons/antibiot.html

BioMedNet
http://www.biomednet.com/

Emerging Infectious Diseases Journal/Centers for Disease Control (CDC)
http://www.cdc.gov/ncidod/eid/index.htm

Heart: An Online Exploration
http://sln.fi.edu/biosci/heart.html

Human Anatomy Online
http://www.innerbody.com/htm/body.html

It's All in the Brain
http://www.hhmi.org/senses/a/a110.htm

Medicine Through Time (BBC)
http://www.bbc.co.uk/education/medicine/

MedicineNet
http://www.medicinenet.com/

MEDtropolis Home Page
http://www.medtropolis.com

Neuroscience for Kids
http://faculty.washington.edu/chudler/neurok.html

NIH Stem Cell Research Guidelines
http://www.nih.gov/news/stemcell/stemcellguidelines.htm

Nutrition and Your Health: Dietary Guidelines for Americans
www.health.gov/dietaryguidelines

Tufts University Nutrition Navigator
http://navigator.tufts.edu/

West Nile Virus Maps
http://nationalatlas.gov/virusmap.html

World Health Organization's Communicable
Disease Surveillance and Response
http://www.who.int/emc/

Paleontology

Coelacanth: The Fish Out of Time
http://www.dinofish.com/

History of Palaeozoic Forests
http://www.uni-muenster.de/GeoPalaeontologie/Palaeo/Palbot/ewald.html

Paleonet
http://www.ucmp.berkeley.edu/Paleonet/

PaleoQuest
http://paleoquest.cet.edu/

Paul Sereno's Dinosaur Web Site
http://dinosaur.uchicago.edu/

So You Want to Be a Paleontologist?
http://www.cisab.indiana.edu/~mrowe/dinosaur-FAQ.html

Willo, the Dinosaur with a Heart
http://www.dinoheart.org/

Physics

American Association of Physics Teachers
http://www.aapt.org

Elemental Data Index
http://physics.nist.gov/PhysRefData/Elements/cover.html

Exploring Gravity
http://www.curtin.edu.au/curtin/dept/phys-sci/gravity/

Gravity Probe B: The Relativity Mission
http://einstein.stanford.edu/

Internet Pilot to Physics
http://physicsweb.org/TIPTOP/

The Particle Adventure
http://www.particleadventure.org

Physicists Find First Direct Evidence for Tau Neutrino at Fermilab
http://www.fnal.gov/directorate/public_affairs/story_neutrino/p1.html

The Physics of...
http://www.kent.wednet.edu/staff/trobinso/physicspages/PhysicsOf.html

Physics Questions/Problems
http://star.tau.ac.il/QUIZ

Physics 2000 (Interactive Journey through Modern Physics)
http://www.colorado.edu/physics/2000/index.pl

PhysicsEd: Physics Education Resources
http://www-hpcc.astro.washington.edu/scied/physics.html

PhysLINK
http://www.physlink.com

Playground Physics
http://www.aps.org/playground.html

Professor Bubbles' Official Bubble Home Page
http://bubbles.org/

Unsolved Mysteries
http://www.pbs.org/wnet/hawking/mysteries/html/myst.html

Visual Quantum Mechanics: Online Interactive Programs
http://phys.educ.ksu.edu/vqm/index.html

Weather, Climate, and Natural Disasters

Building and Fire Research Laboratory: Fire on the Web
http://www.fire.nist.gov/

Dan's Wild Wild Weather Page
http://www.whnt19.com/kidwx/index.html

EPA's Global Warming Site
http://www.epa.gov/globalwarming/

Global Volcanism Program
http://www.volcano.si.edu/gvp/index.htm

Hurricane and Storm Tracking
http://hurricane.terrapin.com/

International Association of Volcanology and Chemistry of the Earth's Interior (IAVCEI)
http://www.iavcei.org/

Lightning Imaging Sensor Data
http://thunder.msfc.nasa.gov/data/lisbrowse.html

The Michigan Technological University (MTU) Volcanoes Page
http://www.geo.mtu.edu/volcanoes/

National Drought Mitigation Center
http://enso.unl.edu/ndmc/

National Earthquake Information Center
http://wwwneic.cr.usgs.gov

National Snow and Ice Data Center
http://www-nsidc.colorado.edu/NSIDC/EDUCATION/SNOW/snow_FAQ.html

Nova Online: Flood!
http://www.pbs.org/wgbh/nova/flood/

Tornado Project Online
http://www.tornadoproject.com/

Tsunami!
http://www.geophys.washington.edu/tsunami/intro.html

Volcano World
http://volcano.und.nodak.edu/

The World-Wide Earthquake Locator
http://www.geo.ed.ac.uk/quakes/quakes.html

Zoology, Entomology, and Veterinary Medicine

AmphibiaWeb
http://elib.cs.berkeley.edu/aw/

Animal Diversity Web
http://animaldiversity.ummz.umich.edu/

Bird Checklists of the United States—NPWRC
http://www.npwrc.usgs.gov/resource/othrdata/chekbird/chekbird.htm

The Bug Club
http://www.ex.ac.uk/bugclub/

Contemporary Herpetology
http://dataserver.calacademy.org/herpetology/herpdocs/index.html

Endangered Species UPDATE
http://www.umich.edu/~esupdate/

The Extinction Files (BBC)
http://www.bbc.co.uk/education/darwin/exfiles/index.htm

FishBase
http://www.fishbase.org/

Ichthyology Web Resources
http://www.biology.ualberta.ca/jackson.hp/IWR/index.php]

Living Links
http://www.emory.edu/LIVING_LINKS/

National Marine Mammal Laboratory's Education Web Site
http://nmml.afsc.noaa.gov/education/

NetVet and the Electronic Zoo
http://netvet.wustl.edu

North American Bird Conservation Initiative (NABCI)
http://www.bsc-eoc.org/nabci.html

Northern Prairie Wildlife Research Center
http://www.npwrc.usgs.gov/

Satellite Tracking of Threatened Species—NASA
http://sdcd.gsfc.nasa.gov/ISTO/satellite_tracking/

2000 IUCN Red List of Threatened Species
http://www.iucn.org/redlist/2000/index.html

Virtual Creatures
http://k-2.stanford.edu/creatures/

Welcome to Coral Forest
http://www.blacktop.com/coralforest/

World Wildlife Fund
http://www.wwf.org

Elizabeth Schafer

Herbert Kroemer and Jack Kilby (below), along with Zhores Alferov (not pictured) shared the the 2000 Nobel Prize in Physics, presented by Carl XVI Gustaf, king of Sweden. (AP/Wide World Photos).

Science Awards in the Year 2000

➤*A variety of science awards are presented annually for researchers exploring scientific and engineering investigations at different levels, ranging from grade school science fair projects to sophisticated theoretical problems posed by experts. Professional societies for every field of science and engineering offer prizes to acknowledge achievements within specific scientific branches. Governmental agencies and private patrons, foundations, corporations, institutions, and schools sponsor science and technology awards.*

This section presents information about the most significant science prizes awarded internationally and in the United States with examples of other representative science prizes to demonstrate the diversity of awards. The awards include those given early in the year 2000, even if for the previous year, and those given in January or early February, 2001, for the year 2000. Web site addresses are provided to locate more information about the history of and criteria for awards and who won them in 2000; these sites also provide information about additional prizes awarded by various groups. This appendix is a sampling, and Internet search engines can be used to find information about other science awards in specific subject areas.

MAJOR AWARDS

The Nobel Prize

http://www.nobel.se

The Nobel Prize is the most prestigious international award that honors scientists and technologists. First presented in 1901, the Nobel Prize was named for dynamite inventor and industrialist Alfred B. Nobel (1833-1896) who bequeathed part of his fortune to establish a trust fund to generate interest to finance scientific awards. Prizes are granted in six categories: literature, chemistry, physics, physiology or medicine, peace, and economics (presented since 1969 and funded by Sweden's national bank). Each winner is selected based on the merit of his or her research or activities in the year before receiving this honor, although winners often are recognized for lifetime achievements. According to Nobel's will, scientists should be chosen according to who achieved the most significant discovery, invention, or application to better society. Nationality is not supposed to be a factor affecting selection of winners. The Swedish Academy of Science grants the chemistry, physics, and economics awards at Stockholm, and the Caroline Medico-Surgical Institute presents the medical or physiological Nobel Prize in that city.

On December 10, the anniversary of Nobel's death, winners are awarded a gold medal, diploma certifying the achievement, and a monetary prize. Each recipient presents a lecture. The Nobel Foundation's board of directors manages the prize fund. Every two years, members of the Swedish Academy of Science and the Caroline Medico-Surgical Institute choose four of the board members, and the Swedish government names the fifth member. Sometimes more than one person is chosen for the annual Nobel Prize in a category, or, occasionally, no winner is named.

2000 Nobel Prize Winners

CHEMISTRY: Alan J. Heeger, Alan G. MacDiarmid, and Hideki Shirakawa

In recognition of their research regarding the modification of plastics to conduct electricity, which enabled the development of conductive polymers to improve materials used for film, television screens, windows, and other items.

ECONOMICS: James J. Heckman and Daniel L. McFadden

For developing methods to study how people make decisions concerning their daily life in areas such as residential location and employment duration, which led to microeconometrics theories using economics and statistics to improve employment training programs and transportation and communication systems.

PHYSICS: Zhores Alferov, Herbert Kroemer, and Jack Kilby

For their contributions to the information revolution. Kilby invented the integrated circuit (computer chip) at Texas Instruments in 1958, and Alferov and Kroemer enhanced satellite and cell phone technology by developing semiconductors. These inventions provided a foundation for information technology by enabling the creation of smaller and efficient devices such as barcode readers, CD players, personal computers, and sophisticated technology to perform both essential and recreational functions.

PHYSIOLOGY OR MEDICINE: Arvid Carlsson, Paul Greengard, and Eric Kandel

For their discoveries about how normal brain cells transmit messages in order to understand how disturbances in signal transmission, which cause Parkinson's disease, depression, and other neurological and psychiatric conditions, can be treated with the application of new drugs, such as pharmaceuticals made from the amino acid dopamine, based on their medical findings.

American Association for the Advancement of Science (AAAS)

http://www.aaas.org/aaas/award.html

Established in 1848 to improve science in the United States, the AAAS encourages scientific research in all fields, promotes scientific education at all levels, and publishes the weekly magazine *Science*.

PHILIP HAUGE ABELSON PRIZE: Leon M. Lederman, Illinois Institute of Technology

For his contributions in advancing scientific education and promoting global scientific collaboration.

INTERNATIONAL SCIENTIFIC COOPERATION AWARD: Kenneth Bridbord, National Institutes of Health

For improving international public health research and training regarding AIDS.

MENTOR AWARD: Lisa A. Pruitt, University of California at Berkeley

For encouraging diversification of the fields of materials and bioengineering.

MENTOR AWARD FOR LIFETIME ACHIEVEMENT: Evelyn L. Hu, University of California, Santa Barbara

For seeking opportunities for female and minority students and faculty in the fields of electrical and computer engineering.

MENTOR AWARD FOR LIFETIME ACHIEVEMENT: William E. Spicer, Stanford University

For his contributions to improve students' self-confidence and support for the inclusion of females and minorities in the fields of electrical engineering, physics, applied physics, and materials science.

NEWCOMB CLEVELAND PRIZE: Gerald M. Rubin and Susan E. Celniker, Drosophilia Genomen Center, University of California at Berkeley, and J. Craig Venter and associated scientists, Celera Genomics

For the review, "The Genome Sequence of Drosophilia melanogaster" which was published in the March 24, 2000, issue of *Science*, which represents the collaboration of academic and industrial researchers to achieve an understanding of the structure of genetic material.

PUBLIC UNDERSTANDING OF SCIENCE AND TECHNOLOGY AWARD: Vaclav Smil, University of Manitoba

For distributing resources to help the public comprehend complex international science issues of immediate concern.

SCIENTIFIC FREEDOM AND RESPONSIBILITY AWARD: Howard Schachman, University of California at Berkeley

For encouraging researchers to perform investigations responsibly and for promoting scientific freedom.

SCIENTIFIC JOURNALISM AWARDS:

Daily Newspapers with a Circulation of 100,000 or More:

Rich Weiss and Deborah Nelson, *The Washington Post*

For their "Gene Therapy" series.

Daily Newspapers with a Circulation of Less Than 100,000:

James B. Erickson, *The Arizona Daily Star*

For his "Signs of Life" series.

Magazine: Mark Schoofs, *The Village Voice*

For his "AIDS: The Agony of Africa" series.

Radio: Moira Rankin and David Barrett Wilson, Soundprint Media Center

For *Gamma Ray Skies*, *Einstein's Blunder*, and *The Fate of the Universe*.

Television: Richard Hudson, Eliene Augenbraun, Ira Flatow, and Gino Del Guercio, Twin Cities Public Television/PBS

For *Transistorized*.

The Franklin Institute

http://www.fi.edu

The Franklin Institute, established in 1824 to encourage scientific investigations and discovery, was named for notable colonial science enthusiast Benjamin Franklin. About one-fifth of Franklin award winners have later won Nobel Prizes for their research.

2000 Benjamin Franklin Medal Winners

PHYSICS: Eric Cornell, Carl Wieman, and Wolfgang Ketterle
For confirming experimentally Satyendra Bose's and Albert Einstein's 1925 theories regarding the condensation of dilute gas.

COMPUTER AND COGNITIVE SCIENCE: John Cocke
For developing reduced instruction set computing (RISC).

EARTH SCIENCE: Eville Gorham
For applying knowledge of plant-environment dynamics to assess the contents of air and the affects on plants and the environment.

CHEMISTRY: Robert Grubbs
For his olefin metathesis research examining the making and breaking of carbon-carbon bonds.

ENGINEERING: Antoine Labeyrie
For inventing speckle interferometry and initiating telescopic interferometry methodology.

ENGINEERING: James Powell and Gordon Danby
For inventing an unique repulsive magnetically levitated train system with superconducting magnets.

2000 Bower Awards

ACHIEVEMENT IN SCIENCE (AWARD AND PRIZE): Alexander Rich
For his discoveries concerning RNA and DNA molecules' structures and functions.

BUSINESS LEADERSHIP: William J. Rutter
To the father of biotechnology because of his efforts to build and enhance that scientific field.

The National Academy of Sciences (NAS)

http://www.nationalacademies.org/

The National Academy of Sciences, incorporated by a U.S. congressional act in 1863, is headquartered in Washington, D.C. American scientists and engineers are eligible for membership in this prestigious society. To recognize their professional expertise, a maximum of fifty Americans and ten international scientists and engineers are elected at the NAS annual meeting. The academy's National Research Council was created in 1916, and members are expected to conduct and report on scientific investigations for federal government agencies without compensation. The NAS presents prizes for achievement in specific scientific and engineering fields, for application of scientific knowledge to societal concerns, for research methodology, and in honor of prominent members. The National Academy of Engineering (NAE; http://www.nae.edu) and Institute of Medicine (IOM; http://www.iom.edu) are included within the National Academy of Sciences.

NAS 2000 Awards

AERONAUTICAL ENGINEERING: Richard T. Whitcomb
BEHAVIORAL RESEARCH RELEVANT TO THE PREVENTION OF NUCLEAR WAR: Philip E. Tetlock
CHEMICAL SCIENCES: K. Barry Sharpless
INITIATIVES IN RESEARCH: Kenneth A. Farley
MATHEMATICS: Ingrid Daubechies
MOLECULAR BIOLOGY: Patrick O. Brown
SCIENTIFIC REVIEWING: Charles F. Stevens
SCIENCES PUBLIC WELFARE MEDAL: Gilbert F. White
DANIEL GIRAUD ELLIOT MEDAL: Geerat J. Vermeij
GILBERT MORGAN SMITH MEDAL: Shirley W. Jeffrey
J. LAWRENCE SMITH MEDAL: George W. Wetherill
JOHN J. CARTY AWARD FOR THE ADVANCEMENT OF SCIENCE: Donald Lynden-Bell
TROLAND RESEARCH AWARDS: Elizabeth Gould

The National Academy of Engineering Awards

THE FOUNDERS AWARD: Charles H. Townes
Honors outstanding engineering achievements; established in 1965.
ARTHUR M. BUECHE AWARD: Charles M. Vest
Honors statesmanship in science and technology; established in 1983.

Institute of Medicine Awards

THE GUSTAV O. LIENHARD AWARD: Philip R. Lee
Recognizes advancements for personal health care services.
WALSH MCDERMOTT MEDAL: Mary Ellen Avery
For service to the Institute of Medicine.
DAVID RALL MEDAL: Stuart Bondurant
For leadership.
ADAM YARMOLINSKY MEDAL: Rashi Fein
For service from a nonhealth and medical sciences IOM member.
RHODA AND BERNARD SARNAT INTERNATIONAL PRIZE IN MENTAL HEALTH: Rosalynn Carter

The President's National Medal of Science

http://www.nsf.gov/nsb/awards/nms/start.htm

The President's National Medal of Science, the United States' highest scientific honor, is administered by the National Science Foundation, an independent governmental agency created by a 1950 congressional act. The foundation, which is responsible for establishing national science policy, funds research, education, and teacher training. The National Medal of Science was established in 1959 and is given to a maximum of twenty people per year who have contributed significantly to science and engineering and who are chosen by the President's Committee on the National Medal of Science.

2000 National Medal of Science Winners

Gary Becker, University of Chicago

For pioneering the economic analysis of racial discrimination and investigating how social forces influence individuals' economic behavior.

Nancy C. Andreasen, University of Iowa Hospitals and Clinics

For combining behavioral science with neuroscience and neuroimaging technologies to examine such mental processes as creativity and memory.

Peter H. Raven, Missouri Botanical Garden and Washington University

For introducing the concept of coevolution as an authority on plant systematics and urging the preservation of global biodiversity.

Carl R. Woese, University of Illinois at Urbana-Champaign

For posing the theory that all living organisms can be classified into three primary evolutionary domains resulting in a quantitative map, also called the universal tree of life, which is used to assess the diversity of life.

John D. Baldeschwieler, California Institute of Technology

For developing ion cyclotron resonance spectroscopy, which is used for chemical and biochemical analyses; this allowed scientists to study molecular structure and reactivity, which enabled the creation of new pharmaceuticals.

Ralph F. Hirschmann, University of Pennsylvania

For developing life-saving machines during his chemistry work in cooperation with Merck and for establishing collaborative university-industry research to achieve biomedical advances.

Yuan-Cheng B. Fung, University of California, San Diego

For posing the theory of aeroelasticity, which defined how aero-structures interacted with aerodynamic flows, advancing aerospace engineering methodology. Considered the founder of biomechanics, he also applied the analytical methods of mechanics to biological tissue examination, introducing novel biomechanics methodologies.

John Griggs Thompson, University of Florida, Gainesville

For classifying all finite simple groups, an act that has been called one of the twentieth century's greatest mathematics achievements.

Karen K. Uhlenbeck, University of Texas, Austin

For contributing to global analysis and gauge theory, which advanced both mathematical physics and the theory of partial differential equations.

Willis E. Lamb, University of Arizona, Tucson

In recognition of his investigations with hydrogen, which led to his discovery of a new relativistic quantum effect that was a foundation for quantum electrodynamics and his establishment of the field of laser physics.

Jeremiah P. Ostriker, Princeton University

For his astrophysics research, which produced revolutionary concepts regarding pulsars, galaxies' sizes and masses, and the nature of matter and how it is distributed in the universe.

Gilbert F. White, University of Colorado, Boulder

For his environmental science research focusing on floodplains, how to lessen flooding damage, and how social and geographical factors should be considered to change public policy and research objectives concerning water use and natural disasters.

National Medals of Technology

http://www.ta.doc.gov/Medal/

The National Medal of Technology was established by Congress in 1980 as a presidential award by the Stevenson-Wydler Technology Act and was first presented in 1985. Similar to the National Medal of Science, this award honors engineering prowess demonstrated by individuals, teams, or companies who create innovative items, methods, or services that involve technology and improve the quality of life, increase employment opportunities, and enhance the United States' global commercial competitiveness. The Tech Museum of Innovation (http://www.thetech.org/nmot) hosts a permanent exhibit for this award.

In 1991, the National Science and Technology Medals Foundation (http://www.asee.org/nstmf/) was formed, and representatives signed a letter of agreement one year later with the National Science Foundation to include the National Medal of Science within its policy mission and to coordinate its activities with the White House Office of Science and Technology Policy in the Executive Office of the President (http://www.ostp.gov).

2000 National Medal of Technology Winners

Douglas C. Engelbart, Bootstrap Institute

For creating personal computing foundations such as the mouse, shared-screen teleconference, hypertext linking, remote collaboration, and continuous, real-time interaction using cathode-ray tube displays.

Dean Kamen, DEKA Research and Development Corporation

For inventing devices that improved international medical care and for promoting the possibilities of science and technology to the American public.

Donald B. Keck and Robert D. Maurer, Corning Incorporated, and Peter C. Schultz, Heraeus Amersil

For inventing low-loss optical fiber, which made the telecommunications revolution possible, and thereby transforming society, creating new industries, expanding future opportunities, and altering how people work, learn, live, and play.

IBM Corporation, Louis V. Gerstner, CEO

For the corporation's forty-year history of innovating and producing hard disk drives and information storage products.

The National Science Foundation

http://www.nsf.gov/nsb/#honors

The U.S. government's science and technology agency presents medals and monetary prizes to selected recipients.

INDIVIDUAL AWARD: Philip and Phylis Morrison

For science reporting and contributions to science education.

GROUP AWARD: Science Service

For publishing *Science News* and sponsoring school science competitions that enhance public awareness of science and technology.

VANNEVAR BUSH AWARD: Norman E. Borlaug, Texas A&M University

For his genetics research with hybrid wheat, which resulted in Third World countries producing sufficient crops to feed native populations and export agricultural goods beginning during the 1960's Green Revolution.

VANNEVAR BUSH AWARD: Herbert F. York, University of California, San Diego
In recognition of his nuclear energy work and leadership in the arms control movement.

ALAN T. WATERMAN AWARD: Jennifer A. Doudna, Yale University
In recognition of her pioneering structural biology investigations of RNA molecules and because she is an outstanding example of a young researcher.

Sigma Xi
http://www.sigmaxi.org/prizes&awards/procter.htm
Sigma Xi is an international science and engineering honorary that recognizes leaders in a variety of research fields and presents numerous awards, grants, and fellowships. The society's most significant awards are as follows:

WILLIAM PROCTER PRIZE FOR SCIENTIFIC ACHIEVEMENT: Francisco J. Ayala
THE JOHN P. MCGOVERN SCIENCE AND SOCIETY AWARD: David Goodstein
YOUNG INVESTIGATOR AWARD: Sherry Yennello
COMMON WEALTH AWARD FOR SCIENCE AND INVENTION: Robert D. Ballard

SCIENTIST OF THE YEAR AWARDS

***Biography Magazine*'s Biography of the Year 2000**
(http://www.biography.com)

J. Craig Venter, CEO of Celera Genomics, and Francis S. Collins, director of the National Human Genome Research Institute
For mapping the human genome.

***Time Magazine*'s Scientist of the Year 2000**
(http://www.time.com)

J. Craig Venter, CEO of Celera Genomics
For genetics research that enabled him to map the human genome.

AWARDS PRESENTED BY SCIENTIFIC ACADEMIES

The following is a selective list; information about other international scientific academies and their awards can be located by using Internet search engines.

Académie des Sciences, Institut de France
http://www.acad-sciences.fr/uk/index_prix.html
GRANDE MÉDAILLE D'OR DE L'ACADÉMIE DES SCIENCES: Robert Langlands
CHARLES LEOPOLD MAYER PRIZE: Robert H. Horvitz

New Zealand Royal Society
http://www.rsnz.govt.nz/

COOPER MEDAL (PHYSICS OR ENGINEERING): Rod White

For his meteorological invention that assesses the accuracy of temperature measurements.

GOLD MEDAL: David Vere-Jones

For his research concerning earthquake forecasting and risk.

HUTTON MEDAL (ANIMAL AND EARTH SCIENCE): Alan Kirton and Hugh Bibby

HECTOR MEDAL (PHYSICAL, MATHEMATICAL AND INFORMATION SCIENCES): Jeff Tallon, Paul Callaghan, and George Seber

THOMSON MEDAL (ORGANIZATION AND APPLICATION OF SCIENCE): Jim Johnston

The Royal Society Medals, Awards, and Prize Lectures
(FRS indicates a Fellow of the Royal Society.)
http://www.royalsoc.ac.uk/royalsoc/am_fr.htm

THE COPLEY MEDAL: Given annually since 1731, it is the Royal Society's most outstanding award for scientific research and is alternated between physical and biological sciences.

Sir Alan (Ruston) Battersby FRS, University of Cambridge.

For his innovative research to determine the biosynthetic pathways to the major families of plant alkaloids, which established the foundation for biosynthetic studies of complex molecules thereby revealing the entire pathway for vitamin B12.

THE RUMFORD MEDAL: Given biennially since 1800 for European researchers' discoveries in the field of thermal or optical properties of matter.

Wilson Sibbett FRS, University of St. Andrews

For his ultrashort pulse laser science and technology research and demonstration of the subpicosecond chronoscopy technique in which cameras act as oscilloscopes. Sibbett's development of self-modelocking methods resulted in the commercialization of subpicosecond pulses over a wide tuning range. He also developed the first all-solidstate optical parametric oscillator.

THE ROYAL MEDALS: Presented annually by the Sovereign to recognize two people for their contributions within the British Commonwealth to the advancement of scientific knowledge and a third person for contributions to applied sciences.

Keith Ingold OC FRS, Steacie Institute for Molecular Sciences, Ontario, Canada.

For developing experimental tools and methodology to study the reaction mechanisms of free radicals.

Geoffrey Burnstock FRS, Director of the Autonomic Neuroscience Institute in the Royal Free and University College Medical School, London.

For his hypotheses that countered previous understanding of autonomic neurotransmission and resulted in improved comprehension of purinergic neurotransmission, which is crucial to several body systems.

Timothy Berners-Lee OBE, Massachusetts Institute of Technology.

For inventing and developing the World Wide Web and designing the universal resource locator (URL), which provides each Web page an unique location, and the two protocols HTTP and HTML, all of which have contributed to an information and communication revolution and increased access to data.

THE DAVY MEDAL: Given annually since 1877 for discoveries in chemistry made in Europe or North America.

Steven Victor Ley FRS, University of Cambridge.

For inventing new synthetic methods that are applied to the synthesis of complex natural products derived from insects, microorganisms, and plants and including the synthesis of avermectin B1a, tetronasin, the milbemycins, indanomycin, and oligosaccharides.

THE DARWIN MEDAL: Given biennially since 1890 for evolutionary biological research.

Brian Charlesworth FRS, University of Edinburgh.

For his research combining theoretical work and experimental analysis by addressing selection in age- structured populations and the theory of the evolution of aging and applying the theories of mutation accumulation and pleiotropy with models incorporating various genetic factors.

THE BUCHANAN MEDAL: Given biennially since 1897 for contributions to medical science.

Sir Stanley Peart FRS, St. Mary's Hospital Medical School in the University of London.

For advancing information about the renin angiotensin system by isolating and analyzing the structure of angiotensin, purifying renin, and investigating the control of renin release.

THE SYLVESTER MEDAL: Given triennially since 1901 for mathematical research.

Nigel James Hitchin FRS, University of Oxford.

For considering modern theories with classical mathematical literature to advance differential geometry by combining it with complex geometry, integrable systems, and mathematical physics.

THE HUGHES MEDAL: Given annually since 1902 for discoveries in the physical sciences, preferably electricity and magnetism or their applications.

Chintamani Nagesa Ramachandra Rao FRS, Jawaharlal Nehru Centre for Advanced Scientific Research, Bangalore, India.

For advancing the field of materials chemistry with his research focusing on the electronic and magnetic properties of transition metal oxides and high temperature superconductors and for inspiring Indian scientists.

THE MULLARD AWARD: Awarded since 1967 and funded by Mullard Limited.

Martin Nicholas Sweeting OBE FRE, Surrey Space Centre, University of Surrey.

For his research and development of low-cost, lightweight satellites that resulted in the creation of the Surrey Satellite Technology, which he directs.

GLAXO WELLCOME PRIZE AND LECTURE: Awarded biennially since 1980 to recognize contributions to medical and veterinary sciences.

David Herman MacLennan FRS, University of Toronto.

For his research of calcium regulatory proteins, which aided the understanding of malignant hyperthermia (MH), central core disease (CCD), Brody disease, and phospholamban and resulted in MacLennan applying these basic science research results concerning the gene in MH in order to diagnose swine disease.

THE MICHAEL FARADAY AWARD: Given since 1986 to honor efforts to educate the public about science.

Lewis Wolpert CBE FRS, Royal Free and University College Medical School.

For assisting public understanding of science by his chairmanship of COPUS, television and radio programs, books, lectures, and newspaper articles, alerting people to scientific topics such as depression.

THE CROONIAN LECTURE: The Royal Society's premier lecture on biological science initiated in 1701.

Peter Nigel Tripp Unwin FRS, Medical Research Council Laboratory of Molecular Biology.

"The Nicotinic Acetylcholine Receptor and the Structural Basis of Synaptic Transmission," October 5, 2000, at University College London.

THE BAKERIAN LECTURE: The Royal Society's premier lecture on physical sciences initiated in 1775.

Robert Stephen John Sparks FRS, University of Bristol.

"How Volcanoes Work," September 7, 2000, at Imperial College London.

THE LEEUWENHOEK LECTURE: The Royal Society's annual lecture on microbiology begun in 1948.

Howard Dalton FRS, University of Warwick.

"The Natural and Unnatural History of Methane-oxidising Bacteria," March 2, 2000, at King's College London.

THE WILKINS LECTURE: The Royal Society's lecture on the history of science originated in 1947.

Roy Porter, The Wellcome Institute for the History of Medicine.

"Reflections on Scientific and Medical Futurology Since the Time of John Wilkins," July 11, 2000, the Royal Society.

Royal Swedish Academy of Science

http://www.kva.se/eng/pg/prizes/crafoord/pressr00.html

CRAFOORD PRIZE: Ravinder N. Maini and Marc Feldmann

Russian Academy of Sciences

http://www.ras.ru/RAS/

ZELDOVICH MEDALS: Sarah T. Gille, Roland Meier, Stephen D. Eckermann, Vassilis Angelopoulos, Takeshi Tsuru, Max P. Bernstein, Jens P. Leypoldt, and Alberto Vecchio

BUSINESS AND INSTITUTIONAL AWARDS

These awards, offered by private foundations, businesses, organizations, and educational institutions, represent some of the more publicized and better known awards. Additional prizes can be located on the Internet.

Clay Mathematics Institute

http://www.claymath.org

CLAY MATHEMATICS RESEARCH AWARD: Laurent Lafforgue and Alain Connes

CMI AMERICAN OLYMPIAD SCHOLAR: Ricky Ini Liu, Newton South High School, Newton Center, Massachusetts

For presenting the most original and mathematically correct solution to problems during the USA Mathematical Olympiad.

Committee on Space Research (COSPAR)

http://www.cospar.itodys.jussieu.fr/Awards/awards.htm

SPACE SCIENCE AWARD: Roger M. Bonnet and Donald M. Hunten

INTERNATIONAL COOPERATION MEDAL: John H. Carver

WILLIAM NORDBERG MEDAL: Kenichi Ijiri

Discover Awards for Technological Innovation

http://www.discover.com

EDITORS' CHOICE AWARD: Harvey Tananbaum

For enabling X-ray vision and space exploration with the creation of the Chandra X-Ray Observatory.

COMPUTING AWARD: Chad Mirkin

For dip-pen nanolithography to etch silicon chips.

ENERGY AWARD: E. Fred Schubert

For photon-recycling semiconductor light-emitting diode.

COMMUNICATIONS AWARD: Randy Giles

For the WaveStar LambdaRouter to make Internet connections more efficient.

AEROSPACE AWARD: Victor F. Petrenko

For a de-icer that transforms ice on aircraft directly from a solid to a gas.

HEALTH AWARD: Todd Golub

For DNA cancer-diagnosis chip that interprets DNA data in order to distinguish leukemia subtypes.

ENTERTAINMENT AWARD: Joseph Paradiso

For expressive shoes that produce sounds according to wearers' movements.

HUMANITARIAN AWARD: Thomas Thundat

For land-mine detector.

TRANSPORTATION AWARD: Gaby Ciccarelli

For RAPTOR, the quiet jackhammer.

CHRISTOPHER COLUMBUS FELLOWSHIP FOUNDATION AWARD: Anthony Atala, tissue engineer and surgeon

For growing organs in laboratories to replace failing organs in humans.

Camille and Henry Dreyfus Foundation

http://www.dreyfus.org/awardslist.shtml

CAMILLE DREYFUS TEACHER-SCHOLAR AWARDS: Scott J. Miller, James L. Leighton, Geoffrey W. Coates, Mark W. Grinstaff, Hilary A. Godwin, Thomas J. Wandless, John P. Toscano, Milan Mrksich, Patrick J. Walsh, Jeffrey R. Long, Carolyn R. Bertozzi, Timothy Deming, Kristi S. Anseth, Todd J. Martinez, James J. Watkins, Marc A. Hillmyer, Deborah G. Evans, Michael R. Gagné, and Uwe H. F. Bunz

NEW FACULTY AWARDS: Shana O. Kelley, Brian M. Stoltz, Brian R. Crane, Stephen L. Craig, Jason M. Haugh, Anatoly B. Kolomeisky, Michael D. McGehee, Victor Munoz, David A. Blank, David M. Leitner, Christian Schafmeister, Daniel T. Chiu, Daesung Lee, T. Daniel Crawford, and J. Patrick Loria

SENIOR SCIENTIST MENTOR INITIATIVE: Reed M. Izatt, John D. Roberts, Vasu Dev, Frank R. Stermitz, Fred W. McLafferty, Luther E. Erickson, Douglas X. West, David M. Schrader, Stuart W. Tanenbaum, Maghar S. Manhas, Robert S. Brodkey, James E. Boggs, Benjamin F. Plummer, August H. Maki, James M. Bobbitt, Paul A. Kitos, Paul H. Gross, and J. Hodge Markgraf

POSTDOCTORAL PROGRAM IN ENVIRONMENTAL CHEMISTRY: Koji Nakanishi, Steven Wofsy, Joseph T. Hupp, Robert F. Curl, Barbara Turpin, Charles T. Driscoll, Ellen R. M. Druffel, and Curtis R. Olson

Gairdner Foundation

http://www.gairdner.org/newwinner.html

GAIRDNER FOUNDATION INTERNATIONAL AWARDS: Jack Hirsh, Roger D. Kornberg, Robert G. Roeder, Alain Townsend, and Emil Unanue

In recognition of medical research achievements.

General Motors

http://www.hhhs.gov/news/speeches/000607.html

GENERAL MOTORS CANCER RESEARCH PRIZE: Monroe Wall, Mansukh Wani, Bert Vogelstein, Avram Hershko, and Alexander Varshavsky

Indian Space Research Organization

http://www.isro.org

VIKRAM SARABHAI MEDAL: Zhen-Xing Liu

Invention Dimension/Massachusetts Institute of Technology

http:/web.mit.edu/invent/index.html

2000 Lemelson-MIT Prize Winners

INVENTION AND INNOVATION: Thomas J. Fogarty

For medical devices such as balloon catheters that revolutionized vascular surgery.

LIFETIME ACHIEVEMENT AWARD: Al Gross

For pioneering wireless communication.

STUDENT PRIZE FOR INVENTION AND INNOVATION: Amy Smith

For technological inventiveness useful for developing countries.

STUDENT TEAM PRIZE: Michael Lim, Jalal Khan, and Thomas Murphy

For research focusing on integrated optical devices.

HIGH SCHOOL INVENTION APPRENTICESHIP AWARD: Charles Johnson

King Faisal Foundation

http://www.kff.net/frameprize.htm

KING FAISAL INTERNATIONAL PRIZE FOR SCIENCE: Edward Osborne Wilson, John Craig Venter, and Cynthia Jane Kenyon

Lamb Medal

http://www.bell-labs.com/news/2000/january/24/2.html

Willis E. Lamb Medal for Laser Science and Quantum Optics: Federico Capasso and Alfred Y. Cho

Lasker Foundation Awards

http://www.laskerfoundation.org

Considered the American Nobel Prize by many scientists.

2000 Albert Lasker Award Winners

Basic Medical Research: Aaron Ciechanover, Avram Hershko, and Alexander Varshavsky

Clinical Medical Research: Michael Houghton and Harvey Alter

Special Achievement in Medical Science: Sydney Brenner

John D. and Catherine T. MacArthur Foundation "Genius Grants"

http://www.macfdn.org/

This foundation has presented $500,000 five-year fellowships annually since 1981 to twenty-five notable scientists, artists, and activists.

"Genius Grants" (science and technology): K. Christopher Beard, Lucy Blake, Peter Hayes, Hideo Mabuchi, Samuel Mockbee, Margaret Murnane, Laura Otis, Matthew Rabin, Carl Safina, Daniel Schrag, Gina Turrigiano, Gary Urton, Deborah Willis, Erik Winfree, and Horng-Tzer Yau

Maria Mitchell Association

http://maria1.mmo.org/

Maria Mitchell Women in Science Award: Catherine Banks and Cinda-Sue C. Davis

National Foundation for Infectious Diseases

http://www.nfid.org

Maxwell Finland Award for Scientific Achievement: R. Gordon Douglas, Jr.

Jimmy and Rosalynn Carter Award for Humanitarian Contributions to the Health of Humankind: Ted Turner

Search for Extraterrestrial Intelligence Institute (SETI)

http://www.seti-inst.edu/science/water_award.html

International Water and Science Award: Natalie Cabrol and Edmond Grin

Sheikh Hamdan Awards for Medical Sciences

http://www.hmaward.org.ae/

Grand Hamdan International Award: Denton A. Cooley

Special Recognition Award: Shahbudin H Rahimtoola

Hamdan Award for Medical Research Excellence Genetics in Diabetes: C. Ronald Kahn

Recombinant Vaccines in Infectious Diseases Winner: Andrew James McMichael

Therapy in Leukemia Winners: Anne Dejean and Pier Paolo Pandolfi de Rinaldis

Hamdan Award for Volunteers in Humanitarian Medical Services: Abdul Sattar Edhi and Red Crescent Society, United Arab Emirates

Templeton Prize for Progress in Religion

http://www.templetonprize.com

Freeman J. Dyson, physicist

For seeking to combine technology and social justice.

Tyler Prize

http://www.usc.edu/admin/provost/tylerprize/

John and Alice Tyler Prize for Environmental Achievement: John P. Holdren

Wolf Foundation Prizes in the Sciences and Arts

http://www.aquanet.co.il/wolf/

Agriculture: Gurdev S. Khush

Chemistry: F. Albert Cotton

Mathematics: Raoul Bott and Jean-Pierre Serre

Physics: Raymond Davis, Jr., and Masatoshi Koshiba

The World Food Prize

http://www.wfpf.org

Recognizes people who better the quality, quantity, and availability of food.

Evangelina Villegas and Surinder Vasal

For developing high-protein corn varieties to feed malnourished populations.

SCIENTIFIC AND TECHNOLOGICAL ASSOCIATION AWARDS

These are some of the better-known professional societies that honor scientists and engineers.

American Astronomical Society

http://www.aas.org/grants/index.htm

Henry Norris Russell Lectureship: Donald Lynden-Bell

Newton Lacy Pierce Prize: Kirpal Nandra

Helen B. Warner Prize: Wayne Hu

Beatrice M. Tinsley Prize: Charles Alcock

Dannie Heineman Prize: Frank H. Shu

Award for Public Service to Science (with the American Mathematical Society and American Physical Society): Harold Varmus, Senator Bill Frist, and Senator Joseph Lieberman

American Chemical Society

http://tungsten.acs.org/awards/2000awards.html

- Arthur W. Adamson Award for Distinguished Service in the Advancement of Surface Chemistry: Alvin W. Czanderna
- Alfred Bader Award in Bioinorganic or Bioorganic Chemistry: Stuart L. Schreiber
- Earle B. Barnes Award for Leadership in Chemical Research Management: George A. Samara
- Herbert C. Brown Award for Creative Research in Synthetic Methods: Samuel J. Danishefsky
- Alfred Burger Award in Medicinal Chemistry: Philip S. Portoghese
- Arthur C. Cope Award: David A. Evans
- James Bryant Conant Award in High School Chemistry Teaching: Frank G. Cardulla, Niles North High School, Skokie, Illinois
- Francis P. Garvan John M. Olin Medal: F. Ann Walker
- The Irving Langmuir Award in Chemical Physics: Richard J. Saykally
- Priestley Medal: Darleane C. Hoffman

American Fisheries Society

http://www.fisheries.org/Awards.shtml

- Award of Excellence: John J. Magnuson
- President's Fishery Conservation Award (AFS Member): John Gunn
- President's Fishery Conservation Award (Non-AFS Member): Columbia River Intertribal Fisheries Commission
- William E. Ricker Resource Conservation Award: Robert J. Behnke
- C.R. Sullivan Fishery Conservation Award: Chesapeake Bay Foundation

American Geophysical Union

http://www.agu.org/inside/awardees.html

- William Bowie Medal: John A. Simpson
- James B. Macelwane Medal: Quentin Williams
- Waldo E. Smith Medal: Rosina Bierbaum
- Walter H. Bucher Medal: James H. Dieterich
- Maurice Ewing Medal: Joseph L. Reid
- John Adam Fleming Medal: John T. Gosling
- Robert E. Horton Medal: M. Gordon Wolman
- Roger Revelle Medal: James R. Holton

American Mathematical Society

http://www.ams.org

- Leroy P. Steele Prize: Isadore M. Singer, John H. Conway, Barry Mazur
- Communications Award: Sylvia Nasar
- Award for Distinguished Public Service: Paul J. Sally, Jr.

American Medical Association

http://www.ama-assn.org/

- Dr. Nathan Davis International Awards: Inge Genefke

Distinguished Service Award: Edwin Lawrence Kendig, Jr.
Scientific Achievement Award: Tom Maniatis
Joseph B. Goldberger Award in Clinical Nutrition: M. Molly McMahon
Medical Executive Achievement Award: Marie D. Zinninger

American Meteorological Society
http://www. www.ametsoc.org/AMS/
Carl-Gustaf Rossby Research Medal: Susan Solomon

American Physical Society
http://www.aps.org/praw/
Maria Goeppart-Mayer Award: Sharon C. Glotzer
Julius Edgar Lilienfeld Prize: Robert J. Birgeneau
Lars Onsager Prize: J. Michael Kosterlitz and David James Thouless
Arthur L. Schawlow Prize in Laser Science: Richard N. Zare
Biological Physics Prize: Paul K. Hansma
Hans A. Bethe Prize: Igal Talmi
Dannie Heineman Prize for Mathematical Physics: Sidney Coleman
Will Allis Prize for the Study of Ionized Gases: John F. Waymouth
Tom W. Bonner Prize in Nuclear Physics: Raymond G. Arnold

American Society for Microbiology
http://www.asmusa.org
The Abbott-ASM Lifetime Achievement Award: Boris Magasanik
The Eli Lilly and Company Research Award: Gisela Storz
The Aventis Pharmaceuticals Award: Richard J. Whitley
Morrison Rogosa Award: Nina Chanishvili
The Procter & Gamble Award in Applied and Environmental Microbiology: Steven E. Lindow
The Chiron Corporation Biotechnology Research Award: Stanley Fields

American Society of Agricultural Engineers
http://www.asae.org
Cyrus Hall McCormick Medal: Stuart O. Nelson
Jerome Case Gold Medal: Dale F. Hermann

Association for Women in Mathematics
http://www.awm-math.org
Emmy Noether Lecture: Margaret H. Wright
Louisa Hay Award: Joan Ferrini-Mundy
Alice T. Schafer Mathematics Prize: Mariana E. Campbell

CAST Center for Applied Special Technology
http://www.cast.org
Charles A. Black Award: Dennis R. Keeney

Institute of Electrical and Electronic Engineers (IEEE)
http://www.ieeeusa.org/awards/
EDISON MEDAL: Junichi Nishizawa

Optical Society of America
http://www.osa.org
FREDERIC IVES MEDAL/JARUS W. QUINN ENDOWMENT: Alexander M. Prokhorov
ESTHER HOFFMAN BELLER AWARD: Henry Stark
EDWIN H. LAND MEDAL: John E. Warnock
ADOLPH LOMB MEDAL: Mikhail Lukin
WILLIAM F. MEGGERS AWARD: Roger Miller

Society for Historical Archaeology
http://www.sha.org
J.C. HARRINGTON AWARD (FOR LIFETIME ACHIEVEMENTS): Roderick Sprague
JOHN L. COTTER AWARD (FOR A SINGLE ACHIEVEMENT): Paul R. Mullins

Society for the History of Technology
http://shot.press.jhu.edu
LEONARDI DA VINCI MEDAL: Silvio A. Bedini

World Meteorological Organization
http://www.wmo.ch/
INTERNATIONAL METEOROLOGICAL ORGANIZATION PRIZE: Edward Norton Lorenz
PROFESSOR DR. VILHO VAISALA AWARD: Barry Goodison and Paul Louie

OTHER SCIENCE AND TECHNOLOGY AWARDS

The National Inventors Hall of Fame
http://www.invent.org
2000 CLASS OF INDUCTEES: Walt Disney, Reginald Fessenden, Alfred Free, Helen Murray Free, J. Franklin Hyde, William Kroll, and Steve Wozniak.

The National Women's Hall of Fame
http://www.greatwomen.org/induct.htm
2000 INDUCTEES WITH SCIENCE-RELATED CAREERS: Faye Glenn Abdellah, Marjory Stoneman Douglas, Sylvia Earle, Frances Kathleen Oldham Kelsey, Mary Walker, Annie Dodge Wauneka

TEACHING AND SCIENCE FAIR AWARDS

National Science Teachers Association
http://www.nsta.org

American Water Works Association Award: Hector Ibarra
Barrick Goldstrike Exemplary Elementary Earth Science Teaching Award: Gail Bushey
Ciba Specialty Chemicals Exemplary Middle Level and High School Science Teaching Award: Melinda Mills (middle level) and Heidi Haugen (high school)
Drug, Chemical, and Allied Trades Association Education Foundation's "Making a Difference" Award: Cynthia Diane Martinez-Bagwill
Distinguished Informal Science Education Award: William Nimke
Distinguished Service to Science Education Award: Rodney Doran
Distinguished Teaching Award: Mark Charles Kroteck
Estes Rocketry/U.S. Space Foundation (USSF) Space Educator Award: Janice Holt
Gustav Ohaus Awards for Innovation in Science Teaching:
Elementary
1st: Jan French
2nd: James Olson
Middle Level
1st: Maureen Barrett
2nd: Wayne Goates
High School
1st: James Rusconi
2nd: Jane Franklin
College
1st: Margaret Johnson
2nd: William Langley
Robert H. Carleton Award: Marvin Druger
Science Screen Report Award: Annette Parrott/SeaWorld/Busch Gardens/NSTA
Environmental Educator Award: Charlene Dindo
The Shell Science Teaching Award: Jim Calaway
Sheldon Exemplary Equipment and Facilities Award: Susan Mitchell

The Collegiate Inventors Competition

http://www.invent.org/collegiate/
This competition is a national contest to test college students' problem-solving and creativity talents in scientific and engineering fields.

2000 Winners

Emilie A. Porter, University of Wisconsin
Invention: Beta-amino acid oligomers for use as an antibiotic
Matthew B. Dickerson and Raymond R. Unocic, Ohio State University
Invention: Ceramic composites processing
Colin A. Bulthaup and Eric J. Wilhelm, Massachusetts Institute of Technology
Invention: Chip fabrication by liquid embossing
Balaji Srinivasan, University of New Mexico
Invention: Fiber lasers
Daniel M. Hartmann, University of California, San Diego
Invention: High performance polymer microlenses

Craftsman/NSTA Young Inventors Awards Program

http://www.nsta.org/programs/craftsman/00regional.htm

2000 Regional Winners

GRADES 3-5: *Region One*

Zachary L. Tuthill, grade 5, Mannington Township School, New Jersey
For a farm tool that allows farmers to remain on their tractors while hooking or unhooking equipment.

Erica L. Atkins, grade 4, Springdale Elementary School, Connecticut
For a kitchen aid that helps small children pour liquids in the glass and not on the countertop.

Alison L. Dadouris, grade 2, Spruce Run School, New Jersey
For a dental device that medicates gums while flossing.

GRADES 3-5: *Region Two*

Rachel Anne Kaminsky, grade 3, Concord Road Elementary School, New York
For a cooking tool that evenly slices onions and prevents users from crying.

Lauren N. Correia, grade 4, Mildred Aitken Elementary, Massachusetts
For a ice-cutting tool that carves small even blocks from larger ice blocks.

Zachary P. Preston, grade 4, Mildred Aitken Elementary, Massachusetts
For a safety brake for skateboards.

GRADES 3-5: *Region Three*

Jordan A. Callender, grade 5, Parkview Baptist School, Louisiana
For a gardening device to plant seedlings evenly.

Andrew Morrow, grade 4, Big A Elementary School, Georgia
For a device that eases the motions of pushing, pulling, lifting, and twisting for disabled people.

Brian J. Tassone, grade 4, Morrisville Year Round Elementary, North Carolina
For a cleaning broom that enables users to sweep at different angles to remove large debris.

GRADES 3-5: *Region Four*

Andrew G. Johannes, grade 5, Webber Elementary School, Michigan
For a tool that compresses plastic one-gallon jugs for disposal.

Joshua A. Colwell, grade 5, Scottsburg Elementary, Indiana
For a kitchen aid that makes dishwashing efficient.

Matthew A. Godfrey, grade 3, River Bend Elementary School, Missouri
For a tool to loosen nuts that are not within arms' reach.

GRADES 3-5: *Region Five*

Timothy M. Wiese, grade 5, Kalamazoo Elementary School, Nebraska
For an invention that enables one person to lift, transport, and stabilize heavy objects.

Melanie C. Anderson, grade 5, Pence Elementary School, Iowa
For a device that permits disabled people to ice skate safely.

Olivia Nicole Wiese, grade 2, Kalamazoo Elementary School, Nebraska
For an invention to transport small animals instead of using a leash.

GRADES 3-5: *Region Six*

Lindsey E. Clement, grade 5, Pine Tree Middle School, Texas
For a tool to collect sweet gum balls from trees.

Shana M. Yamashita, grade 5, Aina Haina School, Hawaii
For an eating utensil created by adding a cutting edge to chopsticks.
Jarhett L. Wirth, grade 5, Shepherd Christian School, California
For a lawn tool that efficiently separates soil from grass and leaves.

GRADES 6-8: *Region One*

Marc L. Shapiro, grade 6, Bernardsville Middle School, New Jersey
For a device that enables trash removal without leaving a building.
Ian R. Ratti, grade 6, Bernardsville Middle School, New Jersey
For an invention that makes exercising pets enjoyable.
Arti L. Kotadia, grade 7, Saddle Brook Middle School, New Jersey
For a tool that easily removes spiders and their webs.

GRADES 6-8: *Region Two*

Brendan Ian Feifer, grade 6, Unqua School, Massapequa Park, New York
For a device that assists people who have lost full functioning of a hand.
Anna Vittoria Frattolillo, grade 6, Dwight School, New York
For a tool that evenly divides pie slices.
Victoria Ann Vitulli, grade 8, St. Boniface School, New York
For a hair-removal device that cleans round brush bristles.

GRADES 6-8: *Region Three*

Kathryn Marie Close, grade 6, Glenridge Middle School, Florida
For a lawn tool that eases back stress during leaf sweeping.
Joe Daniel Richerme, grade 7, Ridgeway Middle School, Tennessee
For a device that transports beach accessories.
Marco Melesio Contreras, grade 7, Walker Middle School, Florida
For a lightbulb changing tool that protects people from bulbs' hot surfaces.

GRADES 6-8: *Region Four*

Mindy Sue Himler, grade 8, Marysville Middle School, Ohio
For a tool to open, adjust, and close high windows.
Lori Beth Bridge, grade 8, Marysville Middle School, Ohio
For a device that assists physically challenged individuals to reach a shower lever.
Malarie Yocum, grade 8, North Middle School, Missouri
For an invention that efficiently moves heavy household appliances.

GRADES 6-8: *Region Five*

Kim N. Brown, grade 6, Virginia Lake Elementary School, Illinois
For a device that changes a bed in one step.
Ashley E. Combs, grade 7, Alton R-IV School, Missouri
For a device that fairly flips a coin or rolls a dice consistently.
Dustin S. Vincke, grade 7, Gentry Middle School, Missouri
For an aid to ease reaching objects stored on high shelves.

GRADES 6-8: *Region Six*

Michael R. Isaacs, grade 6, Hyde Park Middle School, Nevada
For a wheelbarrow that can be both pushed and pulled.
Scott Michael Stedman, grade 7, Holy Name School, California
For a foot-operated device that loosens tire lugs.
Hebah Fotouh, grade 8, Duchesne Academy, Texas
For a kitchen tool to mix condiments inside containers.

Intel International Science and Engineering Fair (ISEF)
http://www.sciserv.org/isef/grnd2000.asp
Ninth- through twelfth-grade high school students who have won a local, regional, state, or national science fair are invited to compete at the Intel ISEF. Almost $1 million is awarded as Intel ISEF prizes, including funds for scholarships and scientific field trips.

Intel Young Scientist Scholarships ($40,000): Karen Kay Powell, Nazanin Jouei, and Jason L. Douglas
Glenn T. Seaborg Nobel Prize Visit Award Trip (to Stockholm, Sweden, in December, 2000, to observe the Nobel Prize festivities): Nazanin Jouei and Garrett J. Young
European Union Contest for Young Scientists (a trip to the Twelfth European Union presented by *Discover Magazine*): Travis Michael Beamish and Avaleigh Nora Milne
Ireland Young Scientist and Technology Exhibit (an all-expense-paid trip to Ireland to attend the Young Scientist and Technology Exhibit presented by: *Discover Magazine*) Joseph E. Pechter and William H. Pechter
Intel Achievement Award (for outstanding work in any field): Stephen Alan Steiner III, Hans Christiansen Lee, Serge A. Tishchenko, Nisha Nagarkatti, Mark Kaganovich, Linda J. Arnade, Daniel Richard Green, and Yuma Takahashi

Intel Science Talent Search
http://www.sciserv.org
The Intel Science Talent Search is considered the United States' most prestigious high school science award. This science fair began in 1941 and was sponsored by Westinghouse until 1999 when Intel assumed sponsorship. The 2000 grand finalist won $100,000, and the 300 semifinalists each received $1,000 scholarships, and their high schools were awarded $1,000 per semifinalist to fund science and math education.

Sometimes referred to as the "Junior Nobel Prize," this competition encourages senior high school students to test original ways to approach scientific problems to increase knowledge and improve standards by undertaking an original scientific research project representing the discipline of their choice. Judged for their research and thinking skills as well as creativity, winners are chosen by leading scientists. A large percentage of past winners have completed science graduate degrees, won major scientific prizes, and been elected to the National Academy of Sciences or the National Academy of Engineering.

2000 Winners

1st: Viviana Risca
For DNA-based Steganography, a computer science project exploring molecular computing and data encryption.

2nd: Jayce Getz
For "Extension of a Theorem of Kiming and Olsson for the Partition Function," mathematics.

3rd: Feng Zhang
For "Genetic Functional Analysis of the Moloney Murine Leukemia Virus

GAG Gene Reveals an Inhibitory Element That Can Be Masked to Control Retroviral Assembly," biochemistry.

THE GLENN T. SEABORG AWARD: Eugene Simuni

For scientific cooperation and communication, winner chosen by Intel finalists.

International Mathematical Olympiad

http://imo.math.ca/

Twelve teams (from more than eighty competing teams) received the highest total scores from the maximum 252 points possible. China (218 points) ranked highest and was followed by Russia (215), United States (184), South Korea (172), Bulgaria (169), Vietnam (169), Belarus (165), Taiwan (164), Hungary (156), Iran (155), Israel (139), and Romania (139).

Invent America!

http://www.inventamerica.org

A student invention competition in which each entry is judged on its usefulness and the level of creativity involved. First place winners receive a $1,000 United States savings bond; the second place award is a $500 savings bond; and the third place prize is a $250 savings bond.

KINDERGARTEN

1st: Bethanie Abbott, Marcellus, New York, The Sit-Up Helper

2nd: Nicholas Ciarrocchi of Drexel Hill, Pennsylvania, Time Teacher

3rd: Shannon Barry, Eagle River, Arkansas, Hand Cuff

FIRST GRADE

1st: Dillan Poe, Hamilton, Texas, Corner Sweeper

2nd: Justine Falcone, Cinnaminson, New Jersey, The Pet Groomer

3rd: Miranda Obert, Big Rapids, Michigan, Paw Cleaner

SECOND GRADE

1st: Alex Patton, Safety Harbor, Florida, Water Saver 2000

2nd: Garrett Kashmer, Cinnaminson, New Jersey, Pick-up Truck Rails

3rd: Colleen Thiersch, N. Brunswick, New Jersey, "Frosty"

THIRD GRADE

1st: Casey Skittle, Presto, Pennsylvania, Self-Extinguishing Safety Candle

2nd: Lauren Lanmon, Hamilton, Texas, ComforTent

3rd: David Hess, Wichita, Kansas, Comet's Leg Cover

FOURTH GRADE

1st: Anna Kieler, Cherry Valley, New York, "Pop It Out" Exacto Brush

2nd: Kristin Shell, Taylorsville, North Carolina, Gluco Ring

3rd: Nathan Froncek, Mesa, Arizona, FreEZeFlo

FIFTH GRADE

1st: Michael Ward, Derby, Kansas, Bike Lift

2nd: Casey Penney, Saltillo, Mississippi, Relax-A-Tub

3rd: Caitlyn Tivy, New Lenox, Illinois, The Lawnilizer

SIXTH GRADE

1st: Walt Gerald, Taylors, South Carolina, Aqua Retriever

2nd: Joanna Jan, Fargo, North Dakota, E-Z Golf Carrier
3rd: Jeffery W. Stier, Eden Prairie, Minnesota, Vari-Bait

Seventh Grade

1st: Charlotte Cantelon, San Diego, California, On the Go Bed for the Handicapped
2nd: Landon Lamb, Hamilton, Texas, N.M.C. Device (No More Cords)
3rd: Robert E. Weidman, Newport News, Virginia, Pug Walker

Eighth Grade

1st: Ericka Fryburg, Warwick, Rhode Island, Reuse-A-Bin
2nd: Claire Kugler, Alexandria, Virginia, Mobile Rabbit Home
3rd: Evan Robert Dufault, Sutton, Massachusetts, Snorer's Solution

Mathcounts
http://mathcounts.org
This national mathematics competition includes timed rounds.

Teams

1st: California
2nd: Virginia
3rd: Indiana
4th: Illinois
5th: Texas
6th: Kansas
7th: Maryland
8th: Florida
9th: Kentucky
10th: Georgia

Individuals

1st: Ruozhou "Joe" Jia, Illinois
2nd: Tiankai Liu, California
3rd: Ho Seung "Paul" Ryu, Kansas
4th: John Shen, Maryland
5th: Ricky Biggs, Virginia
6th: Avinash Kunnath, Florida
7th: Boris Alexeev, Georgia
8th: Adam Rosenfield, Massachusetts
9th: Charley Seelbach, Kentucky
10th: Jack Cackler, Virginia

National Gallery for America's Young Inventors
http://www.pafinc.com/home.htm

2000 Inductees

Edward T. Gemin, Xenia, Ohio
Invention: Heat energy recovery system using Peltier junction modules
Ryan W. Kingsbury, North Fort Myers, Florida
Invention: Thermoelectric-based liquid-cooled personal computer

Ann Lai, Beachwood, Ohio
Invention: Microsensors for monitoring sulfur dioxide emissions
Joseph E. and William H. Pechter, Vero Beach, Florida
Invention: Hybrid text to Speech 2000
Naveen Neil Sinha, Los Alamos, New Mexico
Invention: Multipurpose noninvasive sensor for monitoring contents of closed containers
Spencer Rocco Whale, Pittsburgh, Pennsylvania
Invention: KidKare hospital equipment and supplies.

PhysLINK.com
http://www.physlink.com/ysaward2000_press.cfm
YOUNG SCIENTIST OF THE YEAR AWARD: Jennifer A. Seiler
"The Acoustic Thermometry of Sea Water"

Science Fair Central/Discovery Young Scientist Challenge
http://www.school.discovery.com/sciencefaircentral/dysc/
Forty middle school students who submit research projects and essays and undergo examinations compete for $40,000 in scholarships; the winner receives a $10,000 scholarship.

DYSC 2000 FINALISTS
1st: Shana Matthews, Palm Bay, Florida
"An Investigation of the Factors Affecting Colony Transformation Efficiency Rates"
2nd: Jimmy Yang, Plano, Texas
"Plant, Are You in This Genus?"
3rd: Jonathan-James Eno, Kahului, Hawaii
"Phytoremediation of Hydrocarbon Contaminated Soil"

Science Olympiad
http://www.soinc.org/
A national competition during which high school and junior high school students compete in various scientific categories and earn overall points for their teams.

Top Ten High School Teams in the United States
1. Troy High School, Fullerton, California
2. Centerville High School, Centerville, Ohio
3. Lakeville High School, Lakeville, Minnesota
4. Fayetteville-Manlius High School, Manlius, New York
5. Solon High School, Solon, Ohio
6. Grand Haven High School, Grand Haven, Michigan
7. Harriton High School, Rosemont, Pennsylvania
8. Forest Hills Central High School, Grand Rapids, Michigan
9. State College High School, State College, Pennsylvania
10. Maine-Endwell High School, Endwell, New York

Top Ten Middle School Teams in the United States

1. Booth Middle School, Peachtree City, Georgia
2. Thomas Jefferson Middle School, Valparaiso, Indiana
3. Arendell Parrot Academy, Kinston, North Carolina
4. Malow Junior High School, Shelby Township, Michigan
5. New Mark Middle School, Kansas City, Missouri
6. Rising Starr Middle School, Fayetteville, Georgia
7. South Middle School, Arlington Heights, Illinois
8. Piqua Junior High School, Piqua, Ohio
9. Arthur W. Coolidge Middle School, Reading, Massachusetts
10. Magsig Middle School, Centerville, Ohio

Toshiba ExploraVision Awards Program

Grades K-3

Katherine Finchy Elementary School, Palm Springs, California

For an imagined Good Veggie Machine.

Grades 4-6

Holy Martyrs, Green Bay, Wisconsin

For a proposed Cholestrobot.

Grades 7-9

McLean High School, McLean, Virginia

For suggested Audire Iterum to improve hearing.

Grades 10-12

Don Mills Collegiate Institute, Ontario, Canada

For describing potential Chlorophyll-Photovoltaic Cells (CVCs) for synthesized plant power.

USA Mathematical Olympiad

www.libfind.unl.edu/amc/a-activities/a7-problems/olympiad/

Each contestant solved six problems in six hours, and high scorers won a slot to compete at the International Mathematical Olympiad.

Greitzer-Klamkin Award: Reid W. Barton and Ricky I. Liu

Top Scorers: David G. Arthur, Gabriel D. Carroll, Kamaldeep S. Gandhi, Ian Lee, George Lee, Po-Ru Loh, Po-Shen Loh, Oaz Nir, Paul A. Valiant, and Yian Zhang

Elizabeth Schafer

Metric Conversion Chart

Multiply	by	to get
centimeters	0.0328	feet
centimeters	0.3937	inches
feet	30.48	centimeters
feet	0.3048	meters
gallons	3.785	liters
grams	0.0353	ounces
inches	2.54	centimeters
kilograms	2.205	pounds
kilometers	3280.8	feet
kilometers	0.6214	miles
liters	0.2642	gallons
meters	3.281	feet
miles	1.609	kilometers
ounces	28.349	grams
pounds	0.4536	kilograms
square kilometers	0.3861	square miles
square miles	2.593	square kilometers

Index